Microbe–vector interactions in vector-borne diseases

Several billion people are at daily risk of life-threatening vector-borne diseases such as malaria, trypanosomiasis and dengue, and the dark shadow of plague hovers in the few endemic foci where it waits ready to re-emerge in a deadly pandemic. The abortive attempt to control malaria in the 1960s showed us the problem that we face in eradicating vector-borne diseases. Research into these tropical diseases fell into neglect during the 1960s, but in the 1970s research was once more directed towards vector-borne diseases and more recent initiatives such as the Roll Back Malaria Campaign have kept them in the international spotlight. If we are not to repeat the mistakes of the past it will be necessary to use all of our knowledge of vector biology.

Written by international researchers in the field this volume describes the way in which pathogens interact with the vectors that transmit them. It details the elegant biological adaptations that have enabled pathogens to live with their vectors and in some circumstances control them. This knowledge has led to new control strategies in the form of antibiotics and new vaccines which are targeted not at the pathogen but at its vector. The recent epidemic of West Nile virus infection in the United States and of Nipah virus in Malaysia suggests that vector-borne diseases are of growing concern to everyone. *Microbe–Vector Interactions in Vector-borne Diseases* is essential reading for researchers and clinicians working with these diseases.

Stephen H. Gillespie is Professor of Medical Microbiology in the Department of Infectious Diseases at University College London, UK.

Geoffrey L. Smith is a Wellcome Trust Principal Research Fellow and Head of Department of Virology at Imperial College London, UK.

Anne Osbourn is a Group leader in the Sainsbury Laboratory, John Innes Centre, Norwich, UK.

Symposia of the Society for General Microbiology

Managing Editor: Dr Melanie Scourfield, SGM, Reading, UK
Volumes currently available:

14	Microbial behaviour
23	Microbial differentiation
32	Molecular and cellular aspects of microbial evolution
43	Transposition
45	Control of virus diseases
46	Biology of the chemotactic response
47	Prokaryotic structure and function – a new perspective
51	Viruses and cancer
52	Population genetics of bacteria
53	Fifty years of antimicrobials: past perspectives and future trends
54	Evolution of microbial life
55	Molecular aspects of host–pathogen interaction
56	Microbial responses to light and time
57	Microbial signalling and communication
58	Transport of molecules across microbial membranes
59	Community structure and co-operation in biofilms
60	New challenges to health: the threat of virus infection
61	Signals, switches, regulons and cascades: control of bacterial gene expression
62	Microbial subversion of host cells

DATE DUE FOR RETU

SIXTY-THIRD SYMPOSIUM OF THE
SOCIETY FOR GENERAL MICROBIOLOGY
HELD AT THE UNIVERSITY OF BATH MARCH 2004

Edited by
S. H. Gillespie, G. L. Smith and A. Osbourn

microbe–vector interactions in vector-borne diseases

Published for the Society for General Microbiology

PUBLISHED BY THE PRESS SYNDICATE OF THE UNIVERSITY OF CAMBRIDGE
The Pitt Building, Trumpington Street, Cambridge, United Kingdom

CAMBRIDGE UNIVERSITY PRESS
The Edinburgh Building, Cambridge CB2 2RU, UK
40 West 20th Street, New York, NY 10011-4211, USA
477 Williamstown Road, Port Melbourne, VIC 3207, Australia
Ruiz de Alarcón 13, 28014 Madrid, Spain
Dock House, The Waterfront, Cape Town 8001, South Africa

http://www.cambridge.org

First published 2004

Printed in the United Kingdom at the University Press, Cambridge

Typeface Sabon (Adobe) 10·5/13·5 pt   *System* QuarkXPress™   [SGM]

A catalogue record for this book is available from the British Library

ISBN 0 521 84312 X hardback   ＩＯＯ３８８Ｏ５６Ｘ

Front cover illustration: Xenopsylla cheopis flea, blocked with *Yersinia pestis*, examined immediately after attempting to feed. Fresh blood is present only in the oesophagus and not the midgut, which contains dark-coloured digestion products of previous blood meals and masses of bacteria. Photograph taken from the chapter by B. Joseph Hinnebusch, pp. 331–343.

CONTENTS

Contributors   vii

Editors' Preface   xi

B. W. J. Mahy
Vector-borne diseases   1

S. E. Randolph
Evolution of tick-borne disease systems   19

S. Blanc
Insect transmission of viruses   43

R. Lu, H. Li, W.-X. Li and S.-W. Ding
RNA-based immunity in insects   63

A. Barbour
Specificity of *Borrelia*–tick vector relationships   75

R. M. Elliott and A. Kohl
Bunyavirus/mosquito interactions   91

S. Higgs
How do mosquito vectors live with their viruses?   103

S. C. Weaver, L. L. Coffey, R. Nussenzveig, D. Ortiz and D. Smith
Vector competence   139

P. S. Mellor
Environmental influences on arbovirus infections and vectors   181

N. A. Ratcliffe and M. M. A. Whitten
Vector immunity   199

S. A. MacFarlane and D. J. Robinson
Transmission of plant viruses by nematodes   263

M. J. Taylor
Wolbachia host–symbiont interactions   287

J. A. Carlyon and E. Fikrig
Pathogenic strategies of *Anaplasma phagocytophilum*, a unique bacterium that
colonizes neutrophils   301

B. J. Hinnebusch
Interactions of *Yersinia pestis* with its flea vector that lead to the transmission
of plague   331

P. W. Atkinson and D. A. O'Brochta

Transgenic malaria   345

G. A. T. Targett

Vaccines targeting vectors   363

Index   379

CONTRIBUTORS

Atkinson, P. W.
Department of Entomology, University of California, Riverside, CA 92521, USA

Barbour, A.
Departments of Microbiology and Molecular Genetics and Medicine, University
of California Irvine, B240 Medical Science I, Irvine, CA 92697-4025, USA

Blanc, S.
UMR Biologie et Génétique des Interactions Plante-Parasite (BGPI), CIRAD-INRA-ENSAM,
TA 41/K, Campus International de Baillarguet, 34398 Montpellier Cedex 5, France

Carlyon, J. A.
Section of Rheumatology, Department of Internal Medicine, Yale University School
of Medicine, The Anylan Center for Medical Research and Education, New Haven,
CT 06520-8031, USA

Coffey, L. L.
Center for Biodefense and Emerging Infectious Diseases and Department of Pathology,
University of Texas Medical Branch, Galveston, TX 77555-0609, USA

Ding, S.-W.
Center for Plant Cell Biology, Department of Plant Pathology, University of California,
Riverside, CA 92521, USA

Elliott, R. M.
Division of Virology, Institute of Biomedical and Life Sciences, University of Glasgow,
Church Street, Glasgow G11 5JR, UK

Fikrig, E.
Section of Rheumatology, Department of Internal Medicine, Yale University School
of Medicine, The Anylan Center for Medical Research and Education, New Haven,
CT 06520-8031, USA

Higgs, S.
Department of Pathology, University of Texas Medical Branch, Galveston, TX 77555-0609,
USA

Hinnebusch, B. J.
Laboratory of Human Bacterial Pathogenesis, Rocky Mountain Laboratories,
National Institute of Allergy and Infectious Diseases, National Institutes of Health,
Hamilton, MT 59840, USA

Kohl, A.
Division of Virology, Institute of Biomedical and Life Sciences, University of Glasgow,
Church Street, Glasgow G11 5JR, UK

Li, H.
Center for Plant Cell Biology, Department of Plant Pathology, University of California, Riverside, CA 92521, USA

Li, W.-X.
Center for Plant Cell Biology, Department of Plant Pathology, University of California, Riverside, CA 92521, USA

Lu, R.
Center for Plant Cell Biology, Department of Plant Pathology, University of California, Riverside, CA 92521, USA

MacFarlane, S. A.
Scottish Crop Research Institute, Invergowrie, Dundee DD2 5DA, UK

Mahy, B. W. J.
National Center for Infectious Diseases, Centers for Disease Control and Prevention, Atlanta, GA 30333, USA

Mellor, P. S.
Institute for Animal Health, Pirbright Laboratory, Ash Rd, Pirbright, Woking GU24 0NF, UK

Nussenzveig, R.
Center for Biodefense and Emerging Infectious Diseases and Department of Pathology, University of Texas Medical Branch, Galveston, TX 77555-0609, USA

O'Brochta, D. A.
Center for Biosystems Research, University of Maryland Biotechnology Institute, College Park, MD 20742, USA

Ortiz, D.
Center for Biodefense and Emerging Infectious Diseases and Department of Pathology, University of Texas Medical Branch, Galveston, TX 77555-0609, USA

Randolph, S. E.
Department of Zoology, South Parks Road, Oxford OX1 3PS, UK

Ratcliffe, N. A.
School of Biological Sciences, University of Wales Swansea, Singleton Park, Swansea, UK

Robinson, D. J.
Scottish Crop Research Institute, Invergowrie, Dundee DD2 5DA, UK

Smith, D.
Center for Biodefense and Emerging Infectious Diseases and Department of Pathology, University of Texas Medical Branch, Galveston, TX 77555-0609, USA

Targett, G. A. T.
Department of Infectious and Tropical Diseases, London School of Hygiene & Tropical Medicine, London WC1B 3DP, UK

Taylor, M. J.
Filariasis Research Laboratory, Molecular and Biochemical Parasitology, Liverpool School of Tropical Medicine, Pembroke Place, Liverpool L3 5QA, UK

Weaver, S. C.
Center for Biodefense and Emerging Infectious Diseases and Department of Pathology, University of Texas Medical Branch, Galveston, TX 77555-0609, USA

Whitten, M. M. A.
School of Biological Sciences, University of Wales Swansea, Singleton Park, Swansea, UK

EDITORS' PREFACE

Vector-borne diseases are a major threat to public health throughout the world. They include major human diseases such as malaria, trypanosomiasis and dengue, and others that principally affect animal populations such as theileriosis and West Nile fever. The study of vector-borne diseases has gone through a period when there was little impetus for research. However, through programmes such as the 'Great Neglected Diseases Network' and the Tropical Diseases Research Programme of WHO in the 1980s scientific attention was redirected to vector-borne diseases. More recently initiatives such as the 'Roll Back Malaria' programme have moved the subject on further.

In industrialized countries vector-control programmes have eradicated the important vector-borne diseases although increasing international travel has meant that diseases acquired in the tropics are no longer rare in developed world clinical practice. Climate change, industrialization, changing land use and increasing population have all had a profound impact on vector-borne diseases, resulting in epidemics spread. The recent appearance of West Nile virus in North America and Nipah virus in Malaysia has been responsible for dramatic epidemics in animal and human populations with considerable numbers of human deaths. These epidemics have caught the attention of the public.

Vector-borne infections require an extraordinary integration of host and pathogen. Unravelling these interrelationships is a challenge for biologists. The subtle ways in which parasites control their vectors, and the filaria that contain *Wolbachia* spp., essential to survival and fertility, are examples of elegant and effective co-evolution.

This symposium volume brings together internationally recognized experts to explore the complexity of host–pathogen–vector interactions. The coverage stretches from common endemic diseases such as malaria and dengue through the fortunately rare explosive epidemic diseases such as plague to plant viruses. The focus is on interaction between the components of the life cycle but not one that is abstract. The failure of the malaria eradication campaign demonstrates that it is only by understanding these complex interactions that we can successfully hope to control these diseases.

Microbe–Vector Interactions in Vector-borne Diseases is essential reading for researchers and clinicians working with these diseases. It will prove a treasure trove for those who wish to widen their horizons and study the diversity of biological adaptation.

Stephen H. Gillespie, Geoffrey L. Smith, Anne Osbourn

Vector-borne diseases

Brian W. J. Mahy

National Center for Infectious Diseases, Centers for Disease Control and Prevention, Atlanta, GA 30333, USA

INTRODUCTION

The first discovery of human disease transmission by a vector species was in 1877, when Patrick Manson (who later founded the Royal Society of Tropical Medicine and Hygiene in London) was able to show that transmission of the nematode worms causing lymphatic filariasis involved haematophagous mosquitoes (Manson, 1878). Since that time, hundreds of vector-borne diseases have been described, usually involving arthropod vectors which may transmit not only helminths but also protozoa, bacteria or viruses to cause major epidemics of diseases such as malaria, plague, trypanosomiasis, leishmaniasis, louse-borne typhus, dengue fever, yellow fever, Japanese encephalitis and West Nile fever. The numbers of people who are affected by vector-borne diseases in the world belies the imagination. For example, 750 million people in 76 countries live in areas of endemic filariasis, and an estimated 118 million of them are infected with filariae (World Health Organization, 1995). In India alone, 50 000–100 000 people die each year from visceral leishmaniasis. Malaria, the most important protozoal disease infecting humans, occurs in areas where *Anopheles* mosquitoes are present, and causes some 300–500 million clinical cases and approximately 1·2 million deaths worldwide each year. Malaria is the commonest vector-borne disease imported into the USA, and vector-competent *Anopheles* mosquitoes exist there.

Apart from yellow fever, which can now be controlled effectively by vaccination, most prevention and control programmes have been based on control of the arthropod vector, but a number of factors such as resistance to insecticides and drugs have combined to cause a resurgence of many vector-borne diseases since the 1970s (Gubler, 1998).

SGM symposium 63: Microbe–vector interactions in vector-borne diseases.
Editors S. H. Gillespie, G. L. Smith & A. Osbourn. Cambridge University Press. ISBN 0 521 84312 X

This chapter provides an overview of the most important human diseases that are transmitted by an arthropod vector. There are several diseases which involve rodents or other mammals as the vector, such as haemorrhagic fever with renal syndrome, caused by hantaviruses transmitted by rats, hantavirus pulmonary syndrome, caused by hantaviruses transmitted by deer mice in the Americas, leptospirosis caused by bacteria found in the urine of a wide range of wild and domestic animals, or the South American haemorrhagic fever viruses Junin, Machupo, Guanarito and Sabia, and Lassa fever in West Africa, caused by arenaviruses transmitted by rodents, but these are considered to be outside the scope of this chapter. Nevertheless, pathogens that are maintained in nature by rodents or other mammals and transmitted to humans by an arthropod vector (such as Lyme disease, ehrlichiosis or plague) will be discussed.

VECTOR-BORNE VIRUS DISEASES

Yellow fever

Yellow fever virus (YFV) was introduced into the Americas as a result of the slave trade from Africa in the 17th century. The disease was rife in Cuba, and in the 1880s, Cuban physician Carlos Finlay hypothesized that it was transmitted by mosquitoes. At the conclusion of the Spanish–American war, Cuba, which became a US protectorate in 1903, was chosen as the site for a US Army Commission established to investigate yellow fever. The Commission was headed by Walter Reed and James Carroll, who within 2 years were able to demonstrate that only one species of mosquito, *Aedes aegypti*, was responsible for transmission of the causative agent, which was a filtrable virus (Reed & Carroll, 1902). This was the first human pathogen that was shown to be a virus. YFV causes a clinical disease which starts with sudden onset of acute fever. This is followed by a second phase of viscerotropic (primarily liver) dysfunction and haemorrhage which is fatal in 20–80 % of cases.

Soon after the agent was identified, successful yellow fever control programmes involving controlling the mosquito vector were introduced in both Cuba and Panama in the early 20th century, and later throughout South America, and this eliminated urban yellow fever epidemics there although the virus persisted in jungle areas in a mosquito–monkey cycle known as sylvatic or jungle yellow fever. In this form, the virus infected some 500 unvaccinated forest workers each year in South America, but in Africa sylvatic yellow fever periodically explodes from this endemic cycle to cause major epidemics in urban populations.

In the 1930s, two live attenuated yellow fever vaccines were developed, a neurotropic vaccine obtained by multiple passages in mouse brain, and the better known 17D vaccine strain derived by passaging first in mouse brain and then in chicken tissue

(Barrett, 1997). Both these vaccine strains have lost the ability to cause viscerotropic disease and do not replicate in the mosquito. The 17D strain, derived by Theiler in 1937 (Theiler & Smith, 1937), is now grown in embryonated chicken eggs, and some 20–25 million doses of vaccine for human use are produced annually in 11 different countries, much of it in Brazil and Senegal. This vaccine is one of the safest ever developed, and a single dose probably provides protective immunity for life (Monath, 2001), although revaccination after 10 years is required under International Health Regulations for a valid travel certificate.

Unfortunately, the use of this vaccine in countries where yellow fever is endemic, especially in Africa, has been limited by the extreme poverty of the people living there. In one or two countries, such as Gambia, where high rates of vaccine coverage have been achieved, the disease has been eliminated. However, since the 1980s there has been a resurgence of yellow fever both across Africa and in South America (Robertson *et al.*, 1996). This is also a potential risk to the USA, where the mosquito vector (*Ae. aegypti*) is present in urban areas in the South.

Curiously, epidemic yellow fever has never been documented in Asia (Monath, 2001). The reasons for this are unclear, but may involve a difference in susceptibility of Asian *Ae. aegypti* mosquitoes to YFV as compared to those in Africa or the Americas (Beaty & Aitken, 1979), or perhaps the presence of other flaviviruses such as *Dengue virus* (DENV) and *Japanese encephalitis complex viruses* throughout Asia induces anti-flavivirus antibodies that may provide some cross-protection against YFV (Theiler & Anderson, 1975).

Dengue fever

There are four serotypes of DENV in the genus *Flavivirus*, which together with *Kedougou virus*, an apathogenic virus found in Senegal and the Central African Republic, form the dengue virus serogroup. The four serotypes of DENV (DEN-1 to DEN-4) differ antigenically but cause clinically indistinguishable illnesses in humans. After an incubation period of 5–8 days following the bite of a mosquito, there is sudden onset of fever, severe headache, chills and muscle and joint pains. About 4 days later a maculopapular rash appears on the trunk and later spreads to the face and extremities. Lymphadenopathy and leukopenia occur, and the illness lasts some 5–7 days. A much more severe form of the disease, dengue haemorrhagic fever, which may involve skin haemorrhages, epistaxis, bleeding gums and gastrointestinal haemorrhages, is found, mainly in children under the age of 15 years, in areas where co-circulation of two or more DENV serotypes occurs in a community. The pathogenesis of this severe form of dengue fever seems to involve a process of immune activation when sequential infection with more than one subtype of the virus occurs (Gubler, 1998).

Dengue viruses are found throughout the world in tropical and subtropical areas, with approximately 3 billion persons at risk, and an estimated 50–100 million dengue virus infections each year, several hundred thousand cases of dengue haemorrhagic fever and thousands of deaths (Gubler, 2002b). Despite its importance as a disease of humans, no vaccine exists to control the disease, and control of the mosquito vector is the only option available. Unfortunately, in recent years poor mosquito control, especially in the American tropical regions, has led to a resurgence of the disease.

West Nile virus

West Nile virus (WNV) is a species in the genus *Flavivirus* which was first isolated from a human with febrile illness in Uganda in 1937. It causes a short febrile infection in children, but a more severe disease which can be fatal occurs in elderly people. Normally, a short incubation period of a few days is followed by fever, headache and myalgia, and a rash occurs in about half of the cases. After 3–6 days there is usually complete recovery. The virus can infect birds and many mammals as well as mosquitoes of several different species, though birds are the natural reservoir host.

The virus caused occasional outbreaks of human disease in Israel, France and South Africa over the 50 years following its discovery, but was not considered a serious public health problem. Then in the mid-1990s, the virulence of WNV apparently changed (Hubalek & Halouzka, 1999; Murgue *et al.*, 2001a, b), and epidemics and epizootics of severe neurological disease began to occur in the Mediterranean basin and surrounding countries, finally spreading to the USA in 1999 (Ostlund *et al.*, 2001). When examined by nucleotide sequencing, the virus that was first detected in New York in August 1999 was found to have an identical sequence to a virus isolated from an epizootic of domestic geese in Israel in 1998 (Lanciotti *et al.*, 1999).

How WNV moved from Israel to the USA is unknown, but it might have been introduced in an infected bird, either by migration or as part of the large legal and illegal traffic of exotic birds which enter the USA. Alternatively, it is not impossible that an infected mosquito was brought to the USA on an airliner from the Middle East. Finally, one or more WNV-infected human travellers with a particularly high viraemia might have brought the virus to New York and seeded it there (Gubler, 2002a).

Whatever the mode of introduction, the highly virulent WNV strain rapidly became established within the USA and proved particularly damaging to many bird species, such as the American crow (*Corvus brachyrhyncos*), thousands of which died over the 4 years following introduction of the virus. Since competent mosquito vectors for WNV exist throughout the western hemisphere, a rapid spread of WNV activity occurred south into the Caribbean, Mexico, and Central and South America, north into Canada, and west to virtually every state in the Union by the end of 2003. For example, between January and November 2003, the CDC received reports of 8567 human cases of WNV infection, 199 of which were fatal. In addition, there were reports of 11 350 dead birds and 4146 horses with WNV infection (Centers for Disease Control and Prevention, 2003b).

The public health impact of the epidemic of WNV has been severe, as there is no human vaccine against infection or any specific treatment for the disease even though a killed vaccine for use in horses has recently been licensed. The wide range of mosquito species that appear to be capable of spreading the disease makes vector control difficult, and public education in the use of personal protection against mosquito bites is the only control measure currently available for the human population (Mostashari *et al.*, 2001).

Tick-borne encephalitis virus

Tick-borne encephalitis (TBE) *virus* is a member of the genus *Flavivirus*, and so shares some of the characteristics of yellow fever, dengue and WNV. It causes a meningo-encephalitis in humans that has been recognized in Europe since the 1930s, but the disease was referred to differently as Russian spring-summer encephalitis, Far Eastern encephalitis, Taiga encephalitis, biphasic milk fever, Central European encephalitis, Kumlinge disease and other names, even though the cause and clinical course are the same in each case, the aetiology being any one of three closely related subtypes of TBE virus (European, Siberian and Far-Eastern). The disease in humans is biphasic, with a febrile illness lasting 4–10 days followed by meningitis or meningoencephalitis (Haglund & Gunther, 2003). Mild or inapparent infections occur, but in severe cases there is transient or permanent paralysis. Humans become infected by the bite of a tick, *Ixodes ricinus* or *Ixodes persulcatus*, or occasionally by consumption of milk from infected animals. TBE may also be spread by mites or mosquitoes. The major vertebrate reservoir hosts of TBE virus are rodents, mainly mice of the genera *Apodemus* or *Clethrionomys*, but many other species of domestic animals, birds and even reptiles may be parasitized by *I. ricinus* and could play a role in transmission. For most hosts, TBE virus is apathogenic. The virus replicates in the tick, which becomes infected during feeding, and the tick remains infected for life (Suss, 2003).

As methods of diagnosis of the disease have improved, involving testing of sera or cerebrospinal fluid for the presence of TBE-specific IgM and IgG antibodies using commercially available kits, the extent of the public health problem posed by TBE has been recognized, with 10 000–12 000 hospitalized cases annually. Control of tick populations was attempted in Russia but not found to be effective, and since the 1970s, inactivated vaccines have been developed and are available for human use in Europe and Asia (Barrett *et al.*, 2003). In Austria, a high-risk country for TBE, vaccination rates now exceed 90 % in some areas (Kunz, 2003).

VECTOR-BORNE RICKETTSIAL DISEASES

Typhus and spotted fever

The typhus group of rickettsiae contains two important human pathogens, *Rickettsia prowazekii* (the agent of louse-borne typhus) and *Rickettsia typhi* (the agent of flea-borne murine typhus). The former disease can be spread between humans by lice and does not require an alternative vertebrate host reservoir. It is usually found amongst displaced, crowded populations in which thousands of cases can occur in epidemics with about a 10 % fatality rate. The disease is transmitted by the bite of an infected louse or merely by inhaling aerosols of infectious louse faeces. The primary targets of infection by rickettsiae are vascular endothelial cells, and depending on the extent of vascular damage and inflammation, vasculitis may result clinically in meningo-encephalitis, pneumonitis, myocarditis or focal necrosis of the liver or kidneys. In the louse, which becomes infected by ingesting infected blood, the rickettsiae multiply rapidly within the gut cells and are then discharged in the louse faeces. The lice are killed by the infection, but survive long enough (about 2 weeks) to carry the infection to other individuals (Dasch & Weiss, 1998).

Although modern standards of hygiene have largely eliminated louse-borne typhus from most populations, outbreaks have occurred in refugee populations in Africa, and the disease is endemic in the highlands of Peru and Ethiopia (McDade, 1998). In addition, a sylvatic cycle is maintained in the Eastern USA involving the flying squirrel, *Glaucomys volans*, and its lice and fleas, and occasionally humans may become infected from this reservoir (Reynolds *et al.* 2003).

Murine typhus is a true zoonosis, and the disease is found in areas in which rats are living in close proximity to humans. Infections may cause severe disease but are rarely fatal. Although rat lice may transmit the disease between rats, transmission to humans usually involves the oriental rat flea, *Xenopsylla cheopis*. The rickettsiae grow in the midgut cells of the flea and are excreted in the faeces, but the fleas are not killed by the infection. The distribution of murine typhus is worldwide, wherever the host rats, *Rattus norvegicus* or *Rattus rattus*, are found.

The spotted fever group of rickettsiae contains at least nine pathogens of humans, and probably a few more which are emerging causes of disease. One of the most important of these is *Rickettsia rickettsii*, the cause of Rocky Mountain spotted fever, the aetiology of which was elucidated by Howard Taylor Ricketts in a series of experiments at the beginning of the 20th century (Ricketts, 1909).

The disease is an acute febrile illness, with sudden onset of headache and chills followed by a fever that persists for 2–3 weeks, and a characteristic rash appearing on the trunk and extremities on about the 4th day of the disease. Delirium, shock and renal failure may occur in severe cases, and despite the use of antibiotics, the case-fatality rate has remained at about 5 %. The disease occurs widely throughout South, Central and North America, where it is transmitted by several genera of ixodid ticks (*Dermacentor*, *Ixodes*, *Rhipicephalus* and *Amblyomma*) depending upon the locality. The ticks serve both as reservoirs and vectors of spotted fever group rickettsiae. The infection is maintained in the tick during all stages of the developmental cycle, and most tissues are infected, including the salivary glands. Humans coming into contact with ticks become infected during the feeding process.

Other spotted fever group rickettsiae of note include *Rickettsia conorii*, causing Mediterranean spotted fever, *Rickettsia japonica*, causing Oriental spotted fever, and *Rickettsia sibirica*, causing Siberian tick typhus. Also included in this group is *Rickettsia akari*, the cause of rickettsialpox, but notably the vector for this is a mite, *Liponysoides sanguinus*, that feeds on the house mouse, *Mus muculus* (Kass *et al.*, 1994).

Finally, an important disease of the Asia-Pacific region, spread by chigger mites, particularly of the genus *Leptotrombidium*, is scrub typhus caused by a single species, *Orientia tsutsugamushi*. Transmission of this disease, which generally presents with a fever, rash and eschar at the site of the bite, is limited to the larval stage (chigger) of the mite, which occurs in vegetation and usually feeds on rodents but will also bite humans who come into contact with it. Occasionally the disease can be fatal, with generalized organ failure. Treatment with doxycycline is usually effective, however.

Ehrlichiosis and related diseases

The *Ehrlichieae* are obligate intracellular bacteria that are emerging pathogens of both medical and veterinary importance worldwide. At least two forms of human ehrlichiosis have been described, human monocytic ehrlichiosis (HME) and human granulocytic ehrlichiosis (HGE), with case-fatality ratios in the United States of between 1 and 5 %. The first cases of HME were described in the United States in 1986, and the aetiologic agent was subsequently found to be *Ehrlichia chaffeensis* (Anderson *et al.*, 1991). Since that time several hundred cases of the disease have been reported in the USA (McQuiston *et al.*, 1999) and a retrospective study conducted 10 years later among troops training at Fort Chaffee, Arkansas, revealed that some 15 % were seropositive to *Rickettsia* or *Ehrlichia* species, the major risk factor being tick bites (McCall *et al.*, 2001). Dogs are susceptible to *E. chaffeensis* infection without showing clinical signs and may serve as a natural host (Zhang *et al.*, 2003). White-tailed deer, *Odocoileus virginianus*, are also susceptible, and the tick *Amblyomma americanum* is an important vector in disease transmission. HGE was first described in 1994, and is caused by an agent closely related to *Ehrlichia equi*, which causes equine granulocytic ehrlichiosis, and *Ehrlichia phagocytophila*, which causes disease in ruminants in Europe. In 1999, *Ehrlichia ewingii*, a veterinary pathogen, was shown to cause granulo-cytic ehrlichiosis in humans. Patients infected with *E. ewingii* experience symptoms similar to those observed in other human ehrlichioses, such as fever, headache and myalgia. These ehrlichial agents are all transmitted to humans by ticks and maintained in nature in various vertebrate reservoirs.

Other ehrlichial agents causing human and veterinary diseases include *Neorickettsia sennetsu*, causing senetsu fever in Asia, *Neorickettsia risticii*, which causes equine monocytic ehrlichiosis (Potomac horse fever) in the USA, and *Neorickettsia helminth-oeca*, which causes salmon poisoning in dogs and may also infect cats, found on the West coast of the USA. These agents are all associated with flukes of fish and snails (Dasch & Weiss, 1998; Cohn, 2003) and may have a worldwide distribution, as described in a recent report from China (Wen *et al.*, 2003).

VECTOR-BORNE BACTERIAL DISEASES

Plague

Plague is a disease of animals and humans caused by a single bacterial species, *Yersinia pestis*. The commonest clinical disease in humans is acute febrile lymphadenitis called bubonic plague, but other forms include septicaemic, pneumonic and meningeal plague. The major animal reservoirs of the bacterium are rodents, particularly urban and domestic rats, but squirrels, rabbits and prairie dogs are less common hosts. The vector which transmits the disease between animals and from animals to man is the flea, especially the species *X. cheopis*, the oriental rat flea, although it is estimated that more than 150 species of flea are naturally infected with *Y. pestis* throughout the world. When the flea ingests a blood meal from a bacteraemic animal infected with *Y. pestis*, the coagulase of the organism causes the ingested blood to clot in the foregut of the flea, blocking the flea's swallowing, and the bacteria then multiply in the clotted blood (Bibikova, 1977). The flea may regurgitate thousands of organisms into the animal's skin during attempts to swallow. The inoculated bacteria migrate from cutaneous lymphatics to the regional lymph nodes. The disease in rodents and other animals varies from subclinical bacteraemia to sudden death due to overwhelming septicaemia, and periodically there may be large die-offs of rodents in endemic areas.

Outbreaks of plague still occur throughout the world. WHO receives reports of 1000–3000 cases each year, but in the USA the last urban plague epidemic occurred in Los Angeles in 1924–1925, and since then plague has occurred as mostly scattered cases in rural areas, with some 10–15 cases each year. Prompt treatment with antibiotics is used to control the clinical disease in humans, but if several cases of plague are observed in one locality, there is an immediate need to prevent transmission from animal reservoirs by fleas. This involves both insecticide dusting to kill infected fleas followed by rodent control by trapping and use of rodenticides.

Lyme disease

Lyme disease was named in 1977 when arthritis occurred in a cluster of children living in and around Lyme, Connecticut. A number of factors suggested that it might have a vector-borne origin, and in 1981 the spirochaete causing Lyme disease was isolated from deer ticks (*Ixodes scapularis*) collected from Shelter Island, New York (Burgdorfer *et al.*, 1982). The spirochaete belonged to the genus *Borrelia*, of which there are now 23 species of medical and veterinary significance throughout the world, but the Lyme disease agent *Borrelia burgdorferi sensu stricto* is the only species known to cause Lyme disease in the USA. In Europe, two other species, *Borrelia garinii* and *Borrelia afzelii*, are also causes of Lyme disease in addition to *B. burgdorferi*, and the tick vectors differ between the two continents. In the USA where it is a notionally notifiable disease, Lyme borreliosis is transmitted by *I. scapularis* ticks in the north-east and north-central regions but by *Ixodes pacificus* along the Pacific coast from California to British Columbia and in Nevada and Utah (Steere, 2001). During 1992, 23 763 cases of Lyme disease were reported (Centers for Disease Control and Prevention, 2003a). In western and eastern Europe, Lyme borreliosis is transmitted by *I. ricinus* ticks, and in eastern Europe and Asia it is transmitted by *I. persulcatus* (Hengge *et al.*, 2003). The disease may have a significant cost to public health prevention measures, and a recent estimate of the annual cost of Lyme disease in Scotland alone was £331 000 (about $500 000) (Joss *et al.*, 2003).

Lyme disease in humans is a multisystem disorder which can become a chronic disabling infection if not treated promptly with antibiotics. Classically, about 1 month after the bite of an infected tick, there is a cutaneous response known as erythema migrans. This may expand over the next 4 weeks with associated symptoms of malaise, lethargy, headache and generalized arthralgia. If untreated, there may also be cardiac and neurological symptoms and a variety of musculoskeletal manifestations. There is some evidence that the clinical symptoms of Lyme disease differ in Europe as compared to America (Weber & Pfister, 1994), but the basis for this is unclear.

The principal vertebrate reservoir hosts for Lyme borreliosis may include many species of mammals and birds on which the ticks feed, and deer may play an important role in determining the size of the vector population (Mahy, 2000). However, small rodents such as *Peromyscus leucopus* in North America and *Apodemus* and *Clethrionomys* species in Europe are clearly important hosts for sub-adult forms (larval and nymph stages) of the ticks.

Vaccines for prevention of Lyme disease have been developed, based on recombinant *B. burgdorferi* outer-surface glycoprotein, and these seemed to be safe and effective but were withdrawn from the market due to poor public acceptance (Hengge *et al.*, 2003).

VECTOR-BORNE PROTOZOAN DISEASES

African trypanosomiasis

African trypanosomiasis (sleeping sickness) results from infection by protozoal parasites of the genus *Trypanosoma*. There are two epidemiologically and clinically distinct forms of the disease, caused by distinct subspecies of *Trypanosoma brucei*. *T. brucei rhodesiense* causes East African trypanosomiasis, an acute fulminant febrile illness that occurs in East and Southern Africa, where there are several animal reservoirs of the disease, and humans are incidental hosts. If untreated, East African trypanosomiasis is fatal within 9 months. By contrast, West African trypanosomiasis caused by the subspecies *T. brucei gambiense* has a human reservoir, and has a chronic, progressive clinical course which if untreated may last 4 years before death occurs. Both these agents are transmitted by the bite of the tsetse fly, but the species differ: East African trypanosomiasis is transmitted by *Glossina morsitans*, whereas in West Africa the vector is *Glossina palpalis*.

Trypanosomes appear able to undergo exchange of genetic information, particularly in the salivary glands of the tsetse fly, and this seems to be the basis for a continuing process of antigenic change, which allows the protozoa to resist the host immune response. Variant antigenic types are continually being created by mutations, addition, deletion and recombination during trypanosome growth. Trypanosomiasis, in both animals and humans, is characterized by waves of parasitaemia. Each rise stimulates an immune response by the host to one of the surface coat antigens which may clear the bulk of the parasitaemia, to be followed by a new wave caused by a minor variant in the population which is resistant to the host antibody.

Control of this disease is difficult through vector measures such as dipping cattle in insecticide or the use of insecticide-coated screens, and various approaches to reducing tsetse fly populations by sterilization have been attempted (Langley *et al.*, 1988). For treatment of human cases, two drugs are used: suramin is used in the primary stages, but this drug does not cross the blood–brain barrier, so in later stage disease where the CNS may be involved, melarsoprol is used. There is some evidence that adverse effects of melarsoprol are more common in patients who are also infected with human immunodeficiency virus (HIV). Neither suramin nor melarsoprol are licensed for use in the USA, but occasional imported cases occur, and for these CDC provides clinicians with the drugs for use under an investigational new drug protocol.

New World trypanosomiasis

The Brazilian scientist Carlos Chagas first identified a new protozoan flagellate in the faeces of a large blood-sucking triatomine bug (*Panstrongylus megistus*), and named it after his mentor, Oswaldo Cruz, as *Trypanosoma cruzi*. Chagas' disease, in turn, was named after its discoverer and is now known to be endemic in much of Central and South America, with some 20 million people infected and a further 100 million at risk for the disease. In addition to spread by the bug vector, which is associated with poor housing and economic development, there is increasingly recognized spread of the disease through blood transfusion. Some 10 % of persons who become infected with Chagas' disease will die within a few weeks, and those who survive the acute phase normally go on to develop severe chronic cardiac or gastrointestinal complications. However, the pathogenesis of Chagas' disease is only poorly understood.

The triatomine bug vector may infect numerous reservoir hosts with *T. cruzi*, and at least 150 species of mammals are known to be susceptible. Domestically, dogs are an important host, and in the wild, opossums (*Didelphis* species) are a widespread species with frequent contact with triatomine bugs.

There are five recognized species of bug which transmit *T. cruzi* in the New World. They are *Triatoma infestans*, *Rhodnius prolixus*, *Panstrongylus megistus*, *Triatoma brasiliensis* and *Triatoma dimidiata*. All are adapted to domestic habitats and are responsible for abundant and widespread household infections.

Control of Chagas' disease in the New World has concentrated on the use of insecticides to reduce or eliminate bug populations, and various schemes have been undertaken to this effect (Dias, 1987; Miles, 1992). However, an exciting new approach under development is to attack the problem by reducing the ability of the bugs to transmit trypanosomes by genetic transformation of symbiotic bacteria in the bugs so that they express cecropin A, a peptide which is lethal to trypanosomes and eventually eliminates them (Durvasula *et al.*, 1997). The hope is that this endosymbiont bacterium can be made to spread naturally within the bug population and so provide a novel means of Chagas' disease control, which may eventually also be applicable to other trypanosome diseases (Dotson *et al.*, 2003).

Malaria

Malaria is the most important infectious disease threat for global public health, and more than 40 % of the world population (more than 2 billion people) live in areas where malaria occurs. The malaria parasites are species within the genus *Plasmodium*, and four species are important for human disease: *Plasmodium falciparum*, *Plasmodium vivax*, *Plasmodium malariae* and *Plasmodium ovale*. The infection is caused by the bite of an infected *Anopheles* mosquito. The most serious clinical manifestations of malaria, such as cerebral malaria, severe anaemia and placental malaria, are caused by *P. falciparum*. Following the bite of an infected female mosquito, when the sporozoite stage of the organism is introduced into the bloodstream together with the mosquito's saliva, the sporozoite makes its way to the liver, where it invades a hepatocyte, differentiates and multiplies to produce a schizont, containing several thousand merozoites (10 000–30 000 depending on species). The merozoites invade erythrocytes in the bloodstream, then after multiplication and differentiation an erythrocytic schizont is released containing more merozoites (4–24 depending on species) which, upon rupture of the schizont, in turn invade more erythrocytes. Some of the merozoites then undergo sexual differentiation and gametocytogenesis. The male and female gametocytes remain dormant in the bloodstream waiting to be picked up in the blood meal of a new biting mosquito. Once in the insect midgut fertilization occurs to form an oocyst. This divides by schizogony to produce thousands of sporozoites which migrate to the salivary glands of the mosquito, completing the cycle.

In a non-immune person, there is an incubation period of on average 12–28 days followed by non-specific influenza-like symptoms and usually fever, followed by prostration and evidence of jaundice.

In severe falciparum malaria, in addition to severe anaemia and lactic acidosis, the brain may be involved, and patients may lapse into a coma; the mortality is usually 20 %. The most serious complication is pulmonary oedema, which carries a mortality of more than 50 % (Charoenpan *et al.*, 1990).

The only effective treatment for malaria involves chemotherapy (White, 1988). However, since the parasites have developed resistance to many antimalarial drugs, the choice will depend upon the specific treatment regime recommended for a particular geographical area. Drugs in common use include choroquine, quinine, mefloquine, halofantrine, artesunate and primaquine. Drug resistance of *P. falciparum* to the various antimalarials continues to increase in both intensity and geographical distribution. Control of the *Anopheles* mosquito vector populations has also declined due to insecticide resistance and lack of political will in most affected regions of the world, even though personal protection through the use of pyrethroid-treated bednets has been shown to be a very effective means of malaria control (Meek, 1995).

Cases of malaria occurring in Asia provide a dramatic example of the failure of control measures over the past 40 years (Gubler, 1998). Sri Lanka nearly eliminated malaria in the 1960s, with only 31 and 17 cases reported in 1962 and 1963, respectively. However, by 1967, 3468 cases were reported, there was a major epidemic in 1968 causing 440 644 cases, and in 1969 537 705 cases were reported. Malaria has not been effectively controlled in Sri Lanka since then. A similar situation occurred in India, where there were 7 million cases in 1976. In Africa, as in Asia, the disease is on the increase (Nchinda, 1998), and this led to the 1998 WHO Roll Back Malaria initiative which has the goal to halve the burden of malaria in Africa by 2010. However, this programme has been criticized as having inadequate resources for the task (Narasimhan & Attaran, 2003).

On a worldwide basis, there are now estimated to be more than 300–500 million malaria cases and 1–2 million deaths each year. A great deal of effort is now being put into the development of an effective malaria vaccine, particularly through the Malaria Vaccine Initiative, created with a $50 million grant from the Gates Foundation. There are excellent prospects for success in this area, though the vaccines remain in the predevelopment stage, a long way from their use in humans.

VECTOR-BORNE NEMATODE DISEASES

Lymphatic filariasis

Three species of filarial nematode worms are of major significance to global public health: *Onchocerca volvulus*, *Wuchereria bancrofti* and *Brugia malayi*. The vector for transmission of onchocerciasis worms is the black fly (*Simulium* species), which breeds along fast-flowing rivers, hence the name 'river blindness' given to onchocerciasis. Both *Wuchereria* and *Brugia* are transmitted by mosquitos. Together, these parasites infect an estimated 118 million people worldwide (World Health Organization, 1995).

The disease caused by these worms in humans results from the presence of adult worms in the lymph nodes and vessels, but it may take many months following infection before the adult worms mature, and many years before chronic disease develops. In onchocerciasis, the severity of the symptoms depends upon the length of the infection, and visual damage, the most serious clinical feature of the disease, develops in response to heavy microfilarial loads. The prevalence of visual loss increases with age. In filariasis caused by *Wuchereria* or *Brugia*, it may be several months before adult females in the lymph vessels produce microfilaria, and some individuals may have microfilaraemia for many years without developing disease symptoms. Eventually, attacks of filarial fever lasting 3–15 days accompanied by headache and localized pain in the lymphatic areas in the groin and armpits occur. These attacks may begin as early as 6 years of age and may continue until 25 years of age, when eventually the lymphatics become fibrosed and the nodes calcify. Elephantiasis may develop in the legs, and in males orchitis and hydrocele may occur with inguinal lymphadenopathy.

The possible control of filariasis traditionally has involved chemotherapy with diethylcarbamazine to kill the microfilaria, which is even more effective when used in combination with ivermectin and albendazole (Ottesen & Ramachandran, 1995; Addiss & Dreyer, 2000). However, much attention has been focused recently on the possible use of drugs, especially doxycycline, to attack *Wolbachia* bacteria, ehrlichia-like agents which have been found to be obligate endosymbionts required for the fertility of many species of insects and of female nematode worms (Townson, 2002; Hoereauf, 2003). There is also evidence that *Wolbachia* surface protein may contribute to the inflammatory disease seen in onchocerciasis and filariasis (Taylor, 2003; Punkosdy *et al.*, 2003; Hise *et al.*, 2004). Contrary to this view is evidence that *Wolbachia* symbionts are absent from some pathogenic filaria such as *Loa loa* (Buttner *et al.*, 2003; Grobusch *et al.*, 2003; McGarry *et al.*, 2003). However, the need for additional drugs to aid in the treatment and eventual elimination of filariasis and onchocerciasis has led to clinical trials with doxycycline in Ghana, with encouraging preliminary results (Hoerauf *et al.*, 2003).

Since there is no zoonotic reservoir, the prospects for elimination of filarial nematodes from the human population has led to the designation by WHO of filariasis as a potentially eradicable disease (Taylor & Turner, 1997).

PERSPECTIVE AND CONCLUSIONS

Although more than 100 years have passed since the recognition of the importance of arthropod vectors in the transmission of many serious human pathogens, control of these diseases has not been achieved, and in recent years there has been a resurgence as a result of changes in public health policy, social factors, and the development of insecticide and drug resistance in the vectors and the pathogens they carry. In terms of public health policy, vector control programmes have frequently appeared vulnerable to funding reductions or elimination, especially when the disease in question appears to have been controlled. Societal perceptions have also complicated the control of vector populations, especially by restricting the use of insecticides for vector control. The perceived environmental dangers from the use of DDT outweighed its benefits in the mind of the public, so it was replaced in many countries by new, more expensive, less stable and less effective insecticides. This widespread cessation of DDT use has coincided with a resurgence of dengue fever in the Americas (Gubler, 2002b) and of malaria in many parts of the world (Editorial, 2000). The decision to stop the use of DDT for mosquito control is now being called into question (Attaran *et al.*, 2000).

Few vaccines are available for these vector-borne diseases, though in recent years funds for vaccine development have become available through a number of philanthropic organizations such as the Gates Foundation. Yellow fever is an exception, as an excellent vaccine is available for those at risk, but this also highlights the problem of vaccine application, since many of those needing the vaccine cannot afford it.

As the world population continues to rise, the problems of vector-borne disease control will magnify. New biological methods of vector control, using viruses or bacteria, are under development, and more research is needed on the molecular basis of pathogen–vector–host interactions. The possible control of parasites and their insect vectors by attacking their obligate endosymbiotic bacteria is a promising area for further research, as are prospects for the development of vaccines against parasites (Willadsen, 2001). We must always remain cognizant, however, that the pathogens, as well as their vectors, have in the past, and will in the future, try to find ways to breach our human defences.

ACKNOWLEDGEMENTS

I thank David Addiss and Peter Bloland, colleagues in the Division of Parasitic Diseases at CDC, for helpful advice and discussion.

REFERENCES

Addiss, D. G. & Dreyer, G. (2000). Treatment of lymphatic filariasis. In *Lymphatic Filariasis (Tropical Medicine: Science and Practice)*, pp. 151–199. Edited by T. B. Nutman. London: Imperial College Press.

Anderson, B. E., Dawson, J. E., Jones, D. C. & Wilson, K. H. (1991). *Ehrlichia chaffeensis*, a new species associated with human ehrlichiosis. *J Clin Microbiol* **29**, 2838–2842.

Attaran, A., Roberts, D. R., Curtis, C. F. & Kilama, W. L. (2000). Balancing risks on the backs of the poor. *Nat Med* **6**, 729–731.

Barrett, A. D. T. (1997). Yellow fever vaccines. *Biologicals* **25**, 17–25.

Barrett, P. N., Schober-Bendixen, S. & Ehrlich, H. J. (2003). History of TBE vaccines. *Vaccine* **21**, S1/41–S1/49.

Beaty, B. J. & Aitken, T. H. G. (1979). *In vitro* transmission of yellow fever virus by geographic strains of *Aedes aegypti*. *Mosq News* **39**, 232–238.

Bibikova, V. A. (1977). Contemporary views on the interrelationships between fleas and the pathogens of human and animal diseases. *Annu Rev Entomol* **22**, 23–32.

Burgdorfer, W., Barbour, A. G., Hayes, S. F., Benach, J. L., Grunwaldt, E. & Davis, J. P. (1982). Lyme disease – a tick-borne spirochetosis? *Science* **216**, 1317–1319.

Buttner, D. W., Wanji, S., Bazzocchi, C., Bain, O. & Fischer, P. (2003). Obligatory symbiotic *Wolbachia* bacteria are absent from *Loa loa*. *Filaria J* **2**, 10.

Centers for Disease Control and Prevention (2003a). Notice to readers: final 2002 reports of notifiable diseases. *Morb Mortal Wkly Rep* **52**, 741–750.

Centers for Disease Control and Prevention (2003b). West Nile virus activity – United States, November 20–25, 2003. *Morb Mortal Wkly Rep* **52**, 1160.

Charoenpan, P., Indraprasit, S., Kiatboonsri, S., Suvachittanont, O. & Tanomsup, S. (1990). Pulmonary oedema in severe falciparum malaria. Haemodynamic study and clinicophysiologic correlation. *Chest* **97**, 1190–1197.

Cohn, L. A. (2003). Ehrlichiosis and related infections. *Vet Clin N Am Small Anim Pract* **33**, 863–864.

Dasch, G. A. & Weiss, E. (1998). The Rickettsiae. In *Topley & Wilson's Microbiology and Microbial Infections*, vol. 2, *Systematic Bacteriology*, pp. 853–876. Edited by L. Collier, A. Balows, M. Sussman & B. I. Duerden. London: Arnold.

Dias, J. C. P. (1987). Control of Chagas disease in Brazil. *Parasitol Today* **3**, 336–341.

Dotson, E. M., Plikaytis, B., Shinnick, T. M., Durvasula, R. V. & Beard, C. B. (2003). Transformation of *Rhodococcus rhodnii*, a symbiont of the Chagas disease vector *Rhodnius prolixus*, with integrative elements of the L1 mycobacteriophage. *Infect Genet Evol* **3**, 103–109.

Durvasula, R. V., Gumbs, A., Panackal, A., Kruglov, O., Aksoy, S., Merrifield, R. B., Richards, F. F. & Beard, C. B. (1997). Prevention of insect-borne disease: an approach using transgenic symbiotic bacteria. *Proc Natl Acad Sci U S A* **94**, 3274–3278.

Editorial (2000). Malaria: old challenge, new ideas. *Nat Med* **6**, 1.

Grobusch, M. P., Kombila, M., Autenrieth, I., Mehlhorn, H. & Kremsner, P. G. (2003). No evidence of *Wolbachia* endosymbiosis with *Loa loa* and *Mansonella perstans*. *Parasitol Res* **90**, 405–408.

Gubler, D. J. (1998). Resurgent vector-borne diseases as a global health problem. *Emerg Infect Dis* **4**, 442–450.

Gubler, D. J. (2002a). The global emergence/resurgence of arboviral diseases as public health problems. *Arch Med Res* **33**, 330–342.

Gubler, D. J. (2002b). Epidemic dengue/dengue hemorrhagic fever as a public health, social and economic problem in the 21st century. *Trends Microbiol* **10**, 100–103.

Haglund, M. & Gunther, G. (2003). Tick-borne encephalitis – pathogenesis, clinical course and long-term follow-up. *Vaccine* **21**, S1/11–S1/18.

Hengge, U. R., Tannapfel, A., Tyring, S. K., Erbel, R., Arendt, G. & Ruzicka, T. (2003). Lyme borrelliosis. *Lancet Infect Dis* **3**, 489–500.

Hise, A. G., Gillette-Ferguson, I. & Pearlman, E. (2004). The role of endosymbiotic *Wolbachia* bacteria in filarial disease. *Cell Microbiol* **6**, 97–104.

Hoerauf, A. (2003). Control of filarial infections: not the beginning of the end, but more research is needed. *Curr Opin Infect Dis* **16**, 403–410.

Hoerauf, A., Mand, S., Volkmann, L., Buttner, M., Marfo-Debrekei, Y., Taylor, M., Adjei, O. & Buttner, D. W. (2003). Doxycycline in the treatment of human onchocerciasis: kinetics of *Wolbachia* endobacteria reduction and of inhibition of embryogenesis in female *Onchocerca* worms. *Microbes Infect* **5**, 261–273.

Hubalek, Z. & Halouzka, J. (1999). West Nile fever – a reemerging mosquito-borne viral disease in Europe. *Emerg Infect Dis* **5**, 643–650.

Joss, A. W., Davidson, M. M., Ho-Yen, D. O. & Ludbrook, A. (2003). Lyme disease – what is the cost for Scotland? *Public Health* **117**, 264–273.

Kass, E. M., Szianiawski, W. K., Levy, H., Leach, J., Srinivasan, K. & Rives, C. (1994). Rickettsialpox in a New York City hospital, 1980 to 1989. *N Engl J Med* **331**, 1612–1617.

Kunz, C. (2003). TBE vaccination and the Austrian experience. *Vaccine* **21**, S1/50–S1/55.

Lanciotti, R. S., Roehrig, J. T., Deubel, V. & 21 other authors (1999). Origin of the West Nile virus responsible for an outbreak of encephalitis in the northeastern United States. *Science* **286**, 2333–2337.

Langley, P. A., Felton, T. & Doichi, H. (1988). Juvenile hormone mimics as effective sterilants for the tsetse fly *Glossina morsitans morsitans*. *Med Vet Entomol* **2**, 29–35.

Mahy, B. W. J. (2000). The global threat of emerging infectious diseases. In *Fighting Infection in the 21st Century* (*Society for Applied Microbiology/Society for General Microbiology Symposium*), pp. 1–16. Edited by F. W. Andrew, P. Oyston, G. L. Smith & D. E. Stewart-Tull. Oxford: Blackwell Science.

Manson, P. (1878). On the development of *Filaria sanguis hominis* and on the mosquito considered as a nurse. *J Linn Soc* (*Zool*) **14**, 304–311.

McCall, C. L., Curns, A. T., Rotz, L. D., Singleton, J. A., Jr, Treadwell, T. A., Comer, J. A., Nicholson, W. L., Olson, J. G. & Childs, J. E. (2001). Fort Chaffee revisited: the epidemiology of tick-borne rickettsia and ehrlichial diseases at a natural focus. *Vector Borne Zoonotic Dis* **1**, 119–127.

McDade, J. E. (1998). Rickettsial diseases. In *Topley & Wilson's Microbiology and Microbial Infections*, vol. 3, *Bacterial Infections*, pp. 995–1011. Edited by L. Collier, A. Balows, M. Sussman & W. J. Hausler. London: Arnold.

McGarry, H. F., Pfarr, K., Egerton, G. & 8 other authors (2003). Evidence against *Wolbachia* symbiosis in *Loa loa*. *Filaria J* **2**, 9.

McQuiston, J. H., Paddock, C. D., Holman, R. C. & Childs, J. E. (1999). The human ehrlichioses in the United States. *Emerg Infect Dis* **5**, 635–642.

Meek, S. R. (1995). Vector control in some countries of South-East Asia: comparing the vectors and the strategies. *Ann Trop Med Parasitol* **89**, 135–147.

Miles, M. A. (1992). Disease control has no frontiers. *Parasitol Today* **8**, 221–222.

Monath, T. P. (2001). Yellow fever: an update. *Lancet Infect Dis* **1**, 11–20.

Mostashari, F., Bunning, M. L., Kitsutani, P. T. & 13 other authors (2001). Epidemic West Nile encephalitis, New York, 1999; results of a household-based seroepidemiological survey. *Lancet* **358**, 261–264.

Murgue, B., Murri, S., Zientara, S., Durand, B., Durand, J.-P. & Zeller, H. (2001a). West Nile outbreak in horses in southern France: the return after 35 years. *Emerg Infect Dis* **7**, 692–696.

Murgue, B., Murri, S., Triki, H., Deubel, V. & Zeller, H. G. (2001b). West Nile in the Mediterranean Basin: 1950–2000. *Ann N Y Acad Sci* **951**, 117–126.

Narasimhan, V. & Attaran, A. (2003). Roll Back Malaria? The scarcity of international aid for malaria control. *Malar J* **2**, 8.

Nchinda, T. C. (1998). Malaria: a reemerging disease in Africa. *Emerg Infect Dis* **4**, 398–403.

Ostlund, E. N., Crom, D. D., Pederson, D. J., Johnson, D. J., Williams, W. O. & Schmitt, B. J. (2001). Equine West Nile encephalitis, United States. *Emerg Infect Dis* **7**, 665–669.

Ottesen, E. A. & Ramachandran, C. P. (1995). Lymphatic filariasis and disease control strategies. *Parasitol Today* **11**, 129–131.

Punkosdy, G. A., Addiss, D. G. & Lammie, P. J. (2003). Characterization of antibody responses to *Wolbachia* surface protein in humans with lymphatic filariasis. *Infect Immun* **71**, 5104–5114.

Reed, W. & Carroll, J. (1902). The etiology of yellow fever. *Am Med* **3**, 301–305.

Reynolds, M. G., Krebs, J. W., Comer, J. A. & 9 other authors (2003). Flying squirrel-associated typhus, United States. *Emerg Infect Dis* **9**, 1341–1343.

Ricketts, H. T. (1909). A micro-organism which apparently has a specific relationship to Rocky Mountain spotted fever. *JAMA (J Am Med Assoc)* **52**, 379–380.

Robertson, S. E., Hull, B. P., Tomori, O., Bele, O., LeDuc, J. W. & Esteves, K. (1996). Yellow Fever: a decade of reemergence. *JAMA (J Am Med Assoc)* **276**, 1157–1162.

Steere, A. C. (2001). Lyme disease. *N Engl J Med* **345**, 115–125.

Suss, J. (2003). Epidemiology and ecology of TBE relevant to the production of effective vaccines. *Vaccine* **21**, S1/19–S1/35.

Taylor, M. J. (2003). Wolbachia in the inflammatory pathogenesis of human filariasis. *Ann N Y Acad Sci* **990**, 444–449.

Taylor, M. J. & Turner, P. F. (1997). Control of lymphatic filariasis. *Parasitol Today* **13**, 85–86.

Theiler, M. & Anderson, C. R. (1975). The relative resistance of dengue-immune monkeys to yellow fever virus. *Am J Trop Med Hyg* **24**, 115–117.

Theiler, M. & Smith, H. H. (1937). The use of yellow fever virus modified by *in vitro* cultivation for human immunization. *J Exp Med* **65**, 787–800.

Townson, H. (2002). Wolbachia as a potential tool for suppressing filarial transmission. *Ann Trop Med Parasitol* **96**, S117–S127.

Weber, K. & Pfister, H. W. (1994). Clinical management of Lyme borreliosis. *Lancet* **343**, 1017–1020.

Wen, B., Cao, W. & Pan, H. (2003). Ehrlichiae and ehrlichial diseases in China. *Ann N Y Acad Sci* **990**, 45–53.

White, N. J. (1988). Drug treatment and prevention of malaria. *J Pharmacol* **34**, 1–14.

Willadsen, P. (2001). The molecular revolution in the development of vaccines against ectoparasites. *Vet Parasitol* **101**, 353–368.

World Health Organization (1995). Lymphatic filariasis: ready now for global control. *Parasitol Today* **11**, part 4 (centrefold).

Zhang, X. F., Zhang, J. Z., Long, S. W., Ruble, R. P. & Yu, X. J. (2003). Experimental *Ehrlichia chaffeensis* infection in beagles. *J Med Microbiol* **52**, 1021–1026.

Evolution of tick-borne disease systems

Sarah E. Randolph

Department of Zoology, South Parks Road, Oxford OX1 3PS, UK

INTRODUCTION – BIOLOGICAL CONSTRAINTS ON PATHOGENS' USE OF TICKS AS VECTORS

All vector-borne disease systems involve a triangle of dynamic reciprocal interactions between vector, host and pathogen, which is subject to two major classes of constraints. The intrinsic physical, physiological, cellular barriers determine whether the route of transmission is possible. The pathogen must be adapted to invade, establish and survive in vertebrate and invertebrate tissues alternately, and is also at the mercy of the vector–host interaction upon which depends the passage of fluids (typically vertebrate blood and invertebrate saliva) as a vehicle for infective particles. Transmission is not simply a mechanical matter, but rather a biological process whose complexity at the cellular and molecular level is still the focus of much research aimed at developing blocking agents. These biological barriers, which constitute the innate immunity of host and vector, are not absolute, but become more or less leaky through slow evolutionary change. By comparison, the second class of constraints, ecological and epidemiological hurdles, may change much more rapidly. They depend on the quantitative balance between all the factors determining the rates of transmission: transmission may be biologically possible but may not always occur effectively enough to allow persistent cycles of transmission. Such hurdles are determined to a large extent by the impact of extrinsic multifactorial environmental factors on the vector–host–pathogen triangle, and also by the impact of intrinsic factors such as host acquired immunity.

In this essay, I suggest that the evolution of tick-borne pathogens is *driven* by intrinsic biological barriers, but is *directed* and *constrained* by extrinsic environmental factors.

SGM symposium 63: Microbe–vector interactions in vector-borne diseases.
Editors S. H. Gillespie, G. L. Smith & A. Osbourn. Cambridge University Press. ISBN 0 521 84312 X ©SGM 2004

I shall illustrate the reasoning behind this hypothesis with the RNA flaviviruses of the tick-borne encephalitis (TBE) complex, contrasted with the bacterial spirochaete complex of *Borrelia burgdorferi sensu lato* (*s. l.*), causative agents of Lyme borreliosis. Both are transmitted principally by the ticks *Ixodes ricinus* and *Ixodes persulcatus* through large parts of Europe and Russia, and both appear to have speciated in response to the variable ability of different vertebrate species to act as competent transmission hosts if bitten by infected ticks. The extrinsic constraints on *B. burgdorferi s. l.*, however, are much less intense than those on TBE-complex viruses, largely because the outcome of the intrinsic host–pathogen interaction is more favourable for the spirochaetes than for the viruses. Despite strong stimulation of the adaptive immune system, *B. burgdorferi s. l.* persists in its vertebrate host for very long periods, possibly lifelong under natural conditions (Kurtenbach *et al.*, 2002a). This significantly enhances the transmission potential, which is approximately ten times greater than for TBE virus (Randolph *et al.*, 1996), which remains infective in the vertebrate host for only a few days.

Potential rate of evolutionary change in tick-borne pathogens

A wide variety of different pathogens (including several rickettsial and protozoan species in addition to the viruses and bacteria already mentioned) have overcome the problems associated with utilizing *I. ricinus* as a vector. Nevertheless, all these pathogens face major barriers due to the peculiar biology of ticks as blood feeders, which have a direct impact on the potential rate of evolutionary change. Hard ticks of the family Ixodidae feed only once per life stage, as larvae, nymphs and adults. Each meal by *I. ricinus* takes from about 3 to 10 days to complete, after which the ticks detach from their hosts, drop to the ground and develop to the next stage over a variable period of months depending on ambient temperatures. Therefore, a tick of one stage that picks up pathogens in its blood meal from an infected host cannot transmit them to a new host until it feeds again as the next stage. This introduces the absolute necessity for transstadial maintenance, with the optional extra of transovarial transmission and vertical amplification from one female of one generation via the eggs to many larvae of the next generation. It also introduces a long delay in the transmission process that varies geographically and seasonally depending on climate, during which a high percentage of ticks die, especially in dry summers. Thus seasonal climate that causes ticks to show strong seasonal population dynamics has a significant impact on the pace of transmission, and also on the force of transmission because the longer the delay in transmission, the fewer infected ticks survive (Randolph, 1998). It also determines the very existence of persistent TBE virus cycles (see below).

This protracted mode of transmission by ticks (with each complete cycle measured in months rather than in days as for transmission by repeat-feeding insect vectors) is likely to act as a brake on evolutionary change (Gritsun *et al.*, 1995; Zanotto *et al.*, 1995), despite the potentially high rate common to such RNA viruses (Domingo *et al.*, 1997). Indeed, amongst the flaviviruses, the tick-borne species show an apparent rate of evolution estimated to be about 56 % of the rate seen in the insect-borne species (Zanotto *et al.*, 1996).

TICK-BORNE FLAVIVIRUS EVOLUTION

Driven by intrinsic vertebrate host factors

The evolutionary history of the flaviviruses has been deduced from their molecular phylogeny and has been well described in recent years (Gould *et al.*, 1997, 2001; McGuire *et al.*, 1998; Zanotto *et al.*, 1995, 1996). Division into major clades coincides with different principal vectors (insects, ticks or no known vector), with further subdivision according to vertebrate hosts (Gaunt *et al.*, 2001). Amongst the insect-borne species, neurotropic viruses associated with encephalitic disease in humans (e.g. Japanese encephalitis, West Nile virus) persist in cycles between *Culex* mosquitoes and birds, while non-neurotropic viruses associated with haemorrhagic disease in humans (e.g. dengue, yellow fever) cycle between *Aedes* mosquitoes and primates. Although mosquitoes show species-specific patterns of feeding behaviour this is not absolute, and each genus of mosquitoes may feed on and infect both birds and mammals (Gaunt *et al.*, 2001). Nevertheless, complete transmission cycles appear to be largely confined to one class of vertebrates or the other (*Culex*–birds or *Aedes*–primates) (Gaunt *et al.*, 2001), indicating the dominant role of the vertebrate host in driving the selection process underlying this pattern of clades.

Gould *et al.* (2001) make a clear case for the role of vertebrate hosts in selecting for different genotypes of tick-borne viruses. Early in the evolution of the genus (but apparently not until after the most recent glaciation) a major split occurred into two clades of viruses, the first being transmitted via seabirds [the Tyuleniy (TYU) serogroup] and the second transmitted via mammals (the TBE complex viruses) (Fig. 1). Furthermore, a group of closely related encephalomyelitis viruses all associated with sheep diverged from the other TBE complex viruses that are associated mostly with forest rodents.

The following account is taken from Gould *et al.* (2001). One of the earliest nodes of the tick-borne clades includes an African virus (Kadam), suggesting that during the past 5000 years, infected ticks have been dispersed by seabirds from an African (or Asian) origin, to northwest France (Meaban), to the eastern coast of Russia (TYU) and to the Great Barrier Reef (Saumarez Reef). Corresponding to the migratory behaviour of seabirds, this most ancient lineage of tick-borne viruses shows the greatest

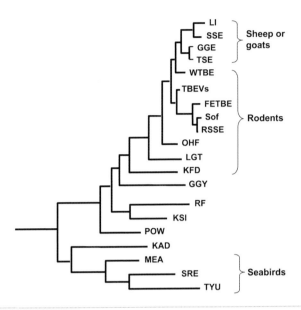

Fig. 1. Phylogenetic tree of the tick-borne clade of flaviviruses: consensus tree based on the 1st and 2nd codon positions for 41 E genes and the NS5 gene sequence, taken from Gould *et al.* (2001) and Gaunt *et al.* (2001). The source of the gene sequences used to construct this tree is given in the original publications. The genus *Flavivirus* contains about 70 distinct antigenically related flaviviruses. These are positive-stranded RNA viruses that consist of three structural proteins (C capsid, M membrane and E envelope) and seven non-structural (NS) proteins. For the geographic distribution of these tick-borne viruses see text: TYU, Tyuleniy; SRE, Saumarez Reef; MEA, Meaban; KAD, Kadam; POW, Powassan; KSI, Karshi; RF, Royal Farm; GGY, Gadget's Gully; KFD, Kyasanur Forest disease; LGT, Langat; OHF, Omsk haemorrhagic fever; RSSE, Russian spring-summer encephalitis; Sof, Sofjin strain of tick-borne encephalitis; FETBE, Far Eastern tick-borne encephalitis; TBEVs, Siberian tick-borne encephalitis; WTBE, Western tick-borne encephalitis; TSE, Turkish sheep encephalitis; GGE, Greek goat encephalitis; SSE, Spanish sheep encephalitis; LI, louping ill. Principal vertebrate host types are shown.

geographic and genetic separation (approx. 60 % amino acid identity compared with 70 % between most of the TBE complex viruses). At the deeper nodes of the tree, the earliest TBE complex viruses are also found at great geographic distance from each other, possibly suggesting similar bird-based introductions to far-flung places (e.g. Gadget's Gully virus, found in ticks under rocks and debris used by penguins and other seabirds on Macquarie Island, several hundred miles off southern Australia). The most recent TBE complex viruses, however, show an asymmetric topology on all phylogenetic trees so far constructed, indicating a continuous rather than interrupted evolution. Moreover, the correlation between the genetic and geographic distances between these viruses reveals that this complex evolved as a cline from east to west across the northern hemisphere (Zanotto *et al.*, 1995). Thus, within the mammal-associated lineage, the viruses of Southeast Asia (Kyasanur Forest disease and Langat) are the ancestors of the viruses of the northern forests of far eastern and central Russia [Omsk haemorrhagic fever, Far Eastern tick-borne encephalitis and Siberian tick-borne encephalitis (TBEVs)], which in turn are the ancestors of the viruses of western Europe [Western tick-borne encephalitis (WTBE), Turkish sheep encephalitis (TSE), Greek goat encephalitis (GGE) and Spanish sheep encephalitis (SSE)], culminating in the most westerly virus [louping ill (LI)] in the British Isles. It is only these last four viruses that are transmitted via sheep, and LI is evidently also transmissible via red grouse (*Lagopus scoticus*) (Hudson *et al.*, 1995) and mountain hares (*Lepus timidus*) (Jones *et al.*, 1997), but not rodents (Gilbert *et al.*, 2000). Despite being found in isolated foci within upland sheep-grazing regions of Turkey (TSE), Bulgaria and Greece (GGE), Spain (SSE), and Ireland, UK and Norway (LI), all are antigenically very closely related, but distinguishable by a unique tripeptide sequence in the envelope (E) gene (Gao *et al.*, 1993; Gould *et al.*, 2001).

Directed by extrinsic environmental factors?
The existence of this cline immediately begs the question: what directed its development? It is not sufficient to point to the extremely limited mobility of ticks. Ticks rely for their displacement on the movement of their hosts, which is also fairly limited in the case of rodents, but not so for birds and ruminants (deer as well as livestock). Clearly, through movement of hosts, either natural or due to man's intervention, TBE complex viruses have been able to colonize a large proportion of the northern hemisphere during the past few thousand years. Yet each virus is characterized by a geographically coherent, focal distribution within the much more continuous ranges of both the competent vector tick species and rodent or ovine (sheep and goats) transmission hosts. The implication is that the chance of each specific virus (within an infected tick or host) successfully invading a new region is rare, and that the distribution of viruses is limited less by their ability to reach a new place than by their ability to survive there. At least for WTBE virus, this limitation is now known to be due to rather specific climatic conditions required for persistent enzootic cycles, which in turn are determined by the biological basis of transmission.

The cellular basis for WTBE virus transmission. Under laboratory conditions, WTBE virus is transmissible by a wide range of Eurasian tick species (Korenberg & Kovalevskii, 1994) and even an African species (Nuttall & Labuda, 1994), and also via a range of natural rodent host species (Labuda *et al.*, 1993b). In natural conditions, however, only the specific ecological association of *I. ricinus* with field mice (*Apodemus flavicollis*) contributes significantly to WTBE virus maintenance (Labuda & Randolph, 1999). The vertebrate specificity is due to a particular route of transmission, now recognized as crucial because of its quantitative impact. A complete transmission cycle, from tick to tick via vertebrates, occurs most efficiently between co-feeding ticks in the absence of a systemic viraemia (Labuda *et al.*, 1993a). This non-systemic route depends on virus replication within particular immunocompetent cells in the skin (notably the

Langerhans cells), and only certain vertebrate species (notably *Apodemus* mice) are susceptible to this (Labuda *et al.*, 1996). The virus has evolved to exploit the very immunological mechanisms by which these vertebrates try to defend themselves against feeding ticks, and also the devices of the tick's counter-attack, as follows. Pharmacologically active substances in tick saliva (Nuttall, 1998; Wikel, 1996) first promote virus replication on delivery to the host's skin (Jones *et al.*, 1992), and then attract virus-infected cells of the host's immune system back from the lymph nodes to the feeding sites of uninfected ticks co-feeding with the infected ticks (Labuda *et al.*, 1996).

Epidemiologically, the significance of this transmission pathway is the slightly longer period over which infected hosts remain infective for ticks (up to 3 rather than 2 days), and the much lower virulence of non-viraemic rather than viraemic infections. The virulence of systemic infections of WTBE virus can be high in certain hosts (e.g. *Pitymys subterraneus*), causing such a high mortality rate that they die before most co-feeding ticks have completed their blood meals. In contrast, *A. flavicollis*, which develops only low viraemias, is much more likely to survive infection (Labuda *et al.*, 1993b). Together, these factors increase the force of transmission by >50 % relative to the systemic pathway. This is crucial to virus survival given the inherent fragility of this system (Randolph *et al.*, 1996).

The environmental context of WTBE virus transmission. A further very important factor essential for WTBE virus survival is that large numbers of infectible larval ticks co-feed on mice alongside any one infected nymphal tick. Only sufficiently large numbers of co-feeding larvae ensure adequate amplification to offset the inevitable high mortality during each transmission cycle. (Adult ticks very rarely feed on rodents.) Empirically, this has been observed within but not outside WTBE foci, and is due to two distinct phenomena. First is the pattern of coincident aggregated tick distributions: both larvae and nymphs show highly aggregated distributions amongst their rodent hosts and, rather than being independent, the distributions of each stage are coincident so that the hosts that feed the most nymphs (a few of which may be infected) also feed large numbers of larvae (Randolph *et al.*, 1999). Second, larvae and nymphs must have synchronous seasons of host-seeking activity with both stages feeding from early to mid-Spring onwards, rather than the larvae showing a delayed onset of activity (Randolph *et al.*, 2000). Amongst the potential tick vectors in Europe only *I. ricinus* has the correct host relationships and the appropriate natural life cycle in certain places to support WTBE virus transmission cycles.

With the help of satellite imagery (which provides uniform information on environmental conditions at sufficiently fine spatial and temporal resolution over continental areas), the critical climatic factors that determine where these biological requirements are satisfied, and therefore where WTBE foci exist, have been identified (Randolph, 2000). First, high humidity at ground level during the summer is necessary for good tick survival and active host-seeking activity, to allow adequate tick populations to exist. These conditions are typically found in deciduous woodlands (Daniel & Kolar, 1990) and also in upland grazing areas with rough vegetation, where large host species such as deer or sheep are also found. These hosts feed adult ticks and therefore maintain the tick population. Second, a particular seasonal profile of land surface temperature characterized by a rapid rate of cooling during the autumn is statistically associated with both synchronous feeding by larvae and nymphs and the presence of WTBE human infections (Randolph *et al.*, 2000), although the precise biological processes that link such temperature and tick patterns have not yet been fully revealed.

So far, WTBE is the only tick-borne flavivirus for which this degree of quantitative explanation of natural enzootic cycles has been achieved. The LI system has also been the subject of intensive field investigations and mathematical modelling. In addition to sheep, red grouse are competent transmission hosts, but their mortality rate when infected (by experimental inoculation with infected mouse brain) was so high (Reid, 1975), and the breeding success of infected birds so low (Reid *et al.*, 1978), that grouse populations are thought to be non-sustainable in LI enzootic areas (Hudson, 1992). In some parts of the Scottish Highlands where sheep are removed from the system by vaccination against the virus or by acaricide treatment against the tick, mountain hares appear to play an important role, both by feeding adult ticks and by transmitting the virus (Hudson *et al.*, 1995; Jones *et al.*, 1997; Norman *et al.*, 1999; Laurenson *et al.*, 2003). Mountain hares, however, do not occur in many parts of England and Wales where LI is present. LI virus thus appears to depend on specific combinations of vertebrate hosts that have not yet been fully identified in all situations (Laurenson *et al.*, 2003).

Very little is known about the natural host relationships of TSE, GGE and SSE viruses beyond their ability to be transmitted via sheep or goats by *I. ricinus* and perhaps also by other tick species.

Gritsun *et al.* (1995) suggest that the mutations in the variable regions of the E gene probably reflect adaptation to wildlife host species. WTBE virus appears to have a considerably narrower effective host range than LI virus, which might explain its greater genetic and antigenic homogeneity across its wide geographic range compared with the apparent strain diversity of LI within the UK (Gould *et al.*, 2001; Guirakhoo *et al.*, 1987; Heinz & Kunz, 1981, 1982). WTBE virus may also be funnelled through vectors of greater than average genetic homogeneity at each transmission cycle, because a large proportion of each cluster of larval ticks co-feeding on any one mouse is likely to come from a single egg batch (groups of larvae quest close to where they emerge).

The limited dispersal of TBE complex viruses associated with sheep. It is clear that natural enzootic cycles of WTBE virus are fragile, lacking the excess slack in the system that would permit invasion into less than optimum biotic or abiotic contexts. Mutations have permitted some tick-borne flaviviruses (TSE, GGE, SSE and LI) to escape from the rodent transmission pathway and to utilize larger vertebrate host species that feed both nymphs and adults of *I. ricinus*. This allows virus transmission between these two tick stages, which much more commonly (always?) show synchronous feeding seasons. These systems, therefore, might be expected to be more robust. Yet, despite the fact that rodents and sheep (and goats) show widespread, overlapping or contiguous distributions throughout Europe, each virus is confined to its own geographically separated, well-defined range. For example, 'hundreds of thousands of sheep' migrate from Bosnia to an area of a natural WTBE focus in Croatia for winter grazing, and take local ticks back with them when they return to Bosnia in the spring, but neither WTBE virus nor any sheep-transmitted viral type has been introduced to Bosnia (Borcic *et al.*, 1999). Isolates of TSE have been recorded only from six villages close to Gebze and to Kirklareli in northwest Turkey (Hartley *et al.*, 1969; Whitby *et al.*, 1993), and of GGE from northeast Greece (Papadopoulos, 1980) and Plodiv district of central Bulgaria (Pavlov, 1968). Their full distributions are unknown, but appear to be somewhat limited. Interestingly (but of no known significance), in the Bulgarian focus, which is several hundred kilometres southwest of known WTBE foci in Romania, GGE was isolated from shrews (*Sorex araneus*) as well as from sheep and ticks.

Once TSE or GGE had established itself in livestock hosts, transportation by humans could easily permit dissemination to other regions where distinct strains have evolved: from Turkey, Greece and Bulgaria to the Basque highlands of Spain (SSE), to Ireland (LI/MA54) and thence to parts of Wales (LI/I), northern England (LI/917 and others), Scotland (LI/31 and others) and finally to south-west England (LI/A and LI/DEV4) (McGuire *et al.*, 1998). Indeed, dates of lineage divergence indicate that LI virus emerged in the British Isles <800 years ago and that most LI virus dispersal occurred during the last 300 years, which has been related to the history of sheep farming (Gould *et al.*, 2001; McGuire *et al.*, 1998). Certain sheep breeds were even imported into Norway from Britain in the 19th century, and LI was isolated from goats in Norway in 1978 and from sheep in 1982 (Ulvund *et al.*, 1983). LI, however, did not spread beyond a limited southwest coastal region closest to Scotland, despite extensive tick populations and other tick-transmitted infections of sheep and wild ruminants along most of the coastal regions of southern Norway as far north as Nordland county (Stuen, 2003).

Why has each virus type remained so isolated? What has prevented it spreading out of its area of presumed introduction, infecting sheep over their much wider distributions?

It appears that each virus may be 'trapped' by a set of peculiar abiotic environmental conditions acting on any one part of the pathogen–vector–host interaction, just as has been identified for WTBE virus. Thus spread throughout Europe is prevented despite the ubiquity of the essential biotic components, competent vectors and hosts.

Testing the role of climate in the evolution of tick-borne flaviviruses

If climate exerts such a strong influence on the successful transmission of tick-borne viruses, it is reasonable to hypothesize that climate has directed and constrained their evolution. One way to investigate whether climatic factors direct the evolution, rather than merely the distribution, of vector-borne flaviviruses is to test whether the eco-climatic spaces occupied by closely related viruses are more or less similar than those occupied by more distantly related ones. If they prove to be very similar one could conclude that climate has been a significant evolutionary constraint, whereas if they are significantly different one could conclude that viruses have been free to leap from one eco-space to another with no climatic constraint.

The idea is to test for matches between virus phylogeny and the environmental conditions in which each virus circulates by seeking correlations between two very different sorts of trees. First are the familiar molecular phylogenetic trees, constructed on the basis of genetic differences as a direct measure of evolutionary distance between related species. These describe evolutionary history. Second are conceptually novel eco-climatic trees, constructed from the statistical distances between the eco-climatic space in which each virus circulates. These trees are purely descriptive, containing no information on evolutionary history.

Constructing phenetic eco-climatic trees for viruses. The basis for these eco-climatic trees is the ability to define the conditions within which each virus can survive (i.e. exists), plotted in multivariate space (illustrated in Fig. 2 for just two variables, temperature and humidity). As made clear above, each virus only survives within a subset of the overlapping range of its hosts and vectors. The complex biological processes that determine the eco-climatic limits of vector-borne pathogens have not yet been completely quantified for any one let alone a whole cladeful. Therefore, statistical pattern-matching methods are currently the most reliable for identifying these limits (Randolph, 2000; Rogers & Randolph, 2000). Correlations are established between the spatial patterns of the distribution of organisms and spatial patterns of environmental conditions as revealed by meteorological satellite imagery (from the NOAA AVHRR sensor) (Hay *et al.*, 2000). Of the variety of statistical methods available for seeking these correlations, discriminant analysis is the most biologically transparent (Green, 1978; Rogers *et al.*, 1996). Furthermore, it has the particular virtue for the present purposes of measuring distance: the Mahalanobis distance is the co-variance-adjusted distance between the geometric centres of multivariate eco-climatic spaces (Fig. 2). It is a way of reducing the many differences in eco-climatic conditions to a single measure of separation between presence and absence, or between the presence of pairs of species. Using forward stepwise selection of up to ten variables, discriminant analysis also assigns relative importance to the variables that define the species distribution, thereby yielding both discrete and distance information.

The distributions of single diseases can now be captured routinely with high accuracy by these statistical methods (reviewed by Hay *et al.*, 2000; Rogers *et al.*, 2002). To create phenetic eco-climatic trees, Mahalanobis distances between species must be derived from the same set of variables. In theory, this could be achieved by putting all the species into a single analysis, but as each disease is added prediction accuracies are inevitably compromised. Nevertheless, to date it has proved possible to capture the distributions of all three most westerly tick-borne flaviviruses (LI, SSE and WTBE) in a single exercise with 80–95 % accuracy [kappa index (Congalton, 1991) = 0·794]. When preliminary data for the Siberian subtype of TBE virus (TBEVs) were added [based on 500 locations within the western distribution of the tick *I. persulcatus* in Latvia, Estonia and western Russia, and from the Perm region of Siberia, with which TBEVs virus effectively, although not absolutely exclusively, coincides (Korenberg, 1994; Lundkvist *et al.*, 2001)], the kappa index was still high, at 0·683. This means that the same ten predictor satellite variables may be used to distinguish areas of presence of each virus from each of the others, and from areas of absence. The identity of the ten predictor variables, however, changes somewhat as new species are added. As more species are added it will be necessary to select a consensus 'top-ten' set of variables that applies across the clade.

Statistically each of the four viruses analysed to date falls within a distinct eco-climatic space defined by factors that are all temporal Fourier variables (i.e. seasonal characteristics) of thermal and moisture conditions. To illustrate this on the page, two of the most significant variables are chosen as axes (Fig. 3). In addition, five points representing the observed locations for GGE and TSE are plotted on the same axes. Even in this simplified bi-variate space, there is clear separation between viral types. Furthermore, the evolutionary sequence from TBEVs to WTBE to GGE/TSE to SSE and finally to LI follows a progressive shift from high to low annual variance in air temperature, with associated changes in annual variance in vapour pressure deficit. Nothing can be deduced from the relative degrees of separation on this graph, partly because of the different numbers of observations and mostly because separation in multivariate space will be differentially increased by other variables. LI and WTBE appear to be close together on this graph, but just as the molecular phylogeny shows that LI virus did

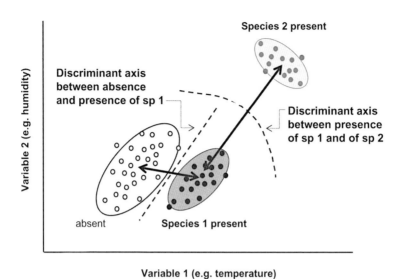

Fig. 2. An illustration of the principle of linear discriminant analysis used to distinguish between the abiotic conditions in which an organism is present or absent, and the conditions occupied by two different species. Mahalanobis distances are represented by the heavy double-ended arrows.

not evolve directly from WTBE virus, so the Fourier-processed satellite imagery of temperature and moisture conditions shows a barrier of seasonally distinct climate in France. Biologically relevant climate factors now explain why WTBE has not reached Britain through France (Randolph *et al.*, 2000), but instead the distinct LI virus entered Britain apparently via Spain and Ireland (Gould *et al.*, 2001; McGuire *et al.*, 1998). These two viruses, however, are now approaching each other geographically via a northern route, as WTBE is spreading westwards through Scandinavia (International Scientific Workgroup on Tick-borne Encephalitis, 2002 – http://www.tbe-info.com/epidemiology/index.html; Skarpass *et al.*, 2002) and LI has been recorded in Norway.

Congruence between phylogenetic and eco-climatic trees? The above results are the first, very important steps towards deriving Mahalanobis distances between all the tick-borne flaviviruses, and so constructing phenetic ecological trees. In many ways, a clade that forms a cline is the least suitable for testing the effects of climate on evolution because of the co-variation between evolutionary change, geographic space and climate. Climate may match phylogeny simply because of the orderly march westwards in this case. The smaller scale of LI virus evolution within the British Isles on the one hand, and the more diffuse geographic patterns of the insect-borne flaviviruses on the other, should provide more clearly interpretable results. Both of these are currently being studied by the Oxford Tick Research Group in Oxford. Whatever degree of congruence between the two types of tree emerges, it will direct the search for

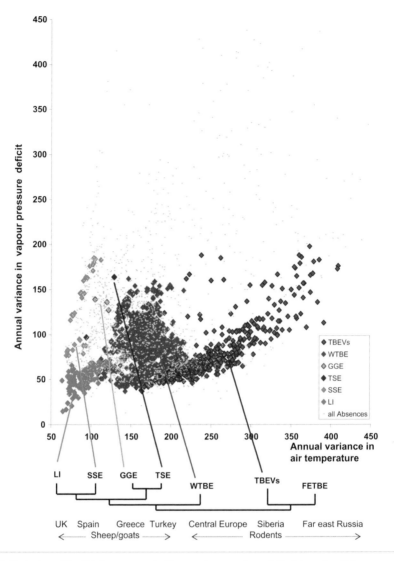

Fig. 3. Each virus of the tick-borne encephalitis complex occupies a distinct 'eco-climatic' space, illustrated here in bi-variate space defined by two of the most significant climatic variables that predict the distribution of each virus. Annual variance is an indicator of the abruptness of the seasonal changes in these climatic variables, especially during spring and autumn. The phylogeny of these viruses, their geographic distribution and principal hosts are shown below.

the processes underlying the evolutionary patterns, leading to new interpretations and understanding of the forces that direct and constrain pathogen evolution. If the trees do not match, we shall have to conclude that forces other than climate must have been more significant in directing the evolution of the flaviviruses. If, on the other hand, the fit is better than random, it will already be clear which particular environmental factors are correlated with the origin of new viruses, and with the shifting incidence of existing vector-borne diseases under the forces of natural and anthropogenic environmental change.

LYME BORRELIOSIS – *B. BURGDORFERI S. L.* EVOLUTION

Driven by intrinsic vertebrate host factors

B. burgdorferi s. l. comprises a genetically diverse complex of spirochaetes. It currently includes 11 named 'genospecies' but new 'species' are continuing to be recognized amongst isolates from around the northern hemisphere (LeFleche *et al.*, 1997; Masuzawa *et al.*, 2001; Postic *et al.*, 1994, 1998). Predictably, those incriminated in causing disease (Lyme borreliosis) were characterized first and have been the subject of intensive research since their recent recognition in the early 1980s. The genetic basis and molecular mechanisms that permit *Borrelia* to escape from the midgut of the tick and infect their vertebrate hosts, where they disseminate and persist despite specific immune responses, appear to hinge on selective expression of outer-surface proteins (Osps) (reviewed by Kurtenbach *et al.*, 2002a). Some, notably OspA, are expressed within the midgut of the unfed tick but not in the vertebrate host. Others, notably OspC, are expressed increasingly as the tick feeds (Schwan *et al.*, 1995), possibly as a prerequisite for escaping the tick's midgut lumen, migrating to the salivary glands and entering the host (de Silva & Fikrig, 1997; Gilmore & Piesman, 2000).

Genetic and phenotypic diversity. Most studies indicate a clonal population structure for *B. burgdorferi s. l.* (Boerlin *et al.*, 1992; Dykhuizen *et al.*, 1993; Wang *et al.*, 1999a). The *ospC* gene, however, is subjected to a high degree of diversifying selection (Ryan *et al.*, 1998) of the sort typical of genes responsible for immunoglobulin synthesis or alleles of the major histocompatibility complex. Intense intraspecific and even interspecific lateral transfers (Dykhuizen & Baranton, 2001; Jauris-Heipke *et al.*, 1995) result in huge diversity among *ospC* sequences (so that phylogenetic trees based on the *ospC* gene do not reflect the 'species'-level taxonomy based on the *ospA* gene) (Lagal *et al.*, 2002), perhaps allowing the emergence of new serotypes to evade effective immunity (Kurtenbach *et al.*, 2002a; Wang *et al.*, 1999b). Nevertheless, host immunity evidently does suppress spirochaetaemia, permitting only low intensity infections to persist within specific tissues, and these tissues (and therefore the clinical symptoms) vary with the genospecies of *B. burgdorferi s. l.* (Van Dam *et al.*, 1993). Thus in humans *Borrelia afzelii* is associated most with skin and cutaneous manifestations (e.g. acrodermatitis chronica atrophicans), *Borrelia garinii* with nervous tissue and paralysis, and *B. burgdorferi sensu stricto* (*s. s.*) with joints and arthritis (reviewed by Stanek *et al.*, 2002).

The species-specificity of the interactions between spirochaetes and human hosts extends also to interactions of spirochaetes with their natural reservoir hosts. Although very many vertebrate species have been shown to play a role in the maintenance of *B. burgdorferi s. l.* cycles, it is now clear that different genotypes show differential transmissibility from ticks to vertebrates depending on the spirochaete–host combination. The general pattern is that mammals, particularly many sorts of rodents and insectivores, are competent reservoir hosts for *B. afzelii*, *Borrelia bissettii*, *Borrelia andersoni*, *Borrelia japonica*, *B. burgdorferi s. s.* and *B. garinii* OspA serotype 4 and the Asian ribotype NT29, but not *B. garinii* OspA serotypes 3, 5, 6, 7 and ribotype 20047 nor *Borrelia valaisiana*, which are transmitted instead via birds (Burkot *et al.*, 2000; Gorelova *et al.*, 1995; Gylfe *et al.*, 1999; Hu *et al.*, 1997, 2001; Humair & Gern, 1998; Humair *et al.*, 1998, 1999; Kurtenbach *et al.*, 1998b, 2002c; Richter *et al.*, 1999, 2000). Only *B. burgdorferi s. s.* appears to be relatively non-specialized, able to infect and be transmitted via both mammals and birds.

The basis for this transmission specificity appears to lie in the vertebrate's alternative complement system, part of the innate immune system. The degree of complement-mediated lysis of cultured strains of *B. burgdorferi s. l.* when exposed *in vitro* to serum from a range of vertebrate species is the mirror image of transmissibility: strains transmissible via mammals are lysed by avian complement, while those transmissible via birds are lysed by mammalian complement (Kurtenbach *et al.*, 1998a, b). This extends to reptiles, rendering American species of lizards non-competent for *B. burgdorferi s. s.* transmission (Kuo *et al.*, 2000; Lane & Quistad, 1998). Interestingly, while *B. burgdorferi s. s.* is fully resistant to most rodent complement, it is only partially resistant to complement from European woodmice, birds and domestic ungulates. Perhaps this explains why sheep do not develop systemic infections but can nevertheless support transmission of *B. burgdorferi s. s.* between co-feeding ticks, which is sufficient on its own to maintain enzootic cycles (Ogden *et al.*, 1997). The results of experimental transmission studies *in vivo* suggest that host complement taken up during the tick's blood meal kills the bacteria in the tick's midgut before they are transmitted to the host (Kurtenbach *et al.*, 2002b, c). Contrary to this model, wild rodents are occasionally found to harbour rodent complement-sensitive strains of *B. garinii* (Kurtenbach *et al.*, 1998b); possibly these *Borrelia* escaped lysis and succeeded in infecting the host because their vectors developed systemic infections outside the midgut before they started to feed (Kurtenbach *et al.*, 2002c), as happens in a small proportion of ticks (Gern *et al.*, 1990).

Clearly there are a number of routes whereby *B. burgdorferi s. l.* may encounter a wide range of host species and, via these, various tick species (e.g. both *I. ricinus* and *Ixodes hexagonus* feed on reservoir-competent hedgehogs; Gern *et al.*, 1997), each of which could select for novel recombinant genotypes. The principal difference of most epidemiological significance between spirochaetes and flaviviruses is the much longer period of infectivity in the vertebrate host for the bacteria. This could promote the evolution of much greater genetic heterogeneity through the wide dissemination of spirochaetes from one source host, first through the greater number and diversity of ticks feeding on any one host during this long period, and secondly through dispersal of the hosts themselves. Indeed, *B. garinii* genotypes show high rates of long-distance dispersal, associated with bird migrations (Gylfe *et al.*, 1999, 2000; Kurtenbach *et al.*, 2002a; Olsen *et al.*, 1995), and *B. garinii* shows the greatest heterogeneity, both genetic (Will *et al.*, 1995) and phenotypic (Hu *et al.*, 2001, and see above). This principle is supported by the recent finding that *B. valaisiana*, also transmitted via birds, shows significant *ospA* sequence heterogeneity indicating three distinct clusters, two from Europe and one from the Far East (Godfroid *et al.*, 2003).

Impact of extrinsic environmental factors

The phylogenetic tree topology for *B. burgdorferi s. l.* differs depending upon which gene is used as the basis (Godfroid *et al.*, 2003; Kurtenbach *et al.*, 2002a; Lagal *et al.*, 2002; Wang *et al.*, 1999a). Until a consensus is reached, conclusions about the impact of environmental factors on the evolution of this complex must rest on general patterns rather than statistical analysis. The high force of transmission of these bacteria (Randolph *et al.*, 1996), and consequently the wide window of opportunity for persistent enzootic cycles, allows *B. burgdorferi s. l.* to occur more or less throughout the ranges of competent tick species under a wide range of ecological (biotic and abiotic) conditions. Therefore, compared with TBE virus, environmental factors are not expected to impose such constraint.

The wide, non-patchy geographic distribution of many *B. burgdorferi s. l.* species appears to confirm this, although there are clear differential ranges that cannot be explained simply by the availability of competent vector or host species (Fig. 4). The more recently described species from eastern Asia are currently known only from the countries where they were originally isolated (*B. japonica*, *Borrelia tanukii* and *Borrelia turdi* from Japan, and *Borrelia sinica* from northwest China), where they are associated with local tick species. This almost certainly reflects an absence of knowledge rather than a knowledge of absence elsewhere. In contrast, where survey work has been much more intensive, *B. afzelii*, *B. garinii* (Eurasian and Asian types) and *B. valaisiana* are found throughout Europe and Asia (Hubálek & Halouzka, 1997; Wang *et al.*, 1999a), but *Borrelia lusitaniae* is known mostly from Portugal and N. Africa and more rarely, but increasingly, from elsewhere in Europe (Jouda *et al.*, 2003). *B. burgdorferi s. s.* and *B. bissettii* are the only species found on both sides of the Atlantic, but both are much more prevalent in N. America than in Europe. In the Old

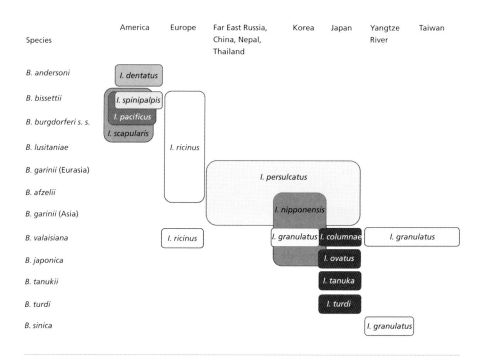

Fig. 4. Broad geographic distributions and principal tick vectors (*Ixodes* species) of each species of the *B. burgdorferi s. l.* complex. Expanded from Miyamoto & Masuzawa (2002), with permission.

World, *B. burgdorferi s. s.* is much more common in western than eastern Europe, where it is transmitted by *I. ricinus* but not in Eurasia by *I. persulcatus*. *B. bissettii*, associated with woodrats and *Ixodes spinipalpis* in the New World, has been recorded from human patients in Slovenia (Picken *et al.*, 1996). *B. andersoni*, associated with cottontail rabbits and *Ixodes dentatus*, is confined to the USA.

The absence of *B. afzelii* and *B. garinii* in America, despite the ability of the widespread American tick species *Ixodes scapularis* to transmit both these species (Dolan *et al.*, 1998), together with the greater diversity of *B. burgdorferi s. l.* in the Old World, suggests that this bacterial complex originated in the Old World (Wang *et al.*, 1999a). The New World, however, may be a (secondary?) centre of endemism. *B. burgdorferi s. s.* isolates from both Europe and N. America are closely related to one another, indicating 'unambiguously … that a constant supply from one continent to the other has occurred' (Lagal *et al.*, 2002). Genetic diversity is evidently higher amongst *B. burgdorferi s. s.* isolates from N. America than amongst isolates of this species from Europe (Foretz *et al.*, 1997; Marti Ras *et al.*, 1997). Furthermore, interspecific lateral transfers have occurred from *B. burgdorferi s. s.* to *B. garinii* and *B. afzelii*, but not in the opposite direction. These facts, it is argued, indicate that no European

B. burgdorferi s. s. isolate has been imported into N. America, only the opposite has occurred (Lagal *et al.*, 2002), i.e. *B. burgdorferi s. s.* originated in America and was introduced into Europe.

Despite the apparent general permissiveness of both biotic and abiotic conditions for circulation of the *B. burgdorferi s. l.* complex of spirochaetes, there are clearly differences which favour the endemic Eurasian species of *B. burgdorferi s. l.* but limit *B. burgdorferi s. s.* in Eurasia, and vice versa in America. Statistical models of the distribution of each species (or subspecies if possible and necessary), incorporating biotic and abiotic predictor variables, would allow some flesh to be put on this obvious statement. Predictive mapping of the risk to humans of Lyme borreliosis (i.e. *B. burgdorferi s. s.*) in America (Dister *et al.*, 1997; Glass *et al.*, 1995; Guerra *et al.*, 2002) has identified conditions that favour tick survival and therefore healthy tick populations, but I suspect that the differences between *Borrelia* species lie in rather more subtle abiotic factors that can determine the precise, quantitative tick–host relationships (Randolph & Storey, 1999). Both the questing behaviour of ticks and the availability of various host species in tick micro-habitats are under environmental determinants. We have a very long way to go on both sides before we can start to test for matches between phylogenetic and eco-climatic trees for *B. burgdorferi s. l.*

CONCLUSIONS

As natural selection acts on the limiting components of biological systems, those features that are limiting are likely to show evolutionary change. This article has described several ways in which tick-borne pathogens are limited by host factors, but these alone do not appear to account for their evolutionary history as inferred from phylogenetic analysis. Climate imposes a further, more fine-grained limitation through its impact on the distribution, abundance, seasonal population dynamics and host relationships of ticks. The precise quantitative identification of these critical climate variables for each tick-borne disease system has only just begun. The quest raises a question of fundamental ecological significance: how have purely *abiotic* factors, climate, directed the responses of three distinct but interacting *biotic* agents – arthropod vectors, vertebrate hosts and micro-parasites – resulting in the evolution and dissemination of new strains of pathogens?

If climate has been a significant force in shaping the past evolutionary pathways of these pathogens, it is likely to be important in future events on both evolutionary and ecological timescales. Identifying the precise nature of these forces will provide a powerful tool for making reliable predictions, not necessarily about the appearance of new pathogens in the future, but about new appearances and spread of existing pathogens. Arguably, the latter may lead to the former – an expansion of the ecological envelope, exposing pathogens to new selective pressures. Because of their higher than average transmission potential, vector-borne, rather than directly transmitted, infections are likely to appear in places where environmental conditions suddenly become permissive through environmental change. But which pathogens are likely to emerge in new places and where will this be?

ACKNOWLEDGEMENTS

These ideas were developed while the author holds a Senior Research Fellowship from the UK Natural Environment Research Council, building on previous work funded by the Wellcome Trust. It is, as ever, a pleasure to thank David Rogers for his valuable advice and comments on the manuscript.

REFERENCES

Boerlin, P., Peter, O., Bretz, A.-G., Postic, D. & Piffaretti, J.-C. (1992). Population genetic analysis of *Borrelia burgdorferi* isolates by multilocus enzyme electrophoresis. *Infect Immun* **60**, 1677–1683.

Borcic, B., Kaic, B. & Krajl, V. (1999). Some epidemiological data on TBE and Lyme borreliosis in Croatia. *Zentbl Bakteriol* **289**, 540–547.

Burkot, T. R., Schneider, B. S., Pieniazek, N. J., Happ, C. M., Rutherford, J. S., Slemenda, S. B., Hoffmeister, E., Maupin, G. O. & Zeidner, N. S. (2000). *Babesia microti* and *Borrelia bissettii* transmission by *Ixodes spinipalpis* ticks among prairie voles, *Microtus ochrogaster*, in Colorado. *Parasitology* **121**, 595–599.

Congalton, R. G. (1991). A review of assessing the accuracy of classifications of remotely sensed data. *Remote Sens Environ* **37**, 35–46.

Daniel, M. & Kolar, J. (1990). Using satellite data to forecast the occurrence of the common tick *Ixodes ricinus* (L.). *J Hyg Epidemiol Microbiol Immunol* **45**, 243–252.

de Silva, A. M. & Fikrig, E. (1997). *Borrelia burgdorferi* genes selectively expressed in ticks and mammals. *Parasitol Today* **13**, 267–270.

Dister, S. W., Fish, D., Bros, S. M., Frank, D. H. & Wood, B. L. (1997). Landscape characterization of peridomestic risk for Lyme disease using satellite imagery. *Am J Trop Med Hyg* **57**, 687–692.

Dolan, M. C., Piesman, J., Mbow, M. L., Maupin, G. O., Peter, O., Brossard, M. & Golde, W. T. (1998). Vector competence of *Ixodes scapularis* and *Ixodes ricinus* (Acari: Ixodidae) for three genospecies of *Borrelia burgdorferi*. *J Med Entomol* **35**, 465–470.

Domingo, E., Escarmis, C., Sevilla, N. & Martinez, M.-A. (1997). Population dynamics and molecular evolution of RNA viruses. In *Factors in the Emergence of Arbovirus Diseases*, pp. 273–278. Edited by J. F. Saluzzo. Paris: Elsevier.

Dykhuizen, D. E. & Baranton, G. (2001). The implications of a low rate of horizontal transfer in *Borrelia*. *Trends Microbiol* **9**, 344–350.

Dykhuizen, D. E., Polin, D. S., Dunn, J. J., Wilske, B., Preac-Mursic, V., Dattwyler, R. J. & Luft, B. J. (1993). *Borrelia burgdorferi* is clonal: implications for taxonomy and vaccine development. *Proc Natl Acad Sci U S A* **90**, 10163–10167.

Foretz, M., Postic, D. & Baranton, G. (1997). Phylogenetic analysis of *Borrelia burgdorferi* sensu stricto by arbitrarily primed PCR and pulsed-field gel electrophoresis. *Int J Syst Bacteriol* **47**, 11–18.

Gao, G. F., Hussain, M. H., Reid, H. W. & Gould, E. A. (1993). Classification of a new member of the TBE flavivirus subgroup by its immunological, pathogenetic and molecular characteristics: identification of subgroup-specific pentapeptides. *Virus Res* **30**, 129–144.

Gaunt, M. W., Sall, A. A., de Lamballerie, X., Falconar, A. K., Dzhivanian, T. & Gould, E. A. (2001). Phylogenetic relationships of flaviviruses correlate with their epidemiology, disease association and biogeography. *J Gen Virol* **82**, 1867–1876.

Gern, L., Zhu, Z. & Aeschlimann, A. (1990). Development of *Borrelia burgdorferi* in *Ixodes ricinus* females during blood feeding. *Ann Parasitol Hum Comp* **65**, 89–93.

Gern, L., Rouvinez, E., Toutoungi, L. N. & Godfroid, E. (1997). Transmission cycles of *Borrelia burgdorferi sensu lato* involving *Ixodes ricinus* and/or *I. hexagonus* ticks and the European hedgehog, *Erinaceus europaeus*, in suburban and urban areas in Switzerland. *Folia Parasitol* **44**, 309–314.

Gilbert, L., Jones, L. D., Hudson, P. J., Gould, E. A. & Reid, H. W. (2000). Role of small mammals in the persistence of Louping-ill virus: field survey and tick co-feeding studies. *Med Vet Entomol* **14**, 277–282.

Gilmore, R. D. & Piesman, J. (2000). Inhibition of *Borrelia burgdorferi* migration from the midgut to the salivary glands following feeding by ticks on OspC-immunized mice. *Infect Immun* **68**, 411–414.

Glass, G. E., Schwarz, B. S., Morgan, J. M., III, Johnson, D. T., Noy, P. M. & Israel, E. (1995). Environmental risk factors for Lyme disease identified with geographic information systems. *Am J Public Health* **85**, 944–948.

Godfroid, E., Hu, C. M., Humair, P.-F., Bollen, A. & Gern, L. (2003). PCR-reverse line blot typing method underscores the genomic heterogeneity of *Borrelia valaisiana* species and suggests its potential involvement in Lyme disease. *J Clin Microbiol* **41**, 3690–3698.

Gorelova, N. B., Korenberg, E. I., Kovaleskii, Y. V. & Shcherbakov, S. V. (1995). Small mammals as reservoir hosts for *Borrelia* in Russia. *Zentbl Bakteriol* **282**, 315–322.

Gould, E. A., Zanotto, P. M. & Holmes, E. C. (1997). The genetic evolution of flaviviruses. In *Factors in the Emergence of Arbovirus Diseases*, pp. 51–63. Edited by B. Dodet. Paris: Elsevier.

Gould, E. A., de Lamballerie, X., Zanotto, P. M. & Holmes, E. C. (2001). Evolution, epidemiology, and dispersal of flaviviruses revealed by molecular phylogenies. *Adv Virus Res* **57**, 71–103.

Green, P. E. (1978). *Analyzing Multivariate Data.* Hinsdale, IL: The Dryden Press.

Gritsun, T. S., Holmes, E. C. & Gould, E. A. (1995). Analysis of flavivirus envelope proteins reveals variable domains that reflect their antigenicity and may determine their pathogenesis. *Virus Res* **35**, 307–321.

Guerra, M., Walker, E., Jones, C. J., Paskewitz, S., Cortinas, M. R., Sancil, A., Beck, L., Bobo, M. & Kitron, U. (2002). Predicting the risk of Lyme disease: habitat suitability for *Ixodes scapularis* in the north central United States. *Emerg Infect Dis* **8**, 289–297.

Guirakhoo, F., Radda, A. C., Heinz, F. X. & Kunz, C. (1987). Evidence for antigenic stability of tick-borne encephalitis virus by analysis of natural isolates. *J Gen Virol* **68**, 859–864.

Gylfe, A., Olsen, B., Strasevicius, D., Ras, N. M., Weihe, P., Noppa, L., Ostberg, Y., Baranton, G. & Bergström, S. (1999). Isolation of Lyme disease *Borrelia* from puffins (*Fratercula arctica*) and seabird ticks (*Ixodes uriae*) on the Faroe Islands. *J Clin Microbiol* **37**, 890–896.

Gylfe, A., Bergström, S., Lundström, J. & Olsen, B. (2000). Reactivation of *Borrelia* infection in birds. *Nature* **403**, 724–725.

Hartley, W. J., Martin, W. B., Hakioglu, F. & Chifney, S. T. E. (1969). A viral encephalomyelitis of sheep in Turkey. *Pendik Inst J* **2**, 89–100.

Hay, S. I., Randolph, S. E. & Rogers, D. J. (editors) (2000). *Remote Sensing and Geographical Information Systems in Epidemiology*. London: Academic Press.

Heinz, F. X. & Kunz, C. (1981). Homogeneity of the structural glycoprotein from European isolates of tick-borne encephalitis viruses: comparison with other viruses. *J Gen Virol* **57**, 263–275.

Heinz, F. X. & Kunz, C. (1982). Molecular epidemiology of tick-borne encephalitis virus: peptide mapping of large non-structural proteins of European isolates and comparison with other flaviviruses. *J Gen Virol* **62**, 271–285.

Hu, C. M., Humair, P.-F., Wallich, R. & Gern, L. (1997). *Apodemus* sp. rodents, reservoir hosts for *Borrelia afzelii* in an endemic area in Switzerland. *Zentbl Bakteriol* **285**, 558–564.

Hu, C. M., Wilske, B., Fingerle, V., Lobet, Y. & Gern, L. (2001). Transmission of *Borrelia garinii* OspA serotype 4 to BALB/c mice by *Ixodes ricinus* ticks collected in the field. *J Clin Microbiol* **39**, 1169–1171.

Hubálek, Z. & Halouzka, J. (1997). Distribution of *Borrelia burgdorferi* sensu lato genomic groups in Europe, a review. *Eur J Epidemiol* **13**, 951–957.

Hudson, P. J. (1992). *Grouse in Space and Time*. Fordingbridge, UK: Game Conservancy Trust.

Hudson, P. J., Norman, R., Laurenson, M. K. & 7 other authors (1995). Persistence and transmission of tick-borne viruses: *Ixodes ricinus* and louping-ill virus in red grouse populations. *Parasitology* **111**, S49–S58.

Humair, P.-F. & Gern, L. (1998). Relationship between *Borrelia burgdorferi* sensu lato species, red squirrels (*Sciurus vulgaris*) and *Ixodes ricinus* in enzootic areas in Switzerland. *Acta Trop* **69**, 213–227.

Humair, P.-F., Postic, D., Wallich, R. & Gern, L. (1998). An avian reservoir (*Turdus merula*) of the Lyme borreliosis spirochetes. *Zentbl Bakteriol* **287**, 521–538.

Humair, P.-F., Rais, O. & Gern, L. (1999). Transmission of *Borrelia afzelii* from *Apodemus* mice and *Clethrionomys* voles to *Ixodes ricinus* ticks: differential transmission pattern and overwintering maintenance. *Parasitology* **118**, 33–42.

Jauris-Heipke, S., Liegl, G., Preac-Mursic, V., Robler, D., Schwab, E., Soutschek, E., Will, G. & Wilske, B. (1995). Molecular analysis of genes encoding outer surface protein C (OspC) of *Borrelia burgdorferi* sensu lato: relationship of *ospA* genotype and evidence of lateral gene exchange of *ospC*. *J Clin Microbiol* **33**, 1860–1866.

Jones, L. D., Kaufman, W. R. & Nuttall, P. A. (1992). Modification of the skin feeding site by tick saliva mediates virus transmission. *Experientia* **48**, 779–782.

Jones, L. D., Gaunt, M., Hails, R. S., Laurenson, K., Hudson, P. J., Reid, H., Henbest, P. & Gould, E. A. (1997). Transmission of louping ill virus between infected and uninfected ticks co-feeding on mountain hares. *Med Vet Entomol* **11**, 172–176.

Jouda, F., Crippa, M., Perret, J.-L. & Gern, L. (2003). Distribution and prevalence of *Borrelia burgdorferi* sensu lato in *Ixodes ricinus* ticks of canton Ticino (Switzerland). *Eur J Epidemiol* **18**, 907–912.

Korenberg, E. I. (1994). Comparative ecology and epidemiology of Lyme disease and tick-borne encephalitis in the former Soviet Union. *Parasitol Today* **10**, 157–160.

Korenberg, E. I. & Kovalevskii, Y. V. (1994). A model for relationships among the tick-borne encephalitis virus, its main vectors, and hosts. *Adv Dis Vector Res* **10**, 65–92.

Kuo, M. M., Lane, R. S. & Giclas, P. C. (2000). A comparative study of mammalian and reptilian alternative pathway of complement-mediated killing of the Lyme disease spirochete (*Borrelia burgdorferi*). *J Parasitol* **86**, 1223–1228.

Kurtenbach, K., Sewell, H., Ogden, N. H., Randolph, S. E. & Nuttall, P. A. (1998a). Serum complement as a key factor in Lyme disease ecology. *Infect Immun* **66**, 1248–1251.

Kurtenbach, K., Peacey, M. F., Rijpkema, S. G. T., Hoodless, A. N., Nuttall, P. A. & Randolph, S. E. (1998b). Differential transmission of the genospecies of *Borrelia burgdorferi* sensu lato by game birds and small rodents in England. *Appl Environ Microbiol* **64**, 1169–1174.

Kurtenbach, K., Schäfer, S. M., de Michelis, S., Etti, S. & Sewell, H.-S. (2002a). *Borrelia burgdorferi* sensu lato in the vertebrate host. In *Lyme Borreliosis: Biology, Epidemiology and Control*, pp. 117–148. Edited by J. S. Gray, O. Kahl, R. S. Lane & G. Stanek. Wallingford, UK: CABI.

Kurtenbach, K., De Michelis, S., Etti, S., Schäfer, S. M., Sewell, H.-S., Brade, V. & Kraiczy, P. (2002b). Host association of *Borrelia burgdorferi* sensu lato – the key role of host complement. *Trends Microbiol* **10**, 74–79.

Kurtenbach, K., Schäfer, S. M., Sewell, H.-S., Peacey, M. F., Hoodless, A. N., Nuttall, P. A. & Randolph, S. E. (2002c). Differential survival of Lyme borreliosis spirochetes in ticks feeding on birds. *Infect Immun* **70**, 5893–5895.

Labuda, M. & Randolph, S. E. (1999). Survival of tick-borne encephalitis virus: cellular basis and environmental determinants. *Zentbl Bakteriol* **288**, 513–524.

Labuda, M., Jones, L. D., Williams, T., Danielová, V. & Nuttall, P. A. (1993a). Efficient transmission of tick-borne encephalitis virus between cofeeding ticks. *J Med Entomol* **30**, 295–299.

Labuda, M., Nuttall, P. A., Kozuch, O., Eleckova, E., Williams, T., Zuffova, E. & Sabo, A. (1993b). Non-viraemic transmission of tick-borne encephalitis virus: a mechanism for arbovirus survival in nature. *Experientia* **49**, 802–805.

Labuda, M., Austyn, J. M., Zuffova, E., Kozuch, O., Fuchsberger, N., Lysy, J. & Nuttall, P. A. (1996). Importance of localized skin infection in tick-borne encephalitis virus transmission. *Virology* **219**, 357–366.

Lagal, V., Postic, D. & Baranton, G. (2002). Molecular diversity of the *ospC* gene in *Borrelia*: impact on phylogeny, epidemiology and pathology. *Wien Klin Wochenschr* **114**, 562–567.

Lane, R. S. & Quistad, G. B. (1998). Borreliacidal factor in the blood of the Western Fence Lizard (*Sceloporus occidentalis*). *J Parasitol* **84**, 29–34.

Laurenson, M. K., Norman, R., Gilbert, L., Reid, H. W. & Hudson, P. J. (2003). Identifying disease reservoirs in complex systems: mountain hares as reservoirs of ticks and louping-ill virus, pathogens of red grouse. *J Anim Ecol* **72**, 177–185.

LeFleche, A., Postic, D., Girardet, K., Peter, O. & Baranton, G. (1997). Characterization of *Borrelia lusitaniae* sp. nov. by 16S ribosomal FDNA sequence analysis. *Int J Syst Bacteriol* **47**, 921–925.

Lundkvist, A., Vene, S., Golovljova, I., Mavtchoutka, V., Forsgren, M., Kalnina, V. & Plyusnin, A. (2001). Characterization of tick-borne encephalitis virus from Latvia: evidence for co-circulation of three distinct sub-types. *J Med Virol* **65**, 730–735.

Marti Ras, N., Postic, D., Foretz, M. & Baranton, G. (1997). *Borrelia burgdorferi* sensu stricto, a bacterial species "made in the *U S A*". *Int J Syst Bacteriol* **47**, 1112–1117.

Masuzawa, T., Takada, N., Fukui, T., Yano, Y., Ishiguro, F., Kawamura, Y., Imai, Y. & Ezaki, T. (2001). *Borrelia sinica* sp. nov., a Lyme disease-related *Borrelia* species isolated in China. *Int J Syst Evol Microbiol* **51**, 1817–1824.

McGuire, K., Holmes, E. C., Gao, G. F., Reid, H. W. & Gould, E. A. (1998). Tracing the origins of louping ill virus by molecular phylogenetic analysis. *J Gen Virol* **79**, 981–988.

Miyamoto, K. & Masuzawa, T. (2002). Ecology of *Borrelia burgdorferi* sensu lato in Japan and East Asia. In *Lyme Borreliosis: Biology, Epidemiology and Control*, pp. 201–222. Edited by J. S. Gray, O. Kahl, R. S. Lane & G. Stanek. Wallingford, UK: CABI.

Norman, R., Bowers, R. G., Begon, M. & Hudson, P. J. (1999). Persistence of tick borne virus in the presence of multiple host species: ticks act as reservoirs and result in parasite mediated competition. *J Theor Biol* **200**, 111–118.

Nuttall, P. A. (1998). Displaced tick-parasite interactions at the host interface. *Parasitology* **116**, S65–S72.

Nuttall, P. A. & Labuda, M. (1994). Tick-borne encephalitides. In *Ecological Dynamics of Tick-borne Zoonoses*, pp. 351–391. Edited by D. E. Sonenshine & T. N. Mather. New York & Oxford: Oxford University Press.

Ogden, N. H., Nuttall, P. A. & Randolph, S. E. (1997). Natural Lyme disease cycle maintained *via* sheep by co-feeding ticks. *Parasitology* **115**, 591–599.

Olsen, B., Duffy, D. C., Jaenson, T. G. T., Gylfe, A., Bonnedahl, J. & Bergström, S. A. (1995). Transhemispheric exchange of Lyme disease spirochaetes by seabirds. *J Clin Microbiol* **33**, 3270–3274.

Papadopoulos, O. (1980). Arbovirus problems in Greece. In *Arboviruses in the Mediterranean Countries* (6th FEMS Symposium), pp. 117–121. Edited by E. Arslanagic. *Zentralblatt für Bakteriologie, Mikrobiologie und Hygiene*. Stuttgart: Gustav Fischer.

Pavlov, P. (1968). Studies of tick-borne encephalitis of sheep and their natural foci in Bulgaria. *Zentbl Bakteriol Parasitenkd Infektkrankh Hyg* **206**, 360–367.

Picken, R. N., Cheng, Y., Strle, F. & Picken, M. M. (1996). Patient isolates of *Borrelia burgdorferi* sensu lato with genotypic and phenotypic similarities of strain 25015. *J Infect Dis* **174**, 1112–1115.

Postic, D., Assous, M., Grimont, P. A. D. & Baranton, G. (1994). Diversity of *Borrelia burgdorferi sensu stricto* as evidenced by restriction length polymorphism of *rrf*(5S)-*rrl*(23S) intergenic spacer amplicon. *Int J Syst Bacteriol* **44**, 743–752.

Postic, D., Marti-Ras, N., Lane, R. S., Hendson, M. & Baranton, G. (1998). Expanded diversity among Californian *Borrelia* isolates and description of *Borrelia bissettii* sp. nov. (formerly *Borrelia* group DN127). *J Clin Microbiol* **36**, 3497–3504.

Randolph, S. E. (1998). Ticks are not insects: consequences of contrasting vector biology for transmission potential. *Parasitol Today* **14**, 186–192.

Randolph, S. E. (2000). Ticks and tick-borne disease systems in space and from space. *Adv Parasitol* **47**, 217–243.

Randolph, S. E. & Storey, K. (1999). Impact of microclimate on immature tick-rodent interactions (Acari: Ixodidae): implications for parasite transmission. *J Med Entomol* **36**, 741–748.

Randolph, S. E., Gern, L. & Nuttall, P. A. (1996). Co-feeding ticks: epidemiological significance for tick-borne pathogen transmission. *Parasitol Today* **12**, 472–479.

Randolph, S. E., Miklisová, D., Lysy, J., Rogers, D. J. & Labuda, M. (1999). Incidence from coincidence: patterns of tick infestations on rodents facilitate transmission of tick-borne encephalitis virus. *Parasitology* **118**, 177–186.

Randolph, S. E., Green, R. M., Peacey, M. F. & Rogers, D. J. (2000). Seasonal synchrony: the key to tick-borne encephalitis foci identified by satellite data. *Parasitology* **121**, 15–23.

Reid, H. W. (1975). Experimental infection of red grouse with louping-ill virus (Flavivirus group). *J Comp Pathol* **85**, 223–229.

Reid, H. W., Duncan, J. S., Phillips, J. D. P., Moss, R. B. & Watson, A. (1978). Studies on louping-ill virus (Flavivirus group) in wild red grouse (*Lagopus lagopus scoticus*). *J Hyg* **81**, 321–329.

Richter, D., Endepols, S., Ohlenbusch, A., Eiffert, H., Spielman, A. & Matuschka, F.-R. (1999). Genospecies diversity of Lyme disease spirochetes in rodent reservoirs. *Emerg Infect Dis* **5**, 291–296.

Richter, D., Spielman, A., Komar, N. & Matuschka, F.-R. (2000). Competence of American robins as reservoir hosts for Lyme disease spirochetes. *Emerg Infect Dis* **6**, 133–138.

Rogers, D. J. & Randolph, S. E. (2000). The global spread of malaria in a future, warmer world. *Science* **289**, 1763–1766.

Rogers, D. J., Hay, S. I. & Packer, M. J. (1996). Predicting the distribution of tsetse flies in West Africa using temporal Fourier processed meteorological satellite data. *Ann Trop Med Parasitol* **90**, 225–241.

Rogers, D. J., Randolph, S. E., Snow, R. W. & Hay, S. I. (2002). Satellite imagery in the study and forecast of malaria. *Nature* **415**, 710–715.

Ryan, J. R., Levine, J. F., Apperson, C. S., Lubke, L., Wirtz, R. A., Spears, P. A. & Orndorff, P. E. (1998). An experimental chain of infection reveals that distinct *Borrelia burgdorferi* populations are selected in arthropod and mammalian hosts. *Mol Microbiol* **30**, 365–379.

Schwan, T. G., Piesman, J., Golde, W. T., Dolan, M. C. & Rosa, P. A. (1995). Induction of an outer surface protein on *Borrelia burgdorferi* during tick feeding. *Proc Natl Acad Sci U S A* **92**, 2909–2913.

Skarpass, T., Sundoy, A., Bruu, A. L., Vene, S., Pedersen, J., Eng, P. G. & Csángó, P. A. (2002). [Tick-borne encephalitis in Norway] (in Norwegian). *Tidsskr Nor Laegeforen* **122**, 30–32.

Stanek, G., Strle, F., Gray, J. & Wormser, G. P. (2002). History and characteristics of Lyme borreliosis. In *Lyme Borreliosis: Biology, Epidemiology and Control*, pp. 1–28. Edited by J. S. Gray, O. Kahl, R. S. Lane & G. Stanek. Wallingford, UK: CABI.

Stuen, S. (2003). *Anaplasma phagocytophilum (formerly Ehrlichia phagocytophila) infection in sheep and wild ruminants in Norway*. PhD thesis, Department of Sheep and Goat Research, Norwegian School of Veterinary Science, Sandnes, Norway.

Ulvund, M. J., Vik, T. & Krogsrud, J. (1983). Louping-ill (tick-borne encephalitis) hos sau i Norge (in Norwegian). *Nor Veterinaertidsskr* **95**, 639–641.

Van Dam, A. P., Kuiper, H., Vos, K., Widjojokusumo, A., Spanjaard, L., De Jongh, B. M., Ramselaar, A. C. P., Kramer, M. D. & Dankert, J. (1993). Different genospecies of *Borrelia burgdorferi* are associated with distinct clinical manifestations in Lyme borreliosis. *Clin Infect Dis* **17**, 708–717.

Wang, G., van Dam, A. P., Schwartz, I. & Dankert, J. (1999a). Molecular typing of *Borrelia burgdorferi* sensu lato: taxonomic, epidemiological, and clinical implications. *Clin Microbiol Rev* **12**, 633–653.

Wang, I. N., Dykhuizen, D. E., Qin, W. G., Dunn, J. J., Bosler, E. M. & Luft, B. J. (1999b). Genetic diversity of *ospC* in a local population of *Borrelia burgdorferi* sensu stricto. *Genetics* **151**, 15–30.

Whitby, J. E., Whitby, S. N., Jennings, A. D., Stephenson, J. R. & Barrett, A. D. T. (1993). Nucleotide-sequence of the envelope protein of a Turkish isolate of tick-borne encephalitis (TBE) virus is distinct from other viruses of the TBE virus complex. *J Gen Virol* **74**, 921–924.

Wikel, S. K. (editor) (1996). *The Immunology of Host-Ectoparasitic Arthropod Relationships*. Wallingford, UK: CABI.

Will, G., Jauris-Heipke, S., Scwab, E., Busch, U., Rössler, D., Soutschek, E., Wilske, B. & Preac-Mursic, V. (1995). Sequence analysis of *ospA* genes shows homogeneity within *Borrelia burgdorferi* sensu stricto and *Borrelia afzelii* strains but reveals major subgroups within the *Borrelia garinii* species. *Med Microbiol Immunol* **184**, 73–80.

Zanotto, P. M., Gao, G. F., Gritsun, T., Marin, M. S., Jiang, W. R., Venugopal, K., Reid, H. W. & Gould, E. A. (1995). An arbovirus cline across the northern hemisphere. *Virology* **210**, 152–159.

Zanotto, P. M., Gould, E. A., Gao, G. F., Harvey, P. H. & Holmes, E. C. (1996). Population dynamics of flaviviruses revealed by molecular phylogenies. *Proc Natl Acad Sci U S A* **93**, 548–553.

Insect transmission of viruses

Stéphane Blanc

UMR Biologie et Génétique des Interactions Plante-Parasite (BGPI), CIRAD-INRA-ENSAM, TA 41/K, Campus International de Baillarguet, 34398 Montpellier Cedex 5, France

INTRODUCTION

Obligate parasites all rely on their host for survival. Transfer from one host to another is thus a key aspect of their life cycle, ensuring both their maintenance (vertical and horizontal transmission) and spread (primarily horizontal transmission) in the environment. Vertical transmission, i.e. the passage of a parasite directly from one host to its descendant(s), is a strategy that does not involve passage outside the host, and so maintains the parasite in an amiable environment. However, the success of this transmission strategy depends on the survival of the host line. Furthermore, it restricts parasite spread within the host population. In fact, the survival of the parasite population is directly related to host fitness. Thus vertical transmission may be expected to lead to parasite populations with lower virulence or to non-pathogenic interactions (Frank, 1996; Day, 2001; Weiss, 2002).

All pathogens have evolved additional (or even exclusive) strategies for horizontal transmission, with an impressive diversity of mechanisms that will be the focus of the present chapter. Pathogens are found among viruses, bacteria, fungi, protozoa and metazoa but the tremendous richness of transmission strategies mentioned above is certainly best illustrated by viruses. Examples of horizontal transmission of organisms other than viruses are described in several sections of this symposium volume. Some viruses have, like other pathogens, developed the ability to transfer from infected to healthy hosts by contact or through passage in the external medium, via a variety of propagating forms (Kuno, 2001). Most viruses, however, exploit an additional organism, itself travelling from host to host, as a transportation device

SGM symposium 63: Microbe–vector interactions in vector-borne diseases.
Editors S. H. Gillespie, G. L. Smith & A. Osbourn. Cambridge University Press. ISBN 0 521 84312 X ©SGM 2004

(referred to as a vector). This chapter presents an overview of the various types of virus–vector interactions that have been reported to successfully promote transmission of viruses of both plants and animals. The vast majority of vectors are nematodes, mites and insects. Interestingly, the strategies controlling virus–insect interactions are very diverse, and include those also reported to promote nematode and mite transmission.

Plant and animal viruses have some very different constraints regarding the transmission step. While animal hosts are mobile and often come into physical contact with each other, plant hosts are immobile and (apart from gametes, which sometimes allow vertical transmission) close contact is rare. Furthermore, plant cells are surrounded by cell walls, thick, solid barriers, made of pectin and cellulose, which a virus cannot penetrate unless a fissure is created either by mechanical stress or by plant-feeding organisms. This presumably explains why nearly all plant viruses are exclusively vector-transmitted in the real world, whereas many animal viruses rely partly or completely on non-vector transmission (Kuno, 2001). As a consequence, a diverse range of specific and intricate interaction strategies has evolved, and these strategies have been the subject of considerable research efforts in the plant virus field. Over the last half century, the classification of these plant-virus–vector interactions has been frequently modified, complemented and updated (Watson & Roberts, 1939; Sylvester, 1956; Kennedy *et al.*, 1962; Harris, 1977; Pirone & Blanc, 1996; Nault, 1997). Vector transmission of animal viruses was originally thought to be of two possible sorts, either 'mechanical' or 'biological'. The former is generally associated with non-specific inoculation of viruses by contaminated mouthparts or body of the vector, while the latter applies to specific associations and requires that the virus can propagate within the vector. Although this simple dichotomy is still in use today (Mellor, 2000; Gray & Banerjee, 1999), it may not reflect the actual diversity of strategies that viruses can develop, particularly for so-called 'mechanical' transmission. Because the classification system developed for the transmission of plant viruses is more detailed, it will be used in the rest of this chapter as a common system applying to both plant and animal viruses.

Recent reviews on virus–vector interactions are available for both animal (Carn, 1996; Weaver, 1997; Mellor, 2000) and plant (Nault, 1997; Harris *et al.*, 2001; Pirone & Perry, 2002) viruses, hence this chapter does not present an exhaustive listing and description of all data available on this tremendously wide subject. Instead, as in previous other reviews (Ammar, 1994; Gray & Banerjee, 1999; Van den Heuvel *et al.*, 1999), the apparent similarities and differences observed between insect transmission of plant and animal viruses will be discussed and illustrated using examples when required. This comparison clearly indicates that vector-transmission research has remained somewhat separate in plant and animal virology and that both fields could profit from each other, especially since there are common questions to be addressed in these areas.

CIRCULATIVE TRANSMISSION

The term 'circulative' was introduced by Sylvester (1956) and Harris (1977). It applies to cases where viruses undergo part of their life cycles within the body of the vector. In this type of relationship the virus is ingested by the vector as it feeds on an infected host. The virus then traverses the gut epithelium at the midgut or hindgut level (for example see Garret *et al.*, 1996), and is released into the haemolymph. The viruses can then enter the salivary glands through various routes and transfer to the saliva, from which they will be inoculated to healthy hosts and initiate new infection. The time required for the virus to complete this cycle is designated the 'incubation period' (for animal viruses) or the 'latent period' (for plant viruses), and varies from several hours to several days depending on the mode followed (see later) and on physical factors such as temperature (reviewed recently by Mellor, 2000). In most cases, once the virus is ingested it will persist and remain infectious for the vector's lifespan.

Obviously, circulative transmission implies that the virus goes through a number of physical barriers, which ultimately condition the efficiency of transmission and the specificity of successful virus–vector interactions. First, viruses are required to pass from the gut lumen to the haemocoel, through the gut epithelium. Pioneering work demonstrating the existence of this gut barrier was conducted by Storey (1933) and Merrill & Tenbroeck (1935). These authors observed that non-vector leaf-hoppers and mosquitoes could transmit *Maize streak virus* (MSV; *Geminiviridae*) and *Eastern equine encephalitis virus* (EEEV; *Togaviridae*), respectively, when their mesenteron was punctured with a needle after they had fed on an infected host. Since then an enormous amount of work involving intra-thoracic injection of viruliferous solutions into vectors has confirmed that this barrier can stop numerous plant and animal viruses and is responsible for inefficient transmission. Second, the virus must then make its way to the haemocoel cavity or through various organs and tissues in order to reach the salivary glands. This transfer can follow different routes (see later) where, again, incompatible interactions can be observed. Third, the final efficient passage through the salivary glands and into the saliva may also be prevented by specific barriers to completion of vector transmission. Details of compatible or incompatible virus–vector interactions at each barrier level are mentioned below and have been exhaustively reviewed for animal viruses (Mellor, 2000). The circulative transmission mode is divided into two subcategories depending on whether the virus can replicate in its vector (circulative-propagative transmission) or not (circulative-non-propagative transmission).

Circulative-propagative transmission

This mode of transmission concerns the vast majority of vector-transmitted vertebrate-infecting viruses, designated arboviruses. Arboviruses and their relationship with vectors have been extensively studied, mostly because they are responsible for a number of severe diseases in humans (yellow fever, dengue, various encephalitides, etc.) and livestock (West Nile encephalomyelitis, Rift Valley fever, vesicular stomatitis, etc.). This mode of transmission also exists in plant viruses, although it is not as frequent as in vertebrates. Indeed, circulative-propagative transmission has been described for the families *Rhabdoviridae*, *Reoviridae* and *Bunyaviridae*, which contain member species associated with both vertebrate and plant hosts [the case of the *Bunyaviridae* is addressed in further detail by Elliott & Kohl (2004) in this volume], and for genera where member species are restricted to plant hosts, such as *Marafivirus* and *Tenuivirus*. Circulative-propagative viruses literally infect their insect vectors, and the consequence of this infection on vector fitness is discussed by Higgs (2004) in this volume.

When viruses reach the mid- or hindgut, the 'gut barrier' (mentioned earlier) may act in different ways (Hardy *et al.*, 1983; Hardy, 1988; Ammar, 1994). Virus–vector recognition is mediated by specific interaction between a motif present on the virus envelope or capsid, and a receptor located on the luminal surface of the epithelial cells. The lack of specificity between these two molecules or any mechanisms preventing their interaction will block the entry of the virus and thus the infection of the vector. The highly specific recognition between viruses and vectors is well described for rhabdoviruses of vertebrates and plants (reviewed recently by Hogenhout *et al.*, 2003). In addition, the replication of many vertebrate viruses is restricted to the mesenteron cells of their mosquito or biting midge vectors (Mellor, 2000). These viruses are incapable of escaping to the haemolymph and of infecting additional organs. Similarly in plants the example of *Tomato spotted wilt virus* (which is transmitted by thrips) demonstrates that even when viruses can enter and infect epithelial cells, they are sometimes unable to escape into the haemocoel (Ullman *et al.*, 1992).

In compatible virus–vector associations, once the gut barrier is passed the viruses can infect numerous organs and tissues of the vector, including the reproductive organs (leading to vertical transmission) and salivary glands (prior to horizontal transmission). The viruses can diffuse in the haemolymph and concomitantly infect different organs (e.g. EEEV) (Scott *et al.*, 1984). Alternatively, they can follow a precise pattern of spread from organ to organ, as demonstrated for *Bluetongue virus* (BTV) (Fu *et al.*, 1999) and rhabdoviruses, where the infection is believed to progress in, and spread from, the central nervous system (Hogenhout *et al.*, 2003). When reproductive organs are infected, the virus can be transmitted vertically to offspring vectors by transovarial passage. While this phenomenon has been described in many different insect species that transmit plant viruses in a circulative-propagative manner (leafhoppers, aphids, whiteflies and planthoppers) (Nault, 1997), in animal viruses transovarial transmission has been described only in virus/mosquito and virus/phlebotomid sandfly associations and not in other haematophagous insects (Mellor, 2000). Another consequence of the infection of the reproductive organs of the vector by viruses is the possibility of venereal transmission from male to female and vice versa. Venereal transmission has been reported in *La Crosse* and *Sindbis viruses* (Thompson & Beaty, 1978; Ovenden & Mahon, 1984) and may also occur in other virus–haematophagous insect interactions. In plants, to the best of our knowledge venereal transmission has been described only once, in the case of the *Tomato yellow leaf curl virus* (TYLCV; *Geminiviridae*), which is transmitted by the white fly *Bemisia tabaci* (Ghanim & Czosnek, 2000). However, it remains unclear whether TYLCV is a circulative-propagative virus (see later).

When viruses have completed the 'diffusion' step, either directly in the haemolymph or through a precise pattern of organ-to-organ passages, the last important barrier that must be overcome in order to achieve transmission is that of the salivary gland(s). To circumvent this barrier, viruses transfer through the basal lamina before infecting the acinar cells, from which they are efficiently released in the terminal acini and join the lumen of the major secretory ducts, together with the saliva. Again, only compatible virus–vector interactions are successful in these processes, and examples of failure of the virus at various stages have been reported in non-vector insects. Viruses may be unable to infect the salivary glands or may simply fail to propagate in the vector body, and it is difficult to distinguish between these possibilities. The presence of virus in various organs and its absence in the salivary glands have been reported for vertebrate viruses such as *Western equine encephalitis virus* (WEEV; *Togaviridae*) in *Culex tarsalis* (Hardy, 1988) and *Bluetongue virus* (*Reoviridae*) in *Culicoides variipennis* (Fu *et al.*, 1999), and for plant viruses such as *Wound tumor virus* (*Reoviridae*) and *Sowthistle yellow vein virus* (*Rhabdoviridae*) in leafhoppers and aphids, respectively (Granados *et al.*, 1967; Behncken, 1973). Cases in which viruses have been detected in the salivary glands but nevertheless failed to escape in the salivary ducts have also been reported. The reasons for this are not clear, and some recent reviews deal with this problem for vertebrate viruses (Mellor, 2000) and for both plant and animal rhabdoviruses (Hogenhout *et al.*, 2003).

Circulative-non-propagative transmission

This mode of transmission concerns solely plant virus species in the families *Luteoviridae* (transmitted by aphids), *Geminiviridae* and *Circoviridae*, genus *Nanovirus* (transmitted by aphids, whiteflies and leafhoppers). Because these viruses are very different (positive ssRNA or ssDNA genomes) and are transmitted by different vectors it seems likely that this transmission strategy has evolved more than once independently.

However, circulative-non-propagative transmission has never been reported in animal viruses. The cycles of circulative-non-propagative viruses in their insect vectors are very similar to those of the propagative viruses described above, the major difference being the total absence of virus replication. The specific implications of this are briefly described below.

Most knowledge about the virus cycle within the vector has been obtained from luteovirus/aphid biological models (Gray & Gildow, 2003). The mechanisms of geminivirus (and nanovirus) relationships with their vectors are usually assumed to be very similar, though no solid experimental data are available to support this. A series of very elegant electron microscopy studies has clearly established that members of the *Luteoviridae* bind to putative receptors of the luminal membrane of the midgut or hindgut epithelial cells of aphids and are internalized in clathrin-coated endocytotic vesicles (reviewed by Gray & Gildow, 2003). The virus does not come into direct contact with the cytoplasm but the virus-containing vesicles converge and fuse with larger cytoplasmic endosomes. From these endosomes, new vesicles bud and are targeted to the basal plasmalemma, where they fuse with the plasma membrane and release their content into the haemocoel cavity. Many luteoviruses, when ingested by non-vector aphid species, are nevertheless able to go through this process and accumulate in the haemocoel (Rochow & Pang, 1961; Gildow & Gray, 1993). However, it has been demonstrated that when the RPV isolate of *Barley yellow dwarf virus* (BYDV) is ingested by the aphid *Metopolophium dirhodum*, the virus is unable to attach to the apical membrane of the epithelial cells (Gildow, 1993). This suggests that at least some degree of specificity exists at this level. The opposite situation has been described for geminiviruses, where the incompatibility of several virus–vector combinations has been shown to be due to the inability of the virus to pass the gut barrier (Hohnle *et al.*, 2001).

A particularly interesting step in circulative-non-propagative transmission concerns transit between the gut and the salivary glands. Indeed, these viruses do not replicate in any of the vector cells and, therefore, they most likely diffuse passively within the haemolymph. The length of time that viruses will remain in suspension in this putatively hostile environment is unknown, but it is believed to be longer for luteo-viruses than for geminiviruses. Indeed, while the haemolymph was hypothesized to be the storage compartment for luteoviruses in the vector, geminiviruses are supposed to be stored (in unknown form) in either the gut or salivary gland cells and are capable of rapid transfer between the two (Lett *et al.*, 2002). Accordingly, the concentration of MSV has been demonstrated by real-time quantitative PCR to be much lower in the haemolymph than in gut cells or salivary glands of the vector *Cicadulina mbila* (Lett *et al.*, 2002). The immune systems of plant-feeding insects are poorly understood but it seems logical to assume that the virus will have to evade these mechanisms, as documented for other vector-transmitted pathogens [see the chapter by Ratcliffe & Whitten (2004) in this volume]. One of the most interesting discoveries in this area was by Van den Heuvel *et al.* (1994, 1997). These authors suggested that a chaperone-like protein designated symbionin, which is produced and secreted by endosymbiotic bacteria of the genus *Buchnera*, was necessary for efficient transmission of luteoviruses and that it protected the virions during transfer into the haemolymph. These authors demonstrated that when aphids were treated with antibiotics (thereby killing the endosymbionts), the transmission efficiency of luteoviruses was dramatically reduced. They later demonstrated that the symbionin could bind *in vitro* to motifs exposed at the surface of the luteovirus particles. However, this binding also occurred between luteoviruses and symbionin of non-vector aphid species, indicating that the role of this protein (if any) is not involved in virus–vector specificity. Interestingly, a symbionin-like protein was also suggested to play a role in the transmission efficiency of TYLCV (*Geminiviridae*) by its whitefly vector (Morin *et al.*, 1999, 2000). Nevertheless, the specific role of symbionin in virus–vector interactions is questionable (Gray & Gildow, 2003) since: (i) the symbionin–virion interaction appears to be relatively non-specific, (ii) it has not been demonstrated *in vivo*, and (iii) the absence of symbionin may affect vector physiology and so modify the vector's capacity to transmit a given virus indirectly.

Circulative-non-propagative viruses are believed to associate specifically with accessory salivary glands. These glands produce a watery saliva that forms the stylet's sheath during insect feeding. The accessory glands are surrounded by an extracellular basal lamina, where transmissible luteoviruses have been shown to accumulate (Gildow & Gray, 1993). This basal lamina represents a specific barrier to luteovirus transmission (Pfeiffer *et al.*, 1997; Gildow *et al.*, 2000). After successfully traversing the basal lamina, some luteoviruses are stopped at the basal plasma membrane of the salivary gland's cells, indicating an additional specific interaction at this level (Pfeiffer *et al.*, 1997). Finally, when a compatible interaction occurs, the virus will translocate to the lumen together with watery saliva. Though the model explaining the intracellular transfer of luteoviruses from the haemolymph to the saliva is similar to that of the transfer through gut cells (see earlier), it may differ in certain details (reviewed recently by Gray & Gildow, 2003).

In members of the families *Geminiviridae* and *Circoviridae* (genus *Nanovirus*), no such details are available. It is, however, noteworthy that a number of studies provided results that could contradict the general assumption of a mode of transmission similar to luteoviruses. For geminiviruses, while no direct proof of virus replication in the insect vector could be obtained, TYLCV was reported to be trans-ovarially transmitted in the whitefly *B. tabaci* for at least two generations (Ghanim *et al.*, 1998). Moreover, cytopathic effects of the virus were observed in the cells of the vector (Rubinstein & Czosnek, 1997), and recent work has demonstrated that, as is often the case for arboviruses (see earlier), TYLCV was also sexually transmitted from male to female vectors or vice versa (Ghanim & Czosnek, 2000). These features are usually associated with replicating viruses, suggesting that more work is required to definitely understand the transmission strategy of geminiviruses. The same is true for nanoviruses, where Franz *et al.* (1999) have identified a helper component (HC) involved in the transmission of *Faba bean necrotic yellows virus* by its aphid vector. This result is puzzling because HC molecules (described in more detail in the next section) have always been associated with cases of non-circulative transmission. Again, more work is needed in order to gain a complete understanding of the so-considered circulative-non-propagative transmission of members of the genus *Nanovirus*.

NON-CIRCULATIVE TRANSMISSION

This mode of transmission encompasses two well-defined strategies of transmission of plant viruses, the 'capsid' and the 'helper' strategies (Pirone & Blanc, 1996), the more obscure beetle-mediated transmission of plant viruses (Sarra & Peters, 2003) and the miscellaneous 'mechanical transmission' of animal viruses (Carn, 1996). The term non-circulative is self-explanatory and stresses the absence of a cycle of the virus within the vector body. Instead, viruses are transported in the mouthparts, or otherwise, by the vector in a manner that may be highly specific and sophisticated or non-specific. There is no latent (or incubation) period associated with this means of transmission and the transmission efficiency of the 'contaminated' vector decreases rapidly with time. In plants, the vast majority of viruses have developed finely tuned molecular mechanisms for regulating the non-circulative virus–vector interaction whereas in animals the situation is unclear and sophisticated molecular models have not been reported. For these reasons, this section will first summarize the data on the strategies adopted by plant viruses and then comment briefly on the data available for animal viruses.

Non-circulative transmission of plant viruses

Two main types of insect vectors can be distinguished for non-circulative viruses of plants: sap-feeding insects, mainly in the orders *Sternorrhyncha* (including aphids and whiteflies) and *Auchenorrhyncha* (including leafhoppers and treehoppers), and plant-'chewing' insects in the order *Coleoptera* (beetles). All non-circulative viruses transmitted by sap-feeding insects have adopted either the helper or the capsid strategy, whereas the mechanism of transmission by beetles is less well understood and it remains unclear whether or not it should be classified as non-circulative.

Non-circulative transmission by sap-feeding insects has long been subdivided into non-persistent and semi-persistent categories (Watson & Roberts, 1939; Sylvester,

1956), depending on the length of time during which the vector remains capable of inoculating the virus after feeding on an infected plant. Although these terms are still frequently used in the literature, a new terminology was more recently proposed based on qualitative rather than quantitative criteria (Pirone & Blanc, 1996), and this is the one preferred below.

The capsid strategy. The capsid strategy is employed by members of the genera *Cucumovirus*, *Alphamovirus* and *Carlavirus*, the best-known example being *Cucumber mosaic virus* (CMV). When feeding on an infected plant, sap-feeding insects ingest contaminating viruses. The virus particles immediately attach to putative receptors associated with the cuticle lining either the inner surface of the food and/or salivary canal in the insect's mouthparts (Martin *et al.*, 1997), or any portion of the anterior gut. The cuticular location of these receptors is demonstrated by the loss of viruses upon insect moult. Viruses that are not retained at this level are lost in the sense that they are never transmitted (for a recent review see Pirone & Perry, 2002).

The ability of the vector to acquire and transmit a virus from purified solution through Parafilm membranes is considered a demonstration that factors in addition to the virus particles themselves are not required for successful transmission. This implies that the sole determinant of the virus–vector interaction is the coat protein (Fig. 1). In a series of elegant experiments Perry and co-workers identified two amino acids in a domain of the CMV coat protein that are involved in the specificity of transmission by different aphid species (Perry *et al.*, 1994, 1998; Pirone & Perry, 2002). The belief that transmission is not simply a non-specific deposition of viruses in plant tissues by the vector resides in the following facts: viruses are not transmitted through contact between plants in the field, and only some insect species transmit a given virus even though the feeding behaviour of non-vector and closely related vector species is often quite similar. Many steps of this process are still unknown and the putative receptors on the insect cuticle, the precise chemical nature of virus–vector interactions and the virus release mechanisms during inoculation into a new host plant remain unclear.

The helper strategy. This strategy has received much attention because a large number of plant viruses are transmitted in this way. In particular, the most important genus of plant viruses, *Potyvirus* (Raccah *et al.*, 2001), and the genus *Caulimovirus*, one of the best-characterized plant virus genera (Blanc *et al.*, 2001), use the helper strategy. The overall relationship between the virus and vector is similar in the capsid and the helper strategies. In the latter, viruses also attach rapidly to the mouthparts' inner cuticle, from where they are released and inoculated upon subsequent feeding of the vector. The critical difference is that in the helper strategy the virus does not bind to the putative receptors directly via its coat protein (Fig. 1). Thus purified virus preparations

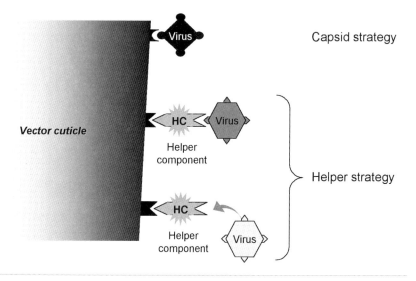

Fig. 1. Two molecular strategies for virus–vector interaction in non-circulative transmission of plant viruses. Both strategies allow the retention of virus particles in the vector mouthparts or foregut on putative receptors located at the surface of the cuticular lining. In the capsid strategy, a motif of the coat protein is able to directly bind to the vector's receptor. In the helper strategy, virus–vector binding is mediated by a virus-encoded non-structural protein, the helper component (HC), which creates a reversible 'molecular bridge' between the two. This mode of action of HC is designated the 'bridge hypothesis'. HC can be acquired alone, prior to virion, and thereby allow HC-transcomplementation. In this case a HC encoded by a genome X (for instance that encapsidated in the dark grey virion) can subsequently assist the transmission of a genome Y of the same population, encapsidated in the light grey virion. This possible sequential acquisition of HC and virion is symbolized by the arrow. It has been demonstrated experimentally that HC and virion can be acquired in different infected cells or even different hosts (Kassanis & Govier, 1971a; Lung & Pirone, 1974). Figure reproduced from Froissart *et al.* (2002) with permission.

are never transmissible. Kassanis and Govier (Kassanis & Govier, 1971a, b; Govier & Kassanis, 1974) and Lung & Pirone (1973, 1974) were the first to demonstrate the involvement of an additional factor, the helper component (HC), in potyviruses and caulimoviruses, respectively. Since then, many studies have characterized the HCs as viral bifunctional molecules capable of interaction on one hand with the putative receptors in the vector's mouthparts, and on the other hand with motifs located at the surface of the virus particles (for review see Raccah *et al.*, 2001; Blanc *et al.*, 2001). A unique feature of the helper strategy is that HC and virions can be acquired sequentially, provided that the HC is acquired first. This then allows the phenomenon of HC-transcomplementation (Froissart *et al.*, 2002). HC-transcomplementation is defined as follows: in a virus population, a HC encoded by a genome X can assist the transmission of a virus particle containing a genome Y, thus allowing co-operation between viral genomes within a population during vector transmission (Fig. 1). HC-transcomplementation is believed to have an important impact on the evolution of viruses that have adopted the helper strategy. This impact has been postulated in the literature (Pirone & Blanc, 1996; Froissart *et al.*, 2002) and will be discussed further later. It is even more clear for the helper strategy than for the capsid strategy that it is a highly specific phenomenon: while purified HC-less virus particles are readily transmitted experimentally by mechanical inoculation, they are not transmitted by vectors unless exogenous functional HC is added in the feeding solution. Again, the putative receptors on the vector cuticle are unknown and the identification of these receptors represents one of the greatest challenges for future research in this field.

Beetle transmission. Transmission of plant viruses by beetles was first classified as purely 'mechanical', meaning non-specific, because only highly physically stable viruses are concerned. A series of work by Gergerich and co-workers (Gergerich & Scott, 1991; Wang *et al.*, 1994a, b; Gergerich, 2001) has suggested a more complex relationship between virus and vector by demonstrating that *Tobacco mosaic virus* (TMV), probably the most stable and resilient plant virus, is never transmitted by beetles. These authors proposed that beetle-transmitted viruses are ingested and regurgitated together with huge amounts of RNase during chewing of the leaf by the vector. The transmissible viruses would be able to translocate into cells distant from the feeding site before replication, therefore escaping the RNase activity, whereas non-transmissible viruses would initiate replication before translocation and be degraded by RNase. No latent or incubation period is required for beetle transmission; however, some viruses were reported in the haemocoel of the insect vector, indicating a possible circulative cycle (Wang *et al.*, 1994b). The correlation between invasion of the haemocoel by a virus and the success of transmission could not be established and it is thus difficult to decide whether these particular cases of vector transmission are circulative, non-circulative or a combination of both.

Non-circulative transmission of animal viruses

This category is termed 'mechanical transmission' for viruses of vertebrates and generally evokes the absence of precise adaptive molecular mechanisms controlling the interaction between virus and insect vectors. Mechanical transmission is most often considered to result from a combination of the ability of a virus to infect a vertebrate orally, through aerosol, by contact or blood transfer, and the commensal, predator or parasitic activity of insects around and within a population of infected hosts. A perfect illustration of these non-specific transmission events is found with *Rabbit hemorrhagic disease virus* (RHDV). Several fly species (*Calliphoridae*) simply transport RHDV, and deposit infectious doses through oral and/or anal excretion that can then be orally transferred to the host (Asgari *et al.*, 1998). Other flies (*Muscidae*) are suspected to directly inoculate the virus to conjunctiva (Gehrmann & Kretzschmar, 1991, quoted in Asgari *et al.*, 1998). Finally, RHDV was shown to be also transmitted by fleas and mosquitoes (Gould *et al.*, 1997). Many vertebrate viruses that are transmitted mechanically by vectors are also transmitted by other means, for example by biological vector transmission or non-vector transmission (Kuno, 2001). It is striking to note that a combination of non-circulative and circulative transmission modes, found for instance in *Dengue virus* (DENV) (Esteva & Vargas, 2000), has never been described for plant viruses.

In other cases such as *Myxoma virus* (Barcena *et al.*, 2000) or *Lumpy skin disease virus* (LSDV) (Chihota *et al.*, 2003), however, mechanical transmission by insects is probably the major mode of horizontal transmission. These viruses are transmitted by haematophagous insects that acquire viruses with their mouthparts during feeding on infected animals and later inoculate healthy hosts, in a manner similar to that of the capsid or helper strategies of plant viruses. One open question is whether these viruses, like plant viruses, have developed specific molecular strategies for interaction with the vector. The case of LSDV is particularly interesting. This virus has long been suspected to be transmitted 'mechanically' by biting insects and intravenous inoculation has been consistently described as the most efficient infection route (Carn & Kitching, 1995). While the stable fly *Stomoxys calcitrans* was identified as the most probable vector (Yeruham *et al.*, 1995), only *Aedes aegypti* could be experimentally demonstrated to efficiently transmit LSDV (Chihota *et al.*, 2001). Interestingly, the stable fly *S. calcitrans*, a vector of other poxviruses (Mellor *et al.*, 1987), failed to transmit LSDV in recent studies (Chihota *et al.*, 2003). In addition, other biting insects, even mosquitoes in the genera *Culex* or *Anopheles*, appear to be non-vectors of LSDV. This specificity for successful virus–vector interaction resembles the cases of non-circulative transmission of plant viruses, and it would be interesting to further investigate the molecular basis of the so-called 'mechanical vector transmission' of poxviruses.

FUTURE RESEARCH AND PROSPECTS

The virus determinants that specifically recognize vector species are the subject of numerous studies. These determinants may be glycoproteins of the viral membranes or capsids in arboviruses of both plants and animals (reviewed by Gray & Banerjee, 1999; Van den Heuvel *et al.*, 1999; Hogenhout *et al.*, 2003), amino acid motifs of the coat protein (*Geminiviridae*) (Hohnle *et al.*, 2001) or coat-associated proteins (*Luteoviridae*) (Gray & Gildow, 2003) in the circulative-non-propagative plant viruses. In non-circulative transmission of plant viruses, specific coat protein amino acids have been identified as determinants of vector specificity in *Cucumoviridae* (Pirone & Perry, 2002), and amino acid motifs of the HCs with similar function have been identified in both *Potyviridae* (Raccah *et al.*, 2001) and *Caulimoviridae* (Blanc *et al.*, 2001). In contrast, very little is known about the receptors in the insect vector that determine the specificity of virus transmission. For insect-infecting arboviruses, many receptors have been identified, presumably corresponding to membrane receptors involved in cell entry in various tissues of the vector (reviewed by Van den Heuvel *et al.*, 1999; Hogenhout *et al.*, 2003). However, receptors that account for the high degree of virus–vector specificity observed in circulative transmission remain to be characterized. Some genetic approaches have investigated the determinants of virus–vector specificity, and, in some cases, only a single vector gene may be involved (reviewed by Mellor, 2000). The development of the international sequencing programmes on mosquitoes and aphids may be beneficial in the near future by allowing genomic approaches similar to that initiated recently in the biting midge *Culicoides sonorensis*, infected with the *Epizootic haemorrhagic disease virus* (Campbell & Wilson, 2002). For *Geminiviridae* and *Luteoviridae*, which do not replicate in their vectors, the situation appears simpler since a less intimate relationship should involve less molecular interaction. However, only one report has been published on a putative receptor of BYDV (luteovirus) in the salivary glands of the aphid vector *Sitobion avenae* (Li *et al.*, 2001). Finally, for non-circulative viruses, even the chemical nature of the putative receptor associated with the cuticle lining the vector mouthparts or anterior gut is unknown. Clearly, identifying and characterizing specific receptors of viruses in insect vectors is the major challenge for studies of vector transmission, since such receptors would be perfect targets for disease control strategies.

An interesting aspect is the modification of vector behaviour that could be either directly or indirectly induced by the virus. In fact, by modifying the feeding behaviour of vectors, parasites can quantitatively and qualitatively increase their chances of transmission. This phenomenon has been studied in parasites such as protozoa or bacteria (Hurd, 2003), but rarely in viruses. It has been shown that when DENV infects the central nervous system of its mosquito vectors, the time length of blood meals increases. This improves the chances of interrupted feeding, favouring immediate re-landing on a new host and increased virus transmission (Platt *et al.*, 1997). In plant luteoviruses, volatile compounds emitted from potato plants infected with *Potato leafroll virus* attract and favour settlement of aphid vectors (Eigenbrode *et al.*, 2002). These rare studies confirm that viruses have indeed evolved strategies for manipulating their insect vectors and open a relatively unexplored field of research in virus–vector relationships.

Experimental studies on the evolution biology of viruses have developed tremendously over the last 15 years. *In vitro* and *in vivo* studies have demonstrated that, at least for RNA viruses and retroviruses, virus populations are structured in a specific way that is sometimes designated quasispecies (Moya *et al.*, 2000; Domingo, 2000). Due to the high error rate of viral polymerases, mutations constantly appear in viral genomes during replication. Rapidly, the population becomes an ensemble of mutant genomes centred on a mean, or consensus, sequence. During repeated and severe population bottlenecks, in asexual populations, the genetic drift is increased and results in an accumulation of mutations, designated Müller's ratchet. Since most mutations are deleterious the mean fitness of virus populations will in consequence dramatically decrease. This debilitating effect of population bottlenecks on virus populations has been experimentally demonstrated many times in bacteria and animal viruses, noticeably in *Vesicular stomatitis virus* (for review see Elena & Lenski, 2003; Garcia-Arenal *et al.*, 2003). Transmission in general and vector transmission in particular has been suggested as a likely step inducing severe population bottlenecks for viruses (Elena *et al.*, 2001; Sacristan *et al.*, 2003). Some studies have estimated the minimum number of virus particles or infectious doses that a vector must ingest in order to transmit either vertebrate (for example see Fu *et al.*, 1999) or plant (Walker & Pirone, 1972; Pirone & Thornbury, 1988) viruses. However, the proof of the existence of such bottlenecks and the evaluation of their actual size, in natural vector transmission, has never been experimentally established. This question is of great interest because, as proposed for the helper strategy (Pirone & Blanc, 1996; Froissart *et al.*, 2002), some strategies of virus–vector interaction may have been selected because they ameliorate the putative severe genetic bottlenecks encountered by virus populations at each round of vector transmission.

ACKNOWLEDGEMENTS

S. Blanc wishes to thank Martin Drucker for critical reading of this manuscript.

REFERENCES

Ammar, E. D. (1994). Propagative transmission of plant and animal viruses by insects: factors affecting vector specificity and competence. *Adv Dis Vector Res* **10**, 289–331.

Asgari, S., Hardy, J. R., Sinclair, R. J. & Cooke, B. D. (1998). Field evidence for mechanical transmission of rabbit haemorrhagic disease virus (RHDV) by flies (*Diptera: Calliphoridae*) among wild rabbits in Australia. *Virus Res* **54**, 123–132.

Barcena, J., Morales, M., Vazquez, B. & 7 other authors (2000). Horizontal transmissible protection against myxomatosis and rabbit hemorrhagic disease by using a recombinant myxoma virus. *J Virol* **74**, 1114–1123.

Behncken, G. M. (1973). Evidence of multiplication of sowthistle yellow vein virus in an inefficient aphid vector, *Macrosiphum euphorbiae*. *Virology* **53**, 405–412.

Blanc, S., Hébrard, E., Drucker, M. & Froissart, R. (2001). Molecular basis of vector transmission: Caulimoviruses. In *Virus-Insect-Plant Interactions*, pp. 143–166. Edited by K. Harris, O. P. Smith & J. E. Duffus. San Diego: Academic Press.

Campbell, C. L. & Wilson, W. C. (2002). Differentially expressed midgut transcripts in *Culicoides sonorensis* (Diptera: ceratopogonidae) following Orbivirus (reoviridae) oral feeding. *Insect Mol Biol* **11**, 595–604.

Carn, V. M. (1996). The role of dipterous insects in the mechanical transmission of animal viruses. *Br Vet J* **152**, 377–393.

Carn, V. M. & Kitching, R. P. (1995). An investigation of possible routes for transmission of lumpy skin disease virus (Neethling). *Epidemiol Infect* **114**, 219–226.

Chihota, C. M., Rennie, L. F., Kitching, R. P. & Mellor, P. S. (2001). Mechanical transmission of lumpy skin disease virus by *Aedes aegypti* (Diptera: Culicidae). *Epidemiol Infect* **162**, 317–321.

Chihota, C. M., Rennie, L. F., Kitching, R. P. & Mellor, P. S. (2003). Attempted mechanical transmission of lumpy skin disease virus by biting insects. *Med Vet Entomol* **17**, 294–300.

Day, T. (2001). Parasite transmission modes and the evolution of virulence. *Evolution Int J Org Evolution* **55**, 2389–2400.

Domingo, E. (2000). Viruses at the edge of adaptation. *Virology* **270**, 251–253.

Eigenbrode, S. D., Ding, H., Shiel, P. & Berger, P. H. (2002). Volatiles from potato plants infected with potato leafroll virus attract and arrest the virus vector, *Myzus persicae* (Homoptera: Aphididae). *Proc R Soc Lond B Biol Sci* **269**, 455–460.

Elena, S. F. & Lenski, R. E. (2003). Evolution experiments with microorganisms: dynamics and genetic basis of adaptation. *Nat Rev Genet* **2**, 137–143.

Elena, S. F., Sanjuan, R., Borkeria, A. V. & Turner, P. E. (2001). Transmission bottlenecks and the evolution of fitness in rapidly evolving RNA viruses. *Infect Genet Evol* **1**, 41–48.

Elliott, R. M. & Kohl, A. (2004). Bunyavirus/mosquito interactions. In *Microbe–vector Interactions in Vector-borne Diseases* (Society for General Microbiology Symposium no. 63), pp. 91–102. Edited by S. H. Gillespie, G. L. Smith & A. Osbourn. Cambridge: Cambridge University Press.

Esteva, L. & Vargas, C. (2000). Influence of vertical and mechanical transmission on the dynamics of Dengue disease. *Math Biosci* **167**, 51–64.

Frank, S. A. (1996). Models of parasite virulence. *Q Rev Biol* **71**, 37–78.

Franz, A. W. E., van der Wilk, F., Verbeek, M., Dullemans, A. M. & Van den Heuvel, J. F. J. M. (1999). Faba bean necrotic yellows virus (genus *Nanovirus*) requires a helper factor for its aphid transmission. *Virology* **262**, 210–219.

Froissart, R., Michalakis, Y. & Blanc, S. (2002). Helper component-transcomplementation in the vector transmission of plant viruses. *Phytopathology* **92**, 576–579.

Fu, H., Leake, C. J., Mertens, P. P. & Mellor, P. S. (1999). The barriers to Bluetongue virus infection, dissemination and transmission in the vector, *Culicoides variipennis* (Diptera: Ceratopogonidae). *Arch Virol* **144**, 747–761.

Garcia-Arenal, F., Fraile, A. & Malpica, J. M. (2003). Variation and evolution of plant virus population. *Int Microbiol* **6**, 225–232.

Garret, A., Kerlan, C. & Thomas, D. (1996). Ultrastructural study of acquisition and retention of potato leafroll luteovirus in the alimentary canal of its aphid vector, *Myzus persicae* Sulz. *Arch Virol* **141**, 1279–1292.

Gergerich, R. C. (2001). Elucidation of transmission mechanisms: mechanism of virus transmission by leaf-feeding beetles. In *Virus-Insect-Plant Interactions*, pp. 133–140. Edited by K. Harris, O. P. Smith & J. E. Duffus. San Diego: Academic Press.

Gergerich, R. C. & Scott, H. A. (1991). Determinants in the specificity of virus transmission by leaf-feeding beetles. *Adv Dis Vector Res* **8**, 1–13.

Ghanim, M. & Czosnek, H. (2000). Tomato yellow leaf curl geminivirus (TYLCV-Is) is transmitted among whiteflies (*Bemisia tabaci*) in a sex-related manner. *J Virol* **74**, 4738–4745.

Ghanim, M., Morin, S., Zeidan, M. & Czosnek, H. (1998). Evidence for transovarial transmission of tomato yellow leaf curl virus by its vector, the whitefly *Bemisia tabac*. *Virology* **240**, 295–303.

Gildow, F. (1993). Evidence for receptor-mediated endocytosis regulating luteovirus acquisition by aphids. *Phytopathology* **83**, 270–277.

Gildow, F. E. & Gray, S. M. (1993). The aphid salivary gland basal lamina as a selective barrier associated with vector-specific transmission of Barley Yellow Dwarf Luteoviruses. *Phytopathology* **83**, 1293–1302.

Gildow, F. E., Reavy, B., Mayo, M. A., Duncan, G. H., Woodford, J. A. T., Lamb, J. W. & Hay, R. T. (2000). Aphid acquisition and cellular transport of Potato leafroll virus-like particles lacking P5 readthrough protein. *Phytopathology* **90**, 1153–1161.

Gould, A. R., Kattenbelt, J. A., Lenghaus, C., Morrissy, C., Chamberlain, T., Collins, B. J. & Westbury, H. A. (1997). The complete nucleotide sequence of rabbit haemorrhagic disease virus (Czech strain V351): use of the polymerase chain reaction to detect replication in Australian vertebrates and analysis of viral population sequence variation. *Virus Res* **47**, 7–17.

Govier, D. A. & Kassanis, B. (1974). A virus induced component of plant sap needed when aphids acquire potato virus Y from purified preparations. *Virology* **61**, 420–426.

Granados, R. R., Himuri, H. & Maramorosch, K. (1967). Electron microscopic evidence for wound tumor virus accumulation in various organs of an inefficient leafhopper vector, *Agalliopsis novella. J Invertebr Pathol* **9**, 147–159.

Gray, S. M. & Banerjee, N. (1999). Mechanisms of arthropod transmission of plant and animal viruses. *Microbiol Mol Biol Rev* **63**, 128–148.

Gray, S. M. & Gildow, F. E. (2003). Luteovirus-aphid interactions. *Annu Rev Phytopathol* **41**, 539–566.

Hardy, J. L. (1988). Susceptibility and resistance of vector mosquitoes. In *The Arboviruses: Epidemiology and Ecology*, pp. 87–126. Edited by T. P. Monath. Boca Raton, FL: CRC Press.

Hardy, J. L., Houk, E. J., Kramer, L. D. & Reeves, W. C. (1983). Intrinsic factors affecting vector competence of mosquitoes for arboviruses. *Annu Rev Entomol* **28**, 229–262.

Harris, K. F. (1977). An ingestion-egestion hypothesis of non circulative virus transmission. In *Aphids as Virus Vectors*, pp. 166–208. Edited by K. F. Harris & K. Maramorosch. New York: Academic Press.

Harris, K. F., Smith, O. P. & Duffus, J. E. (2001). *Virus-Insect-Plant Interactions*. San Diego: Academic Press.

Higgs, S. (2004). How do mosquito vectors live with their viruses? In *Microbe–vector Interactions in Vector-borne Diseases* (Society for General Microbiology Symposium no. 63), pp. 103–137. Edited by S. H. Gillespie, G. L. Smith & A. Osbourn. Cambridge: Cambridge University Press.

Hogenhout, S. A., Redinbaugh, M. G. & Ammar, E. D. (2003). Plant and animal rhabdovirus host range: a bug's view. *Trends Microbiol* **11**, 264–271.

Hohnle, M., Hofer, P., Bedford, I. D., Briddon, R. W., Markham, P. G. & Frischmuth, T. (2001). Exchange of three amino acids in the coat protein results in efficient whitefly transmission of a nontransmissible Abutilon mosaic virus isolate. *Virology* **290**, 164–171.

Hurd, H. (2003). Manipulation of medically important insect vectors by their parasites. *Annu Rev Entomol* **48**, 141–161.

Kassanis, B. & Govier, D. A. (1971a). The role of the helper virus in aphid transmission of potato aucuba mosaic virus and potato virus C. *J Gen Virol* **13**, 221–228.

Kassanis, B. & Govier, D. A. (1971b). New evidence on the mechanism of transmission of potato C and potato aucuba mosaic viruses. *J Gen Virol* **10**, 99–101.

Kennedy, J. S., Day, M. F. & Eastop, V. F. (1962). *A Conspectus of Aphids as Vectors of Plant Viruses*. London: Commonwealth Institute of Entomology.

Kuno, G. (2001). Transmission of arboviruses without involvement of arthropod vectors. *Acta Virol* **45**, 139–150.

Lett, J. M., Granier, M., Hippolyte, I., Grondin, M., Royer, M., Blanc, S., Reynaud, B. & Peterschmitt, M. (2002). Spatial and temporal distribution of geminiviruses in leafhoppers of the genus *Cicadulina* monitored by conventional and quantitative polymerase chain reaction. *Phytopathology* **92**, 65–74.

Li, C., Cox-Foster, D., Gray, S. M. & Gildow, F. E. (2001). Vector specificity of Barley yellow dwarf virus (BYDV) transmission: identification of potential cellular receptors binding BYDV-MAV in the aphid, *Sitobion avenae*. *Virology* **286**, 249–252.

Lung, M. C. Y. & Pirone, T. P. (1973). Studies on the reason for differential transmissibility of cauliflower mosaic virus isolates by aphids. *Phytopathology* **63**, 910–914.

Lung, M. C. Y. & Pirone, T. P. (1974). Acquisition factor required for aphid transmission of purified cauliflower mosaic virus. *Virology* **60**, 260–264.

Martin, B., Collar, J. L., Tjallingii, W. F. & Fereres, A. (1997). Intracellular ingestion and salivation by aphids may cause the acquisition and inoculation of non-persistently transmitted plant viruses. *J Gen Virol* **78**, 2701–2705.

Mellor, P. S. (2000). Replication of arboviruses in insect vectors. *J Comp Pathol* **123**, 231–247.

Mellor, P. S., Kitching, R. P. & Wilkinson, P. J. (1987). Mechanical transmission of capripox virus and African swine fever virus by *Stomoxys calcitrans*. *Res Vet Sci* **43**, 109–112.

Merrill, M. H. & Tenbroeck, C. (1935). The transmission of Equine encephalomyelitis virus by Aedes Aegypti. *J Exp Med* **62**, 687–695.

Morin, S., Ghanim, M., Zeidan, M., Czosnek, H., Verbeek, M. & Van den Heuvel, J. F. J. M. (1999). A GroEL homologue from endosymbiotic bacteria of the whitefly *Bemisia tabaci* is implicated in the circulative transmission of Tomato yellow leaf curl virus. *Virology* **256**, 75–84.

Morin, S., Ghanim, M., Sobol, I. & Czosnek, H. (2000). The GroEL protein of whitefly *Bemisia tabaci* interacts with the coat protein of transmissible and non-transmissible begomoviruses in the yeast two-hybrid system. *Virology* **276**, 404–416.

Moya, A., Elena, S. F., Bracho, A., Miralles, R. & Barrio, E. (2000). The evolution of RNA viruses: a population genetics view. *Proc Natl Acad Sci U S A* **97**, 6967–6973.

Nault, L. R. (1997). Arthropod transmission of plant viruses: a new synthesis. *Ann Entomol Soc Am* **90**, 521–541.

Ovenden, J. R. & Mahon, R. J. (1984). Venereal transmission of Sindbis virus between individuals of *Aedes australis* (Diptera: Culicidae). *J Med Entomol* **21**, 292–295.

Perry, K. L., Zhang, L., Shintaku, M. H. & Palukaitis, P. (1994). Mapping determinants in cucumber mosaic virus for transmission by *Aphis gossypii*. *Virology* **205**, 591–595.

Perry, K. L., Zhang, L. & Palukaitis, P. (1998). Amino acid changes in the coat protein of cucumber mosaic virus differentially affect transmission by the aphids *Myzus persicae* and *Aphis gossypii*. *Virology* **242**, 204–210.

Pfeiffer, M. L., Gildow, F. E. & Gray, S. M. (1997). Two distinct mechanisms regulate luteovirus transmission efficiency and specificity at the aphid salivary gland. *J Gen Virol* **78**, 495–503.

Pirone, T. & Blanc, S. (1996). Helper-dependent vector transmission of plant viruses. *Annu Rev Phytopathol* **34**, 227–247.

Pirone, T. P. & Perry, K. L. (2002). Aphids – non-persistent transmission. In *Advances in Botanical Research*, pp. 1–19. Edited by R. T. Plumb. Academic Press.

Pirone, T. P. & Thornbury, D. W. (1988). Quantity of virus required for aphid transmission of a potyvirus. *Phytopathology* **78**, 104–107.

Platt, K. B., Linthicum, K. J., Myint, K. S., Innis, B. L., Lerdthusnee, K. & Vaughn, D. W. (1997). Impact of Dengue virus infection on feeding behavior of *Aedes aegypti*. *Am J Trop Med Hyg* **57**, 119–125.

Raccah, B., Huet, H. & Blanc, S. (2001). Potyviruses. In *Virus-Insect-Plant Interactions*, pp. 181–206. Edited by K. Harris, O. P. Smith & J. E. Duffus. San Diego: Academic Press.

Ratcliffe, N. A. & Whitten, M. M. A. (2004). Vector immunity. In *Microbe–vector Interactions in Vector-borne Diseases* (Society for General Microbiology Symposium no. 63), pp. 199–262. Edited by S. H. Gillespie, G. L. Smith & A. Osbourn. Cambridge: Cambridge University Press.

Rochow, W. F. & Pang, E. (1961). Aphids can acquire strains of Barley yellow dwarf virus they do not transmit. *Virology* **15**, 382–384.

Rubinstein, G. & Czosnek, H. (1997). Long-term association of tomato yellow leaf curl virus with its whitefly vector *Bemisia tabaci*: effect on the insect transmission capacity, longevity and fecundity. *J Gen Virol* **78**, 2683–2689.

Sacristan, S., Malpica, J. M., Fraile, A. & Garcia-Arenal, F. (2003). Estimation of population bottlenecks during movement of tobacco mosaic virus in tobacco plants. *J Virol* **77**, 9906–9911.

Sarra, S. & Peters, D. (2003). Rice yellow mottle virus is transmitted by cows, donkeys, and grass rats in irrigated rice crops. *Plant Dis* **87**, 804–808.

Scott, T. W., Hildreth, S. W. & Beaty, B. J. (1984). The distribution and development of Eastern equine encephalitis virus in its enzootic mosquito vector, *Culesita melanura*. *Am J Trop Med Hyg* **33**, 300–310.

Storey, H. H. (1933). Investigations of the mechanisms of transmission of plant viruses by insect vectors. *Proc R Soc Lond Ser B* **113**, 463–485.

Sylvester, E. S. (1956). Transmission of Beet yellows virus by the green peach aphid. *J Econ Entomol* **49**, 789–800.

Thompson, W. H. & Beaty, B. J. (1978). Venereal transmission of La Crosse virus from male to female *Aedes triseriatus*. *J Med Entomol* **14**, 499–503.

Ullman, D. E., Cho, J. J., Mau, R. F. L., Hunter, W. B., Westcot, D. M. & Custer, D. M. (1992). Thrips-tomato spotted wilt virus interactions: morphological, behavioral and cellular components influencing thrips transmission. *Adv Dis Vector Res* **9**, 195–240.

Van den Heuvel, J., Verbeek, M. & van der Wilk, F. (1994). Endosymbiotic bacteria associated with circulative transmission of Potato Leafroll Virus by *Myzus persicae*. *J Virol* **75**, 2559–2565.

Van den Heuvel, J., Bruyere, A., Hogenhout, S., Ziegler-Graff, V., Brault, V., Verbeek, M., van der Wilk, F. & Richards, K. (1997). The N-terminal region of the luteovirus readthrough domain determines virus binding to *Buchnera* GroEL and is essential for virus persistence in the aphid. *J Virol* **71**, 7258–7265.

Van den Heuvel, J. F. J. M., Saskia, A., Hogenhout, S. A. & van der Wilk, F. (1999). Recognition and receptors in virus transmission by arthropods. *Trends Microbiol* **7**, 71–76.

Walker, H. L. & Pirone, T. P. (1972). Particle numbers associated with mechanical and aphid transmission of some plant viruses. *Phytopathology* **62**, 1283–1288.

Wang, R. Y., Gergerich, R. C. & Kim, K. S. (1994a). The relationship between feeding and virus retention time in beetle transmission of plant viruses. *Phytopathology* **84**, 995–998.

Wang, R. Y., Gergerich, R. C. & Kim, K. S. (1994b). Entry of ingested plant viruses into the hemocoel of plant virus vector, *Diabrotica undecimpunctata howardi*. *Phytopathology* **84**, 147–153.

Watson, M. A. & Roberts, F. M. (1939). A comparative study of the transmission of hyocyamus virus 3, potato virus Y and cucumber virus by the vector *Myzus persicae* (Sulz), *M. circumflexus* (Buckton) and *Macrosiphum gei* (Koch). *Proc R Soc Lond B* **127**, 543–576.

Weaver, S. C. (1997). Vector biology in arboviral pathogenesis. In *Viral Pathogenesis*, pp. 329–352. Edited by N. Nathanson. Philadelphia: Lippincott-Raven.

Weiss, R. A. (2002). Virulence and pathogenesis. *Trends Microbiol* **10**, 314–317.

Yeruham, I., Braverman, Y., Davidson, M., Grinstein, H., Haymovitch, M. & Zamir, O. (1995). Spread of lumpy skin disease in Israeli dairy herds. *Vet Rec* **137**, 91–93.

RNA-based immunity in insects

Rui Lu, Hongwei Li, Wan-Xiang Li and Shou-Wei Ding

Center for Plant Cell Biology, Department of Plant Pathology, University of California, Riverside, CA 92521, USA

INTRODUCTION

Drosophila has been an excellent model for the mechanistic studies of innate immunity (Hoffmann, 2003). Recently, a new RNA-based antiviral immunity with features of both innate and adaptive immunities has been described in *Drosophila* and *Anopheles* cells (Li *et al.*, 2002, 2004). This RNA-silencing-mediated immunity is characterized by the production of pathogen-derived, 22-nt small RNAs that serve as specificity determinants inside a multi-subunit complex. Similar to innate immunity, however, the new invertebrate antiviral response is capable of a rapid virus clearance in the absence of a virus-encoded suppressor of RNA silencing. The discovery of a new antiviral pathway in insects opens up the possibility of using this pathway to prevent transmission of vector-borne virus pathogens such as dengue and West Nile viruses.

THE RNA-SILENCING PATHWAY

Homology-dependent gene silencing was discovered in transgenic plants in a form of co-suppression between introduced transgenes or between a transgene and its homologous endogenous gene (Matzke *et al.*, 1989; Napoli *et al.*, 1990; Van der Krol *et al.*, 1990). Similar gene-silencing phenomena have subsequently been described in a wide range of eukaryotic organisms such as fungi, worms, flies and mammals (Denli & Hannon, 2003; Fire *et al.*, 1998). A generic term, RNA silencing (Ding, 2000), has been used to describe these related RNA-guided gene regulatory mechanisms variously termed post-transcriptional gene silencing (PTGS) in plants, quelling in fungi and RNA interference (RNAi) in animals.

SGM symposium 63: Microbe–vector interactions in vector-borne diseases.
Editors S. H. Gillespie, G. L. Smith & A. Osbourn. Cambridge University Press. ISBN 0 521 84312 X ©SGM 2004

A core feature of RNA silencing detected in all organisms is the production of 21–26-nt small RNAs from structured or double-stranded RNA (dsRNA) by the endoribonuclease Dicer (Bernstein *et al.*, 2001; Hamilton *et al.*, 2002; Hamilton & Baulcombe, 1999; Hammond *et al.*, 2000; Zamore *et al.*, 2000). These small interfering RNAs (siRNAs) control the specificity of RNA silencing in a homology-dependent manner in an RNA-induced silencing complex (RISC), of which the Argonaute-2 (AGO2) protein is an essential component (Denli & Hannon, 2003; Elbashir *et al.*, 2001; Hammond *et al.*, 2001). RNA silencing in fungi, plants and worms involves a cellular RNA-dependent RNA polymerase (RdRP); however, the multiple-turnover RISC may mediate RNA silencing in the absence of a cellular RdRP in *Drosophila* and mammalian cells (Cogoni & Macino, 1999; Denli & Hannon, 2003; Martinez *et al.*, 2002; Schwarz *et al.*, 2002).

The first indication for a possible antiviral role of RNA silencing came from the observation that virus infection is able to trigger RNA silencing of a homologous virus-derived transgene in transgenic plants (Lindbo *et al.*, 1993). Subsequent work has established that RNA silencing is a natural antiviral response of plants, which is induced upon virus infection and can specifically target the viral and homologous RNAs for degradation (Baulcombe, 1999). This conclusion is further supported by the demonstration in 1998 that HC-Pro and 2b, two essential pathogenic factors encoded by potyviruses and cucumoviruses, are suppressors of RNA silencing (Anandalakshmi *et al.*, 1998; Brigneti *et al.*, 1998; Kasschau & Carrington, 1998; Li *et al.*, 1999; Li & Ding, 2001).

In contrast to the broad recognition of a natural antiviral role for RNA silencing in higher plants, no experimental evidence was available until recently on whether or not RNA silencing plays a similar antiviral role in the animal kingdom. Here we review recent progress in establishing RNA silencing as a novel nucleic-acid-based antiviral immunity in insects and discuss features of this novel antiviral response as compared to the known innate and adaptive immunities.

VIRUSES ARE TARGETS AND INDUCERS OF RNA SILENCING IN INSECTS

Following the first report in *Drosophila* (Kennerdell & Carthew, 1998, 2000; Pal-Bhadra *et al.*, 1997), effective RNAi induced by dsRNA has been demonstrated in many arthropod species (Table 1). These findings indicate that arthropods encode a functional RNA-silencing pathway. The early evidence implicating viruses as targets of RNA silencing in insects came from the use of alphavirus-based expression vectors (Gaines *et al.*, 1996; Olson *et al.*, 1996). Similar to transgenic plants that contain a virus-derived silencing transgene (Baulcombe, 1999), mosquitoes infected with a

Table 1. RNAi in arthropods

Species	Inducer	Means of delivery	Target RNA	Reference*
Fruit fly *Drosophila* *melanogaster*	Transgene dsRNA RNA replication	Transfection Embryo injection Transformation	Cellular and viral	1–5
Medfly *Ceratitis capitata*	dsRNA	Embryo injection	Cellular	6
Milkweed bug *Oncopeltus fasciatus*	dsRNA	Embryo injection	Cellular	7
Mosquito *Anopheles gambiae*	dsRNA Transgene RNA replication	Transfection Adult injection Transformation	Cellular and viral	8–13, 18
Giant silkmoth *Hyalophora cecropia*	dsRNA	Pupa injection	Cellular	14
Red flour beetle *Tribolium castaneum*	dsRNA	Pupa injection Embryo injection	Cellular	15, 16
Noctuid moth *Trichoplusia ni*	dsRNA	Transfection	Viral	17
Silkmoth *Bombyx mori*	RNA replication	Larvae injection	Cellular	18
Tick *Amblyomma* *americanum*	dsRNA	Soaking Adult injection	Cellular	19, 20
Honeybee *Apis mellifera*	dsRNA	Embryo injection Pupa injection	Cellular	21, 22
Spider *Cupiennius salei*	dsRNA	Embryo injection	Cellular	23

*1, Pal-Bhadra *et al.* (1997); 2, Kennerdell & Carthew (1998); 3, Kennerdell & Carthew (2000);
4, Hammond *et al.* (2000); 5, Li *et al.* (2002); 6, Pane *et al.* (2002); 7, Hughes & Kaufman (2000);
8, Billecocq *et al.* (2000); 9, Brown *et al.* (2003); 10, Gaines *et al.* (1996); 11, Hoa *et al.* (2003);
12, Levashina *et al.* (2001); 13, Li *et al.* (2004); 14, Bettencourt *et al.* (2002); 15, Bucher *et al.* (2002);
16, Wheeler *et al.* (2003); 17, Beck & Strand (2003); 18, Uhlirova *et al.* (2003); 19, Aljamali *et al.* (2003);
20, Aljamali *et al.* (2002); 21, Amdam *et al.* (2003); 22, Beye *et al.* (2002); 23, Stollewerk *et al.* (2003).

recombinant alphavirus carrying a heterologous viral sequence also develop a RNA-mediated homology-dependent virus resistance (Adelman *et al.*, 2001; Billecocq *et al.*, 2000; Gaines *et al.*, 1996; Olson *et al.*, 1996). Specific targeting of insect viruses by RNA silencing was confirmed by several recent studies. For example, fruit fly cells transfected with dsRNA corresponding to part of the plus-strand RNA genome of flock house virus (FHV) became resistant to FHV (Li *et al.*, 2002). Similarly, mosquito cells transcribing an inverted-repeat RNA targeting the dengue virus type 2 (DEN-2) RNA genome were unable to support replication of DEN-2 (Adelman *et al.*, 2002). Efficient RNAi of two genes encoded by a double-stranded DNA virus from the *Polydnaviridae* family was also demonstrated in noctuid moth (*Trichoplusia ni*) cells (Beck & Strand, 2003).

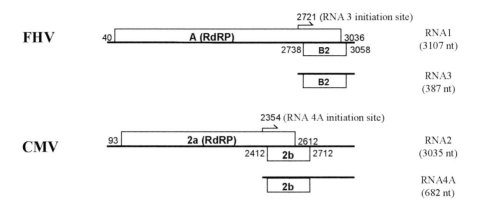

Fig. 1. Genome organization and expression of FHV RNA1 and CMV RNA2. The locations of both start and stop codons in the genomic RNAs for each of the viral genes are indicated. The ORF encoding the viral RdRP, called 2a and A, respectively, is translated from the genomic RNA whereas the smaller overlapping gene (B2 and 2b) is translated from the subgenomic RNA. The conserved GDD motif found in viral RdRPs is encoded by nucleotides 2110–2118 of FHV and nucleotides 1908–1916 of CMV.

Two lines of evidence indicate that specific silencing of the infecting viral RNAs is induced naturally in insect cells upon virus infection. First, infection of either cultured fruit fly cells by FHV or the silkmoth (*Bombyx mori*) larvae by Sindbis virus led to the *in vivo* accumulation of the virus-specific siRNAs (Li *et al.*, 2002; Uhlirova *et al.*, 2003). This indicates that the invertebrate Dicer ribonuclease(s) is able to detect the infecting viruses, most likely by recognizing the viral RNA replicative intermediates as the substrate. Second, increased accumulation of FHV RNAs was observed in the fly cells after the RNA-silencing pathway was made partially defective by RISC depletion through RNAi of AGO2 (Li *et al.*, 2002). Thus a functional RNA-silencing pathway naturally restricts FHV accumulation in infected fruit fly cells. A recent study has further provided evidence for the induction of the RNA-silencing-based antiviral response (RSAR) in both fruit fly and mosquito cells following nodamura virus (NoV) RNA replication (Li *et al.*, 2004) (see later).

INVERTEBRATE VIRAL SUPPRESSORS OF RNA SILENCING

The first invertebrate viral suppressor of RNA silencing, the FHV B2 protein (Li *et al.*, 2002), was discovered following a prediction made in an earlier report (Ding *et al.*, 1995). FHV belongs to the *Alphanodavirus* genus, of which NoV is the type member. FHV was originally isolated from the grass grub *Costelytra zealandica* (Coleoptera) in New Zealand. In the laboratory, FHV also infects *Galleria mellonella* (wax moth) and cultured *Drosophila melanogaster* and mosquito cells and replicates efficiently in plant, yeast and mammalian cells (Ball, 1995; Dasgupta *et al.*, 2003). The nodaviral B2 gene

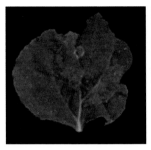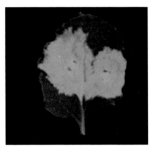

Fig. 2. Cross-kingdom suppression of RNA silencing in plants by an animal viral suppressor of RSAR. The GFP-expressing *Nicotiana benthamiana* leaves were co-infiltrated with a mixture of two *Agrobacterium tumefaciens* strains, as described by Guo & Ding (2002). One directs the expression of GFP and thereby induces GFP RNA silencing, and the other simultaneously expresses the NoV-encoded B2 (right), a plant cucumoviral 2b (middle) or an untranslatable mutant of B2 (left). The leaves were detached and photographed under UV illumination 6 days after infiltration. GFP silencing is visualized in the left leaf as a lack of green fluorescence in the infiltrated patch surrounded by a red colour zone caused by chlorophyll fluorescence.

shares a number of features with the 2b gene of the plant cucumoviruses (Ding *et al.*, 1995), which was demonstrated in 1998 to encode a suppressor of RNA silencing (Brigneti *et al.*, 1998; Li *et al.*, 1999). Both 2b and B2 genes overlap the 3'-end, and occupy the +1 reading frame, of the viral RdRP gene and are translated *in vivo* from a subgenomic RNA, although their encoded proteins do not exhibit any detectable sequence similarities (Fig. 1). Knockout of either B2 in FHV or 2b in cucumber mosaic virus (CMV) resulted in a defect in virus spreading, but neither is required for viral RNA replication (Ball, 1995; Ding *et al.*, 1995).

Using an assay based on transgene RNA silencing in plants, it was shown that the FHV B2 protein indeed is a suppressor of RNA silencing (Li *et al.*, 2002). In this system, B2 suppression was as potent as the 2b protein encoded by tomato aspermy cucumovirus, which is among the strongest plant viral suppressors known (Li *et al.*, 2002). By comparison, the 2b of CMV is a much weaker suppressor that delayed, but did not prevent, the initiation of RNA silencing (Guo & Ding, 2002). A recent study (Li *et al.*, 2004) found that the NoV B2 protein is also a suppressor of RNA silencing in the same *in planta* assay (Fig. 2). Notably, the FHV B2 is able to functionally substitute for 2b of CMV in whole-plant infections, suggesting that the nodaviral B2 is able to inhibit RNA silencing triggered by viral RNA replication (Li *et al.*, 2002).

The nodaviral B2 also suppresses RNA silencing in *Drosophila* cells that is induced by exogenous long dsRNA (Li *et al.*, 2004). Interestingly, RNA-silencing suppression requires B2 expression before dsRNA is introduced into cells to initiate RNAi and suppression was not observed when B2 and dsRNA were simultaneously introduced.

This suggests that B2 suppression may occur before either the production or RISC loading of siRNAs (Li *et al.*, 2004). The fact that the nodaviral B2 protein suppresses RNA silencing in both fruit flies and plants provides the first direct evidence for a conserved RNA-silencing pathway in the plant and animal kingdoms.

SUPPRESSION OF RNA SILENCING IS ESSENTIAL FOR VIRUS INFECTION IN FRUIT FLY CELLS

The important contribution of RNA silencing to invertebrate defence against virus infection was not obvious until the nodavirus-encoded suppressor of RNA silencing was made inactive (Li *et al.*, 2002, 2004). Infection with wild-type FHV rapidly triggers RSAR in infected fruit fly cells, as indicated by the detection of FHV-specific siRNAs 24 h after infection. Nevertheless, FHV accumulation continues to increase in infected cells, suggesting that the induced RSAR fails to provide protection against the nodaviral infection. The FHV mutant containing point mutations that rendered B2 untranslatable, however, does not accumulate to a detectable level in *Drosophila* cells unless it is co-transfected with a B2-expressing plasmid. An efficient rescue of the B2-knockout mutant was also observed when the host machinery RISC was depleted by co-transfection with either a dsRNA or siRNA targeting AGO2 (Li *et al.*, 2002, 2004). These findings show that the B2 deletion did not prevent FHV RNA replication and that the observed essential role of B2 is to suppress the RNA silencing induced by FHV RNA replication that specifically targets the FHV RNAs for destruction. It is important to note that, as occurred in the absence of viral suppression of RNA silencing by B2, the induced RSAR is sufficient to ensure a rapid and complete virus clearance in the challenged *Drosophila* cells.

A mutational analysis on the B2 gene encoded by NoV has been carried out recently (Li *et al.*, 2004). NoV is the only member of the *Alphanodavirus* genus that can lethally infect both insects and mammals in the laboratory and it shares either low or minimal sequence identities with FHV in their encoded proteins (44 % for the viral RdRP and <19 % for B2) (Johnson *et al.*, 2001). The result shows that NoV RNA replication in *Drosophila* cells induces a potent RSAR that is RISC-dependent and capable of eliminating NoV RNAs when its B2 protein is not expressed (Li *et al.*, 2004). In *Drosophila* cells, the RSAR triggered by nodaviral RNA replication can be reciprocally inhibited by heterologous B2 proteins, leading to efficient rescue of either B2-deficient nodavirus mutant (Li *et al.*, 2004). This shows that although targeted destruction of the infecting viral RNAs by RSAR as mediated by siRNAs is specific, viral suppression of invertebrate RSAR is broad-spectrum, implying that the same core pathway is involved in defence against different viruses.

RNA SILENCING AS AN RNA-BASED ANTIVIRAL IMMUNITY IN MOSQUITO CELLS

Mosquito cells support RNAi and homologues of the *Drosophila* RNAi components are encoded by the recently sequenced genome of *Anopheles gambiae*, which transmits both malaria and o'nyong-nyong alphavirus (Christophides *et al.*, 2002; Hoa *et al.*, 2003; Holt *et al.*, 2002; Levashina *et al.*, 2001; Powers *et al.*, 2000). It has been demonstrated recently that replicating virus RNAs are naturally targeted for destruction by RNAi in mosquito cells (Li *et al.*, 2004). This potent RSAR does not require prior activation, e.g. through transformation with an inverted-repeat RNA transgene (Adelman *et al.*, 2002), but it is masked in *A. gambiae* cells by the action of the B2 protein expressed from wild-type NoV RNA1. After its B2 was made untranslatable, the accumulation of NoV RNAs was almost completely abolished in mosquito cells (Li *et al.*, 2004). The NoV RNA1 mutant was rescued by co-transfection with a plasmid expressing B2 of either nodavirus, a dsRNA or siRNA targeting the mRNA of the *A. gambiae* AGO2, but not by an unrelated dsRNA or siRNA. Thus, as has been found in *Drosophila*, NoV RNA replication also triggers RSAR in *A. gambiae* cells which is both AGO2-dependent and sensitive to B2 suppression. These findings establish RSAR as a new RNA-based antiviral immunity in mosquitoes.

FEATURES OF THE INVERTEBRATE RSAR

The RSAR of invertebrates is rapid, effective and adaptive. In contrast to the broad-spectrum innate immunity in *Drosophila* and mosquitoes (Christophides *et al.*, 2002; Hoffmann, 2003), RSAR has features similar to the peptide-based adaptive immunity in vertebrates (Whitton & Oldstone, 2001). The specificity determinants of RSAR are siRNAs, which are derived and processed from the invading virus. After being uploaded into RISC, these virus-specific siRNAs selectively recruit target viral genomic and/or messenger RNAs by base-pairing for RISC-mediated destruction. At the whole-organism level, RNA silencing may also provide a long-term memory, analogous to the lifetime maintenance of specific virus resistance in plants recovered from a virulent primary infection (Covey *et al.*, 1997; Ratcliff *et al.*, 1997, 1999; Xin & Ding, 2003). However, while the peptide-based immunity in vertebrates usually takes more than a week to respond (Whitton & Oldstone, 2001), the RNA-silencing response in fruit flies and mosquitoes is capable of a rapid and complete virus clearance in single cells (Li *et al.*, 2002, 2004), which is similar to the innate immunity. In this regard, it is possible that RNA silencing may contribute to, or be part of, the robust innate immunity occurring at early stages of viral infection in vertebrates (Parham, 2003). This is supported in part by the recent demonstration that a number of mammalian viruses encode suppressors of RNA silencing that were previously shown to inhibit the interferon-regulated innate immunity in vertebrates (Li *et al.*, 2004).

It is known that common components, such as siRNA, Dicer and Argonaute proteins, are involved in PTGS in plants and RNAi in animals (Denli & Hannon, 2003; Tang *et al.*, 2003; Vance & Vaucheret, 2001). Suppression of RSAR in both plants and invertebrates by the same B2 protein provides direct experimental evidence that at least some aspects are conserved in the RNA-silencing pathway between the plant and animal kingdoms (Li *et al.*, 2002; Lindenbach & Rice, 2002). However, it appears that at the single-cell level, RSAR targeting of infecting viruses is not as potent in plants as in insects. For example, most of the plant viral suppressors characterized to date interfere with the cell-to-cell and long-distance spread of RNA silencing, which may explain why plant virus mutants that lack a functional suppressor do not exhibit obvious defect in single cells (Li & Ding, 2001; Silhavy *et al.*, 2002). This is in contrast to an essential role for B2 suppression in nodaviral infection of insect cells (Li *et al.*, 2002, 2004). It is possible that organisms that do not carry a cellular RdRP, such as fruit flies and mosquitoes, have evolved a more efficient multiple-turnover RISC capable of a potent intracellular silencing, which becomes the target of animal viral suppressors.

CONCLUDING REMARKS

There is now compelling evidence supporting RNA silencing as a novel RNA-based antiviral immunity in insects since RNAi was first demonstrated more than 5 years ago. First, virus infection triggers RNA silencing in insect cells that specifically targets the invading viral RNA. Second, invertebrate viruses such as FHV and NoV encode suppressors of RNA silencing essential for infection of insect cells in which the RNA-silencing pathway is not compromised. It is also clear that the insect RSAR is mediated by the RNAi pathway, which is based mostly on the experimental induction of RNA silencing by exogenous dsRNA. This is because the nodaviral B2 protein suppresses RNA silencing in insect cells induced by either dsRNA or viral RNA replication and the insect RSAR involves Dicer recognition of replicating virus RNAs and is RISC-dependent. The demonstration that RNAi naturally protects insects from virus infection opens up the possibility of using this pathway to control insects that are either crop pests or vectors for crop and human pathogens.

ACKNOWLEDGEMENTS

The research projects on the RNA silencing antiviral response in the authors' laboratory are supported by grants from NIH (R01 AI52447), USDA (2001-02678), CRB, CDFA and UC-BIOSTAR.

REFERENCES

Adelman, Z. N., Blair, C. D., Carlson, J. O., Beaty, B. J. & Olson, K. E. (2001). Sindbis virus-induced silencing of dengue viruses in mosquitoes. *Insect Mol Biol* **10**, 265–273.

Adelman, Z. N., Sanchez-Vargas, I., Travanty, E. A., Carlson, J. O., Beaty, B. J., Blair, C. D. & Olson, K. E. (2002). RNA silencing of dengue virus type 2 replication in transformed C6/36 mosquito cells transcribing an inverted-repeat RNA derived from the virus genome. *J Virol* **76**, 12925–12933.

Aljamali, M. N., Sauer, J. R. & Essenberg, R. C. (2002). RNA interference: applicability in tick research. *Exp Appl Acarol* **28**, 89–96.

Aljamali, M. N., Bior, A. D., Sauer, J. R. & Essenberg, R. C. (2003). RNA interference in ticks: a study using histamine binding protein dsRNA in the female tick *Amblyomma americanum*. *Insect Mol Biol* **12**, 299–305.

Amdam, G. V., Simoes, Z. L., Guidugli, K. R., Norberg, K. & Omholt, S. W. (2003). Disruption of vitellogenin gene function in adult honeybees by intra-abdominal injection of double-stranded RNA. *BMC Biotechnol* **3**, 1.

Anandalakshmi, R., Pruss, G. J., Ge, X., Marathe, R., Mallory, A. C., Smith, T. H. & Vance, V. B. (1998). A viral suppressor of gene silencing in plants. *Proc Natl Acad Sci U S A* **95**, 13079–13084.

Ball, L. A. (1995). Requirements for the self-directed replication of flock house virus RNA 1. *J Virol* **69**, 2722.

Baulcombe, D. (1999). Viruses and gene silencing in plants. *Arch Virol Suppl* **15**, 189–201.

Beck, M. & Strand, M. R. (2003). RNA interference silences *Microplitis demolitor* bracovirus genes and implicates glc1.8 in disruption of adhesion in infected host cells. *Virology* **314**, 521–535.

Bernstein, E., Caudy, A. A., Hammond, S. M. & Hannon, G. J. (2001). Role for a bidentate ribonuclease in the initiation step of RNA interference. *Nature* **409**, 363–366.

Bettencourt, R., Terenius, O. & Faye, I. (2002). Hemolin gene silencing by ds-RNA injected into Cecropia pupae is lethal to next generation embryos. *Insect Mol Biol* **11**, 267–271.

Beye, M., Hartel, S., Hagen, A., Hasselmann, M. & Omholt, S. W. (2002). Specific developmental gene silencing in the honey bee using a homeobox motif. *Insect Mol Biol* **11**, 527–532.

Billecocq, A., Vazeille-Falcoz, M., Rodhain, F. & Bouloy, M. (2000). Pathogen-specific resistance to Rift Valley fever virus infection is induced in mosquito cells by expression of the recombinant nucleoprotein but not NSs non-structural protein sequences. *J Gen Virol* **81**, 2161–2166.

Brigneti, G., Voinnet, O., Li, W. X., Ji, L. H., Ding, S. W. & Baulcombe, D. C. (1998). Viral pathogenicity determinants are suppressors of transgene silencing in *Nicotiana benthamiana*. *EMBO J* **17**, 6739–6746.

Brown, A. E., Bugeon, L., Crisanti, A. & Catteruccia, F. (2003). Stable and heritable gene silencing in the malaria vector *Anopheles stephensi*. *Nucleic Acids Res* **31**, e85.

Bucher, G., Scholten, J. & Klingler, M. (2002). Parental RNAi in *Tribolium* (Coleoptera). *Curr Biol* **12**, R85–R86.

Christophides, G. K., Zdobnov, E., Barillas-Mury, C. & 32 other authors (2002). Immunity-related genes and gene families in *Anopheles gambiae*. *Science* **298**, 159–165.

Cogoni, C. & Macino, G. (1999). Gene silencing in *Neurospora crassa* requires a protein homologous to RNA-dependent RNA polymerase. *Nature* **399**, 166–169.

Covey, S. N., Al-Kaff, N. S., Langara, A. & Turner, D. S. (1997). Plants combat infection by gene silencing. *Nature* **385**, 781–782.

Dasgupta, R., Cheng, L. L., Bartholomay, L. C. & Christensen, B. M. (2003). Flock house virus replicates and expresses green fluorescent protein in mosquitoes. *J Gen Virol* **84**, 1789–1797.

Denli, A. M. & Hannon, G. J. (2003). RNAi: an ever-growing puzzle. *Trends Biochem Sci* **28**, 196–201.

Ding, S. W. (2000). RNA silencing. *Curr Opin Biotechnol* **11**, 152–156.

Ding, S. W., Li, W. X. & Symons, R. H. (1995). A novel naturally-occurring hybrid gene encoded by a plant RNA virus facilitates long-distance virus movement. *EMBO J* **14**, 5762–5772.

Elbashir, S. M., Lendeckel, W. & Tuschl, T. (2001). RNA interference is mediated by 21- and 22-nucleotide RNAs. *Genes Dev* **15**, 188–200.

Fire, A., Xu, S. Q., Montgomery, M. K., Kostas, S. A., Driver, S. E. & Mello, C. C. (1998). Potent and specific genetic interference by double-stranded RNA in *Caenorhabditis elegans*. *Nature* **391**, 806–811.

Gaines, P. J., Olson, K. E., Higgs, S., Powers, A. M., Beaty, B. J. & Blair, C. D. (1996). Pathogen-derived resistance to dengue type 2 virus in mosquito cells by expression of the premembrane coding region of the viral genome. *J Virol* **70**, 2132–2137.

Guo, H. S. & Ding, S. W. (2002). A viral protein inhibits the long-range signaling activity of the gene silencing signal. *EMBO J* **21**, 398–407.

Hamilton, A. J. & Baulcombe, D. C. (1999). A species of small antisense RNA in posttranscriptional gene silencing in plants. *Science* **286**, 950–952.

Hamilton, A., Voinnet, O., Chappell, L. & Baulcombe, D. (2002). Two classes of short interfering RNA in RNA silencing. *EMBO J* **21**, 4671–4679.

Hammond, S. M., Bernstein, E., Beach, D. & Hannon, G. J. (2000). An RNA-directed nuclease mediates post-transcriptional gene silencing in *Drosophila* cells. *Nature* **404**, 293–296.

Hammond, S. M., Boettcher, S., Caudy, A. A., Kobayashi, R. & Hannon, G. J. (2001). Argonaute2, a link between genetic and biochemical analyses of RNAi. *Science* **293**, 1146–1150.

Hoa, N. T., Keene, K. M., Olson, K. E. & Zheng, L. (2003). Characterization of RNA interference in an *Anopheles gambiae* cell line. *Insect Biochem Mol Biol* **33**, 949–957.

Hoffmann, J. A. (2003). The immune response of *Drosophila*. *Nature* **426**, 33–38.

Holt, R. A., Subramanian, G. M., Halpern, A. & 120 other authors (2002). The genome sequence of the malaria mosquito *Anopheles gambiae*. *Science* **298**, 129–149.

Hughes, C. L. & Kaufman, T. C. (2000). RNAi analysis of Deformed, proboscipedia and Sex combs reduced in the milkweed bug *Oncopeltus fasciatus*: novel roles for Hox genes in the hemipteran head. *Development* **127**, 3683–3694.

Johnson, K. N., Johnson, K. L., Dasgupta, R., Gratsch, T. & Ball, L. A. (2001). Comparisons among the larger genome segments of six nodaviruses and their encoded RNA replicases. *J Gen Virol* **82**, 1855–1866.

Kasschau, K. D. & Carrington, J. C. (1998). A counterdefensive strategy of plant viruses: suppression of posttranscriptional gene silencing. *Cell* **95**, 461–470.

Kennerdell, J. R. & Carthew, R. W. (1998). Use of dsRNA-mediated genetic interference to demonstrate that frizzled and frizzled 2 act in the wingless pathway. *Cell* **95**, 1017–1026.

Kennerdell, J. R. & Carthew, R. W. (2000). Heritable gene silencing in *Drosophila* using double-stranded RNA. *Nat Biotechnol* **18**, 896–898.

Levashina, E. A., Moita, L. F., Blandin, S., Vriend, G., Lagueux, M. & Kafatos, F. C. (2001). Conserved role of a complement-like protein in phagocytosis revealed by dsRNA knockout in cultured cells of the mosquito, *Anopheles gambiae*. *Cell* **104**, 709–718.

Li, H. W., Lucy, A. P., Guo, H. S., Li, W. X., Ji, L. H., Wong, S. M. & Ding, S. W. (1999). Strong host resistance targeted against a viral suppressor of the plant gene silencing defence mechanism. *EMBO J* **18**, 2683–2691.

Li, H. W., Li, W. X. & Ding, S. W. (2002). Induction and suppression of RNA silencing by an animal virus. *Science* **296**, 1319–1321.

Li, W. X. & Ding, S. W. (2001). Viral suppressors of RNA silencing. *Curr Opin Biotechnol* **12**, 150–154.

Li, W. X., Li, H. W., Lu, R. & 9 other authors (2004). Interferon antagonist proteins of influenza and vaccinia viruses are suppressors of RNA silencing. *Proc Natl Acad Sci U S A* **101**, 1350–1355.

Lindbo, J. A., Silva-Rosales, L., Proebsting, W. M. & Dougherty, W. G. (1993). Induction of a highly specific antiviral state in transgenic plants: implications for regulation of gene expression and virus resistance. *Plant Cell* **5**, 1749–1759.

Lindenbach, B. D. & Rice, C. M. (2002). RNAi targeting an animal virus: news from the front. *Mol Cell* **9**, 925–927.

Martinez, J., Patkaniowska, A., Urlaub, H., Luhrmann, R. & Tuschl, T. (2002). Single-stranded antisense siRNAs guide target RNA cleavage in RNAi. *Cell* **110**, 563–574.

Matzke, M. A., Primig, M., Trnovsky, J. & Matzke, A. J. M. (1989). Reversible methylation and inactivation of marker genes in sequentially transformed tobacco plants. *EMBO J* **8**, 643–649.

Napoli, C., Lemieux, C. & Jorgensen, R. (1990). Introduction of a chimeric chalcone synthase gene into Petunia results in reversible co-suppression of homologous genes in trans. *Plant Cell* **2**, 279–289.

Olson, K. E., Higgs, S., Gaines, P. J., Powers, A. M., Davis, B. S., Kamrud, K. I., Carlson, J. O., Blair, C. D. & Beaty, B. J. (1996). Genetically engineered resistance to dengue-2 virus transmission in mosquitoes. *Science* **272**, 884–886.

Pal-Bhadra, M., Bhadra, U. & Birchler, J. A. (1997). Cosuppression in *Drosophila*: gene silencing of Alcohol dehydrogenase by white-Adh transgenes is Polycomb dependent. *Cell* **90**, 479–490.

Pane, A., Salvemini, M., Delli Bovi, P., Polito, C. & Saccone, G. (2002). The transformer gene in *Ceratitis capitata* provides a genetic basis for selecting and remembering the sexual fate. *Development* **129**, 3715–3725.

Parham, P. (2003). Innate immunity: the unsung heroes. *Nature* **423**, 20.

Powers, A. M., Brault, A. C., Tesh, R. B. & Weaver, S. C. (2000). Re-emergence of Chikungunya and O'nyong-nyong viruses: evidence for distinct geographical lineages and distant evolutionary relationships. *J Gen Virol* **81**, 471–479.

Ratcliff, F., Harrison, B. D. & Baulcombe, D. C. (1997). A similarity between viral defense and gene silencing in plants. *Science* **276**, 1558–1560.

Ratcliff, F. G., MacFarlane, S. A. & Baulcombe, D. C. (1999). Gene silencing without DNA: RNA-mediated cross-protection between viruses. *Plant Cell* **11**, 1207–1215.

Schwarz, D. S., Hutvagner, G., Haley, B. & Zamore, P. D. (2002). Evidence that siRNAs function as guides, not primers, in the *Drosophila* and human RNAi pathways. *Mol Cell* **10**, 537–548.

Silhavy, D., Molnar, A., Lucioli, A., Szittya, G., Hornyik, C., Tavazza, M. & Burgyan, J. (2002). A viral protein suppresses RNA silencing and binds silencing-generated, 21- to 25-nucleotide double-stranded RNAs. *EMBO J* **21**, 3070–3080.

Stollewerk, A., Schoppmeier, M. & Damen, W. G. (2003). Involvement of Notch and Delta genes in spider segmentation. *Nature* **423**, 863–865.

Tang, G., Reinhart, B. J., Bartel, D. P. & Zamore, P. D. (2003). A biochemical framework for RNA silencing in plants. *Genes Dev* **17**, 49–63.

Uhlirova, M., Foy, B. D., Beaty, B. J., Olson, K. E., Riddiford, L. M. & Jindra, M. (2003). Use of Sindbis virus-mediated RNA interference to demonstrate a conserved role of Broad-Complex in insect metamorphosis. *Proc Natl Acad Sci U S A* **100**, 15607–15612.

Vance, V. & Vaucheret, H. (2001). RNA silencing in plants – defense and counterdefense. *Science* **292**, 2277–2280.

Van der Krol, A. R., Mur, L. A., Beld, M., Mol, J. N. & Stuitje, A. R. (1990). Flavonoid genes in petunia: addition of a limited number of gene copies may lead to a suppression of gene expression. *Plant Cell* **2**, 291–299.

Wheeler, S. R., Carrico, M. L., Wilson, B. A., Brown, S. J. & Skeath, J. B. (2003). The expression and function of the achaete-scute genes in *Tribolium castaneum* reveals conservation and variation in neural pattern formation and cell fate specification. *Development* **130**, 4373–4381.

Whitton, J. L. & Oldstone, B. A. (2001). The immune responses to viruses. In *Fields Virology*, 4th edn, pp. 285–320. Edited by D. M. Knipe & P. M. Howley. Philadelphia: Lippincott Williams & Wilkins.

Xin, H. W. & Ding, S. W. (2003). Identification and molecular characterization of a naturally occurring RNA virus mutant defective in the initiation of host recovery. *Virology* **317**, 253–262.

Zamore, P. D., Tuschl, T., Sharp, P. A. & Bartel, D. P. (2000). RNAi: double-stranded RNA directs the ATP-dependent cleavage of mRNA at 21 to 23 nucleotide intervals. *Cell* **101**, 25–33.

Specificity of *Borrelia*–tick vector relationships

Alan Barbour

Departments of Microbiology and Molecular Genetics and Medicine, University of California Irvine, B240 Medical Science I, Irvine, CA 92697-4025, USA

BORRELIA

The genus *Borrelia* comprises diverse species of spirochaetes that are associated with haematophagous arthropods (Paster *et al.*, 1991; Paster & Dewhirst, 2000). Some *Borrelia* species are pathogenic for humans or for livestock. Other spirochaete groups with human pathogens are the treponemes, which include the human-restricted agent of syphilis, and the leptospires, which are mostly free-living spirochaetes that infect a wide variety of animals. The spirochaete phylum also contains a number of species that are symbionts of invertebrates, such as molluscs and termites. *Borrelia* spirochaetes characteristically circulate in the blood of their vertebrate hosts and are transmitted between vertebrates by ticks, with the single, epidemiologically important exception of a louse-borne species. A common strategy of *Borrelia* spp. for prolonging spiro-chaetaemia – thus increasing the probability of vector transmission – is avoidance of the immune response through antigenic variation (Barbour & Restrepo, 2000; Barbour, 2002). Most types of *Borrelia* infections are zoonoses, but humans are the critical reservoirs for at least one species (Barbour & Hayes, 1986; Barbour, 2004).

The number of recognized *Borrelia* species has more than doubled over the last two decades, in part because cultivation methods improved (Barbour, 1988; Cutler *et al.*, 1994) and technologies like PCR allowed identification and taxonomic classification without being able to culture the organism (Anda *et al.*, 1996; Barbour *et al.*, 1996; Kisinza *et al.*, 2003; Scoles *et al.*, 2001). Table 1 is a list of accepted and tentative species designations, as of early 2004. *Borrelia* species have been documented in the Palaearctic, Afro-Tropical, Nearctic, Neotropical and Antarctic ecological regions, and some

SGM symposium 63: Microbe–vector interactions in vector-borne diseases.
Editors S. H. Gillespie, G. L. Smith & A. Osbourn. Cambridge University Press. ISBN 0 521 84312 X ©SGM 2004

Table 1. *Borrelia* species that cause relapsing fever (RF) or Lyme borreliosis (LB) and species with other disease associations

Species	Ecological region	Vector*	Disease†
Relapsing fever group			
B. anserina	?Afro-Tropical‡	A	FS
B. coriaceae	Nearctic	A	?EBA
B. crocidurae	Palaearctic	A	RF
B. duttonii	Afro-Tropical	A	RF
B. hermsii	Nearctic	A	RF
B. hispanica	Palaearctic	A	RF
B. latyschewii	Palaearctic	A	RF
B. lonestari	Nearctic	M	?STARI
B. mazzottii	Nearctic	A	RF
B. miyamotoi	Palaearctic	P	
B. miyamotoi s. l.	Palaearctic, Nearctic	P	
B. parkeri	Nearctic	A	RF
B. persica	Palaearctic	A	RF
B. recurrentis	?Palaearctic‡	L	RF
B. theileri	?Afro-Tropical‡	M	BS
B. turicatae	Nearctic	A	RF
B. venezuelensis	Neotropical	A	RF
Lyme borreliosis group			
B. afzelii	Palaearctic	P	LB
B. andersoni	Nearctic	P	
B. bissettii	Nearctic	P	
B. burgdorferi	Nearctic	P	LB
B. garinii	Palaearctic, Antarctic	P	LB
B. japonica	Palaearctic	P	
B. lusitaniae	Palaearctic	P	
B. tanuki	Palaearctic	P	
B. turdi	Palaearctic	P	
B. valaisiana	Palaearctic	P	

*A, argasid tick; L, human louse; M, metastriate hard tick; P, prostriate hard tick.

†BS, bovine spirochaetosis; FS, fowl spirochaetosis; EBA, epizootic bovine abortion; STARI, Southern tick-associated rash illness.

‡Suspected origins of *B. anserina*, *B. recurrentis* and *B. theileri*, which are global in distribution now.

species that use humans or livestock as reservoirs are now cosmopolitan. There was a single report of a *Borrelia*-like spirochaete in rats in Queensland (Carley & Pope, 1962), but otherwise, indigenous *Borrelia* species have not been documented in the Australasia, Indo-Malayan or Oceania ecological regions. In general, *Borrelia* spp. are not enzootic in rain forests or in highly arid deserts, but most other landscapes with ticks probably support one or more *Borrelia* spp., most of which have yet to be identified.

The vertebrate host ranges of most *Borrelia* spp. are broad. The most common vertebrate reservoirs are rodents, but other host animals include lagomorphs, carnivores, ruminants, primates, bats, insectivores and birds (Barbour, 2004). In the laboratory, one *Borrelia* sp., *Borrelia burgdorferi*, for example, will infect mice, rats, hamsters, guinea pigs, rabbits, dogs, non-human primates and birds. Some *Borrelia* species may have a very limited range of hosts in nature because of the tick's adaptations (see below), but are able to experimentally infect a variety of animals. Some tick vectors of *Borrelia* spp. even feed on lizards, snakes and turtles, but the reservoir competence of these poikilothermic vertebrates has not, with a few exceptions (Lane & Quistad, 1998), been studied. The ability of an unidentified relapsing fever (RF) *Borrelia* isolate to infect the guinea pig was used to discriminate between species (Barbour & Hayes, 1986), but, for the most part, the *Borrelia*–vertebrate host associations were not particularly useful for taxonomic classification or species identification. The apparently specific associations between a *Borrelia* species and a particular group of animals do not appear to be due to the development of specific and restricted affinities between pathogens and their hosts. The lack of infection of some vertebrates, which might serve as reservoirs, more often appears to be due to toxicity of the vertebrate's innate immunity, such as the complement system, for the spirochaete (Kurtenbach *et al.*, 2002; Lane & Quistad, 1998), rather than co-evolution between host and parasite.

There are two major groups of *Borrelia* spp. (Table 1 and Fig. 1). The first group, and the largest in number of identified species, includes several species that cause RF, such as *Borrelia duttonii* in the Afro-Tropical region and *Borrelia hermsii* in the Nearctic region. The second group includes three species documented to cause Lyme borreliosis (LB): *B. burgdorferi*, *Borrelia afzelii* and *Borrelia garinii*. Although the majority of the species of each group may not, after a final, future accounting, be associated with any human disease, the terms 'relapsing fever group' and 'Lyme borreliosis group' are used for ease of presentation, at the risk for some species of false guilt by their associations in a group. The RF group includes species, such as *Borrelia miyamotoi* and some related organisms (*B. miyamotoi sensu lato*), that have yet to be associated with a disease, as well as the species *Borrelia anserina* and *Borrelia theileri*, which are the respective agents of fowl spirochaetosis and bovine borreliosis, two disorders that may be equivalent to RF in their pathogenesis and clinical features but were given other names. This group also includes two species, *Borrelia lonestari* and *Borrelia coriaceae*, that are suspected but not proven to cause disorders of humans (Southern tick-associated rash illness) and domestic animals (epizootic bovine abortion), respectively (Barbour, 2001).

Whereas known LB group species, with the exception of *B. garinii*, which infects sea-birds (Olsen *et al.*, 1995), are limited to the Holoarctic (Palaearctic and Nearctic), RF group species have been found in the Afro-Tropical and Neotropical regions as well as Palaearctic and Nearctic. It is possible that the difference in the distribution of the two groups reflects the time periods and geographic scopes of the collections. RF *Borrelia* spp. have been known for over a century, while the first known example of the LB group was not discovered until 1981 (Barbour & Hayes, 1986). There have only been sporadic, limited searches to date for hitherto unknown *Borrelia* spp. in the Afro-Tropical, Australasia, Oceania and Indo-Malayan ecological regions.

TICK VECTORS AND RESERVOIRS

With the exception of *Borrelia recurrentis*, which is transmitted from person-to-person by the human louse, *Pediculus humanus*, the vectors of *Borrelia* spp. are ticks. Ticks (Ixodidae) are a subgroup of the order of mites (Acari), which are, in turn, members of the larger group Arachnida. Ticks are a form of parasitic mite (Parasitiformes) that is entirely dependent on a periodic blood meal for all its biosynthetic activities and for reproduction. The two major monophyletic clades of ticks are soft (Argasidae) ticks and hard (Ixodidae) ticks. Although most phylogenetic analyses based on morphology or DNA sequences indicate that argasid ticks are basal to ixodid ticks in the evolutionary history of mites (Crampton *et al.*, 1996; Klompen *et al.*, 1997), one molecular phylogenetic analysis did not support this ordering of deep nodes (Black *et al.*, 1997). Fig. 1 on the right shows an unrooted tree without an outgroup of selected tick species.

Humans are more likely to encounter ixodid ticks than argasid ticks and to bring a durably armoured ixodid tick to mind's eye at the mention of 'tick'. Some examples of ixodid ticks are the deer tick (*Ixodes scapularis*), the sheep tick (*Ixodes ricinus*), the taiga tick (*Ixodes persulcatus*), the lone star tick (*Amblyomma americanum*), the cattle tick (*Boophilus microplus*), the rabbit tick (*Haemaphysalis leporispalustris*) and the brown dog tick (*Rhipicephalus sanguineus*). The less-noticeable argasid ticks, which have relatively soft, leathery exoskeletons, tend to feed for less than 30 min each time and often while the host sleeps. Argasid ticks usually feed on a single type of host, such as the chipmunk or wild boar, as fellow residents with the host in its nest, burrow and shelter. Sometimes, especially in environments with human habitation, they are opportunists and feed on whatever type of animal, including humans, happens to sleep in the immediate area. Argasid ticks, because of their spatially restricted range, such as

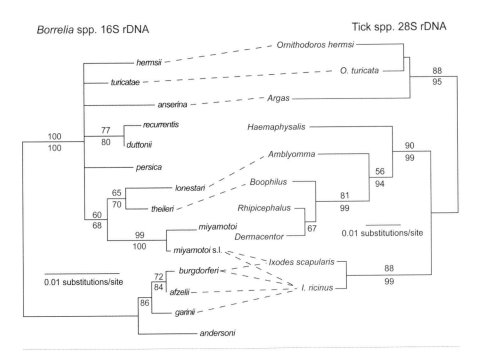

Fig. 1. Micro-organism–vector relationships for selected *Borrelia* spp. (Table 1) and ticks. A nucleotide distance phylogram of partial sequences of 16S mitochondrial rDNA genes of *Borrelia* spp. is shown on the left, and a distance phylogram of partial sequences of 28S rDNA genes of argasid ticks (*Ornithodoros* and *Argas*), metastriate hard ticks (*Haemaphysalis*, *Amblyomma*, *Boophilus*, *Rhipicephalus* and *Dermacentor*) and prostriate hard ticks (*Ixodes*) is shown on the right. The sources of the *Borrelia* sequences are given in Barbour (2004); the source of the tick sequences is described in work submitted for publication by J. Bunikis & A. G. Barbour. The upper numbers near a given node are the percentage bootstrap support of the node from 100 replicates of a heuristic search by maximum-likelihood criterion; the numbers below the branch line are the percentage bootstrap support from 1000 replicates of a neighbour-joining analysis by the distance criterion. The relationship between a given *Borrelia* sp. and its usual vector(s) is shown by a dashed line.

a burrow or hollow log, which may go unoccupied by a vertebrate for long periods, have adapted to survival for several years without attaining a blood meal. The two genera of argasid ticks that include vectors of *Borrelia* are *Argas* and *Ornithodoros*; for example, *Ornithodoros hermsi* is the vector of *B. hermsii*, and *Argas persicus* is a vector of *B. anserina* (Fig. 1).

Ixodid ticks, in contrast to argasid ticks, are more likely to be host-seeking in the environment and not more passively occupy the shelters of their potential hosts. They will ambush an animal as it passes, and once embedded in the skin, they feed on the host for several days, before dropping off to eventually moult to the next stage or, in the case of adult females, to produce and lay eggs. A single feeding of up to several millilitres of blood is usually sufficient for an ixodid tick between moults, but few types of ixodid ticks can survive for as long as argasid ticks in the absence of a blood meal.

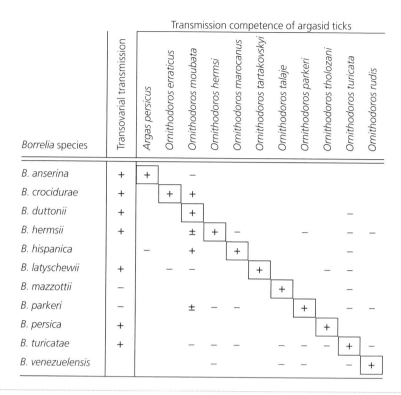

Fig. 2. Transovarial transmission of relapsing fever group *Borrelia* spp. by their usual tick vectors and the transmission competence of other *Ornithodoros* spp. and *Argas* sp. argasid ticks for different *Borrelia* spp. The second column shows the positive (+) or negative (−) result(s) of studies of transovarial transmission of a *Borrelia* sp. in its tick vector. For the columns to the right, the boxes on the diagonal indicate the usual pairing between a *Borrelia* sp. and its argasid tick vector. The results of experimental studies of transmission competence by other tick vectors of a given *Borrelia* sp. are shown as positive (+) or negative (−). In one study's results, which are indicated by the ± symbol, *O. moubata* became infected with and transmitted *B. hermsii* or *B. parkeri*, but the infected ticks were small and had decreased oviposition in comparison to uninfected ticks (Davis, 1942). The references for the transovarial transmission findings are: Adler *et al.* (1937); Aeschlimann (1958); Balashov (1968); Baltazard *et al.* (1954); Bone (1939); Burgdorfer & Mavros (1970); Davis (1943, 1952a, 1956a); Gaber *et al.* (1984); Geigy *et al.* (1954); Pavlovsky & Skvynnik (1958); Zaher *et al.* (1977). Additional references for the transmission competence findings are the following: Baltazard *et al.* (1950, 1952); Brumpt (1933); Brumpt & Brumpt (1939); Burgdorfer & Davis (1954); Davis (1942, 1952b); Davis & Mazzotti (1953); Grün (1950); Herms & Wheeler (1935); Kemp *et al.* (1934); Kirk (1938); Mazzotti (1942, 1943, 1949); Nicolle & Anderson (1927a, b); Nicolle *et al.* (1928).

Ixodid ticks also tend to feed on a different type of animal at each stage; for instance, *I. scapularis* may feed on on a *Peromyscus leucopus* mouse as a larva, on a bird as a nymph, and on a deer as an adult.

The known Ixodidae family is more speciose than Argasidae, perhaps because of the generally greater accessibility of ixodid ticks for would-be collectors. Ixodid tick species have been distributed between two groups, Prostriata and Metastriata

(Hoogstraal & Aeschlimann, 1982). Prostriate ticks comprised species of the genus *Ixodes*, and metastriate ticks included the genera *Amblyomma*, *Boophilus*, *Dermacentor*, *Haemaphysalis* and *Rhipicephalus*. Although Fig. 1 does not show this, other phylogenies based on rRNA sequences suggest that Prostriata is basal to Metastriata in the family Ixodidae (Black *et al.*, 1997; Crampton *et al.*, 1996).

Table 1 and Fig. 1 show that LB group species are associated with the *Ixodes* spp. ticks but not with either metastriate or argasid genera. On the other hand, RF group species have been identified not only in argasid ticks (Hoogstraal, 1985) but in metastriate (Armstrong *et al.*, 1996; Barbour *et al.*, 1996; Rich *et al.*, 2001) and prostriate (Fraenkel *et al.*, 2002; Fukunaga *et al.*, 1995; Scoles *et al.*, 2001) ticks. The three prostriate species, *I. persulcatus*, *I. scapularis* and *I. ricinus*, that are now known to carry *B. miyamotoi* and *B. miyamotoi s. l.* organisms were previously shown to be vectors of the LB agents *B. burgdorferi*, *B. afzelii* and *B. garinii*.

TRANSOVARIAL TRANSMISSION

Another feature distinguishing LB group species from RF group species is the frequency of transovarial transmission of the agent from one generation of ticks to the next (Burgdorfer & Varma, 1967). Transovarial transmission allows the persistence of the micro-organism in a vector population in the absence of sufficient numbers of suitable vertebrate hosts for maintenance. Although some studies suggested that a small percentage of larvae of field-collected *Ixodes* spp. are infected with *B. burgdorferi* or a related LB group species, at least some of these organisms are probably *B. miyamotoi s. l.* spirochaetes and not transovarially transmitted *B. burgdorferi* (Scoles *et al.*, 2001). Experimental studies of vertical transmission of *B. burgdorferi* in *I. scapularis* and *Ixodes pacificus* indicate that it would be insufficient in frequency to maintain *Borrelia* populations for long in the absence of a vertebrate reservoir (Magnarelli *et al.*, 1987; Piesman, 1991; Schoeler & Lane, 1993). When transovarial transmission of the LB group species was demonstrated in the laboratory it was in *Ixodes hexagonus*, a species that is much less frequently infected than the principal disease vector, *I. ricinus*, under natural conditions in western Europe (Toutoungi & Gern, 1993).

In contrast, among the RF group of species transovarial transmission is the rule rather than the exception. Fig. 2 summarizes the results of several studies from five to eight decades ago of this phenomenon in the agents of RF and fowl spirochaetosis. Of the nine species in the reports, only two, *Borrelia parkeri* and *Borrelia mazzottii*, neither of which are common causes of RF (Barbour, 2004), did not demonstrate a high frequency of transovarial transmission. There is also evidence of transovarial transmission of *B. lonestari* in the metastriate tick *Amblyomma americanum* (Stromdahl *et al.*, 2003).

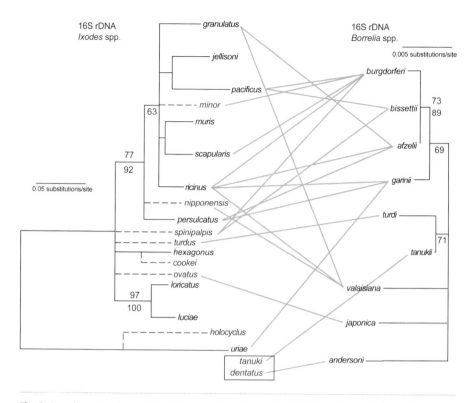

Fig. 3. *Borrelia* sp.–*Ixodes* sp. tick relationships. The tree on the left is based on partial mitochondrial 16S rDNA sequences from *Ixodes* spp., while the unrooted tree on the right is based on 16S rDNA sequences from selected *Borrelia* spp. of the *B. burgdorferi s. l.* group. Both trees were constructed using maximum-likelihood analysis; values above the nodes are bootstrap values based on 100 replicates; values below the nodes are bootstrap values expressed as a percentage of 1000 replicates, generated from a neighbour-joining distance analysis performed using the same dataset. For the analysis of the taxa connected by solid grey lines, the sequences from two regions of the 16S rDNA genes and from the datasets of Black & Piesman (1994) and Xu *et al.* (2003) were used; *Ixodes uriae*, a member with *I. holocyclus* of a sister group to the other *Ixodes* species in the figure, was the outgroup (Klompen *et al.*, 2000). Taxa terminating branches shown in dashed lines occurred in only one of the two datasets, and their placement is more tentative and is shown here without bootstrap support.

In *I. scapularis* there was transovarial transmission of *B. miyamotoi s. l.* but not *B. burgdorferi* (Scoles *et al.*, 2001).

BORRELIA–TICK RELATIONSHIPS

Fig. 1 schematically represents the relationships between selected *Borrelia* and tick species. The figure shows the pathogen–vector pairing between *B. hermsii* and *O. hermsi*, *Borrelia turicatae* and *Ornithodoros turicata*, *B. lonestari* and *Amblyomma americanum*, and *B. theileri* and *Boophilus microplus*, but the occurrence of more than one species of *Borrelia* in *I. scapularis* and *I. ricinus*. The relationships between expanded groups of *Ixodes* spp. and LB group species are shown in Fig. 3. The species are presented as branches of the respective phylograms. Less-well characterized tick species have tentative placements, which are based on less-complete sequence data, in the figure. The branches were manually arranged for more convenient presentation but without the aim of a formal or robust demonstration of co-evolution.

The figure demonstrates the overlapping and complex relationships between the *Ixodes* ticks and the *Borrelia* spirochaetes. *I. ricinus* is the vector of at least five different LB group species: *B. afzelii*, *B. garinii*, *B. burgdorferi* and *Borrelia valaisiana* in Fig. 3, as well as *Borrelia lusitaniae* (Hubalek & Halouzka, 1997). *B. burgdorferi* is probably not even of Palaearctic origin (Marti Ras *et al.*, 1997). It was probably imported to Europe from North America during the last two or three centuries, which is comparatively recent by evolutionary standards. *B. garinii* is transmitted by *I. ricinus* and *I. persulcatus*, which are sister taxa in a clade, and also by *Ixodes uriae*, a tick of seabirds of the Arctic and Antarctic. *I. uriae*, along with *Ixodes holocyclus* and other Australasian region species, comprise a separate clade from other *Ixodes* spp. (Klompen *et al.*, 2000).

The fairly non-stringent specificity between *Ixodes* and *Borrelia* was also demonstrated experimentally. Although some *Ixodes* species, such as *Ixodes cookei* and *I. holocyclus*, were reported as not or poorly competent as vectors of *B. burgdorferi* (Barker *et al.*, 1993; Piesman & Stone, 1991), others, such as *I. hexagonus* (Gern *et al.*, 1991), *I. pacificus* (Lane *et al.*, 1994), *Ixodes spinipalpis* (Dolan *et al.*, 1997), *Ixodes jellisoni* (Lane *et al.*, 1999), *Ixodes angustus* (Peavey *et al.*, 2000), *Ixodes muris* (Dolan *et al.*, 2000) and *Ixodes sinensis* (Sun *et al.*, 2003) were able to transmit LB agent species in the laboratory. [Many studies had demonstrated the incompetence of a variety of metastriate ticks to transmit LB agents (Barbour & Fish, 1993).]

In contrast to *Ixodes* spp. and the LB group of *Borrelia* spp., there appears to be considerably more specificity between RF group *Borrelia* spp. and their argasid vectors in terms of vector competence. A summary of the results of several studies of this phenomenon is presented in Fig. 2. Like the transovarial transmission studies, the tick competence studies date from at least 50 years ago and were carried out by different sets of investigators, but because *Borrelia*–tick associations were the most discriminating means of identifying taxonomically an unknown *Borrelia* sp., these sorts of studies were frequent at the time and, for the most part, well-controlled and carefully described. (The vector competence studies of Gordon E. Davis and his colleagues at the Rocky Mountain Laboratory of the National Institutes of Health in the 1940s and 1950s provide a particularly rich source of information about the biology of these pathogens and their vectors.)

The figure shows that the only tick that was reported to be able to experimentally transmit a *Borrelia* sp. other than its usual pair mate was *Ornithodoros moubata*, the usual vector of *B. duttonii* in sub-Saharan Africa. This tick was able to transmit to laboratory animals *Borrelia crocidurae* and *Borrelia hispanica* as well as *B. duttonii*. Previously, it had been shown that there was cross-immunity between *B. duttonii* and *B. crocidurae* (Schlossberger & Wichmann, 1929). These three species are also highly similar by the criterion of rRNA and other sequences (Marti Ras *et al.*, 1996) and could otherwise be considered as different strains of the same species (A. G. Barbour & J. Bunikis, unpublished findings).

The two other species that *O. moubata* was reported to have transmitted were *B. hermsii* and *B. parkeri*, which are genetically distant from each other and from *B. duttonii* (Marti Ras *et al.*, 1996) (unpublished studies). However, Davis (1942) also noted that the *O. moubata* ticks infected with *B. hermsii* or *B. parkeri* were smaller than uninfected ticks and demonstrated comparatively low oviposition. Thus infection with the non-concordant species may have lowered the fitness of the ticks. *O. moubata* also differs from other known *Ornithodoros* vectors of RF *Borrelia* spp. in transmitting the infection through coxal fluid as well as or instead of through saliva (Varma, 1956b), and thus may offer a different and less discriminating mode of transmission for spirochaetes.

The genetic trait that restricted the transmission by other ticks appears to be dominant. The F_1 generation of mating *O. turicata* and *Ornithodoros parkeri* ticks was able to transmit both *B. turicatae* and *B. parkeri* (Davis, 1942). Possible natural hybrids of these two species of ticks with the ability to transmit both *Borrelia* spp. were also observed. This phenomenon also appears to have been observed in the field (Davis & Burgdorfer, 1955). The putative trait may be expressed in the mouthparts or the midgut of the tick. *B. duttonii* after several passages in mice lost the ability to infect *O. moubata* feeding on a spirochaetaemic mouse, but ticks were rendered infectious when the same *B. duttonii* isolate was injected directly into the haemocoel, thereby bypassing the intestine (Varma, 1956a).

While the studies of Davis and others demonstrated a high degree of specificity between a given *Borrelia* sp. with the species of its vector, Davis did not observe local specificity within species of ticks and *Borrelia* when ticks and organisms of the same species from different locations in the western US were paired in different combinations in transmission studies (Davis, 1956b).

CONCLUDING REMARKS

There appears to have been a set of long-standing relationships between some RF group species and their vector ticks to the extent that the tick is a reservoir as well as a vector of the *Borrelia* sp. and the *Borrelia* sp. has a very narrow host range for vectors. This may have occurred because the ticks have a very limited geographic range and may have only intermittent exposure to a very restricted number of potential hosts. For the *Borrelia* sp. under these conditions the evolutionary strategy may have been to heighten the efficiency of transmission from spirochaetaemic vertebrate to feeding tick, perhaps through high-affinity binding to a mouthpart ligand as the spirochaetes pass or to midgut ligand once the spirochaetes enter the intestine. Another strategy under these conditions would be maintenance of the population size in a given environment through transovarial transmission when suitable vertebrate hosts are lacking.

The LB group of species are considerably less discriminating in their vector choices than the agents of RF. The sole reservoirs for this group of *Borrelia* spp. appear to be vertebrates rather than ticks alone or both ticks and vertebrates. In a vertebrate host, such as a rodent in a hardwood forest in Connecticut or Switzerland, there may be more than one *Ixodes* species that parasitize these animals. Restriction of the *Borrelia* sp. to only one of the ticks would limit opportunities for transmission. The availability of a variety of vertebrate hosts and tick vectors may obviate the need for transovarial transmission, which possibly may have a fitness cost to either the ticks through decreased fecundity or to *Borrelia* through diversion of some organisms to the ova rather than the salivary glands.

The place of the *B. lonestari*–*B. theileri*–*B. miyamotoi* group of organisms in this scheme remains to be determined. The known species in this subgroup are transmitted by ixodid rather than argasid ticks. There is transovarial transmission in at least two of the vectors, but there have been few studies of the specificity of vector competence for these organisms.

ACKNOWLEDGEMENTS

The work from my laboratory was supported by National Institutes of Health grant AI24424 and 37248. I thank Masahito Fukunaga and Joseph Piesman for their helpful comments on the manuscript.

REFERENCES

Adler, S., Theodor, O. & Schieber, H. (1937). Observations on tick-transmitted human spirochaetosis in Palestine. *Ann Trop Med Parasitol* **31**, 25–35.

Aeschlimann, A. (1958). Développement embryonnaire d'*Ornithodorus moubata* (Murray) et transmission transovarienne de *Borrelia duttoni. Acta Trop* **15**, 15–64.

Anda, P., Sanchez-Yebra, W., del Mar Vitutia, M., Perez Pastrana, E., Rodriguez, I., Miller, N. S., Backenson, P. B. & Benach, J. L. (1996). A new *Borrelia* species isolated from patients with relapsing fever in Spain. *Lancet* **348**, 162–165.

Armstrong, P. M., Rich, S. M., Smith, R. D., Hartl, D. L., Spielman, A. & Telford, S. R., 3rd (1996). A new *Borrelia* infecting Lone Star ticks. *Lancet* **347**, 67–68.

Balashov, Y. S. (1968). [Transovarial transmission of the spirochete *Borrelia sogdiana* in *Ornithodoros papillipes* ticks and its effect on biological properties of the agent]. *Parazitologiya* **2**, 198–201.

Baltazard, M., Bahmanyar, M. & Mofidi, C. (1950). Ornithodoros erraticus et fievres recurrentes. *Bull Soc Pathol Exot* **43**, 595–601.

Baltazard, M., Bahmanyar, M., Pournaki, R. & Mofidi, C. (1952). *Ornithodoros tartakovskyi* et *Borrelia latychevi*. *Ann Parasitol Hum Comp* **27**, 311–328.

Baltazard, M., Bahmanyar, M. & Chamsa, M. (1954). Sur l'usage du cobaye pour la differenciation des spirochetes recurrents. *Bull Soc Pathol Exot* **47**, 864–877.

Barbour, A. G. (1988). Laboratory aspects of Lyme borreliosis. *Clin Microbiol Rev* **1**, 399–414.

Barbour, A. G. (2001). *Borrelia*: a diverse and ubiquitous genus of tick-borne pathogens. In *Emerging Infections 5*, pp. 153–174. Edited by W. M. Scheld, W. A. Craig & J. M. Hughes. Washington, DC: American Society for Microbiology.

Barbour, A. G. (2002). Antigenic variation of relapsing fever *Borrelia* and other pathogenic bacteria. In *Mobile DNA II*, pp. 972–994. Edited by N. L. Craig, R. Craigie, M. Gellert & A. Lambowitz. Washington, DC: American Society for Microbiology.

Barbour, A. G. (2004). Relapsing fever. In *Tick-borne Diseases*. Edited by J. Goodman, D. Dennis & D. Sonneshine. Washington, DC: American Society for Microbiology (in press).

Barbour, A. G. & Fish, D. (1993). The biological and social phenomenon of Lyme disease. *Science* **260**, 1610–1616.

Barbour, A. G. & Hayes, S. F. (1986). Biology of *Borrelia* species. *Microbiol Rev* **50**, 381–400.

Barbour, A. G. & Restrepo, B. I. (2000). Antigenic variation in vector-borne pathogens. *Emerg Infect Dis* **6**, 449–457.

Barbour, A. G., Maupin, G. O., Teltow, G. J., Carter, C. J. & Piesman, J. (1996). Identification of an uncultivable *Borrelia* species in the hard tick *Amblyomma americanum*: possible agent of a Lyme disease-like illness. *J Infect Dis* **173**, 403–409.

Barker, I. K., Lindsay, L. R., Campbell, G. D., Surgeoner, G. A. & McEwen, S. A. (1993). The groundhog tick *Ixodes cookei* (Acari: ixodidae): a poor potential vector of Lyme borreliosis. *J Wildl Dis* **29**, 416–422.

Black, W. C. & Piesman, J. (1994). Phylogeny of hard- and soft-tick taxa (Acari: Ixodida) based on mitochondrial 16S rDNA sequences. *Proc Natl Acad Sci U S A* **91**, 10034–10038.

Black, W. C., Klompen, J. S. & Keirans, J. E. (1997). Phylogenetic relationships among tick subfamilies (Ixodida: Ixodidae: Argasidae) based on the 18S nuclear rDNA gene. *Mol Phylogenet Evol* **7**, 129–144.

Bone, G. (1939). La transmission hereditaire du spirochete de Dutton chez *Ornithodoros moubata*. *C R Seances Soc Biol* (*Paris*) **130**, 86–87.

Brumpt, E. (1933). Etude de la fievre recurrente sporadique des Etats-Unis, transmise dans la nature par *Ornithodoros turicata*. *C R Soc Biol* **113**, 1366–1369.

Brumpt, E. & Brumpt, L. (1939). Identite du spirochaete des fievres recurrentes a tiques des plateaux mexicains et du Spirochaeta turicatae agent de la fievre recurrente sporadique des Etats-Unis. *Ann Parasitol Hum Comp* **17**, 287–298.

Burgdorfer, W. & Davis, G. E. (1954). Experimental infection of the African relapsing fever tick, *Ornithodoros moubata* (Murray), with *Borrelia latychevi* (Sofiev). *J Parasitol* **40**, 456–460.

Burgdorfer, W. & Mavros, A. J. (1970). Susceptibility of various species of rodents to the relapsing fever spirochete, *Borrelia hermsii*. *Infect Immun* **2**, 256–259.

Burgdorfer, W. & Varma, M. G. (1967). Trans-stadial and transovarial development of disease agents in arthropods. *Annu Rev Entomol* **12**, 347–376.

Carley, J. G. & Pope, J. H. (1962). A new species of *Borrelia* (*B. queenslandia*) from *Rattus villosissimus* in Queensland. *Aust J Exp Biol Med Sci* **40**, 255–262.

Crampton, A., McKay, I. & Barker, S. C. (1996). Phylogeny of ticks (Ixodida) inferred from nuclear ribosomal DNA. *Int J Parasitol* **26**, 511–517.

Cutler, S. J., Fekade, D., Hussein, K., Knox, K. A., Melka, A., Cann, K., Emilianus, A. R., Warrell, D. A. & Wright, D. J. (1994). Successful in-vitro cultivation of *Borrelia recurrentis*. *Lancet* **343**, 242.

Davis, G. E. (1942). Species unity or plurality of spirochetes. In *Publication No 18*, pp. 41–47. Washington, DC: American Association for the Advancement of Science.

Davis, G. E. (1943). Relapsing fever: the tick *Ornithodoros turicata* as a spirochetal reservoir. *Public Health Rep* **58**, 839–842.

Davis, G. E. (1952a). Biology as an aid to the identification of two closely related species of ticks of the genus *Ornithodoros*. *J Parasitol* **38**, 477–480.

Davis, G. E. (1952b). Observations on the biology of the argasid tick, *Ornithodoros brasiliensis* Aragao, 1923; with the recovery of a spirochete, *Borrelia brasiliensis*, n. sp. *J Parasitol* **38**, 473–476.

Davis, G. E. (1956a). A relapsing fever spirochete, *Borrelia mazzottii* (sp. nov.) from *Ornithodoros talaje* from Mexico. *Am J Hyg* **63**, 13–17.

Davis, G. E. (1956b). The identification of spirochetes from human cases of relapsing fever by xenodiagnosis with comments on local specificity of tick vectors. *Exp Parasitol* **5**, 271–275.

Davis, G. E. & Burgdorfer, W. (1955). Relapsing fever spirochetes: an aberrant strain of *Borrelia parkeri* from Oregon. *Exp Parasitol* **4**, 100–106.

Davis, G. E. & Mazzotti, L. (1953). The non-transmission of the relapsing fever spiro-chete, *Borrelia dugesii* (Mazzotti, 1949) by the argasid tick *Ornithodoros turicata* (Duges, 1876). *J Parasitol* **39**, 663–666.

Dolan, M. C., Maupin, G. O., Panella, N. A., Golde, W. T. & Piesman, J. (1997). Vector competence of *Ixodes scapularis*, *I. spinipalpis*, and *Dermacentor andersoni* (Acari: Ixodidae) in transmitting *Borrelia burgdorferi*, the etiologic agent of Lyme disease. *J Med Entomol* **34**, 128–135.

Dolan, M. C., Lacombe, E. H. & Piesman, J. (2000). Vector competence of *Ixodes muris* (Acari: Ixodidae) for *Borrelia burgdorferi*. *J Med Entomol* **37**, 766–768.

Fraenkel, C. J., Garpmo, U. & Berglund, J. (2002). Determination of novel *Borrelia* genospecies in Swedish *Ixodes ricinus* ticks. *J Clin Microbiol* **40**, 3308–3312.

Fukunaga, M., Takahashi, Y., Tsuruta, Y., Matsushita, O., Ralph, D., McClelland, M. & Nakao, M. (1995). Genetic and phenotypic analysis of *Borrelia miyamotoi* sp. nov., isolated from the ixodid tick *Ixodes persulcatus*, the vector for Lyme disease in Japan. *Int J Syst Bacteriol* **45**, 804–810.

Gaber, M. S., Khalil, G. M., Hoogstraal, H. & Aboul-Nasr, A. E. (1984). *Borrelia crocidurae* localization and transmission in *Ornithodoros erraticus* and *O. savignyi*. *Parasitology* **88**, 403–413.

Geigy, R., Wagner, O. & Aeschlimann, A. (1954). Transmission genitale de *Borrelia duttonii* chez *Ornithodoros moubata*. *Acta Trop* **11**, 81–82.

Gern, L., Toutoungi, L. N., Hu, C. M. & Aeschlimann, A. (1991). *Ixodes* (*Pholeoixodes*) *hexagonus*, an efficient vector of *Borrelia burgdorferi* in the laboratory. *Med Vet Entomol* **5**, 431–435.

Grün, H. (1950). Die experimentelle Übertragung von Rückfallfieber-spirochete durch *Ornithodoros moubata*. *Z Hyg Infektionskr* **131**, 198–218.

Herms, W. B. & Wheeler, C. M. (1935). Tick transmission of California relapsing fever. *J Econ Entomol* **28**, 846–855.

Hoogstraal, H. (1985). Argasid and nuttalliellid ticks as parasites and vectors. *Adv Parasitol* **24**, 135–238.

Hoogstraal, H. & Aeschlimann, A. (1982). Tick-host specificity. *Bull Soc Entomol Suisse* **55**, 5–32.

Hubalek, Z. & Halouzka, J. (1997). Distribution of *Borrelia burgdorferi* sensu lato genomic groups in Europe, a review. *Eur J Epidemiol* **13**, 951–957.

Kemp, H. A., Moursund, W. H. & Wright, D. J. (1934). Relapsing fever in Texas. II. The specificity of the vector, *Ornithodoros turicata*, for the spirochete. *Am J Trop Med* **14**, 159–162.

Kirk, R. (1938). The non-transmission of Abyssinian louse-borne relapsing fever by the tick *Ornithodoros savigny* and certain other blood-sucking arthropods. *Ann Trop Med Parasitol* **32**, 357–365.

Kisinza, W. N., McCall, P. J., Mitani, H., Talbert, A. & Fukunaga, M. (2003). A newly identified tick-borne *Borrelia* species and relapsing fever in Tanzania. *Lancet* **362**, 1283–1284.

Klompen, J. S., Oliver, J. H., Jr, Keirans, J. E. & Homsher, P. J. (1997). A re-evaluation of relationships in the Metastriate (Acari: Parasitformes: Ixodidae). *Syst Parasitol* **38**, 1–24.

Klompen, J. S., Black, W. C. I., Keirans, J. E. & Norris, D. E. (2000). Systematics and biogeography of hard ticks, a total evidence approach. *Cladistics* **16**, 79–102.

Kurtenbach, K., De Michelis, S., Etti, S., Schafer, S. M., Sewell, H. S., Brade, V. & Kraiczy, P. (2002). Host association of *Borrelia burgdorferi* sensu lato – the key role of host complement. *Trends Microbiol* **10**, 74–79.

Lane, R. S. & Quistad, G. B. (1998). Borreliacidal factor in the blood of the western fence lizard (*Sceloporus occidentalis*). *J Parasitol* **84**, 29–34.

Lane, R. S., Brown, R. N., Piesman, J. & Peavey, C. A. (1994). Vector competence of *Ixodes pacificus* and *Dermacentor occidentalis* (Acari: Ixodidae) for various isolates of Lyme disease spirochetes. *J Med Entomol* **31**, 417–424.

Lane, R. S., Peavey, C. A., Padgett, K. A. & Hendson, M. (1999). Life history of *Ixodes* (Ixodes) *jellisoni* (Acari: Ixodidae) and its vector competence for *Borrelia burgdorferi* sensu lato. *J Med Entomol* **36**, 329–340.

Magnarelli, L. A., Anderson, J. F. & Fish, D. (1987). Transovarial transmission of *Borrelia burgdorferi* in *Ixodes dammini* (Acari:Ixodidae). *J Infect Dis* **156**, 234–236.

Marti Ras, N., Lascola, B., Postic, D., Cutler, S. J., Rodhain, F., Baranton, G. & Raoult, D. (1996). Phylogenesis of relapsing fever *Borrelia* spp. *Int J Syst Bacteriol* **46**, 859– 865.

Marti Ras, N., Postic, D., Foretz, M. & Baranton, G. (1997). *Borrelia burgdorferi* sensu stricto, a bacterial species "made in the U.S.A."? *Int J Syst Bacteriol* **47**, 1112–1117.

Mazzotti, L. (1942). Estudio sobre la transmision de *Spirochaeta venezuelensis*. *Rev Inst Salubr Enferm Trop* **3**, 297–301.

Mazzotti, L. (1943). Transmission experiments with *Spirochaeta turicatae* and *S. venezuelensis* with four species of *Ornithodoros*. *Am J Hyg* **38**, 203–206.

Mazzotti, L. (1949). Sobre una nueva espiroqueta de la fiebre recurrente, encontrada en Mexico. *Rev Inst Salubr Enferm Trop* **10**, 277–281.

Nicolle, C. & Anderson, C. (1927a). Transmission du spirochete de la musaraigne par *Ornithodoros moubata* et mecanisme de la transmission des spirochetes recurrents par les tiques. *C R Acad Sci* **185**, 373–375.

Nicolle, C. & Anderson, C. (1927b). Transmission experimentelle du spirochete de la recurrente espagnole par l'*Ornithodoros moubata* et mecanisme de celle transmission. *C R Acad Sci* **185**, 433–434.

Nicolle, C., Anderson, C. & Colas-Belcour, J. (1928). Premiers essais d'adaptation du spirochete des poules a divers Ornithodores. *C R Acad Sci* **187**, 791–792.

Olsen, B., Duffy, D. C., Jaenson, T. G., Gylfe, A., Bonnedahl, J. & Bergstrom, S. (1995). Transhemispheric exchange of Lyme disease spirochetes by seabirds. *J Clin Microbiol* **33**, 3270–3274.

Paster, B. J. & Dewhirst, F. E. (2000). Phylogenetic foundation of spirochetes. *J Mol Microbiol Biotechnol* **2**, 341–344.

Paster, B. J., Dewhirst, F. E., Weisburg, W. G. & 7 other authors (1991). Phylogenetic analysis of the spirochetes. *J Bacteriol* **173**, 6101–6109.

Pavlovsky, Y. N. & Skvynnik, A. N. (1958). [Transovarial transmission of spirochaetes of tick-borne relapsing fever in *Ornithodoros papillipes*]. *Tr Voennomeditsinkoi Akad S M Kirova* **46**, 19–28.

Peavey, C. A., Lane, R. S. & Damrow, T. (2000). Vector competence of *Ixodes angustus* (Acari: Ixodidae) for *Borrelia burgdorferi* sensu stricto. *Exp Appl Acarol* **24**, 77–84.

Piesman, J. (1991). Experimental acquisition of the Lyme disease spirochete, *Borrelia burgdorferi*, by larval *Ixodes dammini* (Acari: Ixodidae) during partial blood meals. *J Med Entomol* **28**, 259–262.

Piesman, J. & Stone, B. F. (1991). Vector competence of the Australian paralysis tick, *Ixodes holocyclus*, for the Lyme disease spirochete *Borrelia burgdorferi*. *Int J Parasitol* **21**, 109–111.

Rich, S. M., Armstrong, P. M., Smith, R. D. & Telford, S. R., 3rd (2001). Lone star tick-infecting borreliae are most closely related to the agent of bovine borreliosis. *J Clin Microbiol* **39**, 494–497.

Schlossberger, H. & Wichmann, F. W. (1929). Experimentelle untersuchungen ueber *Spirochaeta crocidurae* und *Spirochaeta hispanica*. *Z Hyg Infektionskr* **109**, 493–507.

Schoeler, G. B. & Lane, R. S. (1993). Efficiency of transovarial transmission of the Lyme disease spirochete, *Borrelia burgdorferi*, in the western blacklegged tick, *Ixodes pacificus* (Acari: Ixodidae). *J Med Entomol* **30**, 80–86.

Scoles, G. A., Papero, M., Beati, L. & Fish, D. (2001). A relapsing fever group spirochete transmitted by *Ixodes scapularis* ticks. *Vector Borne Zoonotic Dis* **1**, 21–34.

Stromdahl, E. Y., Williamson, P. C., Kollars, T. M., Jr, Evans, S. R., Barry, R. K., Vince, M. A. & Dobbs, N. A. (2003). Evidence of *Borrelia lonestari* DNA in *Amblyomma americanum* (Acari: Ixodidae) removed from humans. *J Clin Microbiol* **41**, 5557–5562.

Sun, Y., Xu, R. & Cao, W. (2003). *Ixodes sinensis*: competence as a vector to transmit the Lyme disease spirochete *Borrelia garinii*. *Vector Borne Zoonotic Dis* **3**, 39–44.

Toutoungi, L. N. & Gern, L. (1993). Ability of transovarially and subsequent transstadially infected *Ixodes hexagonus* ticks to maintain and transmit *Borrelia burgdorferi* in the laboratory. *Exp Appl Acarol* **17**, 581–586.

Varma, M. G. R. (1956a). Comparative studies on the transmission of two strains of *Spirochaeta duttonii* by *Ornithodoros moubata* and of *S. turicatae* by *O. turicata*. *Ann Trop Med Parasitol* **50**, 1–17.

Varma, M. G. R. (1956b). Infections of *Ornithodoros* ticks with relapsing fever spirochaetes, and the mechanisms of their transmission. *Ann Trop Med Parasitol* **50**, 18–31.

Xu, G., Fang, Q. Q., Keirans, J. E. & Durden, L. A. (2003). Molecular phylogenetic analyses indicate that the *Ixodes ricinus* complex is a paraphyletic group. *J Parasitol* **89**, 452–457.

Zaher, M. A., Soliman, Z. R. & Diab, F. M. (1977). An experimental study of *Borrelia anserina* in four species of Argas ticks. 2. Transstadial survival and transovarial transmission. *Z Parasitenkd* **53**, 213–223.

Bunyavirus/mosquito interactions

Richard M. Elliott and Alain Kohl

Division of Virology, Institute of Biomedical and Life Sciences, University of Glasgow, Church Street, Glasgow G11 5JR, UK

INTRODUCTION

Bunyaviruses comprise the largest family of arthropod-transmitted viruses. Of the 300 or so viruses in the family *Bunyaviridae* more than half are transmitted by mosquitoes (Calisher, 1996). A number of these viruses cause disease in man or animals, although thus far many have not been associated with human illness. The family is classified into five genera, *Orthobunyavirus*, *Hantavirus*, *Nairovirus*, *Phlebovirus* and *Tospovirus*, and each genus is associated with a principal arthropod vector, except the hantaviruses, which have no arthropod involvement in their life cycle. Tospoviruses are transmitted by thrips to plants, nairoviruses primarily by ticks, and orthobunyaviruses predominantly by mosquitoes. The *Phlebovirus* genus contains viruses transmitted by phlebotomine flies, ticks and, notably for Rift Valley fever virus, mosquitoes as well. This chapter will deal with the interactions between orthobunyaviruses and mosquitoes.

BUNYAVIRUS CHARACTERISTICS

All bunyaviruses have a tri-segmented single-stranded RNA genome of negative-sense (or a variant of negative-sense, termed ambisense) polarity. The genomic RNA segments are called L (large), M (medium) and S (small). All family members encode four structural proteins – two glycoproteins called Gn and Gc according to their position within the primary gene product, a nucleoprotein, N, that encapsidates the genomic (and antigenomic) RNA segments and an RNA-dependent RNA polymerase, called L protein. The pattern of sizes of the viral proteins and the RNAs is conserved within a genus (Elliott *et al.*, 2000). Viruses in some genera also encode non-structural proteins on their M and/or S RNA segments, termed NSm and NSs, respectively.

SGM symposium 63: Microbe–vector interactions in vector-borne diseases.
Editors S. H. Gillespie, G. L. Smith & A. Osbourn. Cambridge University Press. ISBN 0 521 84312 X ©SGM 2004

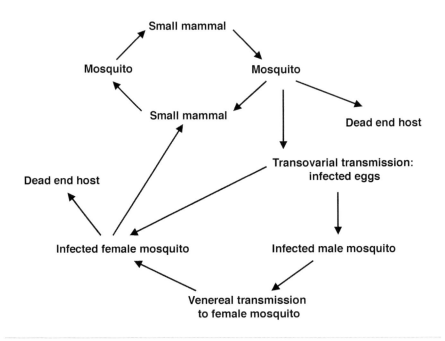

Fig. 1. Generalized life cycle of a mosquito-transmitted orthobunyavirus.

For orthobunyaviruses the NSs open reading frame (ORF) is contained within the N ORF, but is accessed via recognition of an alternate AUG initiation codon in the +1 reading frame with respect to N. The 3′ and 5′ terminal sequences of each genome segment are conserved both within a virus and within a genus, and are complementary. The complementarity of the genome ends results in potential base-pairing to form circular or panhandle forms of the isolated RNA and RNA-N complexes (reviewed in Elliott, 1996).

NATURAL HISTORY OF ORTHOBUNYAVIRUSES

Usually individual viruses are associated with only a few vector species in nature. The maintenance and transmission cycles of orthobunyaviruses involve both horizontal and vertical transmission between a restricted number of mosquito and vertebrate hosts (Fig. 1), and the integrity of such cycles is maintained even when cycles overlap. For example, with California serogroup viruses, individual cycles have been determined involving *Aedes triseriatus* and chipmunks for La Crosse virus; *Aedes communis* and snowshoe hares for snowshoe hare virus; and *Aedes trivittatus* and cottontail rabbits for Trivittatus virus. Although all these viruses have been isolated in the same geographical region, the separate cycles are maintained (Beaty & Calisher, 1991) but the factors responsible for this specificity have yet to be determined.

Horizontal transmission usually refers to the transmission between susceptible vertebrate hosts by the mosquito vector. Following infection the vertebrate host develops a viraemia, and this is a source of virus for another feeding mosquito. The vertebrate therefore acts as an amplifying host. For orthobunyaviruses, horizontal transmission also occurs venereally between mosquitoes (Fig. 1).

Vertical transmission is an important maintenance mechanism for orthobunyaviruses (and Rift Valley fever phlebovirus) in nature, particularly during periods of adverse climatic conditions. Virus can be maintained in mosquito eggs over winter or during periods of drought and when conditions are appropriate mosquito larvae emerge that are infected with the virus. The larvae continue through the normal development process resulting in infected adults.

The ability of a particular mosquito species to transmit a virus to a vertebrate host, vector competence, is dependent on both insect and viral factors. [Further details on this topic can be found in the chapters by Higgs (2004) and Weaver et al. (2004) in this book.] After ingestion of virus in a blood meal, the virus infects the midgut cells of the mosquito. The virus can then be disseminated through the mosquito in the haemocoel, and replicates in essentially all organs. The last organs to be infected are the salivary glands, where extensive virus replication occurs. The virus is then transmitted in saliva when the mosquito feeds. Virus replication in the ovaries leads to transovarial transmission of the virus to progeny. Male offspring develop containing infected gonadal cells, and hence the infected males can pass the virus venereally to uninfected females. (The reader is referred to the following in-depth review articles on the natural history of bunyaviruses: Beaty & Bishop, 1988; Beaty & Calisher, 1991; Beaty et al., 1997; Borucki et al., 2002; Gonzalez-Scarano et al., 1988; Griot et al., 1993.)

Experiments with reassortant bunyaviruses (see later) of the California serogroup indicated that viral factors associated with efficient infection of the mosquito midgut and subsequent transmission mapped to the M segment, encoding Gn, Gc and NSm proteins (Beaty et al., 1981). Furthermore, a La Crosse virus variant containing a neutralizing monoclonal-antibody-resistant mutant of Gc was shown to be deficient in its ability to infect midgut cells following oral infection, but was able to initiate infection when inoculated intrathoracically (Sundin et al., 1987). Others have suggested that Gn is the ligand for binding midgut cell receptors and that Gc is cleaved proteolytically by enzymes in the mosquito midgut (Ludwig et al., 1989, 1991). Taken together the data can be interpreted as cleavage of Gc is required to expose residues on Gn for midgut binding, but proteolytic cleavage is not necessary for disseminated infection after exit from the midgut into the haemocoel.

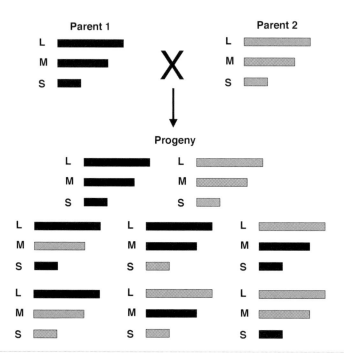

Fig. 2. Generation of reassortant bunyaviruses following a mixed infection of the same cell.

GENOME SEGMENT REASSORTMENT

Like other segmented genome viruses, bunyaviruses can reassort their genomes during the course of a mixed infection (Fig. 2). This has been studied extensively in cell culture (Pringle, 1996) and also in mosquitoes. High-frequency reassortment occurs in mosquitoes infected by two viruses taken together in a blood meal or by separate events that mimic 'interrupted feeding' (when a host's physical reaction to the mosquito disturbs it before it has finished feeding and it seeks another host). However, the time between infections must be less than 2–3 days or else there is resistance to super-infection by the second virus (Beaty *et al.*, 1985).

In slight contrast to the above result on asynchronous infection, female mosquitoes that were infected transovarially have also been shown to produce reassortant virus when given a blood meal containing a different virus (Borucki *et al.*, 1999). In the experimental situations described above, in each case the reassortant viruses generated in the mosquito could be transmitted to susceptible vertebrate hosts.

In contrast to high-frequency reassortment in mosquitoes, attempts to recover reassortant bunyaviruses from dually infected mice were unsuccessful (Beaty & Bishop, 1988). There are a number of reports to indicate that reassortment does occur in nature

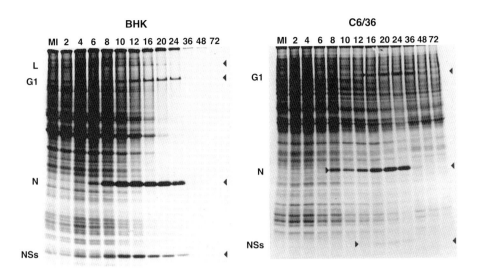

Fig. 3. Protein synthesis in Bunyamwera virus-infected BHK and *Aedes albopictus* C6/36 cells. Cells were either mock-infected (M) or infected with 5 p.f.u. virus per cell, and pulse-labelled for 1 h with [^{35}S]methionine at the indicated times after infection. Viral proteins are indicated at the right. From Scallan & Elliott (1992).

(Pringle, 1996). For instance, the L and S RNA segments of Shark River and Pahoyokee viruses generated almost identical RNA oligonucleotide fingerprints whereas their M segments were distinct (Ushijima *et al.*, 1981). More recently, analysis of the genome of Garissa virus, a bunyavirus associated with human haemorrhagic fever, indicates it to be a reassortant between Bunyamwera virus and an as yet unknown bunyavirus (Bowen *et al.*, 2001). Taken together these data suggest that reassortment in mosquitoes may be the major mechanism of rapid bunyavirus evolution in nature.

EFFECTS OF BUNYAVIRUS REPLICATION ON HOST CELL METABOLISM

One of the striking differences between the replication of orthobunyaviruses in mosquito and mammalian cells is the lack of host cell protein synthesis shut-off in the former (Fig. 3). Bunyaviruses replicate efficiently in cultured mosquito cells, releasing similar titres of infectious virus into the culture medium as from infected mammalian cells. Infected mosquito cells do not show cytopathic effects or evidence of apoptosis (Borucki *et al.*, 2002; Kohl *et al.*, 2003b), in contrast to mammalian cells, where the infection is cytolytic. Mosquito cells in culture become infected persistently and release varying, though low level, amounts of infectious virus into the medium over time (Carvalho *et al.*, 1986; Elliott & Wilkie, 1986; Kascsak & Lyons, 1978; Newton *et al.*, 1981; Rossier *et al.*, 1988; Scallan & Elliott, 1992). In addition, in mosquito cells persistently infected with Bunyamwera virus an over-representation of the S RNA was

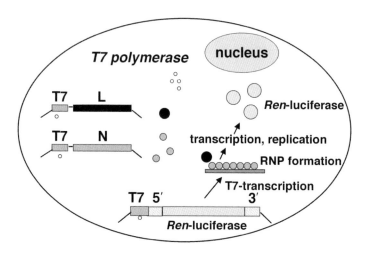

Fig. 4. The bunyavirus minireplicon system. Mosquito cells expressing T7 RNA polymerase constitutively are transfected with plasmids expressing bunyavirus proteins under T7 promoter control and the *Renilla* luciferase minireplicon. The minireplicon is transcribed intracellularly by T7 polymerase to yield a negative-sense transcript containing the complete untranslated regions of a bunyavirus genome segment. These transcripts are encapsidated by the co-expressed N protein to form a functional RNP, leading to transcription and replication of the minireplicon. This gives rise to *Renilla* mRNA and *Renilla* luciferase activity.

observed (Elliott & Wilkie, 1986) and defective RNAs derived from the L segment were detected in long-term cultures (Scallan & Elliott, 1992).

A BUNYAVIRUS MINIREPLICON SYSTEM IN MOSQUITO CELLS
In order to probe the molecular mechanisms that account for the difference in infection outcomes between mammalian and insect cells, recently we have established a 'minireplicon' system in *Aedes albopictus* C6/36 cells (Fig. 4; A. Kohl and others, unpublished). C6/36 cells expressing bacteriophage T7 RNA polymerase (C6-IBT7/3 cells) were selected following transfection with a plasmid encoding the polymerase gene and a blasticidin resistance gene. Bunyamwera virus proteins could be expressed in these cells by transfecting plasmids containing the viral coding sequences under T7 promoter control and baculovirus translational enhancers (Scheper *et al.*, 1997) to ensure efficient expression. The minireplicon plasmid contained a negative-sense *Renilla* luciferase gene flanked by Bunyamwera virus non-coding sequences under control of the T7 promoter. The 3'-end of the minireplicon transcript was generated by hepatitis delta virus ribozyme-mediated self cleavage.

C6-IBT7/3 cells were transfected with various combinations of L and N protein expressing plasmids and the minireplicon plasmid, were incubated at 28 °C for 24 h, and *Renilla* luciferase activity was measured. No *Renilla* luciferase activity was detected in cells transfected with only two of the three plasmids. However, in cells expressing both Bunyamwera virus N and L proteins together with the minireplicon, about 100-fold induction of *Renilla* luciferase activity was detected. This indicates that the minireplicon was transcribed (and by analogy with the mammalian cell system, probably replicated) by the expressed BUN proteins.

The non-coding regions at the 3′- and 5′-ends of bunyavirus genomes or antigenomes interact to form a panhandle in the ribonucleoprotein (RNP), and these non-coding sequences also contain promoter elements that regulate replication levels for each segment. Previous studies in mammalian cells, where minireplicon RNA levels were measured directly, indicated that promoter activity varied between the three segments in the order M>L>S (Barr *et al.*, 2003). Measurement of *Renilla* luciferase activity in cell extracts does not allow discrimination between replication and transcription, but shows that a biologically active RNP is reconstituted and transcribed in the transfected cell. Therefore, we used this as a means to compare promoter activity in mosquito and mammalian cells. T7-expressing cells were transfected with plasmids encoding the L and N proteins, and minireplicons were derived from the L, M or S segments. In addition a minireplicon containing a U-to-G mutation in position 16 of the viral

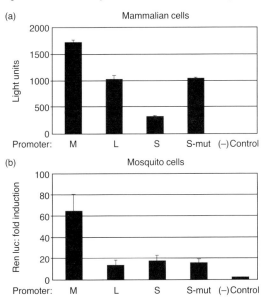

Fig. 5. Analysis of promoter strength in mammalian (BSR-T7/5) or mosquito (C6-IBT7/3) cells expressing T7 RNA polymerase. Cells were transfected with Bunyamwera virus L and N expression vectors and the appropriate minigenome plasmids with different segment promoters. (–) Control, negative control, showing background minigenome activity in the absence of functional L protein. Activities are indicated in light units or as fold induction of *Renilla* luciferase activity. *Ren* luc, *Renilla* luciferase.

genome end, which results in a strong increase in promoter activity in mammalian cells (Kohl *et al*., 2003a), was used. As shown in Fig. 5(a), *Renilla* luciferase activities from the different minireplicons in mammalian cells follow the replicative ability of the L, M and S promoters as deduced by RNA analysis (Barr *et al*., 2003), that is, *Renilla* luciferase activities were strongest in cells transfected with the M segment minireplicon, followed by that from the L and the S segment. As reported previously (Kohl *et al*., 2003a), the mutant S minireplicon gave about three times higher activity than the minireplicon containing the wild-type S sequence. When the same experiment was repeated in mosquito cells the strongest activity was mediated by the M-segment-derived minireplicon. The activities of the other minireplicons, including the mutant-S-derived minireplicon, were similar to each other, giving on average four times less *Renilla* luciferase than the corresponding M sequences (Fig. 5b). Hence using a minireplicon system there appears to be a difference in promoter activities between mammalian and mosquito cells.

Previously it was shown that the NSs protein, whether it is expressed with N from an S-segment-like mRNA or from separate plasmids, strongly inhibits minireplicon activity in mammalian cells (Weber *et al*., 2001). Therefore, we investigated whether the same pertained in mosquito cells. As shown in Fig. 6, expression of both S segment proteins N and NSs compared to expression of N protein alone led to a reduction in *Renilla* luciferase activity in mammalian cells but not in mosquito cells. When expressed from a separate plasmid NSs also had no major effect on overall minireplicon activity in mosquito cells.

EFFECTS OF BUNYAMWERA VIRUS NSs ON HOST PROTEIN SYNTHESIS IN MAMMALIAN AND MOSQUITO CELLS

Recently we described production by reverse genetics of a Bunyamwera virus mutant (BUNdelNSs 9a) that does not express NSs (Bridgen *et al*., 2001). As shown in Fig. 7(a), infection of mammalian cells with wild-type BUN (wtBUN) leads to shut-off of host protein synthesis whereas in cells infected with BUNdelNSs 9a shut-off was markedly reduced. In addition, the N protein was overexpressed in BUNdelNSs 9a-infected cells because of the U-to-G mutation in position 16 of the viral genome (Kohl *et al*., 2003a). In contrast, no shut-off of host protein synthesis was seen in infected C6/36 mosquito cells, and N protein levels were similar between wtBUN- and BUNdelNSs 9a-infected cells (Fig. 7a). The metabolic labelling data suggest that NSs plays a role in shut-off of host protein synthesis. To test this hypothesis, we compared transient expression of a reporter protein in T7-expressing mammalian or mosquito cells also expressing NSs. Cells were transfected with a plasmid that expresses *Renilla* luciferase from either a T7 promoter or a CMV promoter. [The CMV promoter has been shown previously to be functional in other dipteran cell lines (Saraiva *et al*., 2000).] The cells were co-

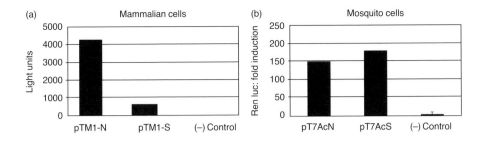

Fig. 6. Effect of NSs on minireplicon activity in mosquito and mammalian cells. Cells were transfected with the L expression and minigenome plasmids, along with plasmids expressing N and NSs (pTM1-BUNS or pT7AcS) or N alone (pTM1-BUNN or pT7AcN) as indicated. (–) Control, negative control without functional L protein. The amounts of DNA transfected were kept constant by adding the corresponding empty plasmid if necessary. Activities are indicated in light units (a; mammalian cells) or as fold induction of *Renilla* luciferase activity (b; mosquito cells).

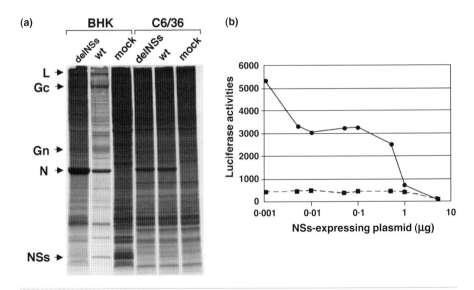

Fig. 7. Effects of NSs expression on protein synthesis. (a) Metabolic labelling of Bunyamwera virus-infected mammalian (BHK-21) and mosquito (C6/36) cells. Cells were infected at an m.o.i. of 5 by wtBUN or BUNdelNSs 9a and labelled with [^{35}S]methionine for 2 h at 18 h post-infection. Cell extracts were analysed by SDS-PAGE (18 % gel). (b) Effects of NSs on transient reporter gene expression in mammalian (BSR-T7, –●–) and mosquito (C6-IBT7/3, ··■··) cells expressing T7 RNA polymerase. Cells were co-transfected with a luciferase reporter plasmid (phRL-CMV; Promega) and various amounts of NSs-expressing plasmid. *Renilla* luciferase reporter gene activities were determined at 16 h post-transfection.

transfected with various amounts of NSs-expressing plasmid. As shown in Fig. 7(b), increasing amounts of NSs expression led to a strong reduction in *Renilla* luciferase activity in mammalian cells, but this was not the case in mosquito cells.

CONCLUSIONS

The mosquito plays a major part in the life cycle of orthobunyaviruses, but there are fundamental differences between virus replication in mosquito and mammalian cells, particularly with the effect of NSs. Perhaps NSs, which also counteracts the interferon response and early induction of apoptosis in mammalian cells (Kohl *et al.*, 2003b; Weber *et al.*, 2002), is more important for replication in mammals, whose immune system seems different in many respects to that of arthropods. Currently we are analysing activation of arthropod immune response pathways after infection by wtBUN or BUNdelNSs 9a to look for a possible role of NSs. The development of these new tools to study orthobunyavirus molecular biology should help us to better understand the differences between the two host systems.

ACKNOWLEDGEMENTS

Work in R. M. E.'s laboratory is supported by Wellcome Trust grants 058356 and 065121.

REFERENCES

Barr, J. N., Elliott, R. M., Dunn, E. F. & Wertz, G. W. (2003). Segment-specific terminal sequences of Bunyamwera bunyavirus regulate genome replication. *Virology* **311**, 326–338.

Beaty, B. J. & Bishop, D. H. L. (1988). Bunyavirus-vector interactions. *Virus Res* **10**, 289–302.

Beaty, B. J. & Calisher, C. H. (1991). Bunyaviridae – natural history. *Curr Top Microbiol Immunol* **169**, 27–78.

Beaty, B. J., Holterman, M., Tabachnick, W., Shope, R. E., Rozhon, E. J. & Bishop, D. H. (1981). Molecular basis of Bunyavirus transmission by mosquitoes: role of the middle-sized RNA segment. *Science* **211**, 1433–1435.

Beaty, B. J., Sundin, D. R., Chandler, L. J. & Bishop, D. H. L. (1985). Evolution of Bunyaviruses by genome reassortment in dually infected mosquitoes (*Aedes triseriatus*). *Science* **230**, 548–550.

Beaty, B. J., Borucki, M. K., Farfan, J. A. & White, D. (1997). Arbovirus-vector interactions: determinants of arbovirus evolution. In *Factors in the Emergence of Arbovirus Diseases*, pp. 23–35. Edited by J. F. Saluzzo & B. Boder. Paris: Elsevier.

Borucki, M. K., Chandler, L. J., Parker, B. M., Blair, C. D. & Beaty, B. J. (1999). Bunyavirus superinfection and segment reassortment in transovarially infected mosquitoes. *J Gen Virol* **80**, 3173–3179.

Borucki, M. K., Kempf, B. J., Blitvich, B. J., Blair, C. D. & Beaty, B. J. (2002). La Crosse virus: replication in vertebrate and invertebrate hosts. *Microbes Infect* **4**, 341–350.

Bowen, M. D., Trappier, S. G., Sanchez, A. J. & 7 other authors (2001). A reassortant bunyavirus isolated from acute hemorrhagic fever cases in Kenya and Somalia. *Virology* **291**, 185–190.

Bridgen, A., Weber, F., Fazakerley, J. K. & Elliott, R. M. (2001). Bunyamwera bunyavirus nonstructural protein NSs is a nonessential gene product that contributes to viral pathogenesis. *Proc Natl Acad Sci U S A* **98**, 664–669.

Calisher, C. H. (1996). History, classification and taxonomy of viruses in the family Bunyaviridae. In *The Bunyaviridae*, pp. 1–17. Edited by R. M. Elliott. New York: Plenum.

Carvalho, M. G. C., Frugulhetti, I. C. & Rebello, M. A. (1986). Marituba (Bunyaviridae) virus replication in cultured *Aedes albopictus* cells and in L-A9 cells. *Arch Virol* **90**, 325–335.

Elliott, R. M. (editor) (1996). *The Bunyaviridae*. New York: Plenum.

Elliott, R. M. & Wilkie, M. L. (1986). Persistent infection of *Aedes albopictus* C6/36 cells by Bunyamwera virus. *Virology* **150**, 21–32.

Elliott, R. M., Bouloy, M., Calisher, C. H., Goldbach, R., Moyer, J. T., Nichol, S. T., Pettersson, R., Plyusnin, A. & Schmaljohn, C. S. (2000). Bunyaviridae. In *Virus Taxonomy. Seventh Report of the International Committee on Taxonomy of Viruses*, pp. 599–621. Edited by M. H. V. van Regenmortel, C. M. Fauquet, D. H. L. Bishop, E. B. Carstens, M. K. Estes, S. M. Lemon, J. Maniloff, M. A. Mayo, D. J. McGeoch, C. R. Pringle & R. B. Wickner. San Diego: Academic Press.

Gonzalez-Scarano, F., Beaty, B., Sundin, D., Janssen, R., Endres, M. & Nathanson, N. (1988). Genetic determinants of the virulence and infectivity of La Crosse virus. *Microb Pathog* **4**, 1–7.

Griot, C., Gonzalez-Scarano, F. & Nathanson, N. (1993). Molecular determinants of the virulence and infectivity of California serogroup Bunyaviruses. *Annu Rev Microbiol* **47**, 117–138.

Higgs, S. (2004). How do mosquito vectors live with their viruses? In *Microbe–vector Interactions in Vector-borne Diseases* (Society for General Microbiology Symposium no. 63), pp. 103–137. Edited by S. H. Gillespie, G. L. Smith & A. Osbourn. Cambridge: Cambridge University Press.

Kascsak, R. J. & Lyons, M. J. (1978). Bunyamwera virus. II. The generation and nature of defective interfering particles. *Virology* **89**, 539–546.

Kohl, A., Bridgen, A., Dunn, E., Barr, J. N. & Elliott, R. M. (2003a). Effects of a point mutation in the 3′ end of the S genome segment of naturally occurring and engineered Bunyamwera viruses. *J Gen Virol* **84**, 789–793.

Kohl, A., Clayton, R. F., Weber, F., Bridgen, A., Randall, R. E. & Elliott, R. M. (2003b). Bunyamwera virus nonstructural protein NSs counteracts interferon regulatory factor 3-mediated induction of early cell death. *J Virol* **77**, 7999–8008.

Ludwig, G. V., Christensen, B. M., Yuill, T. M. & Schultz, K. T. (1989). Enzyme processing of La Crosse virus glycoprotein G1: a Bunyavirus-vector infection model. *Virology* **171**, 108–113.

Ludwig, G. V., Israel, B. A., Christensen, B. M., Yuill, T. M. & Schultz, K. T. (1991). Role of La Crosse virus glycoproteins in attachment of virus to host cells. *Virology* **181**, 564–571.

Newton, S. E., Short, N. J. & Dalgarno, L. (1981). Bunyamwera virus replication in cultured *Aedes albopictus* (mosquito) cells: establishment of a persistent infection. *J Virol* **38**, 1015–1024.

Pringle, C. R. (1996). Genetics and genome segment reassortment. In *The Bunyaviridae*, pp. 189–226. Edited by R. M. Elliott. New York: Plenum.

Rossier, C., Raju, R. & Kolakofsky, D. (1988). La Crosse virus gene expression in mammalian and mosquito cell lines. *Virology* **165**, 539–548.

Saraiva, E., Fampa, P., Cedeno, V., Bergoin, M., Mialhe, E. & Miller, L. H. (2000). Expression of heterologous promoters in *Lutzomyia longipalpis* and *Phlebotomus papatasi* (Diptera: Psychodidae) cell lines. *J Med Entomol* **37**, 802–806.

Scallan, M. F. & Elliott, R. M. (1992). Defective RNAs in mosquito cells persistently infected with Bunyamwera virus. *J Gen Virol* **73**, 53–60.

Scheper, G. C., Vries, R. G., Broere, M., Usmany, M., Voorma, H. O., Vlak, J. M. & Thomas, A. A. (1997). Translational properties of the untranslated regions of the p10 messenger RNA of *Autographa californica* multicapsid nucleopolyhedrovirus. *J Gen Virol* **78**, 687–696.

Sundin, D. R., Beaty, B. J., Nathanson, N. & Gonzalez-Scarano, F. (1987). A G1 glycoprotein-epitope of La Crosse virus: a determinant of infection of *Aedes triseriatus*. *Science* **235**, 591–593.

Ushijima, H., Clerx-van Haaster, C. M. & Bishop, D. H. L. (1981). Analyses of Patois group bunyaviruses: evidence for naturally occurring bunyaviruses and existence of immune precipitable and non-precipitable nonvirion proteins induced in bunyavirus-infected cells. *Virology* **110**, 318–332.

Weaver, S. C., Coffey, L. L., Nussenzveig, R., Ortiz, D. & Smith, D. (2004). Vector competence. In *Microbe–vector Interactions in Vector-borne Diseases* (Society for General Microbiology Symposium no. 63), pp. 139–180. Edited by S. H. Gillespie, G. L. Smith & A. Osbourn. Cambridge: Cambridge University Press.

Weber, F., Dunn, E. F., Bridgen, A. & Elliott, R. M. (2001). The Bunyamwera virus nonstructural protein NSs inhibits viral RNA synthesis in a minireplicon system. *Virology* **281**, 67–74.

Weber, F., Bridgen, A., Fazakerley, J. K., Streitenfeld, H., Kessler, N., Randall, R. E. & Elliott, R. M. (2002). Bunyamwera bunyavirus nonstructural protein NSs counteracts the induction of alpha/beta interferon. *J Virol* **76**, 7949–7955.

How do mosquito vectors live with their viruses?

Stephen Higgs

Department of Pathology, University of Texas Medical Branch, Galveston, TX 77555-0609, USA

INTRODUCTION

The word vector is derived from the Latin and means 'bearer' or 'one who does something', but for the purposes of this chapter, we can use the term as defined by Gordh & Headrick (2001) as an arthropod that carries disease-producing organisms to a vertebrate host. A wide variety of arthropods are involved as vectors in pathogen transmission cycles but the focus of this chapter is the mosquito. To date, there are at least 537 named arthropod-borne viruses (arboviruses) (Karabatsos, 1985), encompassing species from many genera, examples of which are given in Table 1. With one exception (African swine fever virus, transmitted by ticks) all of the viruses regarded as true arboviruses have an RNA genome. The vast majority of arboviruses (e.g. *Togaviridae*, *Flaviviridae*, *Bunyaviridae*) are enveloped, i.e. the nucleocapsid is surrounded by a lipid-containing envelope. A few (e.g. *Reoviridae*) are non-enveloped. Most of these (e.g. bluetongue virus and Colorado tick fever virus) are transmitted by midges and ticks. Fewer than 20 have been isolated from mosquitoes and only lebombo and Orungo have been associated with human infections. As discussed by Mitchell (1983), because arboviruses are typically not pathogenic to their insect hosts, it has been hypothesized that they may have evolved in insects and later became infectious to vertebrates. Tick-borne arboviruses (see review by Nuttall & Labuda, 2003) and the true insect viruses (e.g. nuclear polyhedrosis viruses) that infect and replicate in mosquitoes, but are not transmitted to vertebrate hosts (Miller & Ball, 1998), are not considered in this chapter. Given that almost all mosquito-borne viruses are enveloped and have an RNA genome, unless specified otherwise, this review therefore focuses on this predominant group of arboviruses.

SGM symposium 63: Microbe–vector interactions in vector-borne diseases.
Editors S. H. Gillespie, G. L. Smith & A. Osbourn. Cambridge University Press. ISBN 0 521 84312 X ©SGM 2004

Table 1. Examples of viruses in different families that infect arthropods

Family	Genus	Examples
Arenaviridae	*Arenavirus*	Possibly pichinde and tacharibe
Bunyaviridae	*Bunyavirus*	Akabane, bunyamwera, California encephalitis
	Nairovirus	Crimean-Congo hemorrhagic fever, Nairobi sheep disease
	Phlebovirus	Sandfly fever, Rift Valley fever
Togaviridae	*Alphavirus*	Chikungunya, Semliki Forest, Ross River, Venezuelan equine encephalitis
Flaviviridae	*Flavivirus*	Dengue, yellow fever, tick-borne encephalitis
Nodoviridae	*Nodovirus*	Possibly nodamura
Orthomyxoviridae		Thogoto
Reoviridae	*Coltivirus*	Colorado tick fever, eyach
	Orbivirus	African horse sickness, bluetongue, Kemerovo
Rhabdoviridae	*Rhabdovirus*	Hart Park, Connecticut
	Lyssavirus	Possibly bovine ephemeral fever and rochambeau
	Vesiculovirus	Vesicular stomatitis virus

This chapter is intended for the general microbiologist rather than expert vector biologists or arbovirologists. It aims to present a broad overview of what we currently know and what we need to learn in respect to how mosquitoes can sustain relatively long infections with agents that are often fatal to their vertebrate host. The contents include information from published reviews integrated with more recent information and new interpretations that have added to our understanding of the vector–virus relationship.

Arboviruses have been studied for over 100 years and other arthropod-borne pathogens have been the focus of scientific attention for considerably longer. Following the leads provided by Patrick Manson working on filarial worms and Ronald Ross working on malaria, the involvement of *Aedes aegypti* in the spread of yellow fever was suggested by Carlos Finlay in 1881. By 1901, the team of Walter Reed, James Caroll, Jesse Lazear and Aristides Agramonte, working in Cuba, demonstrated experimentally the infection and transmission of the agent responsible for causing yellow fever by mosquitoes, although the nature of the infective agent was not certain. Despite this uncertainty, the research had an almost immediate impact on public health. With the new knowledge in hand, General William C. Gorgas implemented a mosquito eradication programme in Havana and by September of 1901 yellow fever was eliminated from the city. This success epitomizes the ultimate goal of all researchers who work on vector–pathogen interactions; namely, to reduce the incidence of vector-borne diseases. As described at the end of this chapter, new strategies to accomplish the goal have been developed; however, their success depends upon a thorough understanding of the basic biology of the vector and of vector–pathogen–host interactions.

Despite over 100 years of research we still do not understand fully the complex relationship that exists between arthropods and arboviruses that they transmit. The specifics of the relationship between a particular mosquito species and an arbovirus are influenced by three factors: vector genetics, virus genetics and environmental conditions. Unfortunately, the complexity and interdependence of these factors confounds our knowledge of the relationship and continues to hinder our understanding. For example, the basis of host specificity and why certain mosquitoes are susceptible to infection with a particular virus but are refractory to others is still not understood. There have been several reviews on the interactions between arboviruses and their vectors (Hardy, 1988; Hardy *et al.*, 1983; Leake, 1992; Kramer & Ebel, 2003), and on vector competence (Beerntsen *et al.*, 2000; Gubler *et al.*, 1982; Mitchell, 1983). A good definition of vector competence is 'The combined effect of all of the physiological and ecological factors of vector, host, pathogen, and environment that determine the vector status of a given arthropod population' (McKelvey *et al.*, 1981). This definition emphasizes the complexity of the topic. Despite the recent advances in molecular biology, genetics and imaging technology, the information in these classic reviews remains as relevant today as it was when published about 20 years ago. The vector–virus relationship is a complex one that can be influenced by numerous factors. Since a mosquito may be infected for most of its adult life, a virus may be exposed to many biological processes. Not only does the virus have to survive, replicate and disseminate to different mosquito tissues, but this must be achieved as the female mosquito undergoes complex behavioural, hormonal, biochemical and physiological changes in order to produce and lay eggs. Subsequently, the mosquito must locate and feed upon a vertebrate host in order for the virus to be transmitted horizontally. Hardy *et al.* (1983) provided an excellent flow chart describing the pathway of arboviral infection within a mosquito including hypothesized and conceptual infection and escape barriers. Since the generalities of the process of infection follow a similar pattern in most mosquito–virus relationships, for the purposes of this chapter, I have adopted a chronological approach to discuss how the vector lives with its virus. The basic steps in this process are listed below, and a diagrammatic view of a lateral section through the mosquito is presented in Fig. 1 to familiarize the reader with the basic anatomy that is referred to in the text. Fig. 2 shows representative photomicrographs of different mosquito tissues infected with arboviruses throughout the typical infection process.

STAGES OF ARBOVIRAL INFECTIONS IN MOSQUITOES

For an enveloped arbovirus to be transmitted perorally by a mosquito following an infectious blood meal, a number of criteria must be satisfied. Thus the virus–vector relationship can be regarded as consisting of a number of stages. (1) The mosquito must feed on a viraemic host. (2) An infectious dose of virus (greater than the minimum threshold) must enter the mosquito midgut lumen. (3) Virions must bind to the membrane of midgut epithelial cells. (4) Following endocytosis or fusion of the viral envelope and cellular membranes, the virus genome must enter the cell cytoplasm and replicate to produce infectious virions. (5) Virions must disseminate from the midgut epithelial cells and enter the haemocoel. (6) Virions must infect salivary glands. (7) Virions must be secreted in saliva when the mosquito feeds upon a host.

The successful completion of this transmission process is influenced by several factors, but the overriding consideration is that the virus and species of mosquito must be 'compatible'. From an entomological viewpoint, one would say that the mosquito must be a 'competent' vector for the virus. This 'competence' is determined by factors that allow progress from each of the stages to the next. Consequently there are a number of 'barriers' to this progression that can affect the success/failure and rate of the dissemination and transmission process. Whilst some of these are undoubtedly mosquito-related, others may be virus-dependent. Stages one and two are influenced by host factors and by mosquito behaviour (host preference, time and frequency of feeding, etc.) and are not discussed in detail here.

In brief, a female mosquito typically becomes exposed naturally to a virus as a result of feeding on the blood of a viraemic host. Exceptions are when a mosquito is infected transovarially or venereally, and these are discussed later. Virus enters the midgut with the blood meal and under appropriate conditions disseminates to the haemocoel and subsequently to other organs. The timing of dissemination can be quite variable and sporadic (Romoser et al., 1992), but once virus has entered the haemocoel, numerous tissues can be infected quickly. The time between feeding upon the viraemic host and being infectious (i.e. secreting virus in the saliva and being capable of transmitting it to the vertebrate host) is known as the extrinsic incubation period. This is influenced strongly by temperature and is shorter at higher temperatures. Romoser et al. (1992) reported that the most common sequence of dissemination from the midgut observed for *Culex pipiens* infected with Rift Valley fever virus (RVFV) was: intussuscepted foregut, fat body, salivary glands and thoracic ganglia, epidermis and ommatidia of the compound eyes. Similar sequences have been reported for bunyaviruses (Beaty & Thompson, 1978) and flaviviruses (Kuberski, 1979; Leake & Johnson, 1987; Doi, 1970), although dissemination via the intussuscepted midgut may be specific for RVFV. Alphaviruses are somewhat unusual in infecting muscle tissue (Weaver et al., 1990). Bowers et al. (1995) working with Sindbis virus (SINV) categorized different types of infection in different tissues. 'Persistent infections' were seen in the fat body-haemolymph, hindgut and the tracheole-associated cells (tracheoblasts) around the gut; 'transient infections' occurred in the salivary glands, anterior and posterior midgut, and thoracic muscle; 'cleared infections' characterized those in the head ganglia; whilst ovarioles and Malpighian tubules were designated 'refractory' to infection.

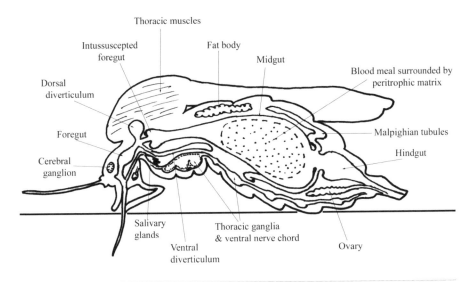

Fig. 1. Diagrammatic longitudinal cross-section through a mosquito, showing the major tissues and organs referred to in the text.

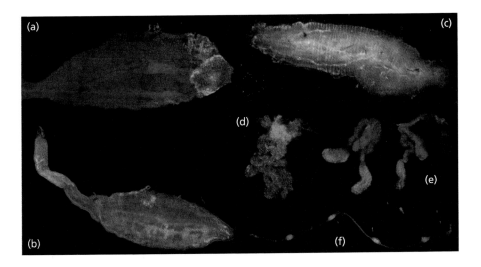

Fig. 2. Mosquito organs infected with viruses as observed by antigen detection using an indirect immunofluorescence assay based on Gould *et al.* (1985). Tissues in the composite are not to scale. Antigen is indicated by green fluorescence, with a contrasting red counterstain. (a) Early infection showing focal Sindbis virus (SINV) infection in epithelial cells of the posterior midgut. (b) Later infection with widely distributed SINV antigen throughout the midgut. (c) Disseminated infection with West Nile virus antigen in the circular and longitudinal muscles of the midgut. (d) Disseminated infection with SINV antigen in the fat body. (e) A pair of salivary glands with SINV antigen in the distal lateral lobes. (f) SINV antigen in the ventral nerve chord and abdominal ganglia.

Infection of the digestive tract

Since a mosquito vector is first exposed to a virus during feeding on a viraemic host the first tissues that are exposed are therefore in the digestive tract and it is the midgut (Lehane & Billingsley, 1996) that constitutes the primary dose-dependent barrier to infection. The blood meal collects in the midgut of the adult mosquito, where it is surrounded by a peritrophic matrix (PM). It has been suggested that this might represent a barrier to infection for some arthropod-borne pathogens (Chamberlain & Sudia, 1961; Stohler, 1961), although other researchers drew a different conclusion (Hardy *et al.*, 1983). To produce a patent infection arboviruses must contact the gut epithelial cells and therefore the potential influence of the PM on the initial establishment of a virus in the vector must be considered. A general review of the PM was published by Peters (1992), with a focus on that of haematophagous insects published by Shao *et al.* (2001). There are two types of peritrophic matrix, PM1 and PM2, which differ with respect to their biochemical properties. Adult mosquitoes, black flies and sandflies produce PM1, whilst the tsetse fly and stable fly produce PM2 (Lehane, 1976). The PM1 is secreted by gut epithelial cells in response to the distension of the midgut epithelium caused by the blood meal, although the exact mechanism involved is uncertain. The PM2 is secreted continuously as a tube by specialized cardia cells at the foregut–midgut junction. In those species producing this type of matrix, the blood meal is therefore effectively always contained within the PM2. For PM1, the time of production and maturation varies according to species. In *Ae. aegypti*, the PM1 has first been observed at 5 h post-feeding (Freyvogel & Stäubli, 1965), but in *Ochlerotatus triseriatus* has been detected within 20–50 min post-feeding (Richardson & Romoser, 1972). The PM is composed of numerous proteins and carbohydrates, including chitin. Maturation to form a membrane takes from minutes to hours, depending on species. Once mature, the PM1 separates the bolus of blood from direct contact with the epithelial cells, but permits enzymic digestion via pores of 20–30 µm diameter. For several reasons, Hardy *et al.* (1983) considered that although the PM might raise the infectious threshold (see later) it does not pose a barrier to viral infection of the mesenteron. Blood coagulation, which is well advanced within 4 h, would, in their opinion, totally immobilize virions within the meal. Based on their observations, they believed that virion attachment occurs soon after the blood meal enters the midgut, before clotting and prior to PM formation.

Given that the PM poses little if any restriction on infection, it is clear that the first significant physical obstacle to infection must be the entry of the virus into the cells of the midgut epithelium. One possible factor that might influence virus entry into the cells of the midgut is the chemical environment within the midgut lumen, which is concerned primarily with digestion of the blood meal and the adsorption of nutrients. Consequently, a number of digestive enzymes are secreted into the lumen, the predominant one being trypsin, which is produced in two phases. Genetic analysis of *Ae. aegypti* and *Anopheles gambiae* indicates that there are at least seven genes encoding trypsins. An early trypsin gene is active briefly immediately following a blood meal and encodes a 32–36 kDa enzyme. It is thought that limited digestion of the meal by this enzyme may stimulate the more abundant secretion of the late trypsin, which digests the bulk of the blood. Within the blood meal, virions will be exposed to the digestive enzymes and this may alter the virion outer surface and influence infectivity. For example, the virion envelope of La Crosse virus (LACV) is composed of two glycoproteins, G1 and G2. The G1 protein is important in binding to vertebrate cells; however, the G2 has this function for mosquito infection. Infection of the mosquito midgut seems to depend on the enzymic cleavage of the G1 protein to expose the G2 protein that can then bind to the epithelial cells (Ludwig *et al.*, 1989, 1991). This observation contradicted those in a previous study (Sundin *et al.*, 1987) using mutant viruses that implicated G1 in binding to midguts. The discrepancy is explicable because the mutation may have been influencing infectivity via changes in the enzyme cleavage site. These observations are an excellent demonstration of how a particular arbovirus has evolved to coexist within the very different host types.

Other environmental factors that might influence infection are the pH of the midgut and the relative charge of the cell surface. Limited studies on the mosquito midgut have demonstrated that prior to feeding, the pH ranges from 7·4 to 7·9 (Clements, 1992). Although manipulating the pH by the addition of buffers with the meal (Hardy *et al.*, 1983) can alter the efficiency of viral infection, there are no data to suggest that this is relevant in the natural process of infection.

Relatively little is known about the gut flora of mosquitoes (Demaio *et al.*, 1996) and there are no data to suggest that this influences the establishment and course of viral infection in the vector. Coinfection with *Plasmodium* spp. seems to have no effect on viral infection; however, the presence of filarial worms may accelerate dissemination from the midgut due to the worms' penetration of the gut during their development (Turrell, 1988).

The midgut is composed of several cell types, but the most abundant is the columnar epithelial cell that rests on the multi-layered porous basement membrane and has a brush border on the luminal surface. In mosquitoes, digestion, absorption and storage are all achieved by a single cell type (see Billingsley, 1990). Recently scanning electron microscopy has revealed a microvillar network that is dispersed over and between the microvilli of the midgut cells of *Ae. aegypti* (Zieler *et al.*, 2000; Shahabbudin, 2002). This appears as a dense mat but its possible influence on midgut infection with pathogens is uncertain. There is some evidence to suggest that certain parasites may infect particular midgut cell types preferentially (Shahabbudin, 2002), but there are no equivalent data for viruses. Based on antigen detection and electron microscopy (EM) it seems clear that the columnar epithelial cells are the first cell type to be infected and, although only a small proportion may be infected initially, there is nothing to suggest that the process is not random. The observation that in some cases infection may predominate in the posterior midgut (Doi, 1970; Doi *et al.*, 1967; Kuberski, 1979) and in others be more widespread may simply reflect the abundance and distribution of virions in the meal. The anterior midgut may be infected less frequently because it is believed to be involved with sugar absorption into the haemocoel (Richardson & Romoser, 1972). The virus abundance can certainly influence the success of an infection and a dose-response phenomenon termed an infection threshold is recognized in which the minimum threshold of infection is vector- and virus species-specific. In cases with a low threshold, i.e. a relatively low viral dose is required to infect a particular species of mosquito, the species is regarded as being highly susceptible to infection with that virus. However, as described later, vector competence is influenced by numerous factors.

The mechanism by which a virus enters a host cell may vary according to both cell and virus type, but generally it is thought that the initial event preceding infection must be virus attachment followed by the fusion of the viral envelope and cellular membranes. The importance of the outer-surface structural proteins in initiating infection has been demonstrated by studies with various viral mutants. The ability of Venezuelan equine encephalitis virus (VEEV) to infect vectors is determined by the E2 protein (Brault *et al.*, 2002) and a single amino acid change restricts its ability to infect and disseminate from the midgut (Woodward *et al.*, 1991). Deletions in the E2 of SINV have a similar effect (Myles *et al.*, 2003). It is supposed that arbovirus infection is a receptor-mediated event, but as yet no specific receptor has been identified. Given that many arboviruses can infect cultured mosquito cells *in vitro*, but infections *in vivo* are often very species specific, it is thought that permissiveness of epithelial cells to infection may be determined by the type of receptors on the cell surface. The so-called midgut infection barrier (MIB) that acts to prevent infection of the epithelial cells may either reflect the lack of appropriate receptors, or may result from a lack of cytoplasmic factors that are necessary for virus replication.

Houk *et al.* (1990) performed binding studies using brush border fragments (BBFs) isolated from mosquito mesenteronal epithelial cells. These workers reported optimal binding at pH 7·2 at 20 °C, based on binding kinetic assays of radiolabelled Western equine encephalomyelitis virus (WEEV) to fragments derived from susceptible and refractory mosquitoes. Although WEEV bound to both susceptible and refractory BBFs, there was significantly greater binding to those derived from susceptible mosquitoes. Binding to susceptible BBFs was specific, but to refractory BBFs was non-specific. Quantitative analysis suggested that there were $1.8–3.5 \times 10^6$ binding sites per cell, but the nature of the receptor was not determined. The role of heparin sulphate as an attachment receptor for SINV, dengue viruses (DENV) and yellow fever virus (YFV) has been demonstrated for vertebrate cells (Germi *et al.*, 2002; Klimstra *et al.*, 1998) but parallel studies have not been performed with mosquito cells. Ludwig *et al.* (1996) described a 32 kDa polypeptide on the surface of C6/36 *Ae. albopictus* cells that bound to the E2 protein of VEEV. They characterized this as a laminin-binding protein, laminin being a multifunctional, extracellular matrix protein present in the basement membranes of epithelial and non-epithelial cells (Sonnenberg, 1992). Also working with C6/36 cells, Salas-Benito & del Angel (1997) identified a 40–45 kDa and a 70–75 kDa protein that bound DEN serotype 4 virus (DEN-4V). A 45 kDa glycoprotein has also been described as a putative receptor for DENV by Mendoza *et al.* (2002). The distribution of this protein on mosquito tissues corresponds with those of known susceptibility to dengue infection. To date, none of the putative receptors described earlier have been implicated unequivocally in the *in vivo* infection process.

An interesting phenomenon that might reflect abundance and specificity of receptors is known as superinfection resistance. In nature, most mosquitoes are unlikely to become infected with more than one arbovirus. However, there are experimental data demonstrating a barrier to infection that is both time- and virus-specific. Coinfection of an individual mosquito with two heterologous viruses can be achieved (Chamberlain & Sudia, 1957; Lam & Marshall, 1968a); however, coinfection with homologous or closely related viruses may be restricted (Sabin, 1952) and only occurs if infectious meals are presented within a relatively short time frame. The phenomenon has been best studied with bunyaviruses (Bishop & Beaty, 1988); however, the mechanism of resistance remains a mystery and it is not clear whether it is virus- or host-mediated. An important consequence of this form of interference is that the opportunities for virus evolution by genome reassortment between viruses within the vector is restricted, although reassortment between bunyaviruses with segmented genomes has been observed (Bishop & Beaty, 1988).

Assuming that the virus binds to the correct receptor, internalization may occur in one of two ways. For some viruses, the viral envelope fuses with the cellular membrane and the nucleocapsid with the viral genome is directly released into the cytoplasm (e.g. for alphaviruses such as WEEV; Hardy, 1988; Hardy *et al.*, 1983). Alternatively, endocytosis results in the internalization of the enveloped virions in an intracytoplasmic vesicle [e.g. flaviviruses such as St. Louis encephalitis virus (SLEV); Whitfield *et al.*, 1973]. Extrapolating from *in vitro* studies, within the acidic environment of the vesicle, uncoating of the virus allows the RNA genome to be released into the cytoplasm.

Virus replication in mosquito cells

During the early stages of infection, progressively more epithelial cells in the midgut become infected. This spread from the primary target cells may either occur directly via the basolateral membranes of neighbouring cells, or may occur indirectly by release into the lumen and infection of distally located cells. Replication of representative arboviruses has been studied in a variety of cell types, including cell culture lines derived from mosquitoes. Based on EM studies, replication and virion production in infected mosquitoes (Murphy *et al.*, 1975; Sriurairatna & Bhamarapravati, 1977) would seem to be the same as observed *in vitro* (Deubel *et al.*, 1981; Gliedman *et al.*, 1975; Grief *et al.*, 1997; Hase *et al.*, 1987; Ng, 1987). Since the general details seem to be similar regardless of the tissues in which observations have been made, the following is a general description based on EM observations.

The rough endoplasmic reticulum (RER) is critical for virion assembly. Sriurairatna & Bhamarapravati (1977) observed that there was generally more replication in cells with abundant RER, commenting that the RER is relatively sparse in the mosquito foregut where virus replication is rarely seen. Deubel *et al.* (1981) observed ER covered with ribosomes and subsequent accumulation of YFV virions in the cisternae. The RER then developed vesicles containing mature virions. Sriurairatna & Bhamarapravati (1977) observed orderly arranged DEN-2V particles in the cisternae of the RER with an increasing involvement over time but with no obvious deleterious effects to the infected cells. They also saw dense threads that they considered as virally directed structures but their role in assembly was not clear. Well-organized structures called convoluted membranes/paracrystalline arrays and vesicle packets are characteristic of the replication complex of alphaviruses and flaviviruses (Friedman *et al.*, 1972; Mackenzie *et al.*, 1999; Uchil & Satchidanandam, 2003). Alphaviruses, for example SINV, characteristically escape from the cell by direct budding (Brown *et al.*, 1972), in contrast to, for example, flaviviruses, whose virions are released following collection in vesicles that coalesce with the cell's outer membrane (Hase *et al.*, 1987). In support of this pattern, Deubel *et al.* (1981) did not observe budding of YFV through the plasma membrane. Murphy *et al.* (1975) observed an SLEV particle at the outer edge of a midgut epithelial cell directly beneath the basement membrane and noted that for both SLEV and Eastern equine encephalitis virus (EEEV) there was a directional effect in which more viruses left the cells on the haemocoel side than left the cells on the luminal border side. Interestingly, Murphy *et al.* (1975) observed that the proportion of infected midgut cells remained relatively constant (20–30 %) and they hypothesized that this indicated a 'limited' or 'homeostatic' infection. This is supported by the observation that the maximum virus titres produced in mosquitoes are not strictly dose-dependent.

The observations described earlier introduce an important concept with respect to the relationship between the mosquito vector and arboviruses, namely that the infection does not harm the mosquito. Generally speaking, this seems to be true, both *in vitro* and *in vivo*; however, as discussed at the end of this chapter, it is not an absolute rule. It is now recognized that there can be a number of effects caused by viral infection. For example, in mosquitoes infected with EEEV, Weaver *et al.* (1988) observed the sloughing of densely staining, heavily infected cells and degeneration within the epithelium, and sometimes the loss of brush border and integrity of the basal lamina. In a later study with the closely related WEEV, lesions were observed in midgut cells, including sloughing of infected cells into the lumen, and also necrosis of cells *in situ* (Weaver *et al.*, 1992).

During the course of a typical infection, the titre of the virus (the quantity of infectious particles) in the mosquito changes. In the early stage of infection, a so-called eclipse phase occurs, during which the infectious virus titre declines to less than the infectious dose that was imbibed. This is the time that is prior to the virus actively replicating in the gut to produce infectious virions.

Although the midgut is the primary location of infection and replication, as described later, virus has been observed in other regions of the digestive tract. Both the foregut and hindgut are lined with a non-cellular chitinous intima that probably makes infection from the lumen relatively inefficient. However, LACV infects the foregut soon after the midgut (Beaty & Thompson, 1978) with large amounts of antigen accumulating early after midgut infection. DEN-2V has been observed in the proventriculus (Kuberski, 1979). In mosquitoes inoculated intrathoracically with virus, the foregut is infected more frequently than the hindgut. Scott *et al.* (1984) considered that the presence of antigen in the hindgut was due to infection with virus in the haemocoel following dissemination from the midgut. Infection has been observed in the diverticula, which, although is concerned with sugar digestion, may also receive a small amount of blood (Romoser *et al.*, 1992; Faran *et al.*, 1988; Hardy *et al.*, 1983).

Virus dissemination from the gut

Since the blood meal is collected and digested in the mosquito midgut, it is assumed that virus disseminates from this region of the digestive tract. The epithelial cells rest on a basement membrane and virions must therefore traverse this to reach the haemocoel. This process has been recorded for SLEV (Whitfield *et al.*, 1973), WEEV (Houk *et al.*, 1985) and VEEV (Weaver, 1986). Whether this is a receptor-mediated event is unknown. The frequency of midgut infection is relatively low when virus is inoculated into mosquitoes intrathoracically (Houk *et al.*, 1985) and Romoser *et al.* (1992) suggested that this may be explained either by a lower density of RVFV-receptor molecules on the haemocoel side of the midgut epithelium, or a basal lamina barrier.

Although direct dissemination from the midgut undoubtedly occurs, at least two alternative routes of escape from the midgut have been suggested. Romoser *et al.* (1987) demonstrated that in *Cx. pipiens* infected with RVFV, dissemination occurs from cells of the intussuscepted foregut (IF) within the cardia at the foregut–midgut junction. Unlike in other regions of the foregut, and also the hindgut, these cells are not protected by the relatively impermeable non-cellular chitinous intima. Virus replicating in these junction cells could spread into cells in the diverticular/oesophagous region and may be released from these tissues into the haemolymph. Consistent with this route of dissemination, infection of the cardiac epithelia was observed by Romoser *et al.* (1992). Lerdthusnee *et al.* (1995) observed large numbers of RVFV virions budding into the basal labyrinth at the periphery of the cardia and into both the non-cellular matrix of the inner epithelial cardial cells and IF cells. The overall significance of dissemination from the IF compared to the accepted route from midgut epithelial cells is uncertain, but this may be an important route for some viruses. Bowers *et al.* (1995) were the first to observe infection of respiratory tissue, when SINV antigen was detected in tracheo-blast cells associated with the tracheoles. They proposed that this extensively branching system could provide an efficient vehicle for the spread of infection to secondary target tissues. It should, however, be noted that their experiments were based on mosquitoes that were infected by intrathoracic inoculation. The possibility of ovarian infection via the tracheal system has also been proposed for RVFV (Romoser *et al.*, 1992) – see later.

The dynamics of virus dissemination in mosquitoes are variable. Romoser *et al.* (1992) calculated a 'viral dissemination index' (DI) that was based on infection of several different tissues, to estimate the extent of dissemination in the haemocoel. They then plotted DIs as a function of time post-oral infection. As reported by others (see Hardy *et al.*, 1983), in some individual mosquitoes disseminated infections can be detected within a few hours of feeding. Weaver (1986) observed VEEV in the fat body adjacent to the midgut within 1 h of feeding. Dissemination without replication in the midgut cells has been observed in several studies (Boorman, 1960; Miles *et al.*, 1973; Weaver, 1986), and although the mechanism of such rapid dissemination is not known, it has been suggested that leakage due to a breach in the gut during feeding may occur (Weaver *et al.*, 1991). Typically dissemination occurs only after replication within the mesenteron, although the efficiency of dissemination may be influenced by dose (Bates & Roca-Garcia, 1946; Chamberlain *et al.*, 1959; Jupp, 1974; Thomas, 1963; Watts *et al.*, 1976).

Apparently some species of mosquitoes can be infected with certain viruses; however, even if the virus replicates within the mesenteron it may fail to disseminate. One explanation for this is the existence of a mesenteronal or midgut escape barrier (MEB) (Kramer *et al.*, 1981). In *Culex tarsalis* the MEB is dose- but not time-dependent for WEEV. In mosquitoes where WEEV failed to disseminate, titres of infectious virus in the mesenteron were 10-fold lower than in competent transmitters (Hardy *et al.*, 1983). Results suggested that nucleocapsids were produced but accumulated along the basal margins of the mesenteronal epithelial cells and were not released. The reasons for the failure of the envelopment process were never determined. A similar barrier has been seen in *Oc. triseriatus* for LACV (Paulson *et al.*, 1989). In some species, for example *Cx. taeniopis*, the MEB cannot be overcome even when high doses of VEEV are ingested (Weaver *et al.*, 1984). For viruses that may cause damage to the infected cells, dissemination may be enhanced by the destruction of the midgut integrity (Weaver *et al.*, 1992). It may be assumed that a MIB and a MEB are a function of vector genetics. Data from elegant quantitative genetic studies with *Ae. aegypti* suggested that for DENV at least two loci were controlling vector competence in the tested laboratory strains under the laboratory conditions of the tests. One locus controlled a MIB and another controlled a MEB (Bosio *et al.*, 1998). The roles of these loci controlling *Ae. aegypti* variation for vector competence in nature remain to be determined and the studies also documented the importance of environmental factors in accounting for most of the observed variation. Furthermore, what is interpreted as a vector-determined barrier may be a consequence of virus-encoded factors. The roles of both vector and virus genetics in mosquito infection are discussed later.

Infection of organs within the haemocoel

The haemocoel and haemolymph. The haemocoel is the insect's body cavity and constitutes an open circulatory system containing the fluid haemolymph. In addition to the organs and muscles, the haemocoel contains cellular components of the insect immune system, for example haemocytes that are distributed in the haemolymph. Sriurairatna & Bhamarapravati (1977) reported replication of DEN-2V in haemocytes but this seems unusual. The biochemical and physiological make-up of the haemolymph can vary considerably with the insect species and, for example, its nutritional and reproductive status. During the dissemination from the gut to other organs enveloped virions must exist in a free state within the haemolymph. A protein with antiviral activity has been described from C6/36 (*Ae. albopictus*) cells that were persistently infected with SINV (Condreay & Brown, 1988). This inhibited total viral RNA synthesis. Although both cellular and humoral components of the mosquito immune system are present in the haemolymph, there is no evidence to suggest that they have antiviral activity. They are, however, active against bacteria and some parasites either directly or via melanization and encapsulation responses. Recently it has been suggested that an intracellular antiviral response similar to RNA interference (RNAi) exists in mosquitoes that is triggered by double-stranded RNA formed during virus replication and leads to the destruction of viral mRNA with sequence identity to the dsRNA trigger (K. E. Olson, personal communication; Sanchez-Vargas *et al.*, 2003).

Fat body. The fat body is a somewhat diffuse tissue (van Heusden, 1996), with some cells surrounding the gut and other organs, whilst others are attached to the integument. It is involved in intermediary metabolism and nutrient storage. Although only a few cells thick, it is surrounded by a permeable basal lamina that allows the rapid diffusion of metabolites between the cells and the haemolymph. The fat body is infected by several arboviruses and for some viruses is the major amplifying tissue (Doi, 1970; Weaver, 1986). For others, for example DEN-2V in *Ae. albopictus*, the fat body is relatively poorly infected (Sriurairatna & Bhamarapravati, 1977).

Skeletal muscle. Despite the abundance of skeletal muscle, particularly in the thorax, with the exception of alphaviruses, this tissue seems relatively unimportant in the dissemination of arboviruses. Visceral muscles associated with the gut may, however, be infected by other viruses including RVFV (Faran *et al.*, 1986).

Nervous tissues. Several viruses infect neural tissue (Hardy & Reeves, 1990; Leake & Johnson, 1987; Larsen & Ashley, 1971) and indeed Leake & Johnson (1987) suggested that the salivary glands may be infected with Japanese encephalitis virus (JEV) via nerve trunks from the infected cerebral ganglia. Miles *et al.* (1973) made a similar suggestion for Whataroa virus. In addition to infection of the ventral nerve chord and ganglia, the brain is also involved, as are the tissues of the eye, including the ommatidia, and the nerves of the antennae, including Johnston's organ. DEN-2V virus particles present in the RER cisternae of *Ae. albopictus* neurons are associated with virally directed dense threads, and although there is an increasing involvement of nervous tissues over time, no obvious deleterious effects of infection have been observed (Sriurairatna & Bhamarapravati, 1977).

Reproductive tissues. The reproductive organs of the female mosquito consist of a pair of ovaries, each joined via lateral oviducts to a common oviduct opening into the genital chamber. The chamber and oviduct are lined with a non-cellular intima. The lateral oviducts penetrate into the ovary to form a chamber – the calyx – from which numerous follicles radiate. The follicles are held together by a muscular ovarian sheath. Each follicle has an outer layer of epithelial cells with a chorionic membrane surrounding the oocyte and nurse cells. For many arboviruses, infection of reproductive tissues occurs rarely, if ever. However, for others, notably members of the *Bunyaviridae*, infection of the female reproductive tract is a critical feature of viral infection that permits overwintering and survival in the absence of vertebrate hosts. The phenomenon of transovarial transmission (TOT) is a mechanism by which virus is transmitted from an infected female to her progeny (see reviews by Leake, 1984; Turrell, 1988) and is dependent on viral infection of the follicles. Vertical transmission of DENV to F_1 progeny following mating between a transovarially infected male and uninfected female does not require infection of the follicles (Rosen, 1987) but may require some replication within the female genital tract. Infection of the egg apparently occurs by virus entering via the micropyle during the fertilization process. Sexual transmission from infected males to uninfected females has also been documented (Rosen, 1987).

Although TOT has been recorded for many viruses (Turrell, 1988), for alphaviruses and flaviviruses it is relatively rare and probably of limited importance in the overall maintenance of the virus in nature. Alphaviruses have been observed in common and lateral oviducts but not in the ovarioles. In contrast, for bunyaviruses such as LACV, TOT is a common component of the life cycle and of primary importance in overwintering and the epidemiology of the virus (Watts et al., 1974). LACV may be detected in the calyx within 2 days of infection, before apparent dissemination from the midgut (Chandler et al., 1998), but typically infection of the ovaries is seen during the second gonadotrophic cycle after an infectious blood meal and occurs infrequently during the first cycle. The intima of the calyx and ovarian ducts may prevent virus from entering the lumen and the ovaries may be infected by disseminated virus via the ovarian sheath. RVFV frequently infects the calyx and has been observed in follicular relics, ovarian and ovariolar ducts – the sheaths – following oral infection. However, this is not seen in mosquitoes infected intrathoracically, leading Romoser et al. (1992) to suggest that the infection of the sheaths may occur as a result of changes in their receptiveness in response to the blood meal. These researchers suggested that RVFV probably enters the ovary without infecting the sheath and proposed that a route involving tracheal cells might be involved. This observation could not be confirmed because of non-specific immunofluorescent staining of muscle, tracheae and tracheoles.

In addition to the direct infection of ovaries following an infectious blood meal, a second mechanism of vertical transmission exists for bunyaviruses. In transovarially infected embryos, the germinal tissues may become infected, thus perpetuating long-term transmission of the virus between generations even in the absence of infected vertebrate hosts (Grimstad, 1988). The presence of virus in both immature and adult stages of the vector allows the virus to be widely distributed in tissues. However, generally virus titres in adult mosquitoes infected in this manner are lower than are seen in mosquitoes infected orally. This and the fact that in the infected embryo virus titres do not change significantly over time shows that there may be vector factors that modulate virus replication. Infected male mosquitoes derived from eggs infected transovarially may infect females venereally during mating (Thompson & Beaty, 1977). From an epidemiological perspective one important factor associated with TOT is that the virus can be transmitted early in the year by the first mosquitoes to eclose, since they do not have to feed upon a viraemic host and no extrinsic incubation period is necessary. Genetic studies by Graham *et al.* (1999) concluded that permissiveness to TOT of LACV by *Ochlerotatus triseriatus* was conditioned by dominant alleles. A subsequent study using quantitative trait loci (QTLs) found alleles at three loci mapping to chromosome II and III that contributed additively to determine the TOT rate. The three loci accounted for 53 % of the observed phenotypic variance in the tested mosquito strain under the laboratory environmental conditions of the experiment (Graham *et al.*, 2003). The roles of these loci in contributing to TOT phenotypic variation in nature remain unknown.

Malpighian tubules. The Malpighian tubules, concerned with ionic balance and secretion, are generally regarded as a 'non-permissive' organ, not typically supporting virus infection. Although both SINV and SLEV have been observed in Malpighian tubules (Muangman *et al.*, 1969; Whitfield *et al.*, 1973) and EEEV may occur in the pericardial cells (Scott *et al.*, 1984), the tubules are not infected by LACV (Beaty & Thompson, 1978), JEV (Doi, 1970; Doi *et al.*, 1967; LaMotte, 1960) or VEEV (Larsen & Ashley, 1971). However, interpretations of these data were confounded by the problems of distinguishing viral antigen by immunofluorescent and immuno-histochemical techniques from the non-specific staining and autofluorescence associated with Malpighian tubules and pericardial cells (Faran *et al.*, 1986).

Salivary glands. With respect to transmission, the mosquito salivary gland (SG) must be considered as the most important organ to become infected. The mosquito has a pair of salivary glands joined by a common duct that allows saliva to flow to the fascicle during feeding. Each of the pairs has three lobes – a median lobe and two lateral lobes that may be subclassified into proximal and distal parts. This morphological distinction is reflected functionally with the median and distal-lateral lobes producing substances concerned primarily with blood feeding, whilst the proximal-lateral lobes secrete enzymes concerned with sugar feeding. There are excellent reviews on the salivary gland and saliva (James, 1994; Ribeiro, 1989, 1995) and recent genetic analysis (Ribeiro & Francischetti, 2002; Valenzuela *et al.*, 2002) has contributed significantly to our understanding of the nature and role of the substances produced by the salivary glands. Although mosquito saliva has no known direct influence on arbovirus phenotype, there are experimental data for several arthropod vectors demonstrating that saliva inoculated into a vertebrate can have a profound effect on the successful infection of the host (see later).

Leake & Johnson (1987) suggested that the salivary glands may be infected with JEV via nerve trunks from the infected cerebral ganglia; however, the salivary gland has a fat body closely associated with it and this too could be the source of infection. To infect the gland, virus must pass through the outer basement membrane on which the salivary acinar cells rest, enter the cell and then replicate within the cytoplasm. The virus then must be shed into the saliva so that it can be transmitted during feeding.

The distribution and abundance of virus in the lobes of salivary glands are variable. Although secretions from the proximal-lateral lobes are not concerned with blood feeding, they can become infected (Janzen *et al.*, 1970). RVFV, for example, may be present in the glands within 48 h of blood feeding (Faran *et al.*, 1988) and, although most frequently seen in the lateral lobes, may occur in all regions (Romoser *et al.*, 1992). Gaidamovitch *et al.* (1973) observed small isolated areas of VEEV antigen in the lateral lobes of *Ae. aegypti* on day 2, in the distal parts by day 5 and in proximal areas by day 10, with antigen still detected on the 52nd day post-feeding. The tendency towards infection of the lobes whose secretions are important for blood feeding is seen with several viruses. Takahashi & Suzuki (1979) reported that JEV was present only in the lateral lobes of *Culex tritaeniorhynchus* and Gubler & Rosen (1976) reported that DENV was only in lateral lobes of *Ae. albopictus*. Sriurairatna & Bhamarapravati (1977) reported that DEN-2V occurred mostly in salivary acinar cells of the distal-lateral lobes. These authors reported the formation of vesicular structures that they considered to be a remarkable change rarely seen in cytoplasm. Characteristic vesicular structures and viral crystalloid aggregates were often seen in the cisternae of the RER. Enveloped DENV was frequently located in the basal part of cells and, although present at the basal lamina, no virions were observed to be penetrating this. Viral aggregates developed in the apical cavities and glandular lumen, together with saliva. An ultrastructural study of SLEV included the secretion process in *Cx. pipiens* and showed crystalline aggregates within the diverticulum of the salivary gland lumen and the subsequent accumulation of virus throughout the length of the chitinous central duct (Murphy *et al.*, 1975; Whitfield *et al.*, 1973).

As with the midgut, both infection and escape barriers have been reported for the salivary gland (Jupp, 1985; Kramer *et al.*, 1981; McClean *et al.*, 1974; Paulson *et al.*, 1989) but the mechanisms involved in producing these are not known.

As mentioned earlier, the function of mosquito saliva is to facilitate blood feeding. Saliva of haematophagous arthropods, including insects and acarids, consists of numerous substances which, in concert, cause vasodilation and prevent clotting. Specific components vary between different species, but functionally they have similar compounds. Recently it has been realized that, in addition to acting directly on vessels and the blood, saliva can also influence the vertebrate host immune system. Although most easily studied in ticks that feed on the host for a long time, research with dipteran vectors has also revealed some unexpected consequences of blood feeding (Gillespie *et al.*, 2000). In the few studies on mosquito saliva, it was observed that mosquito feeding can rapidly and significantly alter the immune response of mice (Zeidner *et al.*, 1999). As a consequence of mosquito feeding, components of the vertebrate immune system such as dendritic cells that may migrate to the feeding site and alterations in cytokine levels, might increase the opportunity for an arbovirus to establish an infection in the vertebrate host and could impact on disease development. Levels of LACV in the vertebrate blood are higher when infection is mediated by mosquito bite than if established by needle inoculation (Osorio *et al.*, 1996) and enhancement of infection in the presence of mosquito saliva was demonstrated recently for both Cache Valley virus (Edwards *et al.*, 1998) and vesicular stomatitis virus (Limesand *et al.*, 2002, 2003). Reisen *et al.* (2000) found no difference between needle inoculation and infectious mosquito bite when infecting avian hosts with WEEV or SLEV.

DELETERIOUS EFFECTS

The classical view of mosquito–arbovirus interactions assumed that the infection was not harmful to the vector. In support of this, it is generally true that whilst arboviruses cause cytopathic effects in vertebrate cell culture, they can produce non-cytolytic, persistent infections in mosquito cells. With the increasing number of detailed studies utilizing a wide range of techniques this dogma has been somewhat dispelled. Clearly some vectors when infected with certain viruses do experience deleterious consequences of the infection. Direct tissue damage may occur: for example, pathologic changes in the midgut including sloughing of infected cells into the lumen and necrosis of cells *in situ* have been observed during infections with WEEV (Weaver *et al.*, 1992); and loss of brush border and basal lamina integrity may occur during EEEV infections (Weaver *et al.*, 1988). SINV infection has been associated with vacuolated cytoplasm and myofilament misalignment of the visceral muscles associated with the midgut in mosquitoes inoculated intrathoracically (Bowers *et al.*, 2003). Semliki Forest virus can cause pathology in salivary glands (Mims *et al.*, 1966; Lam & Marshall, 1968b), as can SINV (Bowers *et al.*, 2003). Less obvious, indirect effects may also be associated with vector infection. For example, Grimstad *et al.* (1980) demonstrated that *Oc. triseriatus* had an extended feeding time following infection with LACV. If virus transmission efficiency is time-dependent, then the more a mosquito probes and the longer it takes to become fully engorged, the greater is the likelihood that the virus is transmitted. One might predict that viruses causing such an effect might be at a selective advantage compared with viruses that did not. However, mosquitoes that spend an extended time feeding on a human host are probably more likely to be detected and killed, which may explain why Putnam & Scott (1995) found no effect of DEN-2V infection on the feeding behaviour of *Ae. aegypti*. *Cx. pipiens* infected with RVFV had a reduced ability to refeed and reduced fecundity (Turrell *et al.*, 1984). Later studies (Faran *et al.*, 1988; Romoser *et al.*, 1992) reported reduced survival and fecundity of *Cx. pipiens* following infection with RVFV and suggested that these effects were due to the extensive tissue and organ infections with a resulting overall energy drain. They suggested that the infection of regulatory tissues and organs, for example the neural ganglia, neurosecretory cells and corpora allata, would probably adversely affect the overall functioning of the mosquito. Eggs of *Oc. triseriatus* that have overwintered and are infected transovarially with LACV terminate diapause more readily than uninfected eggs and the mortality rate was more than double that of the uninfected embryos (McGaw *et al.*, 1998). Apparently selections for high TOT rates in *Oc. triseriatus* result in a decrease in overall fitness because the adults have decreased survival and fecundity (Graham *et al.*, 2003).

Having discussed some of the specifics of the interactions between mosquitoes and arboviruses and how there are very complex parameters that can influence these interactions, it is hoped that many issues will be addressed using new advances in molecular biology and genetics. The elucidation of viral sequences is now almost routine and viral genomes can often be manipulated in the laboratory. The genomic sequences of *An. gambiae* (Holt *et al.*, 2002) and *Plasmodium falciparum* (Gardner *et al.*, 2002) are determined and those for *Ae. aegypti* and *Cx. pipiens* are forthcoming. Although we know that the vector–virus relationship is influenced by three funda-mental components, environmental, vector and virus factors, unfortunately our understanding of the complexity of these factors and their interactions with one another is very rudimentary (Tabachnick, 2003). However, let us now examine the influences of viral and vector genetics with specific examples.

VIRUS GENETICS AND MOSQUITO INFECTION

Undoubtedly the infection and dissemination of arboviruses in mosquitoes is influenced by virally encoded factors. Many virus genomes have been sequenced and can be manipulated using reverse genetics. As a result, it is now possible to study the influence of viral sequence and to begin to identify the molecular mechanisms underlying basic virus–vector interactions. Historically, the differential behaviour of vaccine and wild-type strains of virus, for example YFV, indicated the importance of viral genetics. A genetic variant (V22) of LACV that was selected using a G1 protein-specific monoclonal antibody (mAb) is restricted in its ability to disseminate from the midgut (Ludwig *et al.*, 1989, 1991). Similarly, a variant of VEEV (MARV 1A3B-7) replicated identically to the parent virus when inoculated intrathoracically into *Ae. aegypti*; however, when fed in an artificial blood meal to mosquitoes the variant was significantly less efficient in infection and dissemination from the midgut (Woodward *et al.*, 1991). Characterization of MARV 1A3B-7 determined that there is an isoleucine to phenylalanine substitution at amino acid 207 in the E2^h epitope of E2. This epitope is involved in early virus–cell interactions (attachment) and the mutation may cause a conformational change in E2. Since fusion of the viral and cellular membranes is attributed to the alphavirus E1 protein, the decreased infectivity caused by the E2 mutation is presumably an indirect effect, but further studies are needed to confirm this. For flaviviruses little attention has been paid to the role of virally encoded proteins in vector infection. Sequence analysis has identified distinct variable motifs in the E protein that distinguish between tick-borne and mosquito-borne viruses (Gritsun *et al.*, 1995) but primarily this has been discussed from an evolutionary viewpoint (Kuno *et al.*, 1998; Zanotto *et al.*, 1996) rather than being analysed to identify the mechanism of vector specificity.

WILD-TYPE AND VACCINE STRAINS OF FLAVIVIRUSES IN MOSQUITOES

It has long been known that whilst wild-type YFV can infect, disseminate and be transmitted by *Ae. aegypti* mosquitoes, the 17D strain, which is a live attenuated derivative used for human vaccination, can infect epithelial cells of the midgut but does not disseminate and is not transmitted (Roubaud *et al.*, 1937; Whitman, 1937, 1939; Miller & Adkin, 1988; Jennings *et al.*, 1994). Similarly, the YFV FNV vaccine strain is attenuated in mosquitoes (Davis *et al.*, 1932; Roubaud & Stephanopoulo, 1933; Peltier *et al.*, 1939). When *Ae. aegypti* is presented with virus in a blood meal (virus titres $7 \cdot 2–8 \cdot 3 \log_{10}$ p.f.u. ml^{-1}) dissemination rates ranged from 90 to 100 % for wild-type YFV, but was only 3 % for YFV 17D (Miller & Adkin, 1988). Jennings *et al.* (1994) also observed a lack of dissemination, but reported that midguts became infected with 17D. What seems to be operating to prevent dissemination of YFV 17D could be interpreted as an effective mosquito MEB. Although the MEB might be acting in a virus-strain-specific manner, it seems more likely that the failure of YFV 17D to disseminate in, for example, the RexD strain of *Ae. aegypti* is due to a virally encoded factor. As described later, mosquitoes collected from different geographic locations vary widely with respect to susceptibility to infection (Aitken *et al.*, 1977; Beaty & Aitken, 1979; Tabachnick *et al.*, 1985).

Mosquito experiments have been performed on other flaviviruses to evaluate transmissibility of live vaccine strains. Results have been variable. For candidate DEN-2V vaccine virus (S-1) that was presented orally at a range of doses, a mean of only 16 % of mosquito midguts became infected and none transmitted virus (versus 56 % infection and 14 % transmission rates for the parent; Miller *et al.*, 1982). The trend was consistent regardless of blood meal virus titres, and it was estimated that the vaccine strain was approximately 63 times less efficient than the parent strain for infecting *Ae. aegypti* by the oral route. The authors concluded that although the vaccine strain could replicate in the midgut, generally it was unable to mature and escape into the haemocoel or unable to attach and replicate in secondary organs. Interestingly, a few mosquitoes that were allowed to feed on humans infected with the vaccine strain did become infected but dissemination and transmission did not occur (Bancroft *et al.*, 1982). Both viruses produced similar infection and dissemination rates (98–100 %) following inoculation into mosquitoes. Similar studies with candidate vaccine strains for DEN-1V and DEN-4V viruses (Schoepp *et al.*, 1991) and for DEN-3V (Schoepp *et al.*, 1990) produced variable results. The DEN-4V vaccine candidate (PDK35-TD3 FRhL p3) was attenuated for mosquitoes; however, neither the DEN-1V (45AZ5) nor DEN-3V (CH53489) candidate was attenuated for oral infection and dissemination in *Ae. aegypti*. Based on these observations and data from vertebrate and *in vitro* studies, Schoepp *et al.* (1991) concluded that vector and vertebrate host attenuation are genetically linked; however, they considered that plaque morphology and temperature sensitivity were not reliable markers for attenuation. Whilst these data are interesting, it should be noted that an effective DEN vaccine has not been produced commercially from any of these so-called 'candidates'.

An attenuated JEV vaccine candidate (strain SA14-2-8) also showed reduced replication in mosquitoes compared to the parent virus (SA14) and remained avirulent after mosquito passage (Chen & Beaty, 1982). Similarly, an attenuated infectious clone, DEN-2V 16681 pdk53 (Kinney *et al.*, 1997), is restricted in its ability to infect mosquitoes. A new generation of live vaccine candidates produced as chimeric viruses in a YFV 17D backbone (ChimeriVax-) have also been tested in mosquitoes. Chimeric vaccine candidates for JEV (Bhatt *et al.*, 2000), DENV (Johnson *et al.*, 2002) and West Nile virus (Johnson *et al.*, 2003) are all highly restricted in their ability to infect and be transmitted by mosquitoes and remain genetically stable in the vector.

VIRUS EVOLUTION

The transmission cycle of arboviruses depends upon replication in two physiologically very different biological environments, namely the invertebrate vector and the vertebrate host. *In vitro* experiments have demonstrated that viral phenotype, for example virulence for the vertebrate host, can be changed selectively by passage in, for example, a particular vertebrate culture system. Attenuated vaccines such as the 17D strain of YFV have been derived in this manner. As described earlier, live attenuated vaccine viruses typically are restricted in their ability to infect mosquitoes and the altered phenotype of viruses like YFV 17D reflects genotypic change. Arboviral genotypes are under selective pressures in two radically different environments as they replicate alternately in the vertebrate and the arthropod. The vertebrate's immune response is certainly a selective force that often restricts replication, viraemia and the potential for reinfection. In the arthropod, however, arboviral infections can last for the life of the vector, which for some hibernating species may be several months. This prolonged association between the vector and the virus can be viewed as a time of constant selection to ensure genomic stability (Baldridge *et al.*, 1989) that has resulted in the non-pathogenic infection of the vector as described earlier. Alternatively, the long period of infection can also provide the potential for evolutionary change (Nuttall *et al.*, 1991). Since during their lifetime vectors may feed multiple times, perhaps on hosts infected with different pathogens, there exists the potential for different viruses to co-replicate in the same cells. Although, as described earlier, resistance to superinfection exists, for groups such as the bunyaviruses that have a segmented genome with multiple RNA strands, evolution may occur by the process of reassortment in which RNA segments from different viruses are encapsidated together into a virion (Beaty *et al.*, 1985; Chandler *et al.*, 1990; Turrell *et al.*, 1990). This may be a rare event but in the context of virus evolution may represent the least restrained method by which large genetic changes can occur. Of course, in order to survive, the reassortant virus must be capable of replication and transmission between the vector and vertebrate host.

VECTOR GENETICS

For several vector–virus relationships, it has been shown that different strains of a particular species may vary considerably with respect to susceptibility to infection with many arboviruses (Grimstad *et al.*, 1977; Gubler & Rosen, 1976; Gubler *et al.*, 1979; Hardy *et al.*, 1976; Hayes *et al.*, 1984; Tabachnick, 1994; Tabachnick *et al.*, 1985; Tesh *et al.*, 1976). Laboratory experiments in which susceptibility can be altered selectively via specific breeding strategies (Hardy *et al.*, 1978; Wallis *et al.*, 1985) clearly support a vector genetic basis for susceptibility. This genetic influence of susceptibility is not necessarily a general trait, but can also be virus-specific (Gubler & Rosen, 1976; Tesh *et al.*, 1976). Interestingly, severe modulation of virus titres following infection has been observed by Kramer *et al.* (1998) to be a general alphavirus phenomenon in genetically selected lines of *Cx. tarsalis*.

For *Ae. aegypti*, numerous studies have documented intraspecific variation with respect to susceptibility to infection with YFV and other viruses (Bennett *et al.*, 2002; Gubler *et al.*, 1979; Lorenz *et al.*, 1984; Miller & Mitchell, 1991; Rosen *et al.*, 1985; Tabachnick *et al.*, 1985; Tardieux *et al.*, 1990; Vazeille-Falcoz *et al.*, 1999). Vector competence studies reveal considerable variation between mosquito strains and often poor reproducibility. Susceptibility to YFV is different in populations collected from different geographic locations (Hindle, 1929; Aitken *et al.*, 1977; Beaty & Aitken, 1979; Wallis *et al.*, 1984; Tabachnick *et al.*, 1985; Jupp & Kemp, 2002). In *Ae. aegypti formosus* YFV infected the midgut but failed to disseminate (Miller & Mitchell, 1991). The basis of this refractoriness is unknown but in essence wild-type YFV Asibi in *Ae. aegypti formosus* behaves similarly to YFV 17D in *Ae. aegypti aegypti*. The former is presumably under the control of mosquito genes and, as described earlier , the latter is probably virally encoded. However, they may have a common mechanism. For example, YFV 17D may fail to interact with an as-yet-unidentified mosquito factor and YFV Asibi may fail to disseminate in *Ae. aegypti formosus* because this mosquito lacks that 'factor'.

Colonization affects susceptibility to oral infection (Armstrong & Rico-Hesse, 2001; Lorenz *et al.*, 1984) and genetic selection has been used to produce resistant and susceptible phenotypes (Miller & Mitchell, 1991). These authors hypothesized that virus movement across the midgut is likely to be determined by either a single major gene and modifying minor genes, or a group of closely linked genes. Interestingly, Beaty & Aitken (1979) found that a resistant strain of *Ae. aegypti* (Amphur) from Asia transmitted YFV Asibi at a low rate even when infected by intrathoracic inoculation. This is in contrast to studies with DENV and WEEV, where even the most resistant strains will replicate and transmit virus efficiently when inoculated (Gubler *et al.*, 1982; Kramer *et al.*, 1989). Several studies have demonstrated strain variation in *Ae. aegypti* with respect to susceptibility to DENV infection (Gubler & Rosen, 1976; Gubler *et al.*, 1979; Tardieux *et al.*, 1990). The factor(s) controlling susceptibility to DENV was the same for all serotypes (Gubler *et al.*, 1982) and from crosses between susceptible and resistant phenotypes it was concluded that resistance is dominant over susceptibility. Geographic strain variation in *Ae. albopictus* to infection with Chikungunya virus was shown by Tesh *et al.* (1976) but this did not correlate with variation in susceptibility to dengue. Apparently, multiple factors can operate in a single strain of mosquito to control susceptibility to alphaviruses and flaviviruses.

A review by Beerntsen *et al.* (2000) discussed the genetics of mosquito vector competence. Their conclusion that the future of research in mosquito vector competence is very bright seems somewhat premature. Although it is now possible to put sequence markers on a chromosome map, despite the effort and recent studies (Bosio *et al.*, 1998, 2000; Black *et al.*, 2002), all that one can conclude is that it seems clear that multiple genes are involved and it seems likely that different stages of the infection, for example midgut infection, midgut escape and salivary gland infection, may be influenced by different genes. We have yet to explore the impact of any single gene on the phenotype of susceptibility to infection, almost no work has been done on transmission ability, and the role of the environmental variance in shaping phenotypic variation in nature has not been touched. Tabachnick (2003) reviewed the complexities and challenges ahead in addressing these issues. In the 13 years since Miller and Mitchell concluded that their '...inbred mosquito lines will be useful in discovering the molecular basis for flavivirus resistance in *Ae. aegypti*', we still do not know which specific mosquito genes influenced susceptibility to arboviral infection in the inbred mosquito lines used by Miller and Mitchell, we do not even know how many genes there are, and we certainly do not have any idea which genes are important in natural populations.

The most detailed study on the genetic mechanism controlling susceptibility to arbovirus infection was not accomplished using mosquitoes. The genetic control of phenotypic variation in the susceptibility to infection of a colony of the biting midge *Culicoides sonorensis* with a strain of bluetongue virus was shown to be largely con-trolled by a single genetic locus (Tabachnick, 1991). The locus operated in a complex manner where the dominance recessive relationships between refractoriness and susceptibility were determined by the male parent. In addition, the locus exhibited a maternal effect where the genotype of the mother determined the phenotype of the offspring. This study illustrates that it will be important to couple classical genetic studies using careful family analysis with molecular biological techniques. A purely molecular biological approach without family data would have missed this complex genetic mechanism. Unfortunately, as with all other studies linking genetic control and vector competence, the role of this genetic mechanism in controlling phenotypic variation in nature is unknown. This is despite the considerable technical developments that have occurred since the reviews of the 1980s and 1990s and in particular the recent advances in our knowledge of mosquito genomes. One hopes that we are getting closer to elucidating the relative importance of different mosquito genes in determining vector susceptibility to a specific viral infection but this of course is only part of our battle with mosquito-borne diseases. The real challenge lies in the application of this knowledge in the real world, with the ultimate goal of reducing the impact of mosquito-borne pathogens.

The suggestion that vector-borne diseases might be controlled better by reducing the vectors' ability to transmit pathogens than by reducing vector populations (Curtis, 1968) has been embraced by many scientists and there has certainly been great progress towards being able to test the feasibility of this approach. To this end, we now possess the ability to manipulate several different species of mosquito vectors genetically and with relative ease (Allen *et al.*, 2001; Catteruccia *et al.*, 2000; Coates *et al.*, 1998; Grossman *et al.*, 2001; Handler & James, 2000; Jasinskiene *et al.*, 1998), we have identified some mechanisms by which virus replication in the mosquito can be altered (Higgs *et al.*, 1998; Olson *et al.*, 1996, 2002; Powers *et al.*, 1996), and can manipulate the expression of endogenous genes that might influence pertinent aspects of mosquito biology (Attardo *et al.*, 2003; Shiao *et al.*, 2001). However, the challenge of spreading and establishing mosquitoes with reduced vector competence into wild populations remains (Braig & Yan, 2002; Riehle *et al.*, 2003). It is now important to proceed with fundamental studies on vector transmission of arboviruses and then to study vector capacity traits such as host preference, longevity, density, etc. Such knowledge is critical if the promise for controlling mosquito-borne diseases via genetically reducing vector competence of mosquitoes is to be fulfilled. It will also be important in enabling us to identify new approaches and perhaps unforeseen opportunities by which we may control vector competence and vector populations.

ACKNOWLEDGEMENTS

My thanks to Kate McElroy, Yvette Girard, Laura, D. Kramer, Walter J. Tabachnick, Douglas M. Watts and Dana L. Vanlandingham for their constructive reading of this manuscript and helpful comments, and to Patricia Hamilton for assistance with the production of the photomicrographs. Infection data depicted in the photomicrographs were produced as part of studies supported by NIH grants AI47246 and AI47877 and CDC grant U50/CCU620539.

REFERENCES

Aitken, T. H., Downs, W. G. & Shope, R. E. (1977). *Aedes aegypti* strain fitness for yellow fever virus transmission. *Am J Trop Med Hyg* **26**, 985–989.

Allen, M. L., O'Brochta, D. A., Atkinson, P. W. & Levesque, C. S. (2001). Stable, germ-line transformation of *Culex quinquefasciatus* (Diptera: Culicidae). *J Med Entomol* **38**, 701–710.

Armstrong, P. M. & Rico-Hesse, R. (2001). Differential susceptibility of *Aedes aegypti* to infection by the American and Southeast Asian genotypes of dengue type 2 virus. *Vector Borne Zoonotic Dis* **1**, 159–168.

Attardo, G., Higgs, S., Klingler, K. A., Vanlandingham, D. L. & Raikhel, A. S. (2003). RNAi-mediated knockdown of a GATA factor reveals a molecular mechanism of anautogeny in the yellow fever mosquito, *Aedes aegypti. Proc Natl Acad Sci U S A* **100**, 13374–13379.

Baldridge, G. D., Beaty, B. J. & Hewlett, M. J. (1989). Genomic stability of La Crosse virus during vertical and horizontal transmission. *Arch Virol* **108**, 89–99.

Bancroft, W. H., Scott, R. M., Brandt, W. E., McCown, J. M., Eckels, K. H., Hayes, D. E., Gould, D. J. & Russell, P. K. (1982). Dengue-2 vaccine: infection of *Aedes aegypti* mosquitoes by feeding on viremic recipients. *Am J Trop Med Hyg* **31**, 1229–1231.

Bates, M. & Roca-Garcia, M. (1946). The development of the yellow fever virus in *Haemagogus* mosquitoes. *Am J Trop Med* **26**, 585–605.

Beaty, B. J. & Aitken, T. H. G. (1979). *In vitro* transmission of yellow fever virus by geographic strains of *Aedes aegypti. Mosq News* **39**, 232–238.

Beaty, B. J. & Thompson, W. H. (1978). Tropisms of La Crosse virus in *Aedes triseriatus* (Diptera: Culicidae) following infective blood meals. *J Med Entomol* **14**, 499–503.

Beaty, B. J., Sundin, D. R., Chandler, L. C. & Bishop, D. H. L. (1985). Evolution of bunyaviruses by genome reassortment in dually infected mosquitoes (*Aedes triseriatus*). *Science* **230**, 548–550.

Beerntsen, B. T., James, A. A. & Christensen, B. M. (2000). Genetics of vector competence. *Microbiol Mol Biol Rev* **64**, 115–137.

Bennett, K. E., Olson, K. E., de Lourdes Munoz, M., Fernandez-Salas, I., Farfan-Ale, J. A., Higgs, S., Black, W. C. & Beaty, B. J. (2002). Variation in vector competence for dengue 2 virus among 24 collections of *Aedes aegypti* from Mexico and the United States. *Am J Trop Med Hyg* **67**, 85–92.

Bhatt, T. R., Crabtree, M. B., Guirakhoo, F., Monath, T. P. & Miller, B. R. (2000). Growth characteristics of the chimeric Japanese encephalitis virus vaccine candidate, ChimeriVax-JE (YF/JE SA14-14-2), in *Culex tritaeniorhynchus, Aedes albopictus*, and *Aedes aegypti* mosquitoes. *Am J Trop Med Hyg* **62**, 480–484.

Billingsley, P. F. (1990). The midgut ultrastructure of hematophagous insects. *Ann Rev Entomol* **35**, 219–248.

Bishop, D. H. L. & Beaty, B. J. (1988). Molecular and biochemical studies of the evolution, infection and transmission of insect bunyaviruses. *Philos Trans R Soc Lond B* **321**, 463–483.

Black, W. C. I. V., Bennett, K. E., Gorrochotegui-Escalante, N., Barillas-Mury, C. V., Fernandez-Salas, I., de Lourdes Munoz, M., Farfan-Ale, J. A., Olson, K. E. & Beaty, B. J. (2002). Flavivirus susceptibility in *Aedes aegypti*. *Arch Med Res* **33**, 379–388.

Boorman, J. (1960). Observations on the amount of virus present in the haemolymph of *Aedes aegypti* infected with Uganda S, yellow fever and Semliki Forest viruses. *Trans R Soc Trop Med Hyg* **54**, 362–365.

Bosio, C. F., Beaty, B. J. & Black, W. C. I. V. (1998). Quantitative genetics of vector competence for dengue-2 virus in *Aedes aegypti*. *Am J Trop Med Hyg* **59**, 965–970.

Bosio, C. F., Fulton, R. E., Salasek, M. L., Beaty, B. L. & Black, W. C. I. V. (2000). Quantitative trait loci that control vector competence for dengue-2 virus in the mosquito *Aedes aegypti*. *Genetics* **156**, 687–698.

Bowers, D. F., Abell, B. A. & Brown, D. T. (1995). Replication and tissue tropism of the alphavirus Sindbis in the mosquito *Aedes albopictus*. *Virology* **58**, 81–86.

Bowers, D. F., Coleman, C. G. & Brown, D. T. (2003). Sindbis virus-associated pathology in *Aedes albopictus* (Diptera: Culicidae). *J Med Entomol* **40**, 698–705.

Braig, H. R. & Yan, G. (2002). The spread of genetic constructs in natural insect populations. In *Genetically Engineered Organisms. Assessing Environmental and Human Health Effects*, pp. 251–314. Edited by D. K. Letourneau & B. Elpern Burrows. Boca Raton: CRC Press.

Brault, A. C., Powers, A. M. & Weaver, S. C. (2002). Vector infection determinants of Venezuelan equine encephalitis virus reside within the E2 envelope glycoprotein. *J Virol* **76**, 6387–6392.

Brown, D. T., Waite, M. R. F. & Pfefferkorn, E. R. (1972). Morphology and morphogenesis of Sindbis virus as seen by freeze-etching techniques. *J Virol* **10**, 524–536.

Catteruccia, F., Nolan, T., Loukeris, T. G., Blass, C., Savakis, C., Kafatos, F. C. & Crisanti, A. (2000). Stable germline transformation of the malaria mosquito *Anopheles stephensi*. *Nature* **405**, 959–962.

Chamberlain, R. W. & Sudia, W. D. (1957). Dual infections of eastern and western equine encephalitis viruses in *Culex tarsalis*. *J Infect Dis* **101**, 233–236.

Chamberlain, R. W. & Sudia, W. D. (1961). Mechanism of transmission of viruses by mosquitoes. *Annu Rev Entomol* **6**, 371–390.

Chamberlain, R. W., Sudia, W. D. & Gillett, J. D. (1959). St. Louis encephalitis virus in mosquitoes. *Am J Hyg* **70**, 221–236.

Chandler, L. C., Beaty, B. J., Baldridge, G. D., Bishop, D. H. L. & Hewlett, M. J. (1990). Heterologous reassortment of bunyaviruses in *Aedes triseriatus* mosquitoes and transovarial and oral transmission of newly evolved genotypes. *J Gen Virol* **71**, 1045–1050.

Chandler, L. C., Blair, C. D. & Beaty, B. J. (1998). La Crosse virus infection of *Aedes triseriatus* (Diptera: Culicidae) ovaries before dissemination of virus from the midgut. *J Med Entomol* **35**, 567–572.

Chen, B. Q. & Beaty, B. J. (1982). Japanese encephalitis vaccine (2-8 strain) and parent (SA 14 strain) viruses in *Culex tritaeniorhynchus* mosquitoes. *Am J Trop Med Hyg* **31**, 403–407.

Clements, A. N. (1992). *The Biology of Mosquitoes*, vol. 1, *Development, Nutrition and Reproduction*. London: Chapman & Hall.

Coates, C. J., Jasinskiene, N., Miyashiro, L. & James, A. A. (1998). Mariner transposition and transformation of the yellow fever mosquito, *Aedes aegypti*. *Proc Natl Acad Sci U S A* **95**, 3748–3751.

Condreay, L. D. & Brown, D. T. (1988). Suppression of RNA synthesis by specific antiviral activity in Sindbis-infected *Aedes albopictus* cells. *J Virol* **62**, 346–348.

Curtis, C. F. (1968). A possible genetic method for the control of insect pests, with special reference to tsetse flies (*Glossina spp.*). *Bull Entomol Res* **57**, 509–523.

Davis, N. C., Lloyd, W. & Frobisher, M., Jr (1932). Transmission of neurotropic yellow fever virus by *Stegomyia* mosquitoes. *J Exp Med* **56**, 853–865.

Demaio, J., Pumpuni, C. B., Kent, M. & Beier, J. C. (1996). The midgut bacterial flora of wild *Aedes triseriatus*, *Culex pipiens*, and *Psorophora columbiae* mosquitoes. *Am J Trop Med Hyg* **54**, 219–223.

Deubel, V., Digoutte, J. P., Mattei, X. & Pandare, D. (1981). Morphogenesis of yellow fever virus in *Aedes aegypti* culture cells. *Am J Trop Med Hyg* **30**, 1071–1077.

Doi, R. (1970). Studies on the mode of development of Japanese encephalitis virus in some groups of mosquitoes by the fluorescent antibody technique. *Jpn J Exp Med* **40**, 101–115.

Doi, R., Shiraska, A. & Sasa, M. (1967). The mode of development of Japanese encephalitis virus in the mosquito *Culex tritaeniorhynchus summorosus* as observed by the fluorescent antibody technique. *Jpn J Exp Med* **37**, 227–238.

Edwards, J. F., Higgs, S. & Beaty, B. J. (1998). Mosquito feeding-induced potentiation of Cache Valley Virus (Bunyaviridae) in mice. *J Med Entomol* **35**, 261–265.

Faran, M. E., Romoser, W. S., Routier, R. G. & Bailey, C. L. (1986). Use of the avidin-biotin-peroxidase complex immunocytochemical procedure for detection of Rift valley fever virus in paraffin sections of mosquitoes. *Am J Trop Med Hyg* **35**, 1061–1067.

Faran, M. E., Romoser, W. S., Routier, R. G. & Bailey, C. L. (1988). The distribution of Rift Valley fever virus in the mosquito *Culex pipiens* as revealed by viral titration of dissected organs and tissues. *Am J Trop Med Hyg* **39**, 206–213.

Freyvogel, T. A. & Stäubli, W. (1965). The formation of the peritrophic membrane in Culicidae. *Acta Trop* **22**, 118–147.

Friedman, R. M., Levin, J. D., Grimley, P. M. & Berezesky, I. K. (1972). Membrane-associated replication complex in arbovirus infection. *J Virol* **10**, 504–515.

Gaidamovitch, S. Y., Khutoretskaya, N. V., Lvova, A. I. & Sveshnikova, N. A. (1973). Immunofluorescent staining study of the salivary glands of mosquitoes with group A arboviruses. *Intervirology* **1**, 193–200.

Gardner, M. J., Hall, N., Fung E. & 42 other authors (2002). Genome sequence of the human malaria parasite *Plasmodium falciparum*. *Nature* **419**, 498–511.

Germi, R., Crance, J. M., Garin, D., Guimet, J., Lorat-Jacob, H., Ruigrok, R. W., Zarski, J. P. & Drouet, E. (2002). Heparin sulphate-mediated binding of infectious dengue virus type 2 and yellow fever virus. *Virology* **292**, 162–168.

Gillespie, R. D., Mbow, M. L. & Titus, R. G. (2000). The immunomodulatory factors of bloodfeeding arthropod saliva. *Parasite Immunol* **22**, 319–331.

Gliedman, J. B., Smith, J. F. & Brown, D. T. (1975). Morphogenesis of Sindbis virus in cultured *Aedes albopictus* cells. *J Virol* **16**, 913–926.

Gordh, G. & Headrick, D. H. (2001). *A Dictionary of Entomology*. Wallingford: CABI.

Gould, E. A., Buckley, A. & Cammack, N. (1985). Use of the biotin streptavidin interaction to improve flavivirus detection by immunofluorescence and ELISA tests. *J Virol Methods* **11**, 41–48.

Graham, D. H., Holmes, J. L., Higgs, S., Beaty, B. J. & Black, W. C. I. V. (1999). Selection of refractory and permissive strains of *Aedes triseriatus* (Diptera: Culicidae) for transovarial transmission of La Crosse virus. *J Med Entomol* **36**, 671–678.

Graham, D. H., Holmes, J. L., Beaty, B. J. & Black, W. C. I. V. (2003). Quantitative trait loci conditioning transovarial transmission of La Crosse virus in the eastern treehole mosquito, *Ochlerotatus triseriatus*. *Insect Mol Biol* **12**, 307–318.

Grief, C., Galler, R., Cortes, L. M. C. & Barth, O. M. (1997). Intracellular localization of dengue-2 RNA in mosquito cell culture using electron microscopic in situ hybridization. *Arch Virol* **142**, 2347–2357.

Grimstad, P. R. (1988). California group virus disease. In *The Arboviruses: Epidemiology and Ecology*, vol. II, pp. 99–136. Edited by T. P. Monath. Boca Raton: CRC Press.

Grimstad, P. R., Craig, G. B., Jr, Ross, Q. E. & Yuill, T. M. (1977). *Aedes triseriatus* and La Crosse virus: geographic variation in vector susceptibility and ability to transmit. *Am J Trop Med Hyg* **26**, 990–996.

Grimstad, P. R., Ross, Q. E. & Craig, G. B., Jr (1980). *Aedes triseriatus* (Diptera: Culicidae) and La Crosse virus. II. Modifications of mosquito feeding behavior by virus infection. *J Med Entomol* **17**, 1–7.

Gritsun, T. S., Holmes, E. C. & Gould, E. A. (1995). Analysis of flavivirus envelope proteins reveals variable domains that reflect their antigenicity and may determine their pathogenesis. *Virus Res* **35**, 307–321.

Grossman, G. L., Rafferty, C. S., Clayton, J. R., Stevens, T. K., Mukabayire, O. & Benedict, M. Q. (2001). Germline transformation of the malaria vector, *Anopheles gambiae*, with the *piggy*Bac transposable element. *Insect Mol Biol* **10**, 597–604.

Gubler, D. J. & Rosen, L. (1976). Variation among geographic strains of *Aedes albopictus* in susceptibility to oral infection with dengue viruses. *Am J Trop Med Hyg* **25**, 318–325.

Gubler, D. J., Nalim, S., Tan, R., Saipan, H. & Sulianti Saroso, J. (1979). Variation in susceptibility to oral infection with dengue viruses among geographic strains of *Aedes aegypti*. *Am J Trop Med Hyg* **28**, 1045–1052.

Gubler, D. J., Novak, R. & Mitchell, C. J. (1982). Arthropod vector competence – epidemiological, genetic, and biological considerations. In *Recent Developments in the Genetics of Insect Disease Vectors*, pp. 343–378. Edited by W. W. M. Steiner, W. J. Tabachnick, K. S. Rai & S. Narang. Champaign, IL: Stipes Publishing.

Handler, A. M. & James, A. A. (2000). *Insect Transgenesis. Methods and Applications*. Boca Raton: CRC Press.

Hardy, J. L. (1988). Susceptibility and resistance of vector mosquitoes. In *The Arboviruses: Epidemiology and Ecology*, vol. 1, pp. 87–126. Edited by T. P. Monath. Boca Raton: CRC Press.

Hardy, J. L. & Reeves, W. C. (1990). Experimental studies on infection in vectors. In *Epidemiology and Control of Mosquito-borne Arboviruses in California, 1943–1987*. Edited by W. C. Reeves. Sacramento, CA: California Mosquito and Vector Control Association.

Hardy, J. L., Reeves, W. C. & Sjogren, R. D. (1976). Variation in susceptibility of field and laboratory populations of *Culex tarsalis* to experimental infection with western equine encephalomyelitis virus. *Am J Epidemiol* **103**, 498–505.

Hardy, J. L., Apperson, G., Asman, S. M. & Reeves, W. C. (1978). Selection of a strain of *Culex tarsalis* highly resistant to infection following infection of western equine encephalitis virus. *Am J Trop Med Hyg* **27**, 313–321.

Hardy, J. L., Houk, E. J., Kramer, L. D. & Reeves, W. C. (1983). Intrinsic factors affecting vector competence of mosquitoes for arboviruses. *Annu Rev Entomol* **28**, 229–262.

Hase, T., Summers, P. L., Eckels, K. H. & Baze, W. B. (1987). Maturation process of Japanese encephalitis virus in cultured mosquito cells in vitro and mouse brain cells in vivo. *Arch Virol* **96**, 135–151.

Hayes, C. G., Baker, R. H., Baqar, S. & Ahmed, T. (1984). Genetic variation for West Nile virus susceptibility in *Culex tritaeniorhynchus*. *Am J Trop Med Hyg* **33**, 715–724.

Higgs, S., Rayner, J., Olson, K., Davis, B. S., Beaty, B. J. & Blair, C. D. (1998). Engineered resistance in *Aedes aegypti* to a West African and a South American strain of yellow fever virus. *Am J Trop Med Hyg* **58**, 663–670.

Hindle, E. (1929). An experimental study of yellow fever. *Trans R Soc Trop Med Hyg* **22**, 405–430.

Holt, R. A., Subramanian, G. M., Halpern, A. & 120 other authors (2002). The genome sequence of the malaria mosquito *Anopheles gambiae*. *Science* **298**, 129–159.

Houk, E. J., Kramer, L. D., Hardy, J. L. & Chiles, R. E. (1985). Western equine encephalomyelitis virus: *in vivo* infection and morphogenesis in mosquito mesenteronal epithelial cells. *Virus Res* **2**, 123–138.

Houk, E. J., Arcus, Y. M., Hardy, J. L. & Kramer, L. D. (1990). Binding of western equine encephalomyelitis virus to brush border fragments isolated from mesenteronal epithelial cells of mosquitoes. *Virus Res* **17**, 105–118.

James, A. A. (1994). Molecular and biochemical analyses of the salivary glands of vector mosquitoes. *Bull Inst Pasteur* **92**, 133–150.

Janzen, H. G., Rhoodes, A. J. & Doane, F. W. (1970). Chikungunya virus in salivary glands of *Aedes aegypti* (L.): an electron microscope study. *Can J Microbiol* **16**, 581–586.

Jasinskiene, N., Coates, C. J., Benedict, M. Q., Cornel, A. J., Rafferty, C. S., James, A. A. & Collins, F. H. (1998). Stable transformation of the yellow fever mosquito, *Aedes aegypti*, with the Hermes element from the housefly. *Proc Natl Acad Sci U S A* **95**, 3743–3747.

Jennings, A. D., Gibson, C. A., Miller, B. R. & 12 other authors (1994). Analysis of a yellow fever virus isolated from a fatal case of vaccine-associated human encephalitis. *J Infect Dis* **169**, 512–518.

Johnson, B. W., Chamber, T. V., Crabtree, M. B., Bhatt, T. R., Guirakhoo, F., Monath, T. P. & Miller, B. R. (2002). Growth characteristics of ChimeriVax-DEN2 vaccine virus in *Aedes aegypti* and *Aedes albopictus* mosquitoes. *Am J Trop Med Hyg* **67**, 260–265.

Johnson, B. W., Chambers, T. V., Crabtree, M. B., Arroyo, J., Monath, T. P. & Miller, B. R. (2003). Growth characteristics of the veterinary vaccine candidate ChimeriVax-West Nile (WN) virus in *Aedes* and *Culex* mosquitoes. *Med Vet Entomol* **17**, 235–243.

Jupp, P. G. (1974). Laboratory studies on the transmission of West Nile virus by *Culex* (*Culex*) *univittatus* Theobald; factors influencing the transmission rate. *J Med Entomol* **11**, 455–458.

Jupp, P. G. (1985). *Culex theileria* and Sindbis virus; salivary glands infection in relation to transmission. *J Am Mosq Control Assoc* **1**, 374–376.

Jupp, P. G. & Kemp, A. (2002). Laboratory vector competence experiments with yellow fever virus and five South African mosquito species including *Aedes aegypti*. *Trans R Soc Trop Med Hyg* **96**, 493–498.

Karabatsos, N. (editor) (1985). *International Catalogue of Arboviruses*, 3rd edn. San Antonio, TX: American Society of Tropical Medicine and Hygiene.

Kinney, R. M., Butrapet, S., Chang, G.-J. J., Tsuchiya, K. R., Roehrig, J. T., Bhamarapravati, N. & Gubler, D. (1997). Construction of infectious cDNA clones from Dengue 2 virus: strain 16681 and its attenuated vaccine derivative strain PDK-53. *Virology* **230**, 300–308.

Klimstra, W. B., Ryman, K. D. & Johnston, R. E. (1998). Adaptation of Sindbis virus to BHK cells selects for use of heparin sulphate as an attachment receptor. *J Virol* **72**, 7357–7366.

Kramer, L. D. & Ebel, G. D. (2003). Dynamics of flavivirus infection in mosquitoes. *Adv Virus Res* **60**, 187–232.

Kramer, L. D., Hardy, J. L., Presser, S. B. & Houk, E. J. (1981). Dissemination barriers for Western equine encephalomyelitis virus in *Culex tarsalis* after ingestion of low viral doses. *Am J Trop Med Hyg* **30**, 190–197.

Kramer, L. D., Hardy, J. L., Houk, E. J. & Presser, S. B. (1989). Characterization of the mesenteronal infection with western equine encephalomyelitis virus in an incompetent strain of *Culex tarsalis*. *Am J Trop Med Hyg* **41**, 241–250.

Kramer, L. D., Presser, S. B. & Hardy, J. L. (1998). Characterization of modulation of western equine encephalomyelitis virus by *Culex tarsalis* (Diptera: Culicidae) maintained at 32 °C following parenteral infection. *J Med Entomol* **35**, 289–295.

Kuberski, T. (1979). Fluorescent antibody studies on the development of dengue-2 virus in *Aedes albopictus* (Diptera: Culicidae). *J Med Entomol* **16**, 343–349.

Kuno, G., Chang, G.-J. J., Tsuchiya, K. R., Karabatsos, N. & Cropp, C. B. (1998). Phylogeny of the genus Flavivirus. *J Virol* **72**, 73–83.

Lam, K. S. S. & Marshall, I. D. (1968a). Dual infections of *Aedes aegypti* with arboviruses. I. Arboviruses that have no apparent cytopathic effect in the mosquito. *Am J Trop Med Hyg* **17**, 625–636.

Lam, K. S. S. & Marshall, I. D. (1968b). Dual infections of *Aedes aegypti* with arboviruses. II. Salivary-gland damage by Semliki Forest virus in relation to dual infections. *Am J Trop Med Hyg* **17**, 627–644.

LaMotte, L. C. (1960). Japanese B encephalitis virus in the organs of infected mosquitoes. *Am J Hyg* **72**, 73–87.

Larsen, J. R. & Ashley, R. F. (1971). Demonstration of Venezuelan equine encephalomyelitis virus in tissues of *Aedes aegypti*. *Am J Trop Med Hyg* **20**, 754–760.

Leake, C. J. (1984). Transovarial transmission of arboviruses by mosquitoes. In *Vectors in Virus Biology*, pp. 63–91. Edited by C. A. Mayo & K. A. Harrap. London: Academic Press.

Leake, C. J. (1992). Arbovirus-mosquito interactions and vector specificity. *Parasitol Today* **8**, 123–128.

Leake, C. J. & Johnson, R. T. (1987). The pathogenesis of Japanese encephalitis virus in *Culex tritaeniorhynchus* mosquitoes. *Trans R Soc Trop Med* **81**, 681–685.

Lehane, M. J. (1976). The formation and histochemical structure of the peritrophic membrane of the stable fly, *Stomoxys calcitrans*. *J Insect Physiol* **22**, 1551–1557.

Lehane, M. J. & Billingsley, P. F. (1996). *Biology of the Insect Midgut*. London: Chapman & Hall.

Lerdthusnee, K., Romoser, W. S., Faran, M. E. & Dohm, D. J. (1995). Rift valley fever virus in the cardia of *Culex pipiens*: an immunocytochemical and ultrastructural study. *Am J Trop Med Hyg* **53**, 331–337.

Limesand, K. H., Higgs, S., Pearson, L. D. & Beaty, B. J. (2000). Potentiation of vesicular stomatitis New Jersey virus infection in mice by mosquito saliva. *Parasite Immunol* **22**, 461–467.

Limesand, K. H., Higgs, S., Pearson, L. D. & Beaty, B. J. (2003). The effect of mosquito salivary gland treatment on vesicular stomatitis New Jersey virus replication and interferon α/β expression *in vitro*. *J Med Entomol* **40**, 199–205.

Lorenz, L., Beaty, B. J., Aitken, T. H. G., Wallis, G. P. & Tabachnick, W. J. (1984). The effect of colonization upon *Aedes aegypti* susceptibility to oral infection with yellow fever virus. *Am J Trop Med Hyg* **33**, 690–694.

Ludwig, G. V., Christensen, B. M., Yuill, T. M. & Schultz, K. T. (1989). Enzyme processing of La Crosse virus glycoprotein G1: a bunyavirus-vector infection model. *Virology* **171**, 108–113.

Ludwig, G. V., Isreal, B. A., Christensen, B. M., Yuill, T. M. & Schultz, K. T. (1991). Role of La Crosse virus glycoproteins in attachment of virus to host cells. *Virology* **181**, 564–571.

Ludwig, G. V., Kondig, J. P. & Smith, J. F. (1996). A putative receptor for Venezuelan equine encephalitis virus from mosquito cells. *J Virol* **70**, 5592–5599.

Mackenzie, J. M., Jones, M. K. & Westaway, G. (1999). Marker for *trans*-Golgi membranes and the intermediate compartment localize to induced membranes with distinct replication functions in flavivirus-infected cells. *J Virol* **73**, 9555–9567.

McClean, D. M., Clarke, A. M., Coleman, J. C., Motalbetti, C. A., Skidmore, A. G., Walters, T. E. & Wise, R. (1974). Vector capability of *Aedes aegypti* mosquitoes for California encephalitis and dengue viruses at various temperatures. *Can J Microbiol* **20**, 255–262.

McGaw, M. M., Chandler, L. J., Wasieloski, L. P., Blair, C. D. & Beaty, B. J. (1998). Effect of La Crosse virus infection on overwintering of *Ae. triseriatus*. *Am J Trop Med Hyg* **58**, 168–175.

McKelvey, J. J., Jr, Eldridge, B. F. & Maramorosch, K. (1981). *Vectors of Disease Agents*. New York: Praeger.

Mendoza, M. Y., Salas-Benito, J. S., Lanz-Mendoza, H., Hernandez-Martinez, S. & del Angel, R. M. (2002). A putative receptor for dengue virus in mosquito tissues: localization of a 45-kDa glycoprotein. *Am J Trop Med Hyg* **67**, 76–84.

Miles, J. A. R., Pillai, J. S. & Maguire, T. (1973). Multiplication of Whataroa virus in mosquitoes. *J Med Entomol* **10**, 176–185.

Miller, B. R. & Adkin, D. (1988). Biological characterization of plaque-size variants of yellow fever virus in mosquitoes and mice. *Acta Virol* **32**, 227–234.

Miller, B. R. & Mitchell, C. J. (1991). Genetic selection of a flavivirus-refractory strain of the yellow fever mosquito *Aedes aegypti*. *Am J Trop Med Hyg* **45**, 399–407.

Miller, B. R., Beaty, B. J., Aitken, T. M., Eckels, K. H. & Russell, P. K. (1982). Dengue-2 vaccine: oral infection, transmission, and lack of evidence for reversion in the mosquito, *Aedes aegypti*. *Am J Trop Med Hyg* **32**, 1232–1237.

Miller, L. K. & Ball, L. A. (1998). *The Insect Viruses*. New York: Plenum.

Mims, C. A., Day, M. F. & Marshall, I. D. (1966). Cytopathic effect of Semliki Forest virus in the mosquito *Aedes aegypti*. *Am J Trop Med Hyg* **15**, 775–784.

Mitchell, C. J. (1983). Mosquito vector competence and arboviruses. In *Current Topics in Vector Research*, vol. 1, pp. 63–92. Edited by K. F. Harris. New York: Praeger.

Muangman, D., Frothingham, T. E. & Spielman, A. (1969). Presence of Sindbis virus in anal discharge of infected *Aedes aegypti*. *Am J Trop Med Hyg* **18**, 401–410.

Murphy, F. A., Whitfield, S. G., Sudia, W. D. & Chamberlain, R. W. (1975). Interactions of a vector with a vertebrate pathogenic virus. In *Invertebrate Immunity: Mechanisms of Invertebrate Vector-Parasite Relations*, pp. 25–48. Edited by K. Maramorosch & R. E. Shope. New York: Academic Press.

Myles, K. M., Pierro, D. J. & Olson, K. E. (2003). Deletions in the putative cell-receptor-binding domain of Sindbis virus strain MRE-16 E2 glycoprotein reduce midgut infectivity in *Aedes aegypti*. *J Virol* **77**, 8872–8881.

Ng, M. L. (1987). Ultrastructural studies of Kunjin virus-infected *Aedes albopictus* cell. *J Gen Virol* **68**, 577–582.

Nuttall, P. A. & Labuda, M. (2003). Dynamics of infection in tick vectors and at the tick-host interface. *Adv Virus Res* **60**, 233–272.

Nuttall, P. A., Jones, L. D. & Dacies, C. R. (1991). The role of arthropods in arbovirus evolution. *Adv Dis Vector Res* **8**, 15–45.

Olson, K. E., Higgs, S., Gaines, P. J., Powers, A. M., Davis, B. S., Kamrud, K. I., Carlson, J. O., Blair, C. D. & Beaty, B. J. (1996). Genetically engineered resistance in mosquitoes to dengue virus transmission. *Science* **272**, 884–886.

Olson, K. E., Adelman, Z. N., Travanty, E. A., Sanchez-Vargas, I., Beaty, B. J. & Blair, C. D. (2002). Developing arbovirus resistance in mosquitoes. *Insect Biochem Mol Biol* **32**, 1333–1343.

Osorio, J. E., Godsey, M. S., Defoliart, G. R. & Yuill, T. M. (1996). La Crosse viremias in white-tailed deer and chipmunks exposed by injection or mosquito bite. *Am J Trop Med Hyg* **54**, 338–342.

Paulson, S. L., Grimstad, P. R. & Craig, G. B., Jr (1989). Midgut and salivary gland barriers to La Crosse virus dissemination in mosquitoes of the *Aedes triseriatus* group. *Med Vet Entomol* **3**, 113–123.

Peltier, M., Durieux, C., Jonchere, H. & Arquie, E. (1939). La transmission par piqure de Stegomyia, du virus amaril neurotrope present dans le sang des personnes recemment vaccinees, est-elle possible dans les regions ou ce moustique existe en abondance? *Rev d'Immunol* **5**, 172–195.

Peters, W. (1992). *Peritrophic Membranes*. Berlin: Springer.

Powers, A. M., Kamrud, K. I., Olson, K. E., Higgs, S., Carlson, J. O. & Beaty, B. J. (1996). Molecularly engineered resistance to California serogroup virus replication in mosquito cells and mosquitoes. *Proc Natl Acad Sci U S A* **93**, 4187–4191.

Putnam, J. L. & Scott, T. W. (1995). Blood-feeding behavior of dengue-2 virus-infected *Aedes aegypti*. *Am J Trop Med Hyg* **52**, 225–227.

Reisen, W. K., Chiles, R. E., Kramer, L. D., Martinez, V. M. & Eldridge, B. F. (2000). Method of infection does not alter response of chicks and house finches to western equine encephalomyelitis and St. Louis encephalitis viruses. *J Med Entomol* **37**, 250–258.

Ribeiro, J. M. C. (1989). Vector saliva and its role in parasite transmission. *Exp Parasitol* **69**, 104–106.

Ribeiro, J. M. C. (1995). Blood-feeding arthropods: live syringes or invertebrate pharmacologists? *Infect Agents Dis* **4**, 143–152.

Ribeiro, J. M. C. & Francischetti, I. M. B. (2002). Role of arthropod saliva in blood feeding: sialome and post-sialome perspectives. *Annu Rev Entomol* **48**, 73–88.

Richardson, M. W. & Romoser, W. S. (1972). The formation of the peritrophic membrane in adult *Aedes triseriatus* (Say.) (Diptera: Culicidae). *J Med Entomol* **9**, 495–500.

Riehle, M. A., Srinivasan, P., Moreira, C. K. & Jacobs-Lorena, M. (2003). Towards genetic manipulation of wild mosquito populations to combat malaria: advances and challenges. *J Exp Biol* **206**, 3809–3816.

Romoser, W. S., Faran, M. E. & Bailey, C. L. (1987). Newly recognized route of arbovirus dissemination from the mosquito (Diptera: Culicidae) midgut. *J Med Entomol* **24**, 431–432.

Romoser, W. S., Faran, M. E., Bailey, C. L. & Lerdthusnee, K. (1992). An immuno-cytochemical study of the distribution of Rift Valley fever virus in the mosquito *Culex pipiens*. *Am J Trop Med Hyg* **46**, 489–501.

Rosen, L. (1987). Sexual transmission of dengue viruses by *Aedes albopictus*. *Am J Trop Med Hyg* **37**, 398–402.

Rosen, L., Roseboom, L. E., Gubler, D. J., Lein, J. C. & Chaniotis, B. N. (1985). Comparative susceptibility of mosquito species and strains to oral and parenteral infection with dengue and Japanese encephalitis viruses. *Am J Trop Med Hyg* **34**, 1219–1224.

Roubaud, E. & Stephanopoulo, G. J. (1933). Recherches sur la transmission par la voic stegomyienne du virus neurotrop murin de la fievre jaune. *Bull Soc Pathol Exot* **26**, 305–309.

Roubaud, E., Colas-Belcour, J. & Stefanopoulo, G. J. (1937). Transmission de la fievre jaune par un moustique palearetique repandu dans la region parisienne, l' *Aedes geniculatus Oliv. Clin Rev Acad Sci* **205**, 182.

Sabin, A. B. (1952). Research on dengue during world war II. *Am J Trop Med Hyg* **1**, 30–50.

Salas-Benito, J. S. & del Angel, R. M. (1997). Identification of two surface proteins from C6/36 cells that bind dengue type 4 virus. *J Virol* **71**, 7246–7252.

Sanchez-Vargas, I., Travanty, E. A., Keene, K. M., Franz, A. E., Adelman, Z. N., Beaty, B. J., Blair, C. D. & Olson, K. E. (2003). Arthropods, arboviruses, and RNA interference. *Virus Res* (in press).

Schoepp, R. J., Beaty, B. J. & Eckels, K. H. (1990). Dengue 3 virus infection of *Aedes albopictus* and *Aedes aegypti*: comparison of parent and progeny candidate vaccine viruses. *Am J Trop Med Hyg* **42**, 89–96.

Schoepp, R. J., Beaty, B. J. & Eckels, K. H. (1991). Infection of *Aedes albopictus* and *Aedes aegypti* mosquitoes with dengue parent and progeny candidate vaccine viruses: a possible marker of human attenuation. *Am J Trop Med Hyg* **45**, 202–210.

Scott, T. W., Hildreth, S. W. & Beaty, B. J. (1984). The distribution and development of eastern equine encephalitis virus in its enzootic mosquito vector, *Culiseta melanura*. *Am J Trop Med Hyg* **33**, 300–310.

Shahabbudin, M. (2002). Do *Plasmodium* ookinetes invade a specific cell type in the mosquito midgut? *Trends Parasitol* **18**, 157–161.

Shao, L., Devenport, M. & Jacobs-Lorena, M. (2001). The peritrophic matrix of hema-tophagous insects. *Arch Insect Biochem Physiol* **47**, 119–125.

Shiao, S. H., Higgs, S., Adelman, Z., Christensen, B. M., Liu, S. H. & Chen, C. C. (2001). Effect of prophenoloxidase expression knockout on the melanization of filarial worms in the mosquito, *Armigeres subalbatus*. *Insect Mol Biol* **10**, 315–321.

Sonnenberg, A. (1992). Laminin receptors in the integrin family. *Pathol Biol* **40**, 773–778.

Sriurairatna, S. & Bhamarapravati, N. (1977). Replication of dengue-2 virus in *Aedes albopictus* mosquitoes. An electron microscopic study. *Am J Trop Med Hyg* **26**, 1199–1205.

Stohler, H. R. (1961). The peritrophic membrane in blood sucking Diptera in relation to their role as vectors of blood parasites. *Acta Trop* **18**, 263–266.

Sundin, D. R., Beaty, B. J., Nathanson, N. & Gonzalez-Scarano, F. (1987). A G1 glycoprotein epitope of La Crosse virus: a determinant of infection of *Aedes triseriatus*. *Science* **235**, 591–593.

Tabachnick, W. J. (1991). Genetic control of oral susceptibility to infection of *Culicoides variipennis* for bluetongue virus. *Am J Trop Med Hyg* **45**, 666–671.

Tabachnick, W. J. (1994). Genetics of arthropod vector competence for arboviruses. *Adv Dis Vector Res* **10**, 93–108.

Tabachnick, W. J. (2003). Reflections on the *Anopheles gambiae* genome sequence, transgenic mosquitoes and the prospect for controlling malaria and other vector borne diseases. *J Med Entomol* **40**, 597–606.

Tabachnick, W. J., Wallis, G. P., Aitken, T. H., Miller, B. R., Amato, G. D., Lorenz, L., Powell, J. R. & Beaty, B. J. (1985). Oral infection of *Aedes aegypti* with yellow fever virus: geographic variation and genetic considerations. *Am J Trop Med Hyg* **34**, 1219–1224.

Takahashi, M. & Suzuki, K. (1979). Japanese encephalitis virus in mosquito salivary glands. *Am J Trop Med Hyg* **28**, 122–135.

Tardieux, I., Poupel, O., Lapchin, L. & Rodhain, F. (1990). Variation among strains of *Aedes aegypti* in susceptibility to oral infection with dengue virus type 2. *Am J Trop Med Hyg* **43**, 308–313.

Tesh, R. B., Gubler, D. J. & Rosen, L. (1976). Variation among geographic strains of *Aedes albopictus* in susceptibility to infection with chikungunya virus. *Am J Trop Med Hyg* **25**, 326–335.

Thomas, L. A. (1963). Distribution of the virus of western equine encephalomyelitis in the mosquito vector, *Culex tarsalis*. *Am J Hyg* **78**, 150–165.

Thompson, W. H. & Beaty, B. J. (1977). Venereal transmission of La Crosse (California encephalitis) arbovirus in *Aedes triseriatus* mosquitoes. *Science* **196**, 530–531.

Turrell, M. J. (1988). Horizontal and vertical transmission of viruses by insect and tick vectors. In *The Arboviruses: Epidemiology and Ecology*, vol. 1, pp. 128–152. Edited by T. P. Monath. Boca Raton: CRC Press.

Turrell, M. J., Gargan, T. P., II & Bailey, C. L. (1984). Replication and dissemination of Rift Valley fever virus in *Culex pipiens*. *Am J Trop Med Hyg* **33**, 176–181.

Turrell, M. J., Saluzzo, J.-F., Tammariello, R. F. & Smith, J. F. (1990). Generation and transmission of Rift Valley fever viral reassortants by the mosquito *Culex pipiens*. *J Gen Virol* **71**, 2307–2312.

Uchil, P. D. & Satchidanandam, V. (2003). Architecture of the flavivirus replication complex. *J Biol Chem* **278**, 24388–24398.

Valenzuela, J. G., Pham, V. M., Garfield, M. K., Francischetti, I. M. B. & Ribeiro, J. M. C. (2002). Toward a description of the sialome of the adult female mosquito *Aedes aegypti*. *Insect Biochem Mol Biol* **32**, 1101–1122.

van Heusden, M. C. (1996). Fat body and hemolymph. In *The Biology of Disease Vectors*, pp. 349–355. Edited by W. C. Marquardt & B. J. Beaty. Niwot, CO: University Press of Colorado.

Vazeille-Falcoz, M., Mousson, L., Rhodhain, F., Chungue, E. & Failloux, A. B. (1999). Variation in oral susceptibility to dengue type 2 virus populations of *Aedes aegypti* from the islands of Tahiti and Moorea, French Polynesia. *Am J Trop Med Hyg* **60**, 292–299.

Wallis, G. P., Tabachnick, W. J. & Powell, J. R. (1984). Genetic heterogeneity among Caribbean populations of *Aedes aegypti*. *Am J Trop Med Hyg* **33**, 492–498.

Wallis, G. P., Aitken, T. H. G., Lorenz, L., Amato, G. D., Powell, J. R. & Tabachnick, W. J. (1985). Selection for susceptibility and refractoriness of *Aedes aegypti* to oral infection with yellow fever virus. *Am J Trop Med Hyg* **34**, 1225–1231.

Watts, D. M., Thompson, W. H., Yuill, T. M., DeFoliart, G. R. & Hanson, R. P. (1974). Overwintering of La Crosse in *Aedes triseriatus*. *Am J Trop Med Hyg* **23**, 694–700.

Watts, D. M., DeFoliart, G. R. & Yuill, T. M. (1976). Experimental transmission of trivittatus virus (California virus group) by *Aedes trivittatus*. *Am J Trop Med Hyg* **25**, 173–176.

Weaver, S. C. (1986). Electron microscopic analysis of infection patterns for Venezuelan equine encephalomyelitis virus in the vector mosquito, *Culex (Melanoconion) taeniopus*. *Am J Trop Med Hyg* **35**, 624–631.

Weaver, S. C., Scherer, F. W., Cupp, E. W. & Castello, D. A. (1984). Barriers to dissemination of Venezuelan encephalitis viruses in the Middle American enzootic vector mosquito, *Culex (Melanconion) taenopus*. *Am J Trop Med Hyg* **33**, 953–960.

Weaver, S. C., Scott, T. W., Lorenz, L. H., Lerdthusnee, K. & Romoser, W. S. (1988). Togavirus-associated pathologic changes in the midgut of a natural vector. *J Virol* **62**, 2083–2090.

Weaver, S. C., Scott, T. W. & Lorenz, L. H. (1990). Patterns of infection of *Culiseta melanura* by eastern equine encephalomyelitis virus. *J Med Entomol* **27**, 878–891.

Weaver, S. C., Scott, T. W., Lorenz, L. H. & Repik, P. M. (1991). Detection of eastern equine encephalomyelitis virus deposition in *Culiseta melanura* following ingestion of radiolabeled virus in blood meals. *Am J Trop Med Hyg* **44**, 250–259.

Weaver, S. C., Lorenz, L. H. & Scott, T. W. (1992). Pathologic changes in the midgut of *Culex tarsalis* following infection with western equine encephalomyelitis virus. *Am J Trop Med Hyg* **47**, 691–701.

Whitfield, S. G., Murphy, F. A. & Sudia, W. D. (1973). St. Louis encephalitis virus: an ultrastructural study of infection in a mosquito vector. *Virology* **56**, 70–87.

Whitman, L. (1937). Multiplication of the virus of yellow fever in *Aedes aegypti*. *J Exp Med* **66**, 133–143.

Whitman, L. (1939). Failure of *Aedes aegypti* to transmit yellow fever cultured virus (17D). *Am J Trop Med Hyg* **19**, 19–26.

Woodward, T. M., Miller, B. R., Beaty, B. J., Trent, D. W. & Roehrig, J. T. (1991). A single amino acid change in the E2 glycoprotein of Venezuelan equine encephalitis virus affects replication and dissemination in *Aedes aegypti* mosquitoes. *J Gen Virol* **72**, 2431–2435.

Zanotto, P. M., Gould, E. A., Gao, G. F., Harvey, P. H. & Holmes, E. C. (1996). Population dynamics of flaviviruses revealed by molecular phylogenetics. *Proc Natl Acad Sci U S A* **93**, 548–553.

Zeidner, N. S., Higgs, S., Happ, C. M., Beaty, B. J. & Miller, B. R. (1999). Mosquito feeding modulates Th1 and Th2 cytokines in flavivirus susceptible mice: an effect mimicked by injection of sialokinins, but not demonstrated in flavivirus resistant mice. *Parasite Immunol* **21**, 35–44.

Zieler, H., Garon, C. F., Fischer, E. R. & Shahabuddin, M. (2000). A tubular network associated with the brush-border surface of the *Aedes aegypti* midgut: implications for pathogen transmission by mosquitoes. *J Exp Biol* **203**, 1599–1611.

Vector competence

Scott C. Weaver, Lark L. Coffey, Roberto Nussenzveig, Diana Ortiz and Darci Smith

Center for Biodefense and Emerging Infectious Diseases and Department of Pathology, University of Texas Medical Branch, Galveston, TX 77555-0609, USA

INTRODUCTION TO THE CONCEPT OF VECTOR COMPETENCE

The importance of vector-borne diseases

Infectious diseases remain the leading causes of morbidity and mortality worldwide and arthropod-borne diseases include many of the most important, especially in the tropics and developing countries. For example, malarial parasites infect an estimated 200 million people annually in Africa alone with mortality estimates of about 1 million persons, primarily children (Greenwood, 1999). Vector-borne viruses are also important human pathogens. Approximately 2500 million people (two-fifths of the world's population) are at risk from dengue with about 50 million cases each year (www.who.int/inf-fs/en/fact117.html). Many other vector-borne pathogens cause emerging diseases that have undergone resurgence or threaten to increase in prevalence or distribution in the coming years. However, despite the importance of vector-borne diseases, the mechanisms of transmission of many vector-borne pathogens remain poorly understood.

Here, we review the various factors that contribute to the transmission of pathogens by arthropod vectors. Due to space limitations, we focus primarily on human disease, with emphasis on mosquito-borne pathogens, especially arthropod-borne viruses (arboviruses).

Mechanisms of pathogen transmission by arthropod vectors

Vector-borne pathogens of humans and other animals circulate between their arthropod vectors and vertebrate hosts. Arboviruses represent the most basic form of

SGM symposium 63: Microbe–vector interactions in vector-borne diseases.
Editors S. H. Gillespie, G. L. Smith & A. Osbourn. Cambridge University Press. ISBN 0 521 84312 X ©SGM 2004

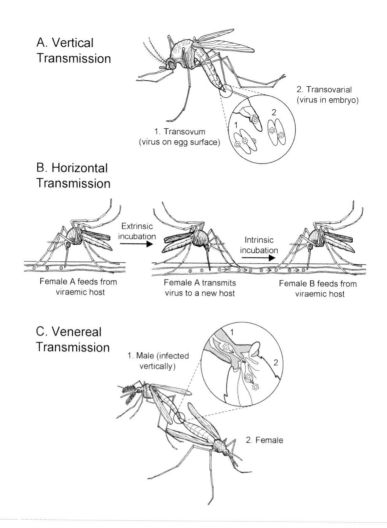

Fig. 1. Transmission of arboviruses by mosquitoes. (A) Vertical transmission occurs when the female mosquito passes the virus to its progeny. Both male and female mosquitoes can be infected via vertical transmission. (B) Horizontal transmission occurs following the infection of a vertebrate host by the bite of an infectious mosquito. After intrinsic incubation during which time the host becomes viraemic, oral infection of susceptible mosquitoes occurs following ingestion of viraemic blood. (C) Venereal transmission is a form of horizontal transmission that occurs when a male that was infected vertically copulates with a female and transmits the virus directly to her with no involvement of a vertebrate.

vector-borne transmission, and undergo horizontal transmission between vertebrate and arthropod hosts (Fig. 1). In some cases they are maintained by vertical transmission from adult arthropod to offspring. Usually a restricted invertebrate host range is observed with only one or few vectors involved in transmission. Within the infected arthropod, the virus must undergo an extrinsic incubation period of variable duration before biological transmission to a susceptible vertebrate host (Fig. 2).

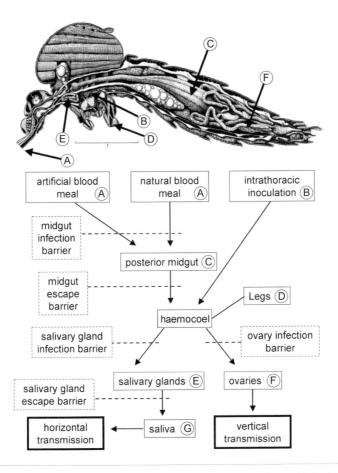

Fig. 2. Illustration of the internal anatomy of the mosquito showing critical sites of infection and dissemination for arbovirus transmission, and flow diagram below showing typical experimental infection approaches and organs/sites assayed. Horizontal, dashed lines show barriers at different stages of the infection process that prevent horizontal or vertical transmission. Circles with enclosed letters show critical organs infected and sampled to determine infection status. Oral infection is initiated by a natural or artificial blood meal (A). Intrathoracic infection (B) typically is used to bypass a midgut infection or escape barrier. Sites used typically for assay to assess vector competence include the midgut (C, often dissected), the legs (D) to sample the haemocoel and assess dissemination from the midgut, the salivary glands (E, often dissected), the saliva (G, collected by inducing salivation into artificial substrates) to assess transmission potential and the ovaries (F). Modified from Jobling & Lewis (1987), with permission from the Wellcome Library, London.

Arbovirus infection is initiated following blood feeding when an ingested virus passes with blood into the arthropod gut, a hostile environment where temperature and pH change, proteolytic enzymes are secreted into the lumen and a chitinous peritrophic matrix forms to isolate the infected blood from the midgut epithelium (Fig. 2). Initiation of virus infection of the mosquito posterior midgut epithelium requires ingestion of sufficient virus to exceed an infection threshold, defined as the minimum

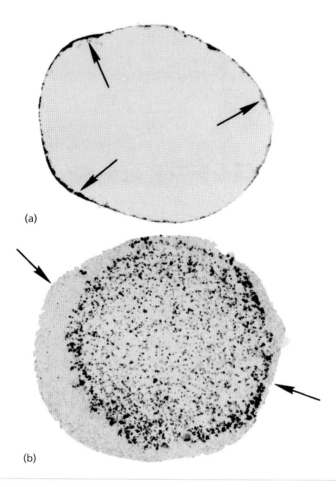

Fig. 3. Transverse section of the abdominal midgut of *Culex tarsalis* mosquitoes following feeding on a viraemic chicken (a) or an artificial blood meal (b) containing ^{3}H-labelled WEEV detected by autoradiography. Brightfield microscopy (mosquito tissues are invisible). Note the concentration of virus particles (silver grains) adjacent to the midgut epithelium (arrows) in the natural blood meal (a) and lack of concentration in the artificial blood meal (b). Adapted from Weaver *et al.* (1993).

titre needed to infect about 1–5 % of individuals (Chamberlain *et al.*, 1954). Exposure to a virus via a blood meal from a viraemic vertebrate generally results in more efficient infection than exposure to artificial blood meals. The major reason for this difference appears to be that natural blood meals clot in the mosquito midgut, resulting in expression of serum along with virus to the periphery where it is concentrated adjacent to the midgut epithelium (Weaver *et al.*, 1993) (Fig. 3). Generally, infection of the posterior midgut epithelial cells is detected first, with a concentration at the posterior end near the hindgut (Fig. 4). Following infection and replication, virus must escape into the haemocoel from the midgut epithelium and replicate in the salivary glands to

Fig. 4. Intact *Ae. aegypti* midgut assayed using immunofluorescence detection of Sindbis virus 2–3 days post-infection. Sindbis virus antigen is restricted to the bulbous posterior portion of the midgut in the centre. Note concentration of antigen on the right adjacent to the hindgut and Malpighian tubules. Magnification, ×110. Adapted from Myles *et al.* (2003).

be transmitted orally, and/or in the reproductive organs for transovarial or venereal transmission. Arthropods transmit virus in their saliva when feeding on a vertebrate. For mosquitoes, only the female ingests blood and transmits horizontally, but for ticks and some others, both males and females transmit. Non-viraemic transmission via simultaneous feeding of both infected and uninfected vectors on the same uninfected host in the absence of detectable viraemia has been reported for several viruses (Labuda *et al.*, 1993; Randolph *et al.*, 1996; Randolph, 1998; Mead *et al.*, 2000). In addition to transovarial transmission (TOT) via infection of the ovaries, seminal fluid from infected males can sometimes infect progeny during fertilization. Arboviruses can also be transmitted venereally (a form of horizontal transmission) from male to female adult mosquito (Woodring *et al.*, 1996).

Several different mechanisms of horizontal transmission can transfer a pathogen from a vector to a vertebrate (Fig. 1, Table 1). The most common, utilized by arboviruses, is through saliva that is injected when an arthropod probes or feeds on blood. Some filarial parasites congregate in the head and mouthparts of vectors and emerge to enter the puncture site actively during blood feeding. The parasite *Trypanosoma cruzi* (aetiologic agent of Chagas' disease) and the bacterium *Rickettsia typhi* (agent of typhus) exit in an arthropod's faeces that are defecated during or after blood feeding. When the host rubs its eyes or scratches its skin, the pathogen gains entrance.

Leishmania spp. and *Yersinia pestis* (plague bacillus) are regurgitated by the arthropod while attempting to feed. These pathogens interfere with feeding, facilitating regurgitation when the vector attempts to clear the obstruction. Other pathogens such as *Rickettsia recurrentis* escape from the vector when it is crushed by the host, gaining entry when the host scratches the affected site. In rare instances, a host can become infected by ingesting an infected arthropod, such as when *T. cruzi* infects rodents that eat the infected vector bug (Edman, 2000). For mechanical transmission, the pathogen does not develop or replicate in the vector, but is transmitted directly, usually by contaminated mouthparts. In the case of arboviruses, after an arthropod feeds on viraemic blood, extracellular virus on the mouthparts is inactivated eventually, so mechanical transmission can occur for a limited time only. Occasionally mechanical transmission is important in virus outbreaks but is not the main contributor to most arboviral cycles (Woodring *et al.*, 1996).

Not all vector species are competent for biological transmission because of physical or replicative barriers to infection and dissemination in the vector (Fig. 2). These include the midgut infection and escape barriers, as well as salivary gland infection and escape barriers (Hardy *et al.*, 1983).

THE CONCEPT OF VECTOR COMPETENCE

For the purposes of our discussion, we define vector competence as *the innate ability of a vector to acquire a pathogen and to successfully transmit it to another susceptible host.*

History of research on vector transmission

Throughout history people have speculated about an association between blood-sucking arthropods and illness. However, the involvement of arthropods in disseminating disease was demonstrated first only about 100 years ago. One of the earliest references to a role for arthropods in disease comes from the 16th century Italian physician Mercurialis, who hypothesized that by feeding on bodily secretions of dying persons, flies transmitted plague by depositing it into human food with their excretions (James & Harwood, 1969; Riley & Johannsen, 1915). Although we now know that this is not the mode of plague transmission, Mercurialis' speculation on a role for flies in mechanical transmission of pathogens was correct. Surprisingly, his observations pre-date van Leeuwenhoek's development of the microscope and his 1676 report of 'animalculus', now thought to have been bacteria in his aqueous samples.

The golden age of medical entomology began in the latter half of the 19th century. In 1848, Josiah Nott suggested that the causes of yellow fever and malaria existed in some form of insect life (Nott, 1848). Around the same time, the French physician

Table 1. Mechanism of pathogen transmission by arthropod vectors

Adapted from Edman (2000).

Transmission category	Specific route of transmission	Pathogen/vector
Vertical	Transovarial (mother to offspring)	*Babesia bigemina*/tick
Vertical	Transstadial (stage to stage)	*Borrelia burgdorferi*/tick
Horizontal	Venereal (male to female)	La Crosse virus/mosquito
Horizontal	Co-feeding	*Borrelia burgdorferi*/tick
Horizontal	Salivation	*Plasmodium* spp./mosquito
Horizontal	Stercorarian	*Trypanosoma cruzi*/triatomid bug
Horizontal	Regurgitation	*Yersinia pestis*/flea
Horizontal	Assisted escape/passive transfer	*Borrelia recurrentis*/louse
Horizontal	Active escape/invasion	*Onchocerca* spp./black fly
Horizontal	Ingestion by host	*Dipylidium caninum*/flea

Beauperthuy suggested that yellow fever and other diseases are transmitted by mosquitoes (Beauperthuy, 1854; Philip & Rozeboom, 1973). In his 1909 review, Boyce called Beauperthuy the father of the doctrine of insect-borne disease. In 1877, Louis Pasteur, after much study of thus far invisible organisms, formulated his 'germ theory'. However, it was not until 1878 that irrefutable evidence of the role for arthropods as vectors of pathogens was established. In the first discovery of primary importance in medical entomology, Manson reported development of the filarial worm *Wuchereria bancrofti* in the mosquito *Culex fatigans* (now called *Culex pipiens*) (Manson, 1878). A series of discoveries further implicating arthropods in dissemination of disease immediately followed Manson's landmark work. Of major significance were: (1) the demonstration that Texas cattle fever is transmitted by the tick *Boophilus annulata* (Smith & Kilbourne, 1893); (2) malaria parasite development in and transmission by mosquitoes (Grassi *et al.*, 1899; Philip & Rozeboom, 1973; Ross, 1897, 1898); and (3) the 1900 report by Reed and colleagues of yellow fever virus (YFV) transmission by *Aedes aegypti* (Philip & Rozeboom, 1973).

Understanding vector competence

Following these historical discoveries, investigators soon recognized that each pathogen is transmitted by only one or a few arthropod species and that each arthropod species could only transmit certain pathogens. This led to the conclusion that genetics played an important role in defining the interrelationships between arthropods (herein termed vectors) and the pathogens that they transmit. Studies on *Cx. pipiens* and its susceptibility to a bird malaria parasite, *Plasmodium cathemerium*, were the first experimental demonstrations of genetic control of vector susceptibility to a pathogen (Huff, 1929, 1931).

Both external or extrinsic and internal or intrinsic factors affect vector competence (Hardy *et al.*, 1983). External factors such as temperature, availability of vertebrate hosts, population density and predator density may affect both the pathogen and the vector. Vector survival and pathogen replication or development rates are critical factors in any transmission scenario. The length of time required for a pathogen to replicate or develop in the vector prior to transmission is known as the extrinsic incubation period. For continuous transmission of any pathogen, the extrinsic incubation period must be shorter than the lifespan of the vector. External factors such as those described here are not governed by the genetic make-up of the vector. Conversely, internal factors controlling vector competence are determined genetically, including behaviour, physiology and metabolism. In-depth discussion of the external and internal factors that affect vector competence follows below.

Models of vector competence

Mathematical modelling of vector-borne diseases began with Hamer (1906) and Ross (1911) modelling malaria transmission, followed by Reed and Frost (cited in Abbey, 1952), who developed a stochastic model of epidemics assuming a binomial distribution. The geographical extent, population at risk for infection and the levels of mortality of malaria make it a prime candidate for mathematical modelling and a vast literature exists on this topic (Bailey, 1982; Bruce-Chwatt, 1976; Dietz *et al.*, 1974; MacDonald, 1957; Molineaux & Gramiccia, 1980; Najera, 1974). Mathematical modelling has also been applied to other vector-borne diseases like schistosomiasis (Macdonald, 1965), filariasis (Webber, 1975; Webber & Southgate, 1981), African (Rogers, 1988) and American (Rabinovich, 1987; Zeledon & Rabinovich, 1981) trypanosomiasis, leishmaniasis (Lysenko & Beljaev, 1987), babesiosis (Mahoney, 1977; Smith, 1983) and arboviruses (Fine & LeDuc, 1978; Kay *et al.*, 1987; Reisen, 1989; Smith, 1987).

The fundamental measure of populations used in modelling is the generation reproductive rate. Its main manifestation in epidemiology is the basic case reproductive number, R_0, the mean number of secondary cases arising from each primary infection. R_0 represents the maximum reproductive rate per generation, unaffected by density-dependent processes like superinfection and immunity in the host population (Dye, 1992). MacDonald (1957) pioneered the use of R_0 in malariology, developing the following relationship between entomological and parasitological variables: $R_0 = ma^2bp^n/-r\ln p$. Most of these factors describe the population dynamics of the adult female vector: m is the number of vectors per person, a is the daily biting rate of an individual female on humans (as opposed to other hosts), b is the fraction of mosquitoes that transmit when feeding on an uninfected host, p is the daily survival rate and n is the number of days between vector infection and ability to transmit (extrinsic

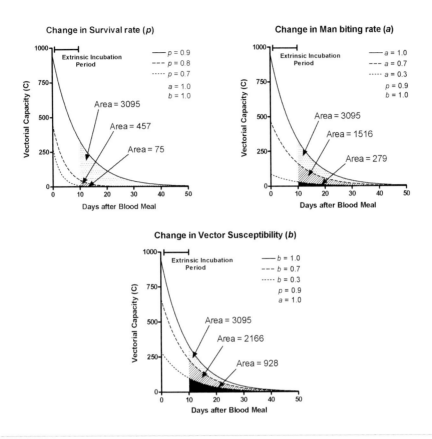

Fig. 5. Graphs demonstrating the effect of varying the survival (*p*, top left panel), human biting rate (*a*, top right panel) and susceptibility (*b*, lower panel) on vectorial capacity. The shaded areas under each line are directly proportional to the capacity of a vector population to transmit a pathogen after the extrinsic incubation, arbitrarily assigned 10 days in this example.

incubation period). The *r* factor is the mean daily recovery rate, or the reciprocal of the duration of infectiousness (Dye, 1992). Note that the biting and survival rates are the only components treated exponentially, implying that changes in their value have a major influence on transmission (Fig. 5).

Parasite populations are in steady state if their reproductive rate equals one (one new host is infected by each case). In malariology, this concept is applied to a human (or other vertebrate host) case as the unit. In other words, one parasitaemic case may generate another case, allowing the perpetuation of transmission. In some situations, a typical case may generate less than one case, triggering a decrease in transmission and usually in disease prevalence. Knowledge of the population dynamics of the reproductive rate of the parasite enables the prediction of epidemics or the evaluation of realistic public health goals for control or eradication. More direct estimates of R_0 can be obtained from epidemiological data where R_0 approximately equals $1/x^*$, where x^* is the equilibrium fraction of susceptible hosts (Dietz, 1975). Derivation of this concept uses data describing the rate of acquisition of infection through time or with age. For vector-borne diseases, the force of infection is often a more sensitive measure of the rate of transmission to humans than crude prevalence or incidence, although it does not lead directly to an estimate of R_0 (Dye, 1992).

Garrett-Jones emphasized the entomological components of McDonald's equation, setting them apart as the vectorial capacity, C (Garrett-Jones & Shidrawi, 1969; Garrett-Jones & Grab, 1964), or the daily rate at which new infections arise from an infective case: $C = ma^2bp^n/-\ln p$. Vectorial capacity can be interpreted as the product of the fraction of vectors capable of transmission (b), the human-biting rate (ma), the human-biting tendency (a) and the longevity factor ($p^n/-\ln p$). The equation emphasizes that the transmission rate is quite sensitive to the two factors treated exponentially: the biting rate (a) because generally a vector has to bite twice to transmit and the daily survival rate (p) because the extrinsic incubation period of the pathogen (n) is usually long compared to the life expectancy of the vector. This latter point played a significant practical contribution to vector control in malaria eradication efforts (Bradley, 1982; Dye, 1992). The relative importance of the biting and survival rates, compared to susceptibility, is illustrated in Fig. 5. Changes in susceptibility from values of 0·3–1·0 have relatively little effect on vectorial capacity (constant a and p), compared to greater changes when survival and biting rates vary.

Vectorial capacity is an indirect way of measuring the transmission rate by a vector population. A direct way is to use the infective biting rate (IBR), which is simply the product of vector biting rate and the proportion of vectors that are infectious (Dye, 1992).

Effects of vector competence on the prevalence of vector-borne diseases

Although the presence of highly susceptible vectors usually is required for vector-borne disease outbreaks, exceptions have been reported. A 1987 Nigerian yellow fever epidemic occurred despite the poor susceptibility of local populations of the principal vector, *Ae. aegypti* (Miller *et al.*, 1989). Although only 7 % of mosquitoes tested were able to transmit, the presence of a high population density of this relatively refractory vector was responsible for initiating and maintaining virus transmission. Thus a marginally susceptible vector is not necessarily benign if it occurs in high densities and has adequate human feeding rates, especially if environmental conditions favour vectorial capacity and pathogen propagation.

INTRINSIC DETERMINANTS OF VECTOR COMPETENCE

Genetic basis of vector susceptibility

The genetic background of individual vectors and populations influences the transmission of vector-borne pathogens. A given pathogen is usually transmitted by a restricted number of vector species and each arthropod has the ability to transmit only certain pathogens; susceptibility can vary inter- as well as intraspecifically, and ranges from complete refractoriness to complete susceptibility (Lehane, 1991). Initial genetic studies of vector competence showed, through selective breeding experiments, that vector susceptibility is genetically based. Ever since Huff (1929) suggested that the susceptibility of *Cx. pipiens* to an avian malaria parasite could be increased through selective mating, considerable research, especially on mosquitoes, has focused on defining vector genes that regulate susceptibility (Beerntsen *et al.*, 2000; Gooding, 1996; Gwadz, 1994; Hardy *et al.*, 1983; Maudlin, 1991; Tabachnick, 1994; Townson & Chaithong, 1991) (Table 1). A complex set of antipathogen reactions have been characterized, including melanotic encapsulation, phagocytosis and the production of antimicrobial compounds in response to bacterial and parasitic infection (Beerntsen *et al.*, 2000). In contrast, factors that regulate infection and/or modulate the replication of arboviruses in vectors are very poorly understood.

Recently, renewed efforts to identify the genetic determinants of vector competence have been directed towards the ultimate goal of developing genetically modified, refractory arthropods for the control of vector-borne diseases. Advances in quantitative and population genetics and DNA-based marker development have also promoted renewed interest in this field. New developments in statistical and genetic research now permit the mapping of loci affecting the expression of quantitative traits, termed quantitative trait loci (QTL). This technique enables the identification of individual genes which often contribute in an additive fashion to multigenic phenotypic traits such as susceptibility to pathogen infection (Severson *et al.*, 2001).

Arboviruses. Initial work on the genetics of *Ae. aegypti* suggested that vector genes are involved in the transmission of arboviruses by mosquitoes (Craig & Hickey, 1967). Numerous studies that followed demonstrated genetic polymorphism in susceptibility to oral infection and the ability to transmit arboviruses among geographical strains of mosquito vectors, including Western equine encephalitis virus (WEEV) in *Culex tarsalis* (Hardy *et al.*, 1976) and *Aedes trivittatus* (Green *et al.*, 1980), chikungunya (CHIKV) and dengue viruses (DENV) in *Aedes albopictus* (Gubler & Rosen, 1976; Tesh *et al.*, 1976), YFV (Aitken *et al.*, 1977) and DENV (Gubler *et al.*, 1979) in *Ae. aegypti*, Japanese encephalitis (JEV) (Takahashi, 1980) and West Nile viruses (WNV) (Ahmed *et al.*, 1979) in *Culex tritaeniorhynchus* and La Crosse encephalitis virus (LACV)

(Grimstad *et al.*, 1977) in *Aedes triseriatus*. Studies of *Ae. triseriatus* mosquitoes suggest selection for resistance to arbovirus infection within endemic regions, which are less capable of transmission than are other populations. Moreover, transmission rates of LACV increased in a colony established from *Ae. triseriatus* mosquitoes collected in the endemic area (Grimstad *et al.*, 1977). *Cx. tritaeniorhynchus* populations from Japan and Pakistan differ in their ability to transmit JEV vertically (Rosen *et al.*, 1989) and *Ae. albopictus* populations from the United States, Japan and Singapore show considerable interpopulation variation in the proportion of females transmitting DEN-1 to their offspring (Bosio *et al.*, 1992).

One technical problem with studies of vector competence is the use of laboratory colonies. Because frequencies of genes involved in vector competence can be influenced by evolutionary forces such as founder effects, selection and genetic drift that accompany colonization, interpopulation differences in susceptibility can occur (Gooding, 1996). This has been demonstrated with biting midges (Mecham & Nunamaker, 1994) and mosquitoes (Beaman & Turell, 1991; Grimstad *et al.*, 1977; Lorenz *et al.*, 1984; Turell & Beaman, 1992). Studies with colonies of *Culicoides variipennis* indicate that the number of individuals used to establish the colony and the magnitude of the initial increase in the census predict whether the colony will be representative of field populations (Jones & Foster, 1978).

The genetics of *Ae. aegypti* susceptibility to flaviviruses, particularly YFV (Tabachnick *et al.*, 1985) and DENV (Gubler *et al.*, 1979), have been studied in detail. Selection of YFV- and DENV-susceptible and refractory colonies suggested that susceptibility is controlled in a polygenic, recessive manner (Miller & Mitchell, 1991). Quantitative genetic studies using two subspecies of *Ae. aegypti* (*aegypti* and *formosus*) infected with DENV determined that the midgut escape barrier is controlled by dominant alleles at three independently segregating loci that generate a susceptibility phenotype. Two QTL control a midgut infection barrier mapped in chromosomes II and III and the other controls a midgut escape barrier located in chromosome III (Bosio *et al.*, 1998, 2000). Susceptibility of *Ae. aegypti* to CHIKV and Sagiyama alphaviruses is linked to the *rosy eye* marker (Mourya *et al.*, 1994) on chromosome III (Munstermann & Craig, 1979). For other arbovirus vectors, however, the genetic determinants of susceptibility may be sex linked. Only one locus for bluetongue virus (BTV) susceptibility in *C. variipennis* has been described; the phenotype is determined by the mother's genotype, with the dominant allele in the mother being the allele inherited from her father. It is possible that other loci are involved or that there are environmental effects, because susceptibility to BTV is often higher in laboratory-reared midges than in field-collected individuals (Tabachnick, 1991).

Mechanisms of virus persistence in vectors include TOT and its genetic determination has been studied. Tesh & Gubler (1975) first described interspecific differences in TOT of LACV in *Ae. albopictus* but not in *Cx. fatigans*. Intraspecific differences in TOT of San Angelo virus (SAV) and Kunjin virus (KUNV) among geographical strains of *Ae. albopictus* were reported by Tesh & Shroyer (1980). In addition, geographical variations in TOT were demonstrated in *Ae. triseriatus* infected with LACV (Grimstad *et al.*, 1977), *Aedes dorsalis* and *Aedes melanimon* infected with California encephalitis virus (CEV) (Turell *et al.*, 1982) and *Ae. aegypti* infected with DENV-2 (Jousset, 1981). Selection for arbovirus susceptibility may be mediated partly by the detrimental effects of arboviruses on vector physiology (Turell & LeDuc, 1983). Grimstad *et al.* (1980) suggested that detrimental effects of arboviruses in mosquitoes may select for a population that is more refractory to infection. Given that mosquitoes infected transovarially only contain 1/10–1/50 of the amount of virus found in individuals infected orally (Tesh & Shroyer, 1980), selection for oral refractoriness may be stronger than that for resistance to TOT in a population with stabilized infections.

Parasites and bacteria. The genetics of vector competence for protozoan and nematode parasites is reviewed by Gooding (1996) and Beerntsen *et al.* (2000). As with arbovirus vectors, intra- and interpopulation variation as well as laboratory colonization effects on susceptibility to parasites have been reported in phlebotomine sandflies (Wu & Tesh, 1990), tsetse flies (Moloo, 1993), ticks (Kubasu, 1992) and mosquitoes (Duhrkopf & Trpis, 1980; Feldmann & Ponnudurai, 1989; McGreevy *et al.*, 1974). The coevolutionary history of parasites and their vectors may differ geographically. For instance, in an *Anopheles gambiae* strain selected for resistance to several species of *Plasmodium*, ookinetes of New World and Asian strains of *Plasmodium falciparum* are killed by melanotic encapsulation, yet *Plasmodium* strains of African origin generally develop normally (Collins *et al.*, 1986). *Anopheles albimanus*, a New World vector of *P. falciparum*, is susceptible to a parasite strain that originated from Central America but not to infection with a strain from Africa (Beerntsen *et al.*, 2000).

Variation has been reported in the specificity of genetic mechanisms that determine susceptibility of vectors to protozoa and nematodes. The recent availability of genetic linkage maps for *Ae. aegypti* and *An. gambiae* have facilitated the determination of chromosomal regions that regulate susceptibility to parasites (Severson *et al.*, 1993). Mosquitoes refractory to *Plasmodium* spp. encapsulate ookinetes, suggesting a common mechanism of pathogen recognition (Collins *et al.*, 1986; Lehane, 1991). Microsatellite linkage maps enabled investigators to determine the genetic background controlling refractoriness to *Plasmodium* spp.; a model using *An. gambiae* and the primate parasite *Plasmodium cynomolgi* B determined that three QTL, located in chromosomes 1, 2 and 3, influence encapsulation and account for 70 % of the variation in the encapsulation response (Collins *et al.*, 1997; Zheng *et al.*, 1997). Three antimicrobial peptide genes (Hao *et al.*, 2001) from the tsetse fly exhibit differential regulation and can discriminate not only between molecular signals specific for bacteria and trypanosome infections but also between different life stages of the trypanosome.

In contrast, mosquitoes refractory to filariae use separate genes to control susceptibility to different parasites. Susceptibility to two *Brugia* spp., which develop in the thoracic musculature, is controlled by a single, sex-linked recessive gene (f^m) located in chromosome 1 of *Ae. aegypti* (Macdonald, 1962, 1965). Susceptibility of the same mosquito to *Dirofilaria immitis*, which develops in the Malpighian tubules, is controlled by a different sex-linked recessive gene (f^t) on chromosome 1 (McGreevy *et al.*, 1974), and susceptibility to *Plasmodium gallinaceum* is controlled by a dominant autosomal allele (*pls*) located in chromosome 2. In *Cx. pipiens*, however, the specific action of genes on particular organ systems may not be apparent. Although both *Brugia pahangi* and *W. bancrofti* follow the same route of infection and both develop in the flight musculature, the sex-linked recessive gene (*sb*) only controls susceptibility of *Cx. pipiens* to *B. pahangi* and not to *W. bancrofti* (Lehane, 1991). Although the genetics are not understood, lectin production in *Rhodnius prolixus*, a vector of Chagas' disease, stimulates the maturation of *T. cruzi* (Pereira *et al.*, 1981). Apparently other lectins produced by *Glossina morsitans* are required for the maturation of two trypanosome species (Maudlin & Welburn, 1988; Welburn & Maudlin, 1989, 1990).

Gooding (1996) pointed out that most vectors appear to have one major gene controlling susceptibility or refractoriness. However, the presence of 'minor' genes that influence susceptibility in an additive manner (Bosio *et al.*, 2000) may be overlooked in studies using colonies, where these alleles may be eliminated by genetic drift. For most models, the expression of genes for refractoriness or susceptibility has been demonstrated in only one genetic background and, in the laboratory, the influence of the environment on these genes has been minimized by unnaturally constant environmental conditions. Genetic mechanisms of susceptibility identified in the laboratory should be confirmed in natural populations.

Behavioural determinants of vector competence

Arthropod behaviours, especially those related to pathogen transmission, are thought to be genetically inherited (Lehane, 1991). Environmental factors such as temperature, relative humidity and precipitation as well as host cues like CO_2 and heat release alter the biochemistry and physiology of the vector, initiating specific behavioural responses. Host preference and seeking, feeding, and oviposition behaviour are probably the most important behavioural determinants for vector competence and, due to their complexity, will be briefly discussed.

Host-seeking behaviour. The difficulty that host-seeking arthropods have in locating their next blood meal depends upon the closeness of their association with the host, their lifestyles (e.g. permanent vs temporary feeders) and mobility (e.g. flying vs crawling/jumping). Sutcliffe (1987) suggested that host-searching behaviour could be divided into three phases: appetitive searching, activation and orientation, and attraction. These behaviours are quite versatile and do not occur necessarily in a strict sequence. Lehane (1991) offers a more complete discussion of these behaviours. Appetence initiates a series of behavioural responses that lead to host contact and successful blood feeding. Passive and active are two basic strategies used by arthropod vectors to find hosts; passive arthropods such as many ticks remain quiescent in their habitat and depend upon contact with the vertebrate animals that invade it, while active arthropods like biting flies or kissing bugs leave their resting environments and fly or walk in search of hosts. In some instances, the release of substances such as an aggregation-attachment pheromone by *Amblyomma variegatum* and *Amblyomma hebraeum* ticks triggers the movement of hungry ticks toward areas of the host already infested. Because these two species have been implicated in the transmission of several animal rickettsioses and one arbovirus (Sonenshine, 1993) this aggregation behaviour may increase the number of ticks that become infected and transmit.

Heat can act as a vector attractant. The louse (*Pediculus humanus*) vector of typhus rarely leaves the host but will do so if body temperature and humidity change drastically, such as following a high fever or death of the host (Kettle, 1983). The louse can then move as a far as 2 metres in search of a new host, using heat as a locating signal. This behaviour increases the chances of encountering a healthy host. In other vectors, such as *Anopheles* mosquitoes, host hyperthermia has no effect on vector feeding (Day & Edman, 1984).

Morphological characteristics of the host like body size can also determine the choice of host. For example, *An. gambiae* seems to prefer human adults over children (Port *et al.*, 1980). Habitat distribution may also influence host selection. Host-seeking (questing) ticks near the ground are mostly exposed to small animals while those questing higher in the vegetation encounter larger animals. These differences are observed in the life stages of the American dog tick, *Dermacentor variabilis*, with immature ticks questing at ground level and parasitizing mice, rats and voles, versus adults questing higher in the vegetation on larger mammals including dogs and humans. These feeding patterns are highly relevant in the epidemiology of Rocky Mountain spotted fever (Sonenshine, 1993).

Host preference and feeding behaviour. One of the most important factors in determining vector–pathogen relationships is host preference and feeding behaviour.

Host choice may be determined by a combination of behavioural, physiological, morphological, ecological, geographical, temporal and genetic factors. Blood feeding by some vector taxa such as mosquitoes is restricted to females, while both males and females of other taxa (e.g. fleas) utilize blood as a source of nutrients for a variety of physiological activities, including reproduction (Lehane, 1991). The source of the blood meal or its quality may influence the course of pathogen infection in the vector. For example, tsetse flies feeding on goats or cattle develop a higher rate of infection with *Trypanosoma vivax* than those fed on mice (Maudlin *et al.*, 1984).

Arthropod blood feeders can be divided into host-specific and opportunistic (Sonenshine, 1993). *Culiseta melanura*, the enzootic vector of Eastern equine encephalitis virus (EEEV), feeds almost exclusively on passerine birds, the reservoir hosts (Scott & Weaver, 1989). *Boophilus microplus*, the tick vector of babesiosis and anaplasmosis of cattle, feeds almost exclusively on cattle; *Ixodes scapularis* and *Ixodes pacificus*, vectors of Lyme disease in North America, feed on a wide variety of vertebrates including large mammals, birds and reptiles. The mosquito *Culex salinarius*, a potential vector of WNV in North America (Sardelis *et al.*, 2001), has a wide host range including birds, equines and canines. This species also takes multiple blood meals during one gonotrophic cycle if feeding is disturbed by defensive behaviours of the host (Cupp & Stokes, 1976). If infected with a pathogen, this behaviour increases the chance of transmission during a short period of time. On the other hand, the competence of the body louse *Pediculus humanus humanus* to transmit epidemic typhus and relapsing fever (Kettle, 1983) relies on its tendency to feed exclusively on humans.

Seasonal variation in host feeding and host preference can enhance transmission of pathogens and initiate disease outbreaks. For instance, *Culex nigripalpus* and *Cx. tarsalis* are vectors of St. Louis encephalitis virus (SLEV) during summer months in Florida and California, respectively. Both species show a marked seasonal change in their feeding patterns, switching from bird feeding in the winter and spring to mammal feeding in the summer when arbovirus transmission occurs (Edman, 1974; Edman & Taylor, 1968; Tempelis & Washino, 1967). These changes in feeding patterns may affect SLEV epidemiology. It has been suggested that the retreat of *Plasmodium malariae* and *P. falciparum* from Europe during the late 19th century was caused by a switch in host preference of several anopheline vectors from humans to domestic animals. Changes in animal husbandry, agricultural practices, housing and a decreased human birth rate may have reduced the number of human hosts available, contributing to reductions in transmission of malaria (Lehane, 1991).

Biological differences in host choice among species complexes have been well documented. *An. gambiae sensu lato* is composed of six sibling species in Africa.

Across its geographical range, one sibling species, *Anopheles arabiensis*, exhibits highly variable host-seeking (from exophilic to endophilic) and feeding (from anthropophilic to zoophilic) behaviours, compared to *An. gambiae sensu stricto*, which is primarily an endophilic species biting man. The sporozooite infection rates of *An. arabiensis* are about 1/15th those in *An. gambiae*, a highly efficient malaria vector. The higher sporozooite rates in *An. gambiae* are not a reflection of higher susceptibility to malaria parasites but arise because *An. gambiae s.s.* lives longer and feeds more often on humans (White, 1974).

Temporary blood feeders (e.g. flies) tend to take relatively large blood meals, thus limiting the number of host contacts and exploiting a blood source that may be available only temporarily. Blood meals taken by tsetse flies are two to three times the mass of their bodies, impairing manoeuvrability during flight from the host and decreasing flying speed significantly (Glasgow, 1961). Since these exothermic flies require a body temperature of 32 °C to enhance take-off, and feed around dusk and dawn, a time without optimum lift temperatures, they have adapted to avoid host defensive responses and escape predators. Immediately after the blood meal, the fly raises its thoracic temperatures towards the optimum by 'buzzing', producing the characteristic sound after which the tsetse is named. By increasing its thoracic temperatures, the fly maximizes its lift and flight speed, allowing it to take a large blood meal and depart from the host (Howe & Lehane, 1986). This behaviour ensures the success of pathogen infection in the vector and transmission to a new host. It may be more disadvantageous for permanent blood feeders closely associated to their hosts to take large blood meals. Many ectoparasites, such as anopluran lice, feed more frequently (every few hours) and take meals that are only 20–30 % of their body mass (Murray & Nicholls, 1965), ensuring the survival of the vector and increasing the chances of further pathogen transmission.

Mechanical transmission of pathogens may be facilitated by the feeding frequency of vectors. Arthropods that take a succession of partial blood meals from several vertebrate hosts are probably the most efficient mechanical vectors. Large insects, such as tabanid flies, are often disturbed by their vertebrate host before they complete a full blood meal. *Tabanus* spp. feed on large mammals usually found in herds, are excellent fliers, and possess sponging mouthparts. Because their painful bite triggers defensive behaviours in the host, a full blood meal is usually accomplished by additional visits to other vertebrate hosts. Moreover, the large amount of blood taken by these vectors increases the chances of pathogen transmission (Soulsby, 1982).

Oviposition behaviour. Environmental modification as a result of human incursions for agriculture or urbanization can affect the epidemiology of vector-borne diseases.

Human habitation and urbanization of enzootic ecosystems has triggered closer associations between human hosts and vectors. Most vector-borne disease transmission is the result of a close relationship between the vector and the affected host or a fortuitous encounter. Because many vectors travel short distances or wait passively during host seeking, the selection of an oviposition site can aid in bringing the vector and the host closer, thus enhancing transmission. Louse-borne typhus epidemics in humans, for example, are associated with overcrowded, unsanitary conditions, especially during times of war, famine and natural disasters (Kettle, 1983). The vector, *P. humanus*, lays its eggs in the thousands in clothing; this microenvironment, along with the conditions encountered during outbreaks, could facilitate the spread of the vector over large numbers of hosts in an explosive manner. Eggs of *Rhodnius* spp., *Panstrongylus* spp. and *Triatoma* spp., vectors of Chagas' disease in South America, are inserted in the rough surfaces, cracks and crevices of houses in rural and urban areas. Because their entire life cycle can occur inside human habitations using a variety of hosts, trypanosome transmission rates in humans are very high inside houses. In contrast, human cases of Chagas' disease occur in the US rarely in spite of the occurrence of the trypanosome, vector and reservoir, because the vectors rarely become domestic (Brenner & Stoka, 1987).

Mosquito oviposition behaviour, particularly site selection, has had a tremendous impact in the emergence and re-emergence of arboviral diseases. The introduction of YFV from Africa to the Neotropics probably occurred during the African slave trade (Chang *et al.*, 1995). Water containers on-board ships crossing the Atlantic from Africa appear to be an ideal breeding site for introduction of *Ae. aegypti* into the Americas. Other mosquito vectors like *Ae. albopictus*, *Ae. triseriatus* and *Culex quinquefasciatus* have adapted to oviposit in peridomestic water containers, bringing arboviruses into closer contact with humans.

PATHOGEN-SPECIFIC DETERMINANTS OF VECTOR TRANSMISSION

Arboviruses

The evolution and genetics of arboviral pathogens are privileged by the flexibility of their genomes, short generation times, high progeny yields and changes in the environment. RNA viruses have the potential to evolve and diverge very rapidly. Mutations in viral genomes are estimated to vary between 0·03 and 2·0 % per position per year (Steinhauer & Holland, 1987). Changes in viral genomes and adaptation to new hosts are the result of high mutation rates, recombination and RNA segment reassortment. Despite these advantages, evolution of arboviruses is generally slower than that of single host RNA viruses, most likely due to the constraints imposed by the alternating host cycle and the requirement for genetic 'trade-offs' not required of more specialist viruses (Weaver *et al.*, 1992, 1999; Novella *et al.*, 1995). Phylogenetic comparison of genomes, within arbovirus families, is characterized by little sequence change in strains from the same geographical area that can differ by years or decades since isolation (Gould & Pritchard, 1990; Weaver *et al.*, 1992; Lepiniec *et al.*, 1994; Meissner *et al.*, 1999; Wilson *et al.*, 2000; Powers *et al.*, 2000; Mutebi *et al.*, 2001). Despite observations of genome stability in nature, arbovirus evolution and divergence has been observed in both the laboratory and field settings. We describe below the evolutionary mechanisms acting on arboviruses, and their relationship to vector competence and preference.

Evolutionary mechanisms driving virus adaptation to vector arthropods

Mutation. RNA virus replication is characterized by a high mutation rate due to error-prone polymerases and these pathogens exist as heterogeneous populations called quasispecies. This heterogeneity and rapid mutation rate confer an important adaptive potential to exploit changes in the environment rapidly, including adaptation to new hosts (Domingo, 2000).

Surface glycoproteins of enveloped viruses include the major epitopes involved in cell entry and infection. The role of these glycoprotein epitopes in vector infection has been examined recently. The E2 structural glycoprotein of alphaviruses has been implicated consistently as a determinant of infection in mosquitoes. Isolation and sequencing of mutant monoclonal antibody-resistant viruses of Venezuelan equine encephalitis virus (VEEV) were used to identify a single amino acid substitution at position 207 in the envelope glycoprotein E2 that prevents dissemination of *Ae. aegypti* infected orally (Woodward *et al.*, 1991). Small numbers of E2 mutations in the E2 protein of natural VEEV strains that accompany emergence of epizootic serotype IAB and IC strains from enzootic, serotype ID progenitors enhance the ability of these viruses to infect *Ochlerotatus taeniorhynchus*, an important epizootic vector (Brault *et al.*, 2002). Subtle differences in E2 from two different strains of Sindbis virus (SINV), TE11 and MRE16, generate distinct midgut infection phenotypes in *Ae. aegypti* (Seabaugh *et al.*, 1998). Studies on LACV have also indicated that a mutant G1 envelope glycoprotein MARV exhibits reduced infectivity and decreased dissemination from the midgut of *Ae. triseriatus* (Sundin *et al.*, 1987).

Comparison of the NS3 gene and protein from South African isolates of BTV and equine encephalosis virus (*Reoviridae*: *Orbivirus*) indicates the existence of discrete virus lineages with different vector infectivity phenotypes (van Niekerk *et al.*, 2003).

Table 2. Infection, dissemination and transmission of La Crosse, snowshoe hare and reassortant viruses in *Aedes triseriatus* mosquitoes

Taken from Beaty & Bishop (1988)

Virus genotype	Percentage of mosquitoes:		
	Infected	**Disseminated**	**Transmitted**
LAC/LAC/LAC SSH/LAC/LAC SSH/LAC/SSH LAC/LAC/SSH	98 (115/117)*	98 (113/115)†	93 (126/136)‡
SSH/SSH/LAC LAC/SSH/LAC LAC/SSH/SSH SSH/SSH/SSH	92 (92/100)*	26 (24/92)†	35 (36/104)‡

*Number with detectable viral antigen in midgut cells/number examined.

†Number with detectable viral antigen in head/number examined.

‡Number transmitting to mice/number with detectable viral antigen in salivary glands.

Recombination. An example of recombination in arboviruses is the New World WEEV. Initially, serological analyses pointed to a close relationship to SINV, while oligonucleotide fingerprinting suggested a close relationship with the New World EEEV. Sequencing of the genome of WEEV revealed it to be a recombinant virus with the envelope glycoproteins E1 and E2 most closely related to SINV and the rest of the genome related to EEEV (Hahn *et al.*, 1988; Weaver *et al.*, 1997). Other closely related viruses including Highlands J and Fort Morgan are also descendants of the same recombinant ancestor.

Modern molecular biology techniques have also been used to generate artificial, recombinant viral chimaeras. For example, SINV strain MRE16 is efficient at oral infection of *Ae. aegypti* mosquitoes, while strain TE11 is not. Recombinant clones containing the E2 envelope protein from strain MRE16 and the remainder of the genome from strain TE11 infect and disseminate in *Ae. aegypti* efficiently (Seabaugh *et al.*, 1998). These experiments demonstrate further the important role of the envelope glycoprotein in mosquito infection and dissemination.

Reassortment. RNA viruses with segmented genomes can exchange genomic segments (reassortment), giving rise to new viral genotypes. In the early 1980s, Beaty and colleagues investigated the genetic basis of bunyavirus infection of *Ae. triseriatus* mosquitoes with LACV. Using LACV and snowshoe hare virus reassortants, the M (middle sized of three) segment was shown to co-segregate with the dissemination and transmission phenotypes (Table 2) (Beaty *et al.*, 1981, 1982). Identification of reassortant field isolates of LACV by oligonucleotide fingerprinting demonstrated that reassortment also occurs in nature (Klimas *et al.*, 1981).

Reassortment of Thogoto virus (*Orthomyxoviridae*: *Thogotovirus*) was demonstrated following superinfection of the tick vector *Rhipicephalus appendiculatus* with two temperature-sensitive strains. A wild-type reassortant was recovered later (Nuttall *et al.*, 1994) and interference studies indicated that the primary site for reassortment was the gut (Jones *et al.*, 1989; Davies *et al.*, 1989).

BTV (*Reoviridae*: *Orbivirus*) is comprised of 23 serotypes with worldwide distribution. *Orbivirus* genomes consist of 10 segments of double-stranded RNA. Mixed oral infections in the primary American vector, *Culicoides variipennis*, with BTV serotypes 10 and 17 resulted in high-frequency reassortment of genome segments (Samal *et al.*, 1987; el Hussein *et al.*, 1989). Phylogenetic comparison of the S3 gene from California isolates versus prototype American strains suggests that this gene segment is formed by different consensus sequences that cocirculate and cannot be grouped by serotype (de Mattos *et al.*, 1996). Phylogenetic results with the prototype American strain of serotype 13 demonstrate further this group to be a result of natural genome reassortment.

Other genetic mechanisms. Analysis of codon usage and base composition among members of the genus *Flavivirus* points to a relationship between base composition and tick versus mosquito specificity (Jenkins *et al.*, 2001). Flaviviruses associated with ticks have a lower G+C content at all amino acid codon positions (de Mattos *et al.*, 1996). Comparison of tick- and mosquito-borne flaviviruses further emphasizes evolutionary divergence in codon usage associated with arthropod adaptation. Moreover, sequence elements on the viral envelope glycoprotein thought to be associated with mosquito receptor recognition are absent in the tick-borne flaviviruses (Bhardwaj *et al.*, 2001).

The molecular determinants of arbovirus infection of arthropod vectors are still incompletely understood, and further investigation is needed. Modern tools for the molecular manipulation of virus and vector genomes have created new opportunities to study the interactions between both organisms.

Parasites

Malaria and *Leishmania* infections represent the best studied examples of parasite determinants of vector transmission. Altered feeding behaviour by infected vectors can increase pathogen transmission, such as an increase in host contact (Molyneux & Jefferies, 1986; Moore, 1993). Mosquitoes infected by malaria parasites exhibit an increase in probing behaviour during blood feeding. Only when the sporozoites have infected the salivary glands are these changes apparent; therefore, the mosquito can take several blood meals before this affected behaviour is apparent. One cause for the increase in probing behaviour may be altered levels of apyrases, ADP-degrading enzymes that inhibit platelet recruitment, counteract host haemostasis, and promote longer blood feeding. Apyrase and sporozooites are both found in the distal regions of the mosquito salivary gland lobes (Sterling *et al.*, 1973). *P. gallinaceum*-infected *Ae. aegypti* has a 25 % reduction in apyrase activity and a threefold increase in the mean time to locate a blood vessel before feeding (Ribeiro *et al.*, 1985; Rossignol *et al.*, 1984). Koella *et al.* (1998) determined that 22 % of *P. falciparum*-infected *An. gambiae* mosquitoes collected from field conditions had bitten more than one person in a night, compared with 10 % of uninfected females, suggesting that this infection increases multiple host contacts and enhances transmission in nature. The mechanism for the decrease in apyrase levels in infected salivary glands is unknown.

The observation that *Leishmania* parasite transmission is enhanced by the feeding behaviour of infected sandflies was made very early (Shortt & Swaminath, 1927). This phenomenon was termed the 'blocked fly hypothesis' to describe the prevention of blood flow into the midgut (Jefferies *et al.*, 1986). Difficulties in feeding lead to an increase in probing behaviour and thereby increased pathogen transmission via increased salivation. The occlusion of the stomodeal valve is caused by a filamentous, gel-like matrix full of infectious metacyclic promastigotes (Rogers *et al.*, 2002). Filamentous promastigote proteophosphoglycan (fPPG), a parasite secretion that forms a major component of the gel-like plug, has been found in at least 10 *Leishmania*–sandfly associations. The fPPG is also thought to prevent unattached promastigotes from being swept into the midgut with the blood meal, enhancing parasite fitness by promoting transmission (Stierhof *et al.*, 1999).

Bacteria

Ticks, mites, fleas, lice and sandflies are the major groups of arthropods that transmit bacterial diseases including several that have caused extensive disease and suffering throughout the world. It has been estimated that plague and typhus have caused more casualties than all wartime-related injuries combined (Zinsser, 1934). After ingestion of plague bacilli (*Y. pestis*) by a competent flea vector, the bacteria multiply in the gut and eventually spread to the proventriculus or anterior midgut (Pollitzer, 1954). As the bacteria multiply they adhere to spines in the flea's proventriculus to form a sticky, gelatinous mass. This mass can inhibit the intake of blood via occlusion. This 'blocked' flea can no longer imbibe a blood meal (Piesman & Gage, 2000) and blockage is a major determinant of transmission (Bacot & Martin, 1914). Blocked fleas begin to starve and therefore continue to attempt feeding on a host. During this process, the flea regurgitates bacteria onto the host in an effort to clear the blockage. Blockage depends partly on the strain of bacteria and is related to haemin storage. The Hms$^+$ phenotype is a pigment-positive strain (binds Congo red in culture) and causes blockage formation in fleas. In contrast, the Hms⁻ phenotype does not cause blockage. The ability to cause blockage is restored in Hms⁻ strains when complemented with a recombinant plasmid containing the three genes that make up the haemin storage locus of *Y. pestis* (Hinnebusch *et al.*, 1996).

Rickettsia prowazekii is the aetiologic agent of louseborne or epidemic typhus. This bacterium is unusual in that it is pathogenic not only for its host, but also for its vector. After lice ingest an infectious blood meal, the pathogen invades the epithelial cells in the gut, destroying cells and eventually killing the vector. Humans become infected when either the infected lice or faeces are scratched or rubbed into the skin or eyes. *R. prowazekii* contrasts with other rickettsiae that have more benign relationships with their vectors. Unlike most pathogens transmitted by vectors, rickettsiae generally do not need vertebrate reservoir hosts to perpetuate their life cycle due to their maintenance through efficient transstadial and TOT in ticks. Non-pathogenic (for the tick) rickettsiae may be selected by TOT (Burgdorfer & Brinton, 1975). Other studies suggest that interspecific competition among closely related rickettsiae may play a role in the vector competence (Azad & Beard, 1998).

ENVIRONMENTAL DETERMINANTS OF VECTOR COMPETENCE

Environmental conditions are primary determinants of vector competence. The vector competence of mosquitoes varies seasonally in field populations as a function of extrinsic environmental factors, influenced largely by rainfall, temperature and other weather variables. Standing water is required for oviposition and/or egg hatching. Ambient air temperature affects adult mosquito survival, the length of the extrinsic incubation period, population dynamics and adult feeding behaviour, as well as the rate of larval development, the speed of virus replication and the expression of genetically controlled determinants of vector competence. Environmentally determined temporal changes in vector competence resulting in increases or decreases in the efficiency of arboviral transmission affect the epidemiology of human pathogens. Understanding environmental determinants of vector competence will be crucial to mosquito control and abatement, especially with increased limitations on the implementation of traditional biological and chemical control agents. Human susceptibility may also be confounded by environmental factors. For example, malnutrition may be caused by climate stress on agriculture and ultraviolet radiation can induce immunosuppression (Patz *et al.*, 1996).

The geographical distribution of many vector-borne diseases is governed by climate and weather variables (Hales *et al.*, 2002). As the earth's climate and other environmental factors change, so will the competence of insect vectors. Global surface temperatures have been increasing since the early 18th century and from the late 1950s this warming trend has accelerated [Intergovernmental Panel on Climate Change, 2001 (http://www.ipcc.ch/pub/spm22-01.pdf); Reiter, 2001]. The lowest 8 kilometres of the atmosphere are strongly influenced by factors subject to human perturbations, including ozone depletion, atmospheric aerosols and the El Niño phenomenon, which affect differentially environmental conditions worldwide [Intergovernmental Panel on Climate Change, 2001 (http://www.ipcc.ch/pub/spm22-01.pdf)]. Here, we review the major environmental determinants of vector competence with pertinent examples being drawn from recent experimental studies. We focus mainly on temperature effects, with a brief review of other important variables that affect vector competence.

Many laboratory studies indicate that the minimum extrinsic incubation under constant temperature varies greatly. The effects of air temperature on measures of vector competence, including virus replication, susceptibility, infectivity, and dissemination and transmission capacities, vary greatly among vector/pathogen interactions. In some cases, peak virus titres, high rates of infectivity, dissemination and transmissibility occur at higher environmental temperatures. In other circumstances, the opposite phenomenon is observed. For example, *Cx. pipiens* infected by intra-thoracic inoculation with Rift Valley Fever virus and held at 26 or 33 °C transmit to naïve hamsters 1 day after inoculation, while mosquitoes held at 13 °C do not transmit until day 14 (Turell *et al.*, 1985b). Fewer *per os* exposed *Cx. pipiens* incubated at 13 °C became infected (31 %) compared to mosquitoes incubated at 26 °C (75 %) or 33 °C (91 %). In the same study, *Oc. taeniorhynchus* exposed orally and held at the three different temperatures did not exhibit significantly different infection rates, although the extrinsic incubation was shorter in mosquitoes held at higher temperatures. A shortened extrinsic incubation with increased holding temperature is a phenomenon common to many arboviruses, including Ockelbo virus infection of *Ae. aegypti* and *Oc. taeniorhynchus* (Turell & Lundstrom, 1990), WNV infection of *Cx. pipiens* (Dohm *et al.*, 2002), DENV-2 infection of *Ae. aegypti* (Watts *et al.*, 1987), EEEV infection of *Ae. triseriatus* (Chamberlain & Sudia, 1955), JEV infection of *Cx. tritaeniorhynchus* (Takahashi, 1976), YFV infection of *Haemagogus capricornii* (Bates, 1946) and Ross River virus infection of *Ochlerotatus vigilax* (Kay & Jennings, 2002). In many of these examples, holding infected mosquitoes at 32 or 33 °C was sufficient to shorten the length of extrinsic incubation to a fraction of the number of days required at 28 °C. Regional, short-term increases in environmental temperature may therefore influence vector competence strongly and can affect arbovirus epidemiology.

The incidence of dengue haemorrhagic fever (DHF) cases in Bangkok, Thailand, parallels temperature-induced variation in the vector competence of *Ae. aegypti* (Watts *et al.*, 1987). DHF epidemics there occur during the hot, dry seasons when infection rates of mosquitoes are maximal. Spring and autumn peaks in the vector competence of *Cx. tarsalis* from California for WEEV correlate with the number of WEE cases; increased resistance of vectors to infection is correlated significantly with the number of summer days when the daily air temperature ranges from approximately 26 to 32 °C (Hardy *et al.*, 1990; Kramer *et al.*, 1983; Reisen *et al.*, 1995, 1996). Increases in WEEV activity and prevalence occur after cool springs when low temperatures maintain high vector competence of *Cx. tarsalis* (Hess *et al.*, 1963). Infection and dissemination rates of *Cx. pipiens* fed on viraemic chickens infected with a WNV isolate recovered from the 1999 New York outbreak were higher for mosquitoes held at warmer temperatures, consistent with those observed immediately preceding and during the outbreak (Dohm *et al.*, 2002). While warmer temperatures promote the intensity of transmission of many arboviruses by shortening the extrinsic incubation period and the gonotrophic cycle of female mosquitoes, an increase in temperature may reduce adult survival (Shope, 1991). High temperatures can exert deleterious effects on mosquitoes by reducing survival (Patz *et al.*, 1996) or decreasing the body size of certain species when larvae develop at high temperature (Schreiber, 2001). However, smaller mosquitoes require a greater number of blood meals to develop eggs (Briegel, 2003), which could enhance feeding on multiple hosts and promote virus transmission. As for many arboviruses, the length of extrinsic incubation for *Plasmodium* parasites decreases with increasing temperatures from 20 to 27 °C (Noden *et al.*, 1995).

However, exceptions to the shortening of the extrinsic incubation period with higher temperature are also described. *Cx. tarsalis* populations from California are more susceptible to WEEV during the spring and autumn months compared to the summer (Hardy *et al.*, 1990; Reisen *et al.*, 1996). The likely explanation for this temporal difference is that this mosquito is a less competent vector when held at 32 °C compared to incubation temperatures of 18 or 25 °C (Kramer *et al.*, 1983). The high temperature does not reduce survival or inhibit initial infection; conversely, infected mosquitoes are able to limit virus replication and/or dissemination at high temperatures. Arbovirus transmission in nature therefore involves the intersection of appropriate environmental temperatures with the intrinsic permissiveness and behavioural determinants of vector competence.

Although adult mosquito incubation temperatures exert a profound influence on susceptibility, larval rearing temperatures have not been shown to be a significant determinant. Kay *et al.* (1989) demonstrated that *Culex annulirostris* larvae and pupae reared at low (approx. 20 °C) temperatures showed no difference in rates of Murray Valley encephalitis virus infection compared to mosquitoes held at 27 °C. Tesh (1980) found that 20–30 °C larval rearing temperatures had no effect on TOT of KUNV (West Nile) or SAV in *Ae. albopictus*.

Changing temperature conditions for diapausing mosquitoes that overwinter can also affect vector competence and have important implications for virus maintenance in temperate regions. *Cx. pipiens* fed on viraemic WNV-infected chickens held at simulated overwintering temperatures of 10 °C for 21–42 days did not have detectable virus in their bodies or legs (Dohm *et al.*, 2002). However, when the mosquitoes were transferred to 26 °C following the simulated winter, infection and dissemination rates increased. It is likely that WNV cannot be detected in some or many infected, overwintering mosquitoes, but when ambient air temperatures rise in the spring, virus replication resumes to detectable levels. Likewise, Kay & Jennings (2002) demonstrated that *Oc. vigilax* susceptibility to Ross River virus is dynamic and can vary even over the course of infection of one mosquito as the holding temperature changes. Temperature-mediated interruptions in egg development can also affect their viability, especially those infected transovarially with arboviruses. While LACV-infected eggs show higher mortality than non-infected eggs overwintering outdoors in Madison, Wisconsin, infected eggs terminate diapause earlier than uninfected eggs, which could confer a temporal advantage in initiating the seasonal transmission cycle (McGaw *et al.*, 1998).

While temperature studies represent the focus of current knowledge regarding environmental determinants of vector competence, other factors such as developmental conditions, food availability, humidity and rainfall, extreme weather events and geography have also been investigated. Catastrophic events, such as flooding or pollution at breeding habitats, reduce larval abundance (Reisen *et al.*, 1989), which indirectly can affect disease prevalence by reducing vector populations. Food shortages in the larval stages have a variable effect on vector competence. *Cx. annulirostris* fed low quantities of a yeast/rat chow diet showed similar susceptibility to Murray Valley encephalitis virus compared to high-diet fed mosquitoes (Kay *et al.*, 1989). Likewise, a similar evaluation of *Cx. tritaeniorhynchus* infection by WNV revealed that suboptimal larval nutrition had minimal effects on susceptibility (Baqar *et al.*, 1980). Conversely, *Cx. tritaeniorhynchus* larvae reared on low-quality diets exhibited higher JEV titres than larvae fed high-quality diets (Takahashi, 1976). Nutritionally deprived *Ae. triseriatus* developed into smaller adults that were more efficient transmitters of LACV, probably because the smaller females consumed larger or more frequent blood meals to fuel egg production (Grimstad & Haramis, 1984).

Rainfall patterns and humidity are often correlated with temperature. In Rwanda, monthly malaria incidence increases at high-altitude locations following record-high temperatures and rainfall (Loevinsohn, 1994). The cool-dry season in Bangkok, Thailand, probably increases the length of extrinsic incubation for mosquitoes infected with DENV as well as reduces overall *Ae. aegypti* population densities (Watts *et al.*, 1987). DENV transmission to humans in many areas peaks in months with abnormally high rainfall and humidity (Gubler *et al.*, 2001). General circulation models predicting climate change linked to models simulating DENV transmission indicate that relatively small increases in temperature in temperate regions, given virus introduction into susceptible human populations, could increase the potential for epidemics (Hales *et al.*, 2002; Jetten & Focks, 1997). Mosquito studies also link population dynamics to climate change; caged populations of *Ae. albopictus* exposed to moderately higher temperatures (30 °C) and continuous precipitation regimes experienced higher survival (Alto & Juliano, 2001). In field conditions, increasing summer temperatures result in higher peak densities of adults. Under forecast climate change scenarios, higher temperatures and increased rainfall may amplify the rate of colonization and spread of certain mosquito species via rapid population growth.

Geographical variations in mosquito populations may also be determinants of vector competence, although the geographical origin of mosquitoes is closely linked to genetic variability. Populations of *Ae. aegypti* representing a worldwide distribution were evaluated for oral susceptibility to YFV using isozyme variability to delineate genetic relatedness (Tabachnick *et al.*, 1985). Populations that showed close relatedness had comparable oral infection rates; the authors conclude that the heterogeneity in susceptibility is probably determined genetically. More recently, a similar phenomenon was observed among *Ae. aegypti* collected from Mexico and the United States (Bennett *et al.*, 2002).

Although it is tempting to draw generalized conclusions regarding environmental determinants of vector competence, many of the experiments cited from laboratory studies or the sampling of a single field population must be interpreted with caution because of differences in experimental design and methods of vector competence assessment. A constant laboratory temperature is not necessarily representative of fluctuating field conditions. Similarly, many of these studies used colony mosquitoes with unmentioned degrees of genetic relatedness. Mosquitoes sampled from a single field location can be polymorphic in their susceptibility to pathogens and susceptibility can vary temporally. Nevertheless, the literature indicates that environmental variables are a major determinant of vector competence. Fig. 6 reviews the effects of environmental determinants on the prevalence of vector-borne disease, indicating the complex set of interactions between environmental variables and the transmission dynamics of arthropod-borne pathogens.

EVOLUTION OF VECTOR–PATHOGEN INTERACTIONS

The advantages to a pathogen of vector transmission are obvious; for example, the pathogen can increase its mobility without relying on host dispersal and the presence of the pathogen in the blood alone can ensure its transmission. These kinds of advantages

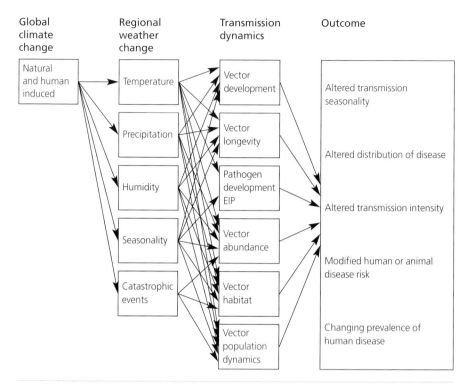

Fig. 6. Environmental influence on the transmission dynamics and outcome of arboviral disease. Arrows indicate the influence of one factor upon others. Note that transmission dynamics are each influenced by a variety of variables. EIP, extrinsic incubation period. Adapted from Gubler *et al.* (2001).

provide a clear means of selection for pathogens that are capable of infection and transmission by arthropod vectors.

Evolution of virulence for vertebrate hosts

Ewald (1994) has argued that the evolution of high vertebrate virulence is favoured for many vector-borne pathogens because immobilization associated with severe disease favours vector feeding on hosts with impaired defensive behaviour. However, many arboviruses, despite their virulence for humans and domesticated animals, which represent dead-end hosts, have very limited virulence for their natural reservoir hosts (Weaver, 1997). Limits on virulence to the vertebrate host have invoked the reasoning that host, and hence parasite, mortality selects against more virulent strains. However, selection for low virulence within clones of the rodent malaria parasite, *Plasmodium chabaudi*, failed to attenuate the pathogen despite strong selection that mimicked 50–75 % host mortality. Selection on between-host differences in virulence was also unable to counteract selection for increased virulence caused by within-host selection processes (Mackinnon & Read, 1999). In other studies employing parasite clones selected for virulence, Mackinnon *et al.* (2002) showed that the total number of gametocytes produced during the infection, a measure of parasite fitness, was fourfold higher in mice that survived than in those which died. Among mice that survived, total gametocyte production was greatest in the host genotype that suffered intermediate levels of morbidity (anaemia and weight loss). Thus transmission costs of high virulence were partly due to host mortality, but perhaps also due to some factor related to high morbidity. Other studies examining the impact of host immunity on malaria parasite virulence predict that anti-disease vaccines will select for higher virulence in those microparasites for which virulence is linked to transmission (Mackinnon & Read, 2003).

Evolution of virulence for vectors

Any effect of infection and transmission on the fitness of the vector was for many years discounted because obvious effects like high mortality of infected arthropods were not observed. However, recent work has demonstrated that vector infection can result in frank pathology in vectors, along with reduced fecundity and other measures of fitness. A few examples for arboviruses and *Plasmodium* spp. are highlighted below.

Several studies have shown that arbovirus infection of vectors, particularly mosquitoes, can reduce survival and fecundity (Weaver, 1997). Rift Valley fever virus decreases survival of *Cx. pipiens* mosquitoes (Faran *et al.*, 1987) and also reduces blood feeding success and fecundity (Turell *et al.*, 1985a). LACV also reduces the ability of *Ae. triseriatus* to obtain blood meals (Grimstad *et al.*, 1980). Although most tick-borne viruses appear to cause no detrimental effects on their vectors (Nuttall *et al.*, 1994), African swine fever virus increases mortality rates of infected ticks, including nymphs of *Ornithodoros coriaceus* (Groocock *et al.*, 1980), third instar *Ornithodoros puerto-ricensis* (Endris *et al.*, 1992a), adult female *Ornithodoros moubata* and *Ornithodoros serraticus* and several stages of *Ornithodoros marocanus* (Endris *et al.*, 1992b). Tick mortality can lead to clearance of African swine fever virus from colonies (Hess *et al.*, 1989).

Other arboviruses, including YFV, KUNV, SAV and CEV delay larval development of mosquito vectors infected transovarially and also increase mortality rates (Turell, 1988). However, LACV apparently causes no detectable deleterious effects on mosquitoes infected transovarially (Patrican & DeFoliart, 1985). Flaviviruses are not known to cause any deleterious effects on their insect or tick vectors (Ludwig & Iacono-Connors, 1993).

Pathological lesions have also been detected within tissues of mosquitoes infected with several alphaviruses. Semliki Forest virus causes cytopathic changes in the salivary glands of *Ae. aegypti* (Lam & Marshall, 1968; Mims *et al.*, 1966), although this

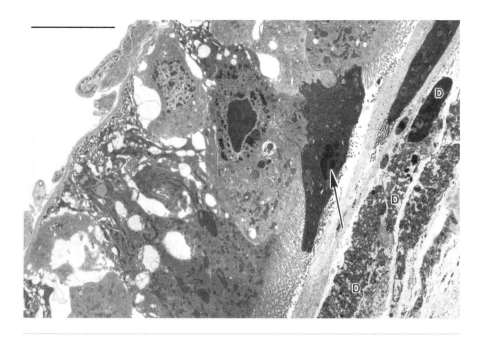

Fig. 7. Transmission electron micrograph of a *Culiseta melanura* midgut 4 days after infection with EEEV. The lumen (right) contains bands of degenerating epithelial cells (D). The electron-dense cell in the centre (arrow) extends from the epithelium into the lumen, suggesting a sloughing process. Bar, 10 µm. Taken from Weaver *et al.* (1988) with permission.

mosquito is not known to be a natural vector of this virus. Both EEEV (Weaver *et al.*, 1988) and WEEV (Weaver *et al.*, 1988, 1992) cause cytopathic changes in midgut epithelial cells of their natural mosquito vectors (Fig. 7). Infection with blood meals containing high virus titres results in sloughing of epithelial cells into the midgut lumen, which may decrease viral loads and modulate the infection. Midgut lesions sometimes compromise the integrity of the abdominal midgut, allowing blood meal components and virus direct access to the haemocoel (Weaver *et al.*, 1988, 1992). These pathological changes may therefore enhance virus dissemination to the salivary glands, providing some selective advantage for mosquito virulence (see below). The lack of arbovirus-induced pathological lesions outside the midgut of most natural mosquito vectors may reflect modulation of virus replication during later stages of infection. However, salivary glands generally produce large amounts of virus without apparent cytopathology.

Considerable investigation into the effects of malaria parasites on mosquito vectors has also been undertaken. However, the question of whether malaria parasites kill their mosquito vectors remains open. Some studies have reported *Plasmodium*-induced mortality, while others have not. Meta-analyses (Ferguson & Read, 2002) show that, overall, malaria parasites do reduce mosquito survival. However, mortality effects are more commonly detected in unnatural vector–parasite combinations and in studies of longer duration.

More detailed studies have demonstrated interesting effects of mixed populations of malaria parasites on vectors. When infected with *P. chabaudi*, *Anopheles stephensi* fecundity is reduced 20 % with a mixture of two genotypes, compared to infection by either genotype alone. Mixed genotype infections are associated with high gametocyte loads and anaemia in the vertebrate host (mice), both of which were correlated with smaller blood meals. Higher parasite loads result in a higher percentage of mosquitoes taking a meal. Overall, Ferguson *et al.* (2003) conclude that mosquitoes may prefer highly parasitaemic hosts that will most impair their fecundity. These and other studies indicate the complexity of selective forces operating on pathogen populations within both vertebrate hosts and vectors and the need for more experimental work employing many different vector-borne pathogens to understand the virulence we face today in human vector-borne diseases, as well as potential changes in virulence as we humans continue to alter the selective landscape.

REFERENCES

Abbey, H. (1952). An examination of the Reed-Frost theory of epidemics. *Human Biol* **24**, 201–233.

Ahmed, T., Hayes, C. G. & Baqar, S. (1979). Comparison of vector competence for West Nile virus of colonized populations of *Culex tritaeniorhynchus* from southern Asia and the Far East. *Southeast Asian J Trop Med Public Health* **10**, 498–504.

Aitken, T. H., Downs, W. G. & Shope, R. E. (1977). *Aedes aegypti* strain fitness for yellow fever virus transmission. *Am J Trop Med Hyg* **26**, 985–989.

Alto, B. W. & Juliano, S. A. (2001). Precipitation and temperature effects on populations of *Aedes albopictus* (Diptera: Culicidae): implications for range expansion. *J Med Entomol* **38**, 646–656.

Azad, A. F. & Beard, C. B. (1998). Rickettsial pathogens and their arthropod vectors. *Emerg Infect Dis* **4**, 179–186.

Bacot, A. W. & Martin, C. J. (1914). Observation on the mechanism of the transmission of plague by fleas. *J Hyg (Cambridge) Plague Suppl 3*, 423–439.

Bailey, N. T. J. (1982). *The Biomathematics of Malaria: the Biomathematics of Disease.* 1. London: Griffin.

Baqar, S., Hayes, C. G. & Ahmed, T. (1980). The effect of larval rearing conditions and adult age on the susceptibility of *Culex tritaeniorhynchus* to oral infection with West Nile virus. *Mosq News* **40**, 165–172.

Bates, M. R.-G. M. (1946). The development of the virus of yellow fever in Haemagogus mosquitoes. *Am J Trop Med*

Beaman, J. R. & Turell, M. J. (1991). Transmission of Venezuelan equine encephalomyelitis virus by strains of *Aedes albopictus* (Diptera: Culicidae) collected in North and South America. *J Med Entomol* **28**, 161–164.

Beaty, B. J. & Bishop, D. H. (1988). Bunyavirus-vector interactions. *Virus Res* **10**, 289–301.

Beaty, B. J., Holterman, M., Tabachnick, W., Shope, R. E., Rozhon, E. J. & Bishop, D. H. (1981). Molecular basis of bunyavirus transmission by mosquitoes: role of the middle-sized RNA segment. *Science* **211**, 1433–1435.

Beaty, B. J., Miller, B. R., Shope, R. E., Rozhon, E. J. & Bishop, D. H. (1982). Molecular basis of bunyavirus per os infection of mosquitoes: role of the middle-sized RNA segment. *Proc Natl Acad Sci U S A* **79**, 1295–1297.

Beauperthuy, L. D. (1854). Transmission of yellow fever and other diseases by mosquito. *Gazeta Oficial de Cumana* May, 23.

Beerntsen, B. T., James, A. A. & Christensen, B. M. (2000). Genetics of mosquito vector competence. *Microbiol Mol Biol Rev* **64**, 115–137.

Bennett, K. E., Olson, K. E., Munoz Mde, L., Fernandez-Salas, I., Farfan-Ale, J. A., Higgs, S., Black, W. C. & Beaty, B. J. (2002). Variation in vector competence for dengue 2 virus among 24 collections of *Aedes aegypti* from Mexico and the United States. *Am J Trop Med Hyg* **67**, 85–92.

Bhardwaj, S., Holbrook, M., Shope, R. E., Barrett, A. D. & Watowich, S. J. (2001). Biophysical characterization and vector-specific antagonist activity of domain III of the tick-borne flavivirus envelope protein. *J Virol* **75**, 4002–4007.

Bosio, C. F., Thomas, R. E., Grimstad, P. R. & Rai, K. S. (1992). Variation in the efficiency of vertical transmission of dengue-1 virus by strains of *Aedes albopictus* (Diptera: Culicidae). *J Med Entomol* **29**, 985–989.

Bosio, C. F., Beaty, B. J. & Black, W. C. (1998). Quantitative genetics of vector competence for dengue-2 virus in *Aedes aegypti*. *Am J Trop Med Hyg* **59**, 965–970.

Bosio, C. F., Fulton, R. E., Salasek, M. L., Beaty, B. J. & Black, W. C. (2000). Quantitative trait loci that control vector competence for dengue-2 virus in the mosquito *Aedes aegypti*. *Genetics* **156**, 687–698.

Boyce, R. W. (1909). *Mosquito or Man? The Conquest of the Tropical World*. London: John Murray.

Bradley, D. J. (1982). Epidemiological models and reality. In *Population Dynamics of Infectious Diseases*, pp. 320–333. Edited by R. M. Anderson. London: Chapman & Hall.

Brault, A. C., Powers, A. M. & Weaver, S. C. (2002). Vector infection determinants of Venezuelan equine encephalitis virus reside within the E2 envelope glycoprotein. *J Virol* **76**, 6387–6392.

Brenner, R. R. & Stoka, A. (1987). *Chagas' Disease Vectors: Taxonomic, Ecological, and Epidemiological Aspects*. Boca Raton, FL: CRC Press.

Briegel, H. (2003). Physiological bases of mosquito ecology. *J Vector Ecol* **28**, 1–11.

Bruce-Chwatt, L. J. (1976). Swellengrebel oration: mathematical models in the epidemiology and control of malaria. *Trop Geogr Med* **28**, 1–8.

Burgdorfer, W. & Brinton, L. P. (1975). Mechanisms of transovarial infection of spotted fever Rickettsiae in ticks. *Ann N Y Acad Sci* **266**, 61–72.

Chamberlain, R. W. & Sudia, W. D. (1955). The effects of temperature upon the extrinsic incubation of Eastern equine encephalitis in mosquitoes. *Am J Hyg* **62**, 295–300.

Chamberlain, R. W., Sikes, R. K., Nelson, D. B. & Sudia, W. D. (1954). Studies on the North American arthropod-borne encephalitides. VI. Quantitative determinations of the virus-vector relationships. *Am J Hyg* **60**, 278–285.

Chang, G. J., Cropp, B. C., Kinney, R. M., Trent, D. W. & Gubler, D. J. (1995). Nucleotide sequence variation of the envelope protein gene identifies two distinct genotypes of yellow fever virus. *J Virol* **69**, 5773–5780.

Collins, F. H., Sakai, R. K., Vernick, K. D., Paskewitz, S., Seeley, D. C., Miller, L. H., Collins, W. E., Campbell, C. C. & Gwadz, R. W. (1986). Genetic selection of a *Plasmodium* refractory strain of the malaria vector *Anopheles gambiae*. *Science* **234**, 607–610.

Collins, F. H., Zheng, L., Paskewitz, S. M. & Kafatos, F. C. (1997). Progress in the map-based cloning of the *Anopheles gambiae* genes responsible for the encapsulation of malarial parasites. *Ann Trop Med Parasitol* **91**, 517–521.

Craig, G. B., Jr & Hickey, W. A. (1967). Genetics of *Aedes aegypti*. In *Genetics of Insect Vectors of Disease*, pp. 794. Edited by J. W. Wright & R. Pal. Amsterdam: Elsevier.

Cupp, E. W. & Stokes, G. M. (1976). Feeding patterns of *Culex salinarius* Coquillett in Jefferson parish, Louisiana. *Mosq News* **36**, 332–335.

Davies, C. R., Jones, L. D. & Nuttall, P. A. (1989). Viral interference in the tick, *Rhipicephalus appendiculatus*. I. Interference to oral superinfection by Thogoto virus. *J Gen Virol* **70**, 2461–2468.

Day, J. F. & Edman, J. D. (1984). The importance of disease induced changes in mammalian body temperature to mosquito blood feeding. *Comp Biochem Physiol A* **77**, 447–452.

de Mattos, C. C., de Mattos, C. A., MacLachlan, N. J., Giavedoni, L. D., Yilma, T. & Osburn, B. I. (1996). Phylogenetic comparison of the S3 gene of United States prototype strains of bluetongue virus with that of field isolates from California. *J Virol* **70**, 5735–5739.

Dietz, K. (1975). Transmission and control of arbovirus diseases. In *Epidemiology*, pp. 104–121. Edited by D. Ludwig & K. L. Cooke. Philadelphia: Society for Industrial and Applied Mathematics.

Dietz, K., Molineaux, L. & Thomas, A. (1974). A malaria model tested in the African savannah. *Bull W H O* **50**, 347–357.

Dohm, D. J., O'Guinn, M. L. & Turell, M. J. (2002). Effect of environmental temperature on the ability of *Culex pipiens* (Diptera: Culicidae) to transmit West Nile virus. *J Med Entomol* **39**, 221–225.

Domingo, E. (2000). Viruses at the edge of adaptation. *Virology* **270**, 251–253.

Duhrkopf, R. E. & Trpis, M. (1980). The degree of susceptibility and levels of infection in ten different strains of *Aedes polynesiensis* Marks infected with subperiodic *Brugia malayi* and *Brugia pahangi*. *Am J Trop Med Hyg* **29**, 815–819.

Dye, C. (1992). The analysis of parasite transmission by bloodsucking insects. *Annu Rev Entomol* **37**, 1–19.

Edman, J. D. (1974). Host-feeding patterns of Florida mosquitoes. 3. *Culex* (Culex) and *Culex* (Neoculex). *J Med Entomol* **11**, 95–104.

Edman, J. D. (2000). Arthropod transmission of vertebrate parasites. In *Medical Entomology*, pp. 151–163. Edited by B. F. Eldridge & J. D. Edman. Dordrecht: Kluwer.

Edman, J. D. & Taylor, D. J. (1968). *Culex nigripalpus*: seasonal shift in the bird-mammal feeding ratio in a mosquito vector of human encephalitis. *Science* **161**, 67–68.

el Hussein, A., Ramig, R. F., Holbrook, F. R. & Beaty, B. J. (1989). Asynchronous mixed infection of *Culicoides variipennis* with bluetongue virus serotypes 10 and 17. *J Gen Virol* **70**, 3355–3362.

Endris, R. G., Haslett, T. M. & Hess, W. R. (1992a). African swine fever virus infection in the soft tick, *Ornithodoros* (*Alectorobius*) *puertoricensis* (Acari: Argasidae). *J Med Entomol* **29**, 990–994.

Endris, R. G., Hess, W. R. & Caiado, J. M. (1992b). African swine fever virus infection in the Iberian soft tick, *Ornithodoros* (*Pavlovskyella*) *marocanus* (Acari: Argasidae). *J Med Entomol* **29**, 874–878.

Ewald, P. W. (1994). *Evolution of Infectious Disease*. New York: Oxford University Press.

Faran, M. E., Turell, M. J., Romoser, W. S., Routier, R. G., Gibbs, P. H., Cannon, T. L. & Bailey, C. L. (1987). Reduced survival of adult *Culex pipiens* infected with Rift Valley fever virus. *Am J Trop Med Hyg* **37**, 403–409.

Feldmann, A. M. & Ponnudurai, T. (1989). Selection of *Anopheles stephensi* for refractoriness and susceptibility to *Plasmodium falciparum*. *Med Vet Entomol* **3**, 41–52.

Ferguson, H. M. & Read, A. F. (2002). Why is the effect of malaria parasites on mosquito survival still unresolved? *Trends Parasitol* **18**, 256–261.

Ferguson, H. M., Rivero, A. & Read, A. F. (2003). The influence of malaria parasite genetic diversity and anaemia on mosquito feeding and fecundity. *Parasitology* **127**, 9–19.

Fine, P. E. & LeDuc, J. W. (1978). Towards a quantitative understanding of the epidemiology of Keystone virus in the United States. *Am J Trop Med Hyg* **27**, 322–338.

Garrett-Jones, C. & Grab, B. (1964). The assessment of insecticidal impact on the malaria mosquito vectorial capacity, from data on the proportion of parous females. *Bull W H O* **31**, 531–545.

Garrett-Jones, C. & Shidrawi, G. R. (1969). Malaria vectorial capacity of a population of *Anopheles gambiae*: an exercise in epidemiological entomology. *Bull W H O* **40**, 531–545.

Glasgow, J. P. (1961). The feeding habits of *Glossina swynnertoni*. *J Anim Ecol* **30**, 77–85.

Gooding, R. H. (1996). Genetic variation in arthropod vectors of disease-causing organisms: obstacles and opportunities. *Clin Microbiol Rev* **9**, 301–320.

Gould, A. R. & Pritchard, L. I. (1990). Relationships amongst bluetongue viruses revealed by comparisons of capsid and outer coat protein nucleotide sequences. *Virus Res* **17**, 31–52.

Grassi, B., Bignami, A. E. & Bastianelli, G. (1899). Ciclo evolutivo delle semilune nell' Anopheles claviger ed altri studi sulla malaria dall' ottobre 1898 al maggio 1899. *Atti Soc Studi Malaria* **1**, 14–27.

Green, D. W., Rowley, W. A., Wong, Y. W., Brinker, J. P., Dorsey, D. C. & Hausler, W. J., Jr (1980). The significance of western equine encephalomyelitis viral infections in *Aedes trivittatus* (Diptera: Culicidae) in Iowa. I. Variation in susceptibility of *Aedes trivittatus* to experimental infection with three strains of western equine encephalomyelitis virus. *Am J Trop Med Hyg* **29**, 118–124.

Greenwood, B. (1999). Malaria mortality and morbidity in Africa. *Bull W H O* **77**, 617–618.

Grimstad, P. R. & Haramis, L. D. (1984). *Aedes triseriatus* (Diptera: Culicidae) and La Crosse virus. III. Enhanced oral transmission by nutrition-deprived mosquitoes. *J Med Entomol* **21**, 249–256.

Grimstad, P. R., Craig, G. B., Jr, Ross, Q. E. & Yuill, T. M. (1977). *Aedes triseriatus* and La Crosse virus: geographic variation in vector susceptibility and ability to transmit. *Am J Trop Med Hyg* **26**, 990–996.

Grimstad, P. R., Ross, Q. E. & Craig, G. B., Jr (1980). *Aedes triseriatus* (Diptera: Culicidae) and La Crosse virus. II. Modification of mosquito feeding behavior by virus infection. *J Med Entomol* **17**, 1–7.

Groocock, C. M., Hess, W. R. & Gladney, W. J. (1980). Experimental transmission of African swine fever virus by *Ornithodoros coriaceus*, an Argasid tick indigenous to the United States. *Am J Vet Res* **41**, 591–594.

Gubler, D. J. & Rosen, L. (1976). Variation among geographic strains of *Aedes albopictus* in susceptibility to infection with dengue viruses. *Am J Trop Med Hyg* **25**, 318–325.

Gubler, D. J., Nalim, S., Tan, R., Saipan, H. & Sulianti Saroso, J. (1979). Variation in susceptibility to oral infection with dengue viruses among geographic strains of *Aedes aegypti*. *Am J Trop Med Hyg* **28**, 1045–1052.

Gubler, D. J., Reiter, P., Ebi, K. L., Yap, W., Nasci, R. & Patz, J. A. (2001). Climate variability and change in the United States: potential impacts on vector- and rodent-borne diseases. *Environ Health Perspect* **109** (Suppl. 2), 223–233.

Gwadz, R. W. (1994). Genetic approaches to malaria control: how long the road? *Am J Trop Med Hyg* **50** (6 Suppl.), 116–125.

Hahn, C. S., Lustig, S., Strauss, E. G. & Strauss, J. H. (1988). Western equine encephalitis virus is a recombinant virus. *Proc Natl Acad Sci U S A* **85**, 5997–6001.

Hales, S., de Wet, N., Maindonald, J. & Woodward, A. (2002). Potential effect of population and climate changes on global distribution of dengue fever: an empirical model. *Lancet* **360**, 830–834.

Hamer, W. H. (1906). Epidemic diseases in England. *Lancet* **1**, 733.

Hao, Z., Kasumba, I., Lehane, M. J., Gibson, W. C., Kwon, J. & Aksoy, S. (2001). Tsetse immune responses and trypanosome transmission: implications for the development of tsetse-based strategies to reduce trypanosomiasis. *Proc Natl Acad Sci U S A* **98**, 12648–12653.

Hardy, J. L., Reeves, W. C. & Sjogren, R. D. (1976). Variations in the susceptibility of field and laboratory populations of *Culex tarsalis* to experimental infection with western equine encephalomyelitis virus. *Am J Epidemiol* **103**, 498–505.

Hardy, J. L., Houk, E. J., Kramer, L. D. & Reeves, W. C. (1983). Intrinsic factors affecting vector competence of mosquitoes for arboviruses. *Annu Rev Entomol* **28**, 229–262.

Hardy, J. L., Meyer, R. P., Presser, S. B. & Milby, M. M. (1990). Temporal variations in the susceptibility of a semi-isolated population of *Culex tarsalis* to peroral infection with western equine encephalomyelitis and St. Louis encephalitis viruses. *Am J Trop Med Hyg* **42**, 500–511.

Hess, A. D., Cherubin, C. E. & LaMotte, L. C. (1963). Relation of temperature to activity of western and St. Louis encephalitis viruses. *Am J Hyg* **12**, 657–667.

Hess, W. R., Endris, R. G., Lousa, A. & Caiado, J. M. (1989). Clearance of African swine fever virus from infected tick (Acari) colonies. *J Med Entomol* **26**, 314–317.

Hinnebusch, B. J., Perry, R. D. & Schwan, T. G. (1996). Role of the *Yersinia pestis* hemin storage (hms) locus in the transmission of plague by fleas. *Science* **273**, 367–370.

Howe, M. A. & Lehane, M. J. (1986). Post-feed buzzing in the tsetse, *Glossina morsitans morsitans*, is an endothermic mechanism. *Physiol Entomol* **11**, 279–286.

Huff, C. G. (1929). The effects of selection upon susceptibility to bird malaria in *Culex pipiens* Linn. *Ann Trop Med Parasitol* **23**, 427–439.

Huff, C. G. (1931). The inheritance of natural immunity to *Plasmodium cathemerium* in two species of *Culex*. *J Prev Med* **5**, 249–259.

James, M. T. & Harwood, R. F. (1969). *Herm's Medical Entomology*, 6th edn. London: Macmillan.

Jefferies, D., Livesey, J. L. & Molyneux, D. H. (1986). Fluid mechanics of bloodmeal uptake by *Leishmania*-infected sandflies. *Acta Trop* **43**, 43–53.

Jenkins, G. M., Pagel, M., Gould, E. A., Zanotto, P. M. & Holmes, E. C. (2001). Evolution of base composition and codon usage bias in the genus Flavivirus. *J Mol Evol* **52**, 383–390.

Jetten, T. H. & Focks, D. A. (1997). Potential changes in the distribution of dengue transmission under climate warming. *Am J Trop Med Hyg* **57**, 285–297.

Jobling, B. & Lewis, D. J. (1987). *Anatomical Drawings of Biting Flies*. London: British Museum of Natural History.

Jones, L. D., Davies, C. R., Booth, T. F. & Nuttall, P. A. (1989). Viral interference in the tick, *Rhipicephalus appendiculatus*. II. Absence of interference with Thogoto virus when the tick gut is by-passed by parenteral inoculation. *J Gen Virol* **70**, 2469–2473.

Jones, R. H. & Foster, N. M. (1978). Relevance of laboratory colonies of the vector in arbovirus research – *Culicoides variipennis* and bluetongue. *Am J Trop Med Hyg* **27**, 168–177.

Jousset, F. X. (1981). Geographic *Aedes aegypti* strain and Dengue-2 virus: susceptibility and ability to transmit to vertebrates and transovarial transmission. *Ann Virol* (*Inst Pasteur*) **132E**, 357–370.

Kay, B. H. & Jennings, C. D. (2002). Enhancement or modulation of the vector competence of *Ochlerotatus vigilax* (Diptera: Culicidae) for ross river virus by temperature. *J Med Entomol* **39**, 99–105.

Kay, B. H., Saul, A. J. & McCullagh, A. (1987). A mathematical model for the rural amplification of Murray Valley encephalitis virus in southern Australia. *Am J Epidemiol* **125**, 690–705.

Kay, B. H., Edman, J. D., Fanning, I. D. & Mottram, P. (1989). Larval diet and the vector competence of *Culex annulirostris* (Diptera: Culicidae) for Murray Valley encephalitis virus. *J Med Entomol* **26**, 487–488.

Kettle, D. S. (1983). *Medical and Veterinary Entomology*. New York & Toronto: Wiley.

Klimas, R. A., Thompson, W. H., Calisher, C. H., Clark, G. G., Grimstad, P. R. & Bishop, D. H. (1981). Genotypic varieties of La Crosse virus isolated from different geographic regions of the continental United States and evidence for a naturally occurring intertypic recombinant La Crosse virus. *Am J Epidemiol* **114**, 112–131.

Koella, J. C., Sorensen, F. L. & Anderson, R. A. (1998). The malaria parasite, *Plasmodium falciparum*, increases the frequency of multiple feeding of its mosquito vector, *Anopheles gambiae*. *Proc R Soc Lond B Biol Sci* **265**, 763–768.

Kramer, L. D., Hardy, J. L. & Presser, S. B. (1983). Effect of temperature of extrinsic incubation on the vector competence of *Culex tarsalis* for western equine encephalomyelitis virus. *Am J Trop Med Hyg* **32**, 1130–1139.

Kubasu, S. S. (1992). The ability of *Rhipicephalus appendiculatus* (Acarina: Ixodidae) stocks in Kenya to become infected with *Theileria parva*. *Bull Entomol Res* **82**, 349–353.

Labuda, M., Danielova, V., Jones, L. D. & Nuttall, P. A. (1993). Amplification of tick-borne encephalitis virus infection during co-feeding of ticks. *Med Vet Entomol* **7**, 339–342.

Lam, K. S. K. & Marshall, I. D. (1968). Dual infections of *Aedes aegypti* with arboviruses. II. Salivary gland damage by Semliki Forest virus in relation to dual infections. *Am J Trop Med Hyg* **17**, 637–644.

Lehane, M. J. (1991). *Biology of Blood-sucking Mosquitoes*. London: Academic Press.

Lepiniec, L., Dalgarno, L., Huong, V. T., Monath, T. P., Digoutte, J. P. & Deubel, V. (1994). Geographic distribution and evolution of yellow fever viruses based on direct sequencing of genomic cDNA fragments. *J Gen Virol* **75**, 417–423.

Loevinsohn, M. E. (1994). Climatic warming and increased malaria incidence in Rwanda. *Lancet* **343**, 714–718.

Lorenz, L., Beaty, B. J., Aitken, T. H., Wallis, G. P. & Tabachnick, W. J. (1984). The effect of colonization upon *Aedes aegypti* susceptibility to oral infection with yellow fever virus. *Am J Trop Med Hyg* **33**, 690–694.

Ludwig, G. V. & Iacono-Connors, L. C. (1993). Insect-transmitted vertebrate viruses: Flaviviridae. *In Vitro Cell Dev Biol* **29A**, 296–309.

Lysenko, A. J. & Beljaev, A. E. (1987). Quantitative approaches to epidemiology.

In *Leishmaniases in Biology and Medicine*, pp. 263–290. Edited by W. Peters & R. Killick-Kendrick. London: Academic Press.

MacDonald, G. (1957). *The Epidemiology and Control of Malaria*. London: Oxford University Press.

Macdonald, W. W. (1962). The genetic basis of susceptibility to infection with semi-periodic *Brugia malayi* in *Aedes aegypti*. In *The Molecular Biology of Insect Disease Vectors: a Methods Manual*, pp. 337–345. Edited by J. M. Crampton, C. B. Beard & C. Louis. New York: Chapman & Hall.

Macdonald, W. W. (1965). The influence of the gene f^m (filarial susceptibility, *Brugia malayi*) on the susceptibility of *Aedes aegypti* to seven strains of *Brugia*, *Wuchereria*, and *Dirofilaria*. *Ann Trop Med Parasitol* **59**, 64–73.

Mackinnon, M. J. & Read, A. F. (1999). Selection for high and low virulence in the malaria parasite *Plasmodium chabaudi*. *Proc R Soc Lond B Biol Sci* **266**, 741–748.

Mackinnon, M. J. & Read, A. F. (2003). The effects of host immunity on virulence-transmissibility relationships in the rodent malaria parasite *Plasmodium chabaudi*. *Parasitology* **126**, 103–112.

Mackinnon, M. J., Gaffney, D. J. & Read, A. F. (2002). Virulence in rodent malaria: host genotype by parasite genotype interactions. *Infect Genet Evol* **1**, 287–296.

Mahoney, D. F. (1977). Babesia of domestic animals. *Parasitic Protozoa* **4**, 1–43.

Manson, P. (1878). On the development of *Filaria sanguinis hominis*, and on the mosquito considered as a nurse. *Nature* **1878**, 439.

Maudlin, I. (1991). Transmission of African trypanosomiasis: interactions among tsetse immune system, symbionts, and parasites. *Adv Dis Vector Res* **7**, 117–148.

Maudlin, I. & Welburn, S. C. (1988). The role of lectins and trypanosome genotype in the maturation of midgut infections in *Glossina morsitans*. *Trop Med Parasitol* **39**, 56–58.

Maudlin, I., Kabayo, J. P., Flood, M. E. & Evans, D. A. (1984). Serum factors and the maturation of *Trypanosoma congolense* infections in *Glossina morsitans*. *Z Parasitenkd* **70**, 11–19.

McGaw, M. M., Chandler, L. J., Wasieloski, L. P., Blair, C. D. & Beaty, B. J. (1998). Effect of La Crosse virus infection on overwintering of *Aedes triseriatus*. *Am J Trop Med Hyg* **58**, 168–175.

McGreevy, P. B., McClelland, G. A. & Lavoipierre, M. M. (1974). Inheritance of susceptibility to *Dirofilaria immitis* infection in *Aedes aegypti*. *Ann Trop Med Parasitol* **68**, 97–109.

Mead, D. G., Ramberg, F. B., Besselsen, D. G. & Mare, C. J. (2000). Transmission of vesicular stomatitis virus from infected to noninfected black flies co-feeding on nonviremic deer mice. *Science* **287**, 485–487.

Mecham, J. O. & Nunamaker, R. A. (1994). Complex interactions between vectors and pathogens: *Culicoides variipennis sonorensis* (Diptera: Ceratopogonidae) infection rates with bluetongue viruses. *J Med Entomol* **31**, 903–907.

Meissner, J. D., Huang, C. Y., Pfeffer, M. & Kinney, R. M. (1999). Sequencing of prototype viruses in the Venezuelan equine encephalitis antigenic complex. *Virus Res* **64**, 43–59.

Miller, B. R. & Mitchell, C. J. (1991). Genetic selection of a flavivirus-refractory strain of the yellow fever mosquito *Aedes aegypti*. *Am J Trop Med Hyg* **45**, 399–407.

Miller, B. R., Monath, T. P., Tabachnick, W. J. & Ezike, V. I. (1989). Epidemic yellow fever caused by an incompetent mosquito vector. *Trop Med Parasitol* **40**, 396–399.

Mims, C. A., Day, M. F. & Marshall, I. D. (1966). Cytopathic effects of Semliki Forest virus in the mosquito, *Aedes aegypti*. *Am J Trop Med Hyg* **15**, 775–784.

Molineaux, L. & Gramiccia, G. (1980). *The Garki Project: Research on the Epidemiology and Control of Malaria in the Sudan Savannah of West Africa*. Geneva: World Health Organization.

Moloo, S. K. (1993). A comparison of susceptibility of two allopatric populations of *Glossina pallidipes* for stocks of *Trypanosoma congolense*. *Med Vet Entomol* **7**, 369–372.

Molyneux, D. H. & Jefferies, D. (1986). Feeding behaviour of pathogen-infected vectors. *Parasitology* **92**, 721–736.

Moore, J. (1993). Parasites and the behavior of biting flies. *J Parasitol* **79**, 1–16.

Mourya, D. T., Gokhale, M. D., Malunjkar, A. S., Bhat, H. R. & Banerjee, K. (1994). Inheritance of oral susceptibility of *Aedes aegypti* to Chikungunya virus. *Am J Trop Med Hyg* **51**, 295–300.

Munstermann, L. & Craig, G. B., Jr (1979). Genetics of *Aedes aegypti*: updating the linkage map. *J Hered* **70**, 291–296.

Murray, M. D. & Nicholls, D. G. (1965). Studies on the ectoparasites of seals and penguins. I. The ecology of the louse *Lepidophthirus macrorhini* Enderlain on the southern elephant seal, *Mironga leonina* (L.). *Aust J Zool* **13**, 437–454.

Mutebi, J. P., Wang, H., Li, L., Bryant, J. E. & Barrett, A. D. (2001). Phylogenetic and evolutionary relationships among yellow fever virus isolates in Africa. *J Virol* **75**, 6999–7008.

Myles, K. M., Pierro, D. J. & Olson, K. E. (2003). Deletions in the putative cell receptor-binding domain of Sindbis virus strain MRE16 E2 glycoprotein reduce midgut infectivity in *Aedes aegypti*. *J Virol* **77**, 8872–8881.

Najera, J. A. (1974). A critical review of the field application of a mathematical model of malaria eradication. *Bull W H O* **50**, 449–457.

Noden, B. H., Kent, M. D. & Beier, J. C. (1995). The impact of variations in temperature on early *Plasmodium falciparum* development in *Anopheles stephensi*. *Parasitology* **111**, 539–545.

Nott, J. C. (1848). On the origins of yellow fever. *New Orleans Med Sci J* **4**, 563–601.

Novella, I. S., Clarke, D. K., Quer, J. & 7 other authors (1995). Extreme fitness differences in mammalian and insect hosts after continuous replication of vesicular stomatitis virus in sandfly cells. *J Virol* **69**, 6805–6809.

Nuttall, P. A., Jones, L. D., Labuda, M. & Kaufman, W. R. (1994). Adaptations of arboviruses to ticks. *J Med Entomol* **31**, 1–9.

Patrican, L. A. & DeFoliart, G. R. (1985). Lack of adverse effect of transovarially acquired La Crosse virus infection on the reproductive capacity of *Aedes triseriatus* (Diptera: Culicidae). *J Med Entomol* **22**, 604–611.

Patz, J. A., Epstein, P. R., Burke, T. A. & Balbus, J. M. (1996). Global climate change and emerging infectious diseases. *JAMA (J Am Med Assoc)* **275**, 217–223.

Pereira, M. E., Andrade, A. F. & Ribeiro, J. M. (1981). Lectins of distinct specificity in *Rhodnius prolixus* interact selectively with *Trypanosoma cruzi*. *Science* **211**, 597–600.

Philip, C. B. & Rozeboom, L. E. (1973). *Medico-Veterinary Entomology: a Generation of Progress*. Palo Alto, CA: Annual Reviews.

Piesman, J. & Gage, K. L. (2000). Bacterial and rickettsial diseases. In *Medical Entomology*, pp. 377–413. Edited by B. F. Eldridge & J. D. Edman. Dordrecht: Kluwer.

Pollitzer, R. (1954). *Plague*. Geneva: World Health Organization.

Port, G. R., Bateham, P. F. L. & Bryan, J. H. (1980). The relationship of the host size to feeding by mosquitoes of the *Anopheles gambiae* complex (Diptera: Culicidae). *Bull Entomol Res* **70**, 133–144.

Powers, A. M., Brault, A. C., Tesh, R. B. & Weaver, S. C. (2000). Re-emergence of chikungunya and o'nyong-nyong viruses: evidence for distinct geographical lineages and distant evolutionary relationships. *J Gen Virol* **81**, 471–479.

Rabinovich, J. E. (1987). Mathematical modeling in the study and control of triatominae populations. In *Chagas' Disease Vectors*, vol. 1, *Taxonomic, Ecological, and Epidemiological Aspects*, pp. 119–146. Edited by R. R. Brenner & A. Stoka. Boca Raton, FL: CRC Press.

Randolph, S. E. (1998). Ticks are not insects: consequences of contrasting vector biology for transmission potential. *Parasitol Today* **14**, 186–192.

Randolph, S. E., Gern, L. & Nutthall, P. A. (1996). Co-feeding ticks: epidemiological significance for tick-borne pathogen transmission. *Parasitol Today* **12**, 472–479.

Reisen, W. K. (1989). Mosquito vector and vertebrate host interaction: the key to maintenance of certain arboviruses. In *Ecology and Physiology of Parasites*, pp. 223–230. Edited by A. M. Fallis. Toronto.

Reisen, W. K., Meyer, R. P., Shields, J. & Arbolante, C. (1989). Population ecology of preimaginal *Culex tarsalis* (Diptera: Culicidae) in Kern County, California. *J Med Entomol* **26**, 10–22.

Reisen, W. K., Lothrop, H. D. & Hardy, J. L. (1995). Bionomics of *Culex tarsalis* (Diptera: Culicidae) in relation to arbovirus transmission in southeastern California. *J Med Entomol* **32**, 316–327.

Reisen, W. K., Hardy, J. L., Presser, S. B. & Chiles, R. E. (1996). Seasonal variation in the vector competence of *Culex tarsalis* (Diptera:Culicidae) from the Coachella Valley of California for western equine encephalomyelitis and St. Louis encephalitis viruses. *J Med Entomol* **33**, 433–437.

Reiter, P. (2001). Climate change and mosquito-borne disease. *Environ Health Perspect* **109** (Suppl. 1), 141–161.

Ribeiro, J. M. C., Rossignol, P. A. & Spielman, A. (1985). Salivary gland apyrase determines probing time in anopheline mosquitoes? *J Insect Physiol* **31**, 689–692.

Riley, W. M. & Johannsen, O. A. (1915). *Handbook of Medical Entomology*, 1st edn. Ithaca, NY: The Comstock Publishing Company.

Rogers, D. J. (1988). A general model for the African trypanosomiases. *Parasitology* **97**, 193–212.

Rogers, M. E., Chance, M. L. & Bates, P. A. (2002). The role of promastigote secretory gel in the origin and transmission of the infective stage of *Leishmania mexicana* by the sandfly *Lutzomyia longipalpis*. *Parasitology* **124**, 495–507.

Rosen, L., Lien, J. C., Shroyer, D. A., Baker, R. H. & Lu, L. C. (1989). Experimental vertical transmission of Japanese encephalitis virus by *Culex tritaeniorhynchus* and other mosquitoes. *Am J Trop Med Hyg* **40**, 548–556.

Ross, R. (1897). On some peculiar pigmented cells found in two mosquitoes fed on malarial blood. *Br Med J* 1786–1788.

Ross, R. (1898). Report on the cultivation of Proteosoma, Labbe, in grey mosquitoes. *Indian Med Gaz* **33**, 401–408.

Ross, R. (1911). *The Prevention of Malaria*. London: Murray.

Rossignol, P. A., Ribeiro, J. M. & Spielman, A. (1984). Increased intradermal probing time in sporozoite-infected mosquitoes. *Am J Trop Med Hyg* **33**, 17–20.

Samal, S. K., el-Hussein, A., Holbrook, F. R., Beaty, B. J. & Ramig, R. F. (1987). Mixed infection of *Culicoides variipennis* with bluetongue virus serotypes 10 and 17: evidence for high frequency reassortment in the vector. *J Gen Virol* **68**, 2319–2329.

Sardelis, M. R., Turell, M. J., Dohm, D. J. & O'Guinn, M. L. (2001). Vector competence of selected North American *Culex* and *Coquillettidia* mosquitoes for West Nile virus. *Emerg Infect Dis* **7**, 1018–1022.

Schreiber, K. V. (2001). An investigation of relationships between climate and dengue using a water budgeting technique. *Int J Biometeorol* **45**, 81–89.

Scott, T. W. & Weaver, S. C. (1989). Eastern equine encephalomyelitis virus: epidemiology and evolution of mosquito transmission. *Adv Virus Res* **37**, 277–328.

Seabaugh, R. C., Olson, K. E., Higgs, S., Carlson, J. O. & Beaty, B. J. (1998). Development of a chimeric sindbis virus with enhanced per Os infection of *Aedes aegypti*. *Virology* **243**, 99–112.

Severson, D. W., Mori, A., Zhang, Y. & Christensen, B. M. (1993). Linkage map for *Aedes aegypti* using restriction fragment length polymorphisms. *J Hered* **84**, 241–247.

Severson, D. W., Brown, S. E. & Knudson, D. L. (2001). Genetic and physical mapping in mosquitoes: molecular approaches. *Annu Rev Entomol* **46**, 183–219.

Shope, R. (1991). Global climate change and infectious diseases. *Environ Health Perspect* **96**, 171–174.

Shortt, H. E. & Swaminath, C. S. (1927). The method of feeding of *Phlebotomus argentipes* with relation to its bearing on the transmission of Kala-azar. *Indian J Med Res* **15**, 827–836.

Smith, C. E. (1987). Factors influencing the transmission of western equine encephalomyelitis virus between its vertebrate maintenance hosts and from them to humans. *Am J Trop Med Hyg* **37**(3 Suppl.), 33S–39S.

Smith, R. D. (1983). *Babesia bovis*: computer simulation of the relationship between the tick vector, parasite, and bovine host. *Exp Parasitol* **56**, 27–40.

Smith, T. & Kilbourne, F. L. (1893). Investigations into the nature, causation and prevention of Texas or southern cattle fever. *US Dep Agric Bur Ann Ind Bull* **1**, 301.

Sonenshine, D. E. (1993). *Biology of Ticks*, vol. 2. New York & Oxford: Oxford University Press.

Soulsby, E. J. L. (1982). *Helminths, Arthropods, and Protozoa of Domesticated Animals*. London: Baillière Tindall.

Steinhauer, D. A. & Holland, J. J. (1987). Rapid evolution of RNA viruses. *Annu Rev Microbiol* **41**, 409–433.

Sterling, C. R., Aikawa, M. & Vanderberg, J. P. (1973). The passage of *Plasmodium berghei* sporozoites through the salivary glands of *Anopheles stephensi*: an electron microscope study. *J Parasitol* **59**, 593–605.

Stierhof, Y. D., Bates, P. A., Jacobson, R. L., Rogers, M. E., Schlein, Y., Handman, E. & Ilg, T. (1999). Filamentous proteophosphoglycan secreted by *Leishmania* promastigotes forms gel-like three-dimensional networks that obstruct the digestive tract of infected sandfly vectors. *Eur J Cell Biol* **78**, 675–689.

Sundin, D. R., Beaty, B. J., Nathanson, N. & Gonzalez-Scarano, F. (1987). A G1 glycoprotein epitope of La Crosse virus: a determinant of infection of *Aedes triseriatus*. *Science* **235**, 591–593.

Sutcliffe, J. F. (1987). Distance orientation of biting flies to their hosts. *Insect Sci Appl* **8**, 611–616.

Tabachnick, W. J. (1991). Genetic control of oral susceptibility to infection of *Culicoides variipennis* with bluetongue virus. *Am J Trop Med Hyg* **45**, 666–671.

Tabachnick, W. J. (1994). Genetics of insect vector competence for arboviruses. *Adv Dis Vector Res* **10**, 93–108.

Tabachnick, W. J., Wallis, G. P., Aitken, T. H., Miller, B. R., Amato, G. D., Lorenz, L., Powell, J. R. & Beaty, B. J. (1985). Oral infection of *Aedes aegypti* with yellow fever virus: geographic variation and genetic considerations. *Am J Trop Med Hyg* **34**, 1219–1224.

Takahashi, M. (1976). The effects of environmental and physiological conditions of *Culex tritaeniorhynchus* on the pattern of transmission of Japanese encephalitis virus. *J Med Entomol* **13**, 275–284.

Takahashi, M. (1980). Variation in susceptibility among colony strains of *Culex tritaeniorhynchus* to Japanese encephalitis virus infection. *Jpn J Med Sci Biol* **33**, 321–329.

Tempelis, C. H. & Washino, R. K. (1967). Host-feeding patterns of *Culex tarsalis* in the Sacramento Valley, California, with notes on other species. *J Med Entomol* **4**, 315–318.

Tesh, R. B. (1980). Experimental studies on the transovarial transmission of Kunjin and San Angelo viruses in mosquitoes. *Am J Trop Med Hyg* **29**, 657–666.

Tesh, R. B. & Gubler, D. J. (1975). Laboratory studies of transovarial transmission of La Crosse and other arboviruses by *Aedes albopictus* and *Culex fatigans*. *Am J Trop Med Hyg* **24**, 876–880.

Tesh, R. B. & Shroyer, D. A. (1980). The mechanism of arbovirus transovarial transmission in mosquitoes: San Angelo virus in *Aedes albopictus*. *Am J Trop Med Hyg* **29**,1394– 1404.

Tesh, R. B., Gubler, D. J. & Rosen, L. (1976). Variation among geographic strains of *Aedes albopictus* in susceptibility to infection with chikungunya virus. *Am J Trop Med Hyg* **25**, 326–335.

Townson, H. & Chaithong, U. (1991). Mosquito host influences on development of filariae. *Ann Trop Med Parasitol* **85**, 149–163.

Turell, M. J. (1988). Horizontal and vertical transmission of viruses by insect and tick vectors. In *The Arboviruses: Epidemiology and Ecology*, vol. I, pp. 127–152. Edited by T. P. Monath. Boca Raton, FL: CRC Press.

Turell, M. J. & Beaman, J. R. (1992). Experimental transmission of Venezuelan equine encephalomyelitis virus by a strain of *Aedes albopictus* (Diptera: Culicidae) from New Orleans, Louisiana. *J Med Entomol* **29**, 802–805.

Turell, M. J. & LeDuc, J. W. (1983). The role of mosquitoes in the natural history of California serogroup viruses. *Prog Clin Biol Res* **123**, 43–55.

Turell, M. J. & Lundstrom, J. O. (1990). Effect of environmental temperature on the vector competence of *Aedes aegypti* and *Ae. taeniorhynchus* for Ockelbo virus. *Am J Trop Med Hyg* **43**, 543–550.

Turell, M. J., Reeves, W. C. & Hardy, J. L. (1982). Evaluation of the efficiency of transovarial transmission of California encephalitis viral strains in *Aedes dorsalis* and *Aedes melanimon*. *Am J Trop Med Hyg* **31**, 382–388.

Turell, M. J., Gargan, T. P. & Bailey, C. L. (1985a). *Culex pipiens* (Diptera: Culicidae) morbidity and mortality associated with Rift Valley fever virus infection. *J Med Entomol* **22**, 332–337.

Turell, M. J., Rossi, C. A. & Bailey, C. L. (1985b). Effect of extrinsic incubation temperature on the ability of *Aedes taeniorhynchus* and *Culex pipiens* to transmit Rift Valley fever virus. *Am J Trop Med Hyg* **34**, 1211–1218.

van Niekerk, M., Freeman, M., Paweska, J. T., Howell, P. G., Guthrie, A. J., Potgieter, A. C., van Staden, V. & Huismans, H. (2003). Variation in the NS3 gene and protein in South African isolates of bluetongue and equine encephalosis viruses. *J Gen Virol* **84**, 581–590.

Watts, D. M., Burke, D. S., Harrison, B. A., Whitmire, R. E. & Nisalak, A. (1987). Effect of temperature on the vector efficiency of *Aedes aegypti* for dengue 2 virus. *Am J Trop Med Hyg* **36**, 143–152.

Weaver, S. C. (1997). Vector biology in viral pathogenesis. In *Viral Pathogenesis*, pp. 329–352. Edited by N. Nathanson. New York: Lippincott-Raven.

Weaver, S. C., Scott, T. W., Lorenz, L. H., Lerdthusnee, K. & Romoser, W. S. (1988). Togavirus-associated pathologic changes in the midgut of a natural mosquito vector. *J Virol* **62**, 2083–2090.

Weaver, S. C., Lorenz, L. H. & Scott, T. W. (1992). Pathologic changes in the midgut of *Culex tarsalis* following infection with Western equine encephalomyelitis virus. *Am J Trop Med Hyg* **47**, 691–701.

Weaver, S. C., Lorenz, L. H. & Scott, T. W. (1993). Distribution of western equine encephalomyelitis virus in the alimentary tract of *Culex tarsalis* (Diptera: Culicidae) following natural and artificial blood meals. *J Med Entomol* **30**, 391–397.

Weaver, S. C., Kang, W., Shirako, Y., Rumenapf, T., Strauss, E. G. & Strauss, J. H. (1997). Recombinational history and molecular evolution of western equine encephalomyelitis complex alphaviruses. *J Virol* **71**, 613–623.

Weaver, S. C., Brault, A. C., Kang, W. & Holland, J. J. (1999). Genetic and fitness changes accompanying adaptation of an arbovirus to vertebrate and invertebrate cells. *J Virol* **73**, 4316–4326.

Webber, R. H. (1975). Theoretical considerations in the vector control of filariasis. *Southeast Asian J Trop Med Public Health* **6**, 544–548.

Webber, R. H. & Southgate, B. A. (1981). The maximum density of anopheline mosquitoes that can be permitted in the absence of continuing transmission of filariasis. *Trans R Soc Trop Med Hyg* **75**, 499–506.

Welburn, S. C. & Maudlin, I. (1989). Lectin signalling of maturation of *Trypanosoma congolense* infections in tsetse. *Med Vet Entomol* **3**, 141–145.

Welburn, S. C. & Maudlin, I. (1990). Haemolymph lectin and the maturation of trypanosome infections in tsetse. *Med Vet Entomol* **4**, 43–48.

White, G. B. (1974). *Anopheles gambiae* complex and disease transmission in Africa. *Trans R Soc Trop Med Hyg* **68**, 278–301.

Wilson, W. C., Ma, H. C., Venter, E. H., van Djik, A. A., Seal, B. S. & Mecham, J. O. (2000). Phylogenetic relationships of bluetongue viruses based on gene S7. *Virus Res* **67**, 141–151.

Woodring, J. L., Higgs, S. & Beaty, B. J. (1996). Natural cycles of vector-borne pathogens. In *The Biology of Disease Vectors*, pp. 51–72. Edited by B. J. Beaty & W. C. Marquardt. Niwot, CO: University Press of Colorado.

Woodward, T. M., Miller, B. R., Beaty, B. J., Trent, D. W. & Roehrig, J. T. (1991). A single amino acid change in the E2 glycoprotein of Venezuelan equine encephalitis virus affects replication and dissemination in *Aedes aegypti* mosquitoes. *J Gen Virol* **72**, 2431–2435.

Wu, W. K. & Tesh, R. B. (1990). Selection of *Phlebotomus papatasi* (Diptera: Psychodidae) lines susceptible and refractory to *Leishmania major* infection. *Am J Trop Med Hyg* **42**, 320–328.

Zeledon, R. & Rabinovich, J. E. (1981). Chagas' disease: an ecological appraisal with special emphasis on its insect vectors. *Annu Rev Entomol* **26**, 101–133.

Zheng, L., Cornel, A. J., Wang, R., Erfle, H., Voss, H., Ansorge, W., Kafatos, F. C. & Collins, F. H. (1997). Quantitative trait loci for refractoriness of *Anopheles gambiae* to *Plasmodium cynomolgi* B. *Science* **276**, 425–428.

Zinsser, H. (1934). *Rats, Lice and History*. Boston: Little, Brown & Company.

Environmental influences on arbovirus infections and vectors

P. S. Mellor

Institute for Animal Health, Pirbright Laboratory, Ash Rd, Pirbright, Woking GU24 0NF, UK

INTRODUCTION

In 430 BC, Hippocrates said 'Whoever would study medicine aright must learn of the following subjects. First, he must consider the effect of the seasons of the year and the differences between them. Second, he must study the warm and cold winds, both those which are common to every country and those peculiar to a particular locality.'

From this we can see that the influence of climate and environmental variables on infectious diseases has been a topic of interest to medicine for over 2000 years. Recently there has been an increase in this interest spurred on by the various climate change scenarios that have been put forward and highlighted by the fact that 9 of the 10 warmest years on record have occurred since 1990, thereby supporting the concept of ongoing climate change. Furthermore, recent climate models project an increase in global mean temperature of between 1·4 and 5·8 °C during the 21st century (Intergovernmental Panel on Climate Change, 2001). It is further predicted that maximum warming will occur at high latitudes and during the winter, and that night-time temperatures will increase more than daytime temperatures (Karl *et al.*, 1993; Kukla & Karl, 1993; McMichael *et al.*, 1996).

In the context of vector-borne diseases, these predicted temperature increases, and accompanying climatic changes in rainfall patterns and extreme weather events (for example increased windiness), are likely to have a variable but significant effect. By definition, vector-borne diseases possess a vector stage, usually an insect, tick, crustacean or mollusc, which is poikilothermic and therefore peculiarly liable to be

SGM symposium 63: Microbe–vector interactions in vector-borne diseases.
Editors S. H. Gillespie, G. L. Smith & A. Osbourn. Cambridge University Press. ISBN 0 521 84312 X ©SGM 2004

profoundly influenced by climatic variables, especially temperature. The components of climate that are likely to have a substantial and direct effect on vector biology include the following (Hack, 1955; Sellers, 1980; Sellers & Mellor, 1993; Mellor, 1996; Wittmann & Baylis, 2000; Mellor & Leake, 2000; Mellor & Wittmann, 2002).

(a) *Temperature*. A rise in temperature will increase a vector's metabolic rate. Blood-feeding vectors, therefore, will be likely to feed more frequently and hence their biting rates will increase, leading to enhanced egg production and, potentially, a greater population size. However, the daily survival rate of individual vectors may decrease as temperature rises. Temperature may also affect the geographical range of many vectors since this tends to be controlled by minimum and maximum temperature (and by humidity).

(b) *Humidity*. High relative humidity favours most (terrestrial) vector metabolic processes so that at higher temperatures a high humidity (that is, a low saturation deficit) will tend to prolong survival, although increased susceptibility to fungal and bacterial pathogens may offset this to a variable degree. Low humidity (that is, a high saturation deficit) causes a decrease in the daily survival rate of many vectors because of dehydration but in some cases it may also cause an increase in the blood-feeding rate in an attempt to compensate for the high levels of water loss.

(c) *Precipitation*. Precipitation is important to most aquatic vectors but also to many blood-feeding groups of insects, including *Culicidae*, *Simuliidae* and *Culicoides*, since these all have aquatic or semi-aquatic immature stages. Precipitation, therefore, frequently determines the presence, absence, size and endurance of suitable breeding sites. The precise impact of precipitation on a particular type of breeding site will depend upon local evaporation rates (saturation deficits), soil type, slope of the terrain and, possibly, proximity to large bodies of water. Many insect vectors tend to breed in minor seepages of water from drinking trough overflows, irrigation pipe leaks, leaky taps and residual water from earlier rains, and in these situations very heavy precipitation may disrupt the sites and wash the immatures away or drown them.

(d) *Wind*. Since winds can contribute to the passive dispersal of many flying insects, wind direction and wind speeds may affect vector distribution. Some insect vectors, including species of *Culicidae*, *Simuliidae*, *Phlebotominae* and *Culicoides*, can be dispersed for hundreds of kilometres in this way.

In addition to these direct effects, climatic variables may also have important indirect effects upon vector abundance and distribution, and hence disease occurrence. For example, one vector species may be displaced by another with a different vectorial capacity in response to changes such as deforestation, expansion in irrigation or increase in brackish water breeding sites due to sea level rise (McMichael *et al.*, 1996).

INSECT VECTORS

Palaeoclimatic records show that most shifts in the distribution of insect taxa have been associated mainly with temperature change and that these shifts have apparently occurred much more rapidly than shifts in the distribution of vegetation and higher animals (Elias, 1995). Considering the high mobility of insects, particularly winged ones, this should come as no surprise and suggests that the current distribution of many insect species that act as vectors will change rapidly following any changes in climate. In this context, Peters (1993) has estimated that a 1 °C rise in temperature will correspond to an extension equivalent to 90 km of latitude and 150 m of altitude in the range of a vector, which provides a measure of the likely impact of such changes.

Confirmation that changes in the distribution of some insect vectors is already occurring can be inferred from newspaper reports relating the apparent extension of *Phlebotomus* sandflies (vectors of *Leishmania*) from mainland France into the Channel Islands (see *Today* 19th August 1994 edition), but there is a paucity of more convincing evidence on vector species. However, with more 'visible' insects, like butterflies, such reliable information is readily available. For example, it has recently been shown for 52 non-migratory butterfly species in Europe that the northern boundaries of 65 % of them have shifted northwards during the late 20th century (Parmesan *et al.*, 1999).

It is not, of course, obligatory that insects extend their range into new areas of suitable climate solely under their own power: human activities can also play a part. *Aedes albopictus* (the Asian 'tiger' mosquito), an efficient vector of yellow fever and dengue viruses, was introduced into Italy in 1989 via its immature stages, in water contained in second-hand car tyres imported from the USA. This mosquito is now found widely over Italy having been recorded from over 10 regions and 19 provinces (Knudsen, 1996). In 1999 its immature stages were also identified near Caen in northern France (Schaffner & Karsh, 1999) and in 2003 it was found in southern Switzerland (A. Y. Cagienard, personal communication). Since the distribution of *Ae. albopictus* is bounded conservatively by the mean 0 °C isotherm in the coldest month of the year (January), this suggests that in Europe, where this isotherm runs as far north as southern Scandinavia, all or most of the continent is open to colonization. In the USA where *Ae. albopictus* was introduced in 1985, it already occupies 26 southeastern states, ranging as far north as Nebraska and Iowa (McMichael *et al.*, 1996). It also expanded southwards through Florida at the rate of 40 miles per year and now occupies all of the counties in that State. Should climate change result in the temperature increases projected, then the range of *Ae. albopictus* is likely to extend to include most of the populated areas of the eastern USA as well as the southern tip of Canada (Nawrocki & Hawley, 1987). In Europe the projected increases could see *Ae. albopictus* occupying virtually all of western and central areas of the continent up to and including southern Scandinavia. However, such increases in distribution by a vector do not necessarily imply an equivalent increase in areas 'at risk' to disease transmission by that vector. In respect of dengue, for example, the lower limiting temperatures for transmission of this virus seem to be about 20 °C (Halstead, 1990). It is likely, therefore, that transmission of dengue virus would be possible only over the warmer parts of its vector's range. In this context it should be remembered that 'warmer' should not be interpreted only in terms of latitude. An increase in mean temperature of 4 °C would also permit transmission of dengue at the altitude of Mexico City.

Effect of temperature

Temperature is probably the most important extrinsic variable affecting arbovirus transmission via the vector and is therefore likely to have the most significant effect on the epidemiology of many arboviral diseases. However, its precise effects upon the dynamics of arbovirus transmission may be difficult to predict and are likely to vary with each virus–vector combination. Generally speaking, at higher temperatures a vector may blood-feed more frequently and the rate of virogenesis within a vector is usually faster, leading to an enhanced probability of transmission. On the other hand, increased environmental temperature may shorten the lifespan of the vector, which would lessen the transmission potential. As temperature decreases, virogenesis usually slows and at some point (specific for each arbovirus) may cease altogether; however, at lower temperatures the lifespan of the vector may be extended. The likelihood of arbovirus transmission by arthropod vectors is therefore a function of the interaction of these trends (McMichael *et al.*, 1996). Additionally, there may be other rather ill-defined but more subtle effects of temperature upon the ability of arthropods to transmit arboviruses; for example upon the ability of the vector to modulate virus replication within its cells (Hardy *et al.*, 1983; Kramer *et al.*, 1983) or on the threshold or efficacy levels of the various barriers to virus replication and dissemination that exist within a variable proportion of most vector species. Some of these effects of temperature and other climatic variables are considered in more detail using the Office International des Epizooties (OIE) List 'A' viral diseases African horse sickness (AHS) and bluetongue (BT) as examples.

Effect of temperature on African horse sickness virus (AHSV) and bluetongue virus (BTV) infection of vector *Culicoides* spp. AHSV and BTV are

transmitted by certain species of *Culicoides* biting midge, which are competent biological vectors of these viruses (Meiswinkel *et al.*, 1994).

In the case of AHSV, Wellby *et al.* (1996) and Mellor & Wellby (1999) have shown that replication occurs in a high proportion of vector midges *Culicoides sonorensis* (=*variipennis*) kept at a constant temperature of 25 °C and the virus can be transmitted by 9 days post-infection (d.p.i.). However, at this temperature the survival rate of the insects was low, and most died within 11 days. As the ambient temperature was reduced the proportion of vectors infected decreased and the time to transmission extended dramatically, but conversely survival rates increased significantly. At constant temperatures lower than 15 °C, AHSV failed to replicate in, or to be transmitted by, any vector. From these findings it can be seen that the likelihood of transmission of AHSV is a function of the interaction of these two opposing trends, and the optimum temperature for transmission must fall somewhere between the two extremes. Under variable temperature regimes, chosen to simulate nocturnal–diurnal variations, the infection rates and the virus replication rates were shown to be proportional to the time spent at temperatures permitting virogenesis, and the cumulative total of time spent at these temperatures was the major factor controlling transmission potential. Interestingly, Wellby *et al.* (1996) also showed that no detectable AHS virogenesis occurred in vectors kept at 10 °C, and by 13 d.p.i. virtually all insects tested were free from virus. However, when surviving vectors that had been kept at 10 °C for 35 days were placed at 25 °C for 3 days, the infection rate increased from apparently zero to 15·5 %. This suggests that at low temperatures AHSV does not replicate in the vector, but it may persist in some midges at a level below that detectable by traditional assay systems. Then when the temperature subsequently rises to permissive levels, 'latent' virus begins to replicate and transmission may become possible.

Mullens *et al.* (1995) showed a similar lower temperature requirement for the replication of BTV in *C. sonorensis*, and Wittmann (2000) found that within the range of temperatures over which virus replication could occur, the proportion of infected vectors tended to increase linearly with temperature, suggesting an increase in susceptibility rate with temperature.

In parallel with these studies it has been demonstrated that *Culicoides imicola*, the major Old World field vector of both AHSV and BTV, is active at temperatures as much as 3 °C below the minimum required for their replication (Wellby *et al.*, 1996; Sellers & Mellor, 1993; R. Meiswinkel, personal communication). In practical terms this suggests that *C. imicola* may be capable of AHSV and BTV transmission only over the 'warmer' parts of its range, while in cooler areas, even if the vector is present, transmission will either be impossible or possible only at restricted times of the year (in summer for example) or in climatically favourable localities. Such findings may have a considerable influence on our understanding of the epidemiology of AHS and BT and of other arboviral diseases, particularly in relation to the ability of virus to survive

Table 1. AHSV: an overwintering mechanism

---

1. The virus seems to require temperatures ≥15 °C in order to replicate in and be transmitted by vector *Culicoides*.

2. Adults of the AHSV vector *C. imicola* are active at temperatures as low as 12 °C and can survive even cooler temperatures in an inactive state.

3. Since the virus requires a higher minimum temperature than the vector this suggests that replication and transmission may be possible over only part of the range of the vector.

4. However, during cold periods virus can survive for extended periods of time at 'undetectable' levels in adult vectors, the lifespan of which is extended at these temperatures to up to 90 days in some cases.

5. When temperatures rise to permissive levels (≥15 °C) virus replication in the vector recommences and transmission may become possible.

6. This set of events could constitute an overwintering mechanism for AHSV in the absence of vertebrate involvement or transmission and also in the absence of detectable virus in the vectors.

---

periods of adverse climatic conditions. Indeed, on the basis of this work, an overwintering mechanism for AHSV, in the absence of infected vertebrates, has recently been postulated; see Table 1 (Mellor & Leake, 2000).

Effect of temperature on AHSV and BTV infection of non-vector *Culicoides* spp.

The biting midge *Culicoides nubeculosus* has an oral susceptibility rate for AHSV of less than 1 % when its immature stages are reared at 25 °C and when the adults are fed on virus at a standard titre. Unlike the proven biological vector *C. sonorensis*, selective breeding from susceptible parents does not in *C. nubeculosus* enhance this rate of oral susceptibility (Mellor *et al.*, 1998). However, if the rearing temperature is raised to 30–35 °C, which is close to the upper lethal limit for this species, the oral susceptibility rate increases to over 10 % and virus replicates to levels suggesting that transmission is possible (Mellor & Wellby, 1999).

Similar results were obtained with *C. nubeculosus* and BTV (Wittmann *et al.*, 1998; Wittmann, 2000). Wittmann (2000) also showed that exposure to 'hot' conditions for the duration of the immature stages is not essential to induce vector competence since exposure of the pupae alone for as little as 2 days had a similar effect.

Since selective breeding has no effect on the *C. nubeculosus* oral susceptibility rate, this is clearly not a genetically controlled, heritable trait as is exhibited by most biological vectors of arboviruses. Instead, Mellor *et al.* (1998) have suggested that adults reared from immature stages maintained under these 'hot' conditions may develop a 'leaky midgut' where virus can leak directly into the haemocoel, bypassing the midgut barriers to infection (see Mellor, 2000). Once in the haemocoel the virus can replicate and be transmitted even by what is normally considered to be a non-vector species. Such a sequence of events may be envisaged as a hybrid mechanism whereby infection is initiated by a mechanical event (that is, passage of virus from the lumen of the midgut through the gut wall and into the haemocoel, without replication in the gut cells) but transmission is the result of a series of subsequent biological events (that is, virus infection of the salivary gland cells, replication in these cells and release of progeny virus particles into the salivary ducts). Such a phenotypic effect is unlikely to be restricted to *C. nubeculosus*, and other non-vector species of *Culicoides* are equally likely to exhibit a similar response when reared at temperatures close to their upper lethal limit.

Effect of other climatic variables (precipitation and wind) on AHSV and BTV transmission by vector *Culicoides* spp.

Where temperatures are suitable, precipitation can have a major effect upon the distribution and abundance of vector species of *Culicoides*. For example, *C. imicola*, the major Old World vector of BTV and AHSV, breeds in wet or damp, but not flooded, organically enriched soil (Walker & Davies, 1971; Braverman *et al.*, 1974; Lubega & Khamala, 1976; Walker, 1977; Braverman, 1978). In Africa it tends to occur in areas with rainfall of 300–700 mm per year (Meiswinkel & Baylis, 1998). Areas with higher rainfall levels are probably unsuitable because the pupae and larvae of *C. imicola* are unable to swim or float in free water and so drown or are washed away when the breeding sites are flooded (Nevill, 1967). Other potential vector species of *Culicoides* (for example the *Culicoides obsoletus* group, the *Culicoides pulicaris* group, *C. sonorensis*) also breed in similar semi-aquatic sites.

The precise level of impact of a given amount of rain upon *Culicoides* breeding sites will of course also depend upon a series of other linked variables. These include local evaporation rates, soil type, slope of terrain and the proximity to large bodies of water (for example rivers, lakes, ponds). The bottom line is that all breeding sites must persist for long enough to be detected by the vector, for eggs to be laid and for the immature stages to complete their development under the prevailing temperature conditions (Mellor & Leake, 2000).

Wind speed and direction can also affect *Culicoides* distribution by passively dispersing the adults. Due to their small size *Culicoides* are exceptionally prone to dispersal in this way. Sellers (1992) suggests that in winds of 10–40 km h^{-1}, at heights of <1·5 km and at temperatures between 12 and 35 °C *Culicoides* may be carried as aerial plankton for up to 700 km, and Hayashi *et al.* (1979) have recovered wind-borne *Culicoides* over 100 km out at sea. This long-distance movement can result in species colonizing new areas (Dyce, 1982) and in the introduction of virus (should the *Culicoides* be infected) into regions remote from the source. Indeed, the majority of BTV and AHSV outbreaks have been attributed to carriage on the wind of infected adult *Culicoides*

(Sellers *et al.*, 1977, 1978; Boorman & Wilkinson, 1983; Mellor, 1987; Anonymous, 1999a, b). As one may imagine, spread of disease in this way can cause major problems in terms of designing and implementing effective control strategies.

In the light of these effects of temperature and other climatic variables upon the vectors and hence transmission of AHSV and BTV, and in the face of ongoing climate change, are there any examples where such diseases can be seen to alter their epi-demiological patterns or distributions, apparently in response to environmental influences? The timing of epizootics of AHS in southern Africa and the epidemiological pattern of the current outbreak of BT in the Mediterranean Basin may provide two such examples.

AHSV

AHSV is a dsRNA virus that causes an infectious, non-contagious, insect-borne disease of equids. Nine serotypes of the virus exist and these are transmitted between their vertebrate hosts by certain species of *Culicoides* biting midge. As the name implies, AHSV originated in Africa and is enzootic in tropical and sub-tropical areas south of the Sahara (Mellor, 1994).

AHS and ENSO in South Africa

In South Africa major epizootics of AHS occur every 10–15 years but the reason for this pattern is unclear. However, a strong link between the timing of these epizootics and the warm (El Niño) phase of the El Niño/Southern Oscillation (ENSO) has recently been found. Baylis *et al.* (1999) have suggested that the association is mediated by the combination of heavy rainfall and drought brought to South Africa by ENSO. The link between this climatic phenomenon and AHS is via the vector *C. imicola*, which, as mentioned earlier, breeds in saturated soil. In years of heavy rain, breeding sites proliferate greatly and the abundance of this insect can increase by over 200-fold. Baylis *et al.* (1999) showed that since 1803, 13 of the 14 major epizootics of AHS in South Africa have coincided with a warm-phase ENSO. However, many ENSOs have also occurred since 1803 (42) when epizootics of AHS did not develop. Baylis *et al.* (1999) explain this apparent anomaly by showing that in South Africa, warm-phase ENSOs typically bring heavy rain followed by drought; these were not the years when AHS epizootics occurred. However, there is a subset of ENSO years where the pattern is reversed, pronounced drought occurring first, followed by heavy rain. These were the years when the 13 AHS epizootics occurred. The reasons why this pattern is conducive to AHS remain unclear but its recognition means that prediction of future AHS epizootics in South Africa will be made significantly easier. Similar links between excessive rainfall and vector-borne disease prevalence have also been reported for many other diseases including Rift Valley fever (Mellor & Leake, 2000).

BT IN THE MEDITERRANEAN BASIN 1998–2003

BTV and BT epidemiology

BTV is a dsRNA virus that exists as 24 serotypes and is transmitted between its ruminant hosts almost entirely by certain species of *Culicoides* biting midge. In consequence its world distribution is restricted to areas where these vector species occur and transmission is limited to those times of the year when the adult vectors are active. In epizootic zones this usually means during the late summer and autumn since vector populations tend to peak at these times. In general the global distribution of BTV and, therefore, also of its competent vectors lies approximately between latitudes 35° S and 40° N. Within these areas the virus has a virtually worldwide distribution, being found in North, Central and South America, the Middle East, the Indian subcontinent, China, SE Asia and Australia. The virus has at times made incursions into southern Europe although in the past it has not been able to establish itself (Mellor & Boorman, 1995). Prior to the 1990s the most recent incursion into Europe involved the Greek islands of Rhodes and Lesvos during the years 1979–1980 (Mellor & Boorman, 1995).

The major Old World vector of BTV

The major Old World vector of BTV is, as for AHSV, the midge *C. imicola*. This is an Afro-Asiatic species and occurs throughout most of Africa, the Middle East and SE Asia. It was first identified in Europe in 1982 and a series of surveys carried out in this continent between 1982 and 1996 had recorded its presence throughout most of Portugal, SW Spain and certain Greek islands adjacent to the Anatolian Turkish coast (Lesvos, Rhodes and Chios) (Mellor *et al.*, 1983, 1985; Boorman, 1986; Rawlings *et al.*, 1997, 1998; Ortega *et al.*, 1998; Boorman & Wilkinson, 1983; Capela *et al.*, 1993; Mellor & Wittmann, 2002). It had been looked for but not found in mainland Greece, eastern Spain, Sicily, parts of southern mainland Italy and Bulgaria (Mellor *et al.*, 1984; Scaramozzino *et al.*, 1996; Gallo *et al.*, 1984; Glouchova *et al.*, 1991; Dilovski *et al.*, 1992).

Start of the 1998–2003 European outbreak

In October 1998, after an absence from Europe of 20 years, BTV was confirmed on four Greek islands (Rhodes, Leros, Kos, Samos) adjacent to the Anatolian Turkish coast. The virus was identified as BTV serotype 9 and this was the first occasion that this serotype had been recorded in Europe though it had been identified serologically during 1979–1980 in western Anatolia (Taylor & Mellor, 1994). As BTV was shown, retrospectively, to be circulating in western Anatolia from early 1998 (A. Ozkul, personal communication) and as no other country in the area reported the presence of this virus, Turkey was presumed to be the source of the 1998 Greek incursions. In the absence of animal movements from Anatolia to any of the affected Greek islands,

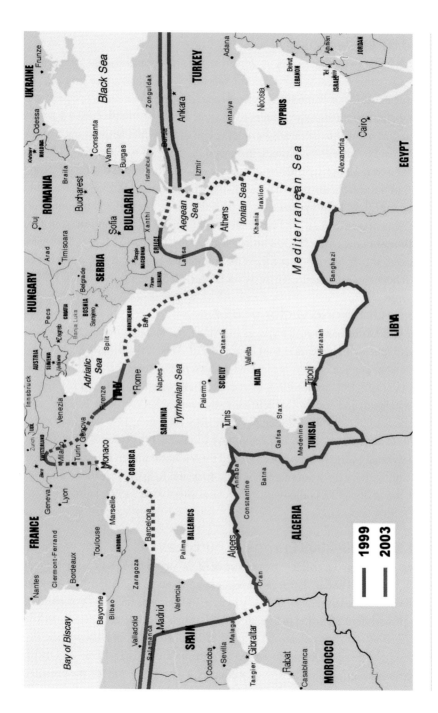

Fig. 1. *Culicoides imicola* distribution, known northern limits at 1999 and 2003.

and none were reported, the mode of incursion was assumed to be by movement of infected vectors on the wind. This assumption was supported by the fact that populations of the major BTV vector, *C. imicola*, were known to be present in western Anatolia (Jennings *et al.*, 1983) and outbreaks on the Greek islands invariably commenced in those areas closest to the Turkish coast, and then moved in a westerly direction (D. Panagiotatos, personal communication). The distances from the Greek islands to the Turkish coast vary from less than 2 to about 15 km, which is well within the known dispersal range of *Culicoides* on the wind (see above).

Similarly, in August 2000, a second and separate incursion of BTV (serotype 2) occurred into Europe. The first cases were reported from the Italian island of Sardinia, but later on in the year the virus also spread to include Sicily, southern mainland Italy, Corsica and two of the Balearic Islands. The original source of infection was Tunisia and/or Algeria in North Africa where BTV serotype 2 had been known to be circulating since December 1999 (Mellor & Wittmann, 2002). Again, in the absence of animal movements between the infected areas, and none were documented, it is likely that the mode of incursion was via wind-borne infected *Culicoides*. In the case of the Balearic Islands this implies a movement of some 300–400 km from the nearest other infected regions (Algeria and Sardinia), which approaches the sort of distances predicted by Sellers (1992).

In general, both BTV incursions into Europe, once established, moved in a westerly direction with the prevailing winds across the region (Mellor & Wittmann, 2002).

Expansion in range of the major vector

Prior to 1999, the northernmost position ever reached by BTV in Europe exactly paralleled the known distribution of *C. imicola* in the same region (Fig. 1), which suggested not only that this species was the major vector in Europe but that it was the only major vector. However, in the period 1999–2002, outbreaks of BT occurred in mainland Greece, European Turkey, Bulgaria, several Balkan countries, Italy, Corsica and the Balearic Islands (Mellor & Wittmann, 2002). These are all areas beyond the '*imicola* line'. The immediate question at the time, therefore, was to ask whether *C. imicola* had expanded its range northwards and westwards into these new areas, possibly in response to climate change, and the answer was a partial yes. Surveys carried out in the affected areas during 2000–2003 have recorded the presence of *C. imicola* in many locations on mainland Greece, mainland Italy, Sicily, Sardinia, Corsica and the Balearics so that the new '*imicola* line' is as shown in Fig. 1 (Mellor & Wittmann, 2002; P. S. Mellor, unpublished). The presence of this known major vector in many of these areas is clearly the major reason why BTV was able to gain entry and be transmitted there. Furthermore, as most of these new areas had been surveyed, without success, for

C. imicola in the 1980s this provides strong evidence that the insect is expanding its range and taking the threat of disease with it. However, this is not the whole picture. Despite repeated surveys carried out in northern Greece, European Turkey during the period 1999–2002 and Bulgaria during the period 1991–2002, *C. imicola* has not been recovered from any of these areas and yet BTV was transmitted very efficiently in these locations during this same period (Mellor & Wittmann, 2002). So what might be the vectors?

Novel vectors

In the areas where BTV was transmitted but where *C. imicola* is absent, *C. obsoletus* and *C. pulicaris* groups of midges comprise over 90 % of the *Culicoides* population (Mellor & Wittmann, 2002). Both of these species groups have long been considered as suspect vectors mainly on the basis of BTV isolations made from *C. obsoletus* in Cyprus (Mellor & Pitzolis, 1979) and AHSV isolations made from mixed pools of *C. obsoletus* and *C. pulicaris* in Spain (Mellor *et al.*, 1990). In this context it should be borne in mind that BTV and AHSV tend to utilize the same *Culicoides* species as vectors (Mellor, 1993). However, vector competence studies on British populations of *C. obsoletus* and *C. pulicaris* have recorded oral susceptibility rates of less than 2 % in comparison with a known major vector, *C. sonorensis* (=*variipennis*) (19·5 %) (Mellor & Jennings, 1988), which suggested that they were likely to be of only minor importance as vectors. This was a judgement supported by the fact that BT in Europe had never previously been reported in the absence of *C. imicola*. Nevertheless, the high abundance and high survival rates of these suspect vectors, as were exhibited in the affected areas, could compensate for low levels of vector competence (Mellor & Wittmann, 2002). Precisely this situation exists in Australia, where *Culicoides brevitarsis* is considered to be the most important BTV vector on the basis of its high abundance and prevalence even though its vector competence levels are extremely low (Standfast *et al.*, 1985). Furthermore, expression of competence for a particular virus is an hereditary trait and populations of a vector species with high, low or intermediate levels of competence can be derived by selective breeding (Jones & Foster, 1974; Wellby *et al.*, 1996; Fu *et al.*, 1999). In the case of *C. obsoletus* and *C. pulicaris*, until very recently the only populations tested for BTV competence were from southern England (Mellor & Jennings, 1988). Other populations across Europe remain untested and their levels of competence are therefore completely unknown. This situation has been and is a matter of real concern because *C. obsoletus* and *C. pulicaris* are probably the commonest species of *Culicoides* across the whole of central and northern Europe including the UK. It is therefore of major importance that their susceptibility to BTV be determined so informed risk assessments can be made for northern and central Europe. Indeed, work has now commenced in several areas spurred on by recent multiple isolations of the virus reported from *C. obsoletus* and *C. pulicaris* in many parts of

Italy during 2002 (Savini *et al.*, 2003; Caracappa *et al.*, 2003; C. de Liberato and others, personal communication). Field studies from the UK support the importance of this work, with preliminary results showing susceptibility rates of almost 11 % for some populations of *C. obsoletus* in England – levels that are more than five times higher than any previously reported (S. Carpenter, personal communication).

CONCLUSIONS

Climate change is now accepted as fact by most authorities. This phenomenon is already bringing higher temperatures (projected to be between 1·4 and 5·8 °C) and an increased frequency of extreme weather events (for example windiness) to many areas. In respect of vector-borne diseases like BT and AHS which are strongly affected by environmental variables, this is likely to have the following effects.

The temperature rise will cause:
- Extension in the range of some vectors (for example *C. imicola* by >500 km)
- Increased abundance of the vectors, leading to enhanced disease transmission
- A longer vector season, leading to a longer disease season
- Increased competence of existing vectors, leading to enhanced disease transmission
- Extension of competence to additional *Culicoides* species, leading to enhanced disease transmission.

Extra windiness may cause:
- Enhanced dispersion of vectors, aiding colonization of new habitats
- Enhanced spread of infected vectors, causing increased numbers of outbreaks.

The long-term outlook for BT and AHS, therefore, is for more disease over a wider area for a long period of time, a picture that is likely to be repeated with many vector-borne diseases.

REFERENCES

Anonymous (1999a). Bluetongue returns to Greece. *Animal Pharm* **410**, 7.
Anonymous (1999b). Bulgaria confirms bluetongue outbreaks. *Animal Pharm* **426**, 9.
Baylis, M., Mellor, P. S. & Meiswinkel, R. (1999). Horse sickness and ENSO in South Africa. *Nature* **397**, 574.
Boorman, J. (1986). Presence of bluetongue virus vectors on Rhodes. *Vet Rec* **118**, 21.
Boorman, J. P. T. & Wilkinson, P. J. (1983). Potential vectors of bluetongue in Lesbos, Greece. *Vet Rec* **113**, 395–396.
Braverman, Y. (1978). Characteristics of *Culicoides* (Diptera: Ceratopogonidae) breeding places near Salisbury, Rhodesia. *Ecol Entomol* **3**, 163–170.

Braverman, Y., Galun, R. & Ziv, M. (1974). Breeding sites of some *Culicoides* species (Diptera: Ceratopogonidae) in Israel. *Mosq News* **34**, 303–308.

Capela, R., Sousa, C., Pena, I. & Caeiro, V. (1993). Preliminary note on the distribution and ecology of *Culicoides imicola* in Portugal. *Med Vet Entomol* **7**, 23–26.

Caracappa, S., Torina, A., Guercio, A., Vitale, F., Calabro, A., Purpari, G., Ferrantelli, V., Vitale, M. & Mellor, P. S. (2003). Identification of a novel bluetongue virus vector species of *Culicoides* in Sicily. *Vet Rec* **153**, 71–74.

Dilovski, M., Nedelchev, N. & Petkova, K. (1992). Studies on the species composition of *Culicoides* – potential vectors of bluetongue virus in Bulgaria. *Vet Sci* **26**, 52–56.

Dyce, A. L. (1982). Distribution of *Culicoides* (*Avaritia*) spp. (Diptera: Ceratopogonidae) west of the Pacific Ocean. In *Arbovirus Research in Australia, Proceedings of the 3rd Symposium,* pp. 35–43. Edited by T. D. St George & B. H. Kay. Brisbane: Commonwealth & Industrial Research Organization & Queensland Institute of Medical Research.

Elias, S. A. (1995). Insects and climate change: fossil evidence from the Rocky Mountains. *Bioscience* **41**, 552–559.

Fu, H., Leake, C. J., Mertens, P. P. C. & Mellor, P. S. (1999). The barriers to bluetongue virus infection, dissemination and transmission in the vector, *Culicoides variipennis* (Diptera: Ceratopogonidae). *Arch Virol* **144**, 747–761.

Gallo, C., Guercio, V., Caracappa, S., Boorman, J. P. T. & Wilkinson, P. J. (1984). Indagine siero-entomologica sulla possibile presenza del virus bluetongue nei bovini in Sicilia. *Atti Soc Ital Buiatria* **16**, 393–398.

Glouchova, V. M., Nedelchev, N. K., Rousev, I. & Tanchev, T. (1991). On the fauna of blood-sucking midges of the genus *Culicoides* (Diptera: Ceratopogonidae) in Bulgaria. *Vet Sci* **25**, 63–66.

Hack, W. H. (1955). Estudio sobre la biologica del *Triatoma infesten. An Inst Med Reg Tucuman Argentina* **4**, 125–147.

Halstead, S. B. (1990). Dengue. In *Tropical and Geographic Medicine,* 2nd edn, pp. 675–685. Edited by K. Warren & A. A. F. Mahmoud. New York: McGraw-Hill.

Hardy, J. L., Houk, G. J., Kramer, L. D. & Reeves, W. C. (1983). Intrinsic factors affecting vector competence of mosquitoes for arboviruses. *Annu Rev Entomol* **28**, 229–262.

Hayashi, K., Suzuki, H. & Makino, Y. (1979). Notes on the transoceanic insects captured on the East China Sea in 1976-1978. *Trop Med* **21**, 1–10.

Intergovernmental Panel on Climate Change (2001). Projections of future climate change. In *Climate Change 2001: Projections of Future Climate Change,* pp. 527–578. Edited by J.-W. Kim & J. Stone. Cambridge: Cambridge University Press.

Jennings, M., Boorman, J. P. T. & Ergun, H. (1983). *Culicoides* from western Turkey in relation to bluetongue disease of sheep and cattle. *Rev Elev Med Vet Pays Trop* **36**, 67–70.

Jones, R. H. & Foster, N. M. (1974). Oral infection of *Culicoides variipennis* with bluetongue virus: development of susceptible and resistant lines from a colony population. *J Med Entomol* **11**, 316–323.

Karl, T. R., Jones, P. D., Knight, R. W. & 7 other authors (1993). A new perspective on recent global warming; asymmetric trends of daily maximum and minimum temperature. *Bull Am Meteorol Soc* **74**, 1007–1023.

Knudsen, A. B. (1996). Occurrence and spread in Italy of *Aedes albopictus* with implications for its introduction into other parts in Europe. *J Am Mosq Control Assoc* **12**, 177–183.

Kramer, L. D., Hardy, J. L. & Presser, S. B. (1983). Effect of temperature of extrinsic incubation on the vector competence of *Culex tarsalis* for western equine encephalomyelitis virus. *Am J Trop Med Hyg* **32**, 1130–1139.

Kukla, G. & Karl, T. R. (1993). Night-time warming and the greenhouse effect. *Environ Sci Technol* **8**, 1468–1474.

Lubega, R. & Khamala, C. P. M. (1976). Larval habitats of common *Culicoides* Latreille (Diptera: Ceratopogonidae) in Kenya. *Bull Entomol Res* **66**, 421–425.

McMichael, A. J., Haines, A., Slooff, R. & Kovats, S. (1996). *Climate Change and Human Health.* Geneva: World Health Organization.

Meiswinkel, R. & Baylis, M. (1998). Morphological confirmation of the separate species status of *Culicoides (Avaritia) nudipalpis* Delfinado 1961 and *C. (A.) imicola* Keiffer 1913 (Diptera: Ceratopogonidae). *Onderstepoort J Vet Res* **65**, 9–16.

Meiswinkel, R., Nevill, E. M. & Venter, G. J. (1994). Vectors: *Culicoides* spp. In *Infectious Diseases of Livestock with Special Reference to Southern Africa,* vol. 1, pp. 68–89. Edited by J. A. W. Coetzer, G. R. Thomson & R. C. Tustin. Cape Town: Oxford University Press.

Mellor, P. S. (1987). Factors limiting the natural spread of bluetongue virus in Europe. In *Agriculture, Bluetongue in the Mediterranean Region*, pp. 69–74. Edited by W. P. Taylor. Brussels: Commission of the European Communities.

Mellor, P. S. (1993). African horse sickness: transmission and epidemiology. *Ann Rech Vet* **24**, 199–212.

Mellor, P. S. (1994). Epizootiology and vectors of African horse sickness virus. *Comp Immunol Microbiol Infect Dis* **17**, 287–296.

Mellor, P. S. (1996). *Culicoides*: vectors, climate change and disease risk. *Vet Bull* **66**, 301–306.

Mellor, P. S. (2000). Replication of arboviruses in insect vectors. *J Comp Pathol* **123**, 231–247.

Mellor, P. S. & Boorman, J. (1995). The transmission and geographical spread of African horse sickness and bluetongue viruses. *Ann Trop Med Parasitol* **89**, 1–15.

Mellor, P. S. & Jennings, D. M. (1988). The vector potential of British *Culicoides* species for bluetongue virus. In *Orbiviruses and Birnaviruses, Proceedings of the Double-Stranded RNA Virus Symposium, Oxford 9–13 Sept 1986*, pp. 12–21. Edited by P. Roy, C. E. Schore & B. I. Osburn. Oxford: NERC Institute Oxford.

Mellor, P. S. & Leake, C. J. (2000). Climatic and geographical influences on arboviral infections and vectors. *OIE Sci Tech Rev* **19**, 41–54.

Mellor, P. S. & Pitzolis, G. (1979). Observations on breeding sites and light trap collections of *Culicoides* during an outbreak of bluetongue in Cyprus. *Bull Entomol Res* **69**, 229–234.

Mellor, P. S. & Wellby, M. P. (1999). Effect of temperature on African horse sickness virus serotype 9 infection of vector species of *Culicoides* (Diptera: Ceratopogonidae). *Equine Infect Dis* **8**, 2246–2251.

Mellor, P. S. & Wittmann, E. J. (2002). Bluetongue virus in the Mediterranean Basin (1998–2001). *Vet J* **164**, 20–37.

Mellor, P. S., Boorman, J. P. T., Wilkinson, P. J. & Martinez-Gomez, F. (1983). Potential vectors of bluetongue and African horse sickness viruses in Spain. *Vet Rec* **112**, 229–230.

Mellor, P. S., Jennings, M. & Boorman, J. P. T. (1984). *Culicoides* from Greece in relation to the spread of bluetongue virus. *Rev Elev Med Vet Pays Trop* **37**, 286–289.

Mellor, P. S., Jennings, M., Wilkinson, P. J. & Boorman, J. P. T. (1985). *Culicoides imicola*, a bluetongue virus vector in Spain and Portugal. *Vet Rec* **116**, 589–590.

Mellor, P. S., Boned, J., Hamblin, C. & Graham, S. (1990). Isolations of African horse sickness virus from vector insects made during the 1988 epizootic in Spain. *Epidemiol Infect* **105**, 447–454.

Mellor, P. S., Rawlings, P., Baylis, M. & Wellby, M. P. (1998). Effect of temperature on African horse sickness virus infection in *Culicoides*. *Arch Virol* (Suppl.) **14**, 155–163.

Mullens, B. A., Tabachnick, W. J., Holbrook, F. R. & Thompson, L. H. (1995). Effects of temperature on virogenesis of bluetongue virus serotype 11 in *Culicoides variipennis sonorensis*. *Med Vet Entomol* **9**, 71–76.

Nawrocki, S. J. & Hawley, W. A. (1987). Estimation of the northern limits of distribution of *Aedes albopictus* in North America. *J Am Mosq Control Assoc* **3**, 314–317.

Nevill, E. M. (1967). *Biological studies on some South African Culicoides species (Diptera: Ceratopogonidae) and the morphology of their immature stages*. MSc Thesis, University of Pretoria.

Ortega, M. D., Mellor, P. S., Rawlings, P. & Pro, M. J. (1998). The seasonal and geographical distribution of *Culicoides imicola*, *C. pulicaris* group and *C. obsoletus* group biting midges in central and southern Spain. *Arch Virol* (Suppl.) **14**, 85–92.

Parmesan, C., Ryrholm, N., Stefanescu, C. & 10 other authors (1999). Poleward shifts in geographical ranges of butterfly species associated with regional warming. *Nature* **399**, 579–583.

Peters, R. L. (1993). Conservation of biological diversity in the face of climate change. In *Global Warming and Biological Diversity*, pp. 15–30. Edited by R. L. Peters & T. E. Lovejoy. New Haven: Yale University Press.

Rawlings, P., Pro, M. J., Pena, I., Ortega, M. D. & Capela, R. (1997). Spatial and seasonal distribution of *Culicoides imicola* in Iberia in relation to the transmission of African horse sickness virus. *Med Vet Entomol* **11**, 49–57.

Rawlings, P., Capela, R., Pro, M. J., Ortega, M. D., Pena, I., Rubio, C., Gasca, A. & Mellor, P. S. (1998). The relationship between climate and the distribution of *Culicoides imicola* in Iberia. *Arch Virol* (Suppl.) **14**, 93–102.

Savini, G., Goffredo, M., Monaco, F., de Santis, P. & Meiswinkel, R. (2003). Transmission of bluetongue virus in Italy. *Vet Rec* **152**, 119.

Scaramozzino, P., Boorman, J., Vitale, F., Semproni, G. & Mellor, P. S. (1996). Entomological survey on Ceratopogonidae in central-southern Italy. *Parassitologia* **38**, 1–2.

Schaffner, F. & Karsh, S. (1999). *Aedes albopictus* discovered in France. *Soc Vector Ecol Newsl* **30**, 11.

Sellers, R. F. (1980). Weather, host and vector – their interplay in the spread of insect-borne animal virus diseases. *J Hyg* **85**, 65–102.

Sellers, R. F. (1992). Weather, *Culicoides* and the distribution and spread of bluetongue and African horse sickness viruses. In *Bluetongue, African Horse Sickness and Related Orbiviruses, Proceedings of the 2nd International Symposium*, pp. 284–290. Edited by T. E. Walton & B. I. Osburn. Boca Raton: CRC Press.

Sellers, R. F. & Mellor, P. S. (1993). Temperature and the persistence of viruses in *Culicoides* during adverse conditions. *OIE Sci Tech Rev* **12**, 733–755.

Sellers, R. F., Pedgley, D. E. & Tucker, M. R. (1977). Possible spread of African horse sickness on the wind. *J Hyg* **79**, 279–298.

Sellers, R. F., Pedgley, D. E. & Tucker, M. R. (1978). Possible windborne spread of bluetongue to Portugal June-July 1956. *J Hyg* **81**, 189–196.

Standfast, H. A., Dyce, A. L. & Muller, M. J. (1985). Vectors of bluetongue virus in Australia. In *Bluetongue and Related Orbiviruses*, pp. 177–186. Edited by T. L. Barber & M. M. Jochim. New York: A. R. Liss.

Taylor, W. P. & Mellor, P. S. (1994). Bluetongue virus distribution in Turkey 1978–1981. *Epidemiol Infect* **112**, 623–633.

Walker, A. R. (1977). Seasonal fluctuations of *Culicoides* species (Diptera: Ceratopo-gonidae) in Kenya. *Bull Entomol Res* **67**, 217–233.

Walker, A. & Davies, F. C. (1971). A preliminary survey of the epidemiology of bluetongue in Kenya. *J Hyg* **69**, 47–61.

Wellby, M. P., Baylis, M., Rawlings, P. & Mellor, P. S. (1996). Effect of temperature on the rate of virogenesis of African horse sickness virus in *Culicoides* (Diptera: Ceratopo-gonidae) and its significance in relation to the epidemiology of the disease. *Med Vet Entomol* **86**, 715–720.

Wittmann, E. J. (2000). *Temperature and the transmission of arboviruses by Culicoides biting midges*. PhD thesis, University of Bristol.

Wittmann, E. J. & Baylis, M. (2000). Climate change: effects on *Culicoides*-transmitted viruses and implications for the UK. *Vet J* **160**, 107–117.

Wittmann, E. J., Baylis, M. & Mellor, P. S. (1998). Higher immature rearing temperatures induce vector competence for bluetongue virus in *Culicoides nubeculosus* Meigen (Ceratopogonidae). In *Proceedings of the 4th International Congress on Dipterology (Abstracts Volume) 6–13 Sept*, pp. 248–249. Oxford: Oxford International.

Vector immunity

Norman A. Ratcliffe and Miranda M. A. Whitten

School of Biological Sciences, University of Wales Swansea, Singleton Park, Swansea, UK

INTRODUCTION

Invertebrates, particularly insects, act as vectors of the most debilitating diseases in many already socially and economically compromised populations of the developing world. The diseases vectored include malaria, sleeping sickness, Chagas' disease, leishmaniasis, lymphatic filariases and river blindness, dengue and yellow fever, and schistosomiasis. Only the latter is transmitted by non-insectan invertebrates, i.e. snails of the genus *Biomphalaria*. Hurd (2003) recently pointed out that of the ten most important tropical diseases affecting the poorer nations, eight of these are transmitted by invertebrate vectors (Table 1). In addition, and often underestimated in importance since they are probably second only to mosquitoes as vectors of human infectious diseases (Parola & Raoult, 2001), are the ticks. In fact, in the USA ticks transmit more vector-borne diseases than any other vector (US Centers for Disease Control and Prevention, 1999). The diseases transmitted by ticks include Lyme borreliosis, tick-borne encephalitis, ehrlichiosis and babesiosis (Gratz, 1999).

The reason that insects are particularly widespread vectors undoubtedly reflects the success of this group, occupying almost every habitat on earth in vast numbers. In addition, their power of flight, their extraordinarily well developed sense organs and their haematophagous habit have made them ideal vehicles for transmitting human blood-borne diseases. Since many insects live in places infested with pathogens and parasites they can only do so due to the extreme efficiency of their immune defences so that any invading parasite must counteract these defences to survive.

SGM symposium 63: Microbe–vector interactions in vector-borne diseases.
Editors S. H. Gillespie, G. L. Smith & A. Osbourn. Cambridge University Press. ISBN 0 521 84312 X ©SGM 2004

Table 1. Major human diseases vectored by arthropods with an indication of those newly emergent or resurging

Disease	Pathogen	Vector	Locations*
Arboviral diseases			
Chikungunya	Alphavirus (Togaviridae)	Mosquito	Africa, India, S.E. Asia
Equine encephalomyelitis (E. and W. Venezuelan)	Alphavirus (Togaviridae)	Mosquito	Americas
Japanese encephalitis	Flavivirus	Mosquito (*Culex*)	Asia
Crimean–Congo haemorrhagic fever	Bunyavirus	Ixodid (hard) tick	C. Asia, C. Europe, Africa
Phlebotomous fever	Bunyavirus	Sandfly	Americas, Europe, Africa, Middle East, Russia
St Louis encephalitis†	Flavivirus	Mosquito	USA, S. Canada
Barmah Forest and Ross River virus (epidemic polyarthritis)†	Alphaviruses (Togaviridae)	Mosquito	Australia
West Nile fever†	Flavivirus	Mosquito	E. USA, Canada, Africa, Middle East, Russia
Kyasanur forest disease†	Flavivirus	Ixodid (hard) tick	W. India
O'nyong-nyong fever‡	Alphavirus (Togaviridae)	Mosquito	*E. Africa*
Yellow fever‡	Flavivirus	Mosquito (*Aedes*)	*Peru, W. Africa, S. and C. America*
Dengue and dengue haemorrhagic fever‡	Flavivirus	Mosquito (*Aedes*)	*India, S. America, E. USA*
Rift Valley fever‡	Bunyavirus	Mosquito	*W. Africa, Saudi Arabia*
Parasitic diseases			
Chagas' disease (S. American trypanosomiasis)	*Trypanosoma cruzi*	Reduviid bug	S. and C. America
Loiasis	*Loa loa*	Horsefly	C. Africa
Leishmaniasis	*Leishmania* spp.	Sandfly	All tropics
Schistosomiasis (bilharzia)	*Schistosoma mansoni*	Aquatic snail	Most tropics
Elephantiasis	*Brugia* sp., *Dirofilaria* sp.	Mosquito	Asia
Filariasis	*Wuchereria* sp. *Mansonella* sp.	Mosquito Midge and blackfly	All tropics Africa, S. America
Onchocerciasis (river blindness)	*Onchocerca* sp.	Blackfly	S. and C. America, Middle East, C. Africa
Human babesiosis†	*Babesia* sp.	Ixodid (hard) tick	USA
Malaria‡	*Plasmodium* spp.	Mosquito	All tropics, *Europe, W. Asia, Korea*
African trypanosomiasis‡ (sleeping sickness)	*Trypanosoma brucei, Trypanosoma vivax*	Tsetse fly	*Africa*

Table 1. cont.

Disease	Pathogen	Vector	Locations*
Bacterial diseases			
Scrub typhus	*Rickettsia tsutsugamushi*	Mite	India, Asia, Australia
Epidemic (louse-borne) typhus	*Rickettsia prowazekii*	Body louse	Africa
Murine typhus	*Rickettsia typhi*	Flea	India, S.E. Asia, USA
Epidemic relapsing fever	*Borrelia recurrentis*	Body louse	E. Africa
Endemic relapsing fever†	*Borrelia* sp.	Argasid (soft) tick	W. USA, most continents
Lyme borreliosis†	*Borrelia burgdorferi*	Ixodid (hard) tick	Europe, USA
Lyme disease-like borreliosis†	*Borrelia bissettii*	Ixodid (hard) tick	Slovenia
Tularaemia†	*Francisella tularensis*	Tick and deerfly	USA, Europe, Russia
Human ehrlichioses† (monocytic and granulocytic)	*Ehrlichia* sp.	Ixodid (hard) tick	Americas, Asia, Europe
Rocky Mountain spotted fever†	*Rickettsia rickettsii*	Ixodid (hard) tick	USA
Cat flea typhus†	*Rickettsia felis*	Flea	USA, UK, S. America, S. Europe
Oriental spotted fever‡	*Rickettsia japonica*	Ixodid (hard) tick	*Japan*
Plague‡	*Yersinia pestis*	Flea	*S. America, Africa, Asia*
Trench fever‡	*Bartonella quintana*	Body louse	*USA, Europe*

*In the case of resurging diseases, locations in italics refers to areas of spreading disease.

†Emerging diseases (bold type).

‡Resurging diseases (bold type).

Thus, most parasites are not simply passively transported from human to human by their vectors, but interact intimately with their invertebrate hosts, usually undergoing significant biochemical and molecular modifications to survive, differentiate and multiply in their vectors. The parasites may then manipulate the vector to optimize transmission to new hosts. This manipulation may, for example, alter the feeding behaviour of the female vector as with infected sandflies and mosquitoes that exhibit enhanced probing of the host during feeding (Hurd, 2003).

This review aims to outline the work undertaken on the immune systems of the most important invertebrate vectors of diseases. Due to the constraints of time and space, we have relied heavily on review articles and apologise to those whose original papers, often culminating after years of work, have not been included.

The present attention to vector immunity is justified as only by understanding the many processes involved will it be possible to determine the survival strategies of the parasites and identify possible new molecular targets for parasite control. Research in this area has taken on a new urgency for several reasons. First, due to the increasing resistance of insect vectors to the main groups of insecticides (Hemingway & Ranson, 2000), as well as to the resistance of the parasites themselves to chemotherapy (Collins & Paskewitz, 1995). Second, there is the problem of the emergence and resurgence of many vector-borne diseases (Table 1) associated with global warming, increased human travel and ecological changes (Gratz, 1999; Snow, 2000; Wittmann & Baylis, 2000; Lounibos, 2002). There is, for example, the real possibility of the reappearance of malaria in Britain due to introduced exotic mosquitoes, although the risk of establishment is regarded as small (Snow, 2000). The situation with tick-borne diseases is particularly worrying as their incidences continue to increase. For example, in the USA, Lyme disease has shown 20–40 % increases in reported cases in some years while new tick-borne diseases, such as ehrlichiosis, have recently been discovered (Gratz, 1999).

The identification of key vector immune factors determining the development of the parasite will also assist the quest for infection-blocking genes for use in the production of refractory transgenic insects or even transmission-blocking vaccines. One possible example is the transgene for the antibacterial peptide defensin, which has been reported by Kokoza *et al*. (2000) to be expressed in *Aedes aegypti* following a blood meal but which has yet to be shown to reduce vectorial capability (Ghosh *et al*., 2002). In addition, many genes have now been identified in mosquitoes that are associated with the innate immune response to parasites. These include genes that control melanotic encapsulation and the refractory status of some *Anopheles* mosquitoes (detailed below; Dimopoulos *et al*., 2001; Dimopoulos, 2003). Studies on other parasite–host immune interactions have also yielded potential candidates for manipulation, for example the 28 kDa glutathione *S*-transferase of *Schistosoma mansoni*, which has been fully sequenced (McNair *et al*., 1993) and now tested as a possible target for a vaccine antigen (Riveau *et al*., 1998; Capron *et al*., 2002). Lastly, the gene to be targeted may not originate in the vector but could be derived from elsewhere. Thus, in *Rhodnius prolixus*, normally symbiotic bacteria, such as *Rhodo- coccus rhodnii*, have been isolated, genetically transformed *in vitro* with an appropriate refractory gene and reintroduced into the vector insect to deliver their engineered molecule. This paratransgenic approach has been used to introduce the gene for L-cecropin A, originally from *Hyalophora cecropia* moths, into *R. prolixus*, and showed that the Chagas' disease agent, *Trypanosoma cruzi*, can be successfully eliminated in the majority of treated insects (reviewed by Beard *et al*., 2002).

PARASITE LIFE CYCLE AND VECTOR INTERACTIONS

Before considering the details of the vector immune system, it is important to understand aspects of the parasite life cycle in the invertebrate vector. Although not confirmed with most parasites, it is probably true that, as with the vertebrate phase of the life cycle, the immune potential of particular sites in the body is an important determinant of the differing locations colonized by parasites in the host. For example, it has recently been shown for two closely related trypanosome species, *T. cruzi* and *Trypanosoma rangeli*, that have markedly different life cycles in their insect vector, *R. prolixus*, that the ability of *T. rangeli*, but not *T. cruzi*, to invade the haemocoel is correlated with its ability to counteract the reactive oxygen and nitrogen radicals generated in this location (Whitten *et al.*, 2001; M. M. A. Whitten, G. A. Schaub, F. Sun, A. J. Nappi & N. A. Ratcliffe, unpublished). It is a reasonable assumption that the vector and its antimicrobial armoury were evolved before the parasite invaded the vector tissues, although some significant modifications would subsequently have occurred in the host innate immune system due to the selective pressure exerted by the parasite itself, and similarly the host environment would have forced modifications in parasites (see discussion on parasite evolution in Kreier & Baker, 1987).

The main diseases covered in this review each have unique life cycles in their vectors but will potentially encounter a similar set of innate immune defences if they pass through identical regions of their hosts. Each compartment in the vector will therefore pose a set of problems to be solved by the parasite (Fig. 1). Interestingly, different parasites in the same vector tissue seem to have adapted to or avoided the innate defences in a variety of ways. Thus, various parasites of the vector gut may adapt to the conditions, differentiate and multiply therein, as with *T. cruzi* in *R. prolixus*, the African trypanosomes in the tsetse fly, *Glossina*, and *Leishmania* spp. in sandflies (*Phlebotomus* spp.). Alternatively, development is initiated in the gut, but the parasites leave this site and invade the haemocoel via the midgut wall to complete their life cycles, as with *T. rangeli* in *R. prolixus* as well as *Plasmodium*, arboviruses and filarial worms in mosquitoes (see below for details).

VECTOR COMPARTMENTS AND BARRIERS TO INFECTION

A. Physico-chemical barriers

What then are the main compartments and barriers encountered by parasites during their invasion of the vector? First, there are the physico-chemical barriers that in insects and ticks include the tough cuticular exoskeleton, that also lines the foregut, the hindgut, the tracheae and the genital ducts, together with the peritrophic membrane of the midgut (Fig. 1). The gastropod snails also have a horny and calcified shell with the soft

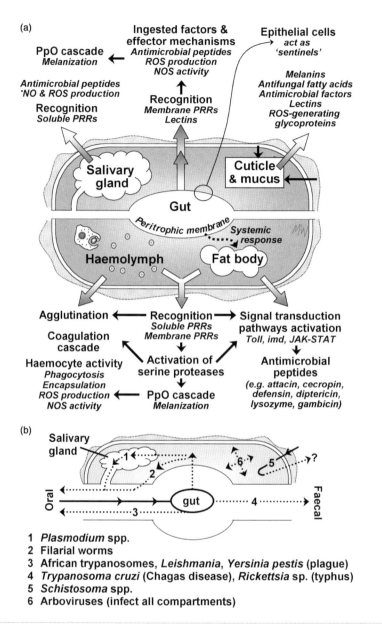

Fig. 1. Overview of arthropod innate immune defences. (a) Immune defences and their locations within a model vector. (b) Compartments infected by major vectored pathogens and routes of transmission. Most pathogens exploit the oral route to infect the host, although some species of trematode (e.g. *Schistosoma*) invade and exit directly through the cuticle. Parasites such as *T. cruzi* pass out via the faeces, while others exit through the mouth (e.g. *Yersinia*). If the gut barrier is traversed and the haemocoel is infected, parasites may either pass through the salivary glands (e.g. malaria) or break directly through the mouthparts (filarial worms). Abbreviations: •NO, nitric oxide; PpO, prophenoloxidase; PRR, pattern recognition receptor; ROS, reactive oxygen species.

tissues of aquatic forms, like *Biomphalaria*, protected by mucus. These tough external barriers include antimicrobial chemical components of the innate immune system that impart protection against would-be invaders. Thus, the arthropod exoskeleton contains phenols and quinones, as components of the prophenoloxidase (PpO) cascade responsible for hardening the cuticle (Asano & Ashida, 2001). In addition, cuticular extracts from insects have been reported to include aldehydes and fatty acids with both antifungal and antibacterial activities. Many insects also actively discharge various antimicrobial hydrocarbons on to the body surface via exocrine glands (Gross *et al.*, 1998). The importance of these externalized secretions should not be underestimated. They may, for example, explain how larval blowflies can thrive on infected human wounds, removing and killing human pathogens, such as methicillin-resistant *Staphylococcus aureus* (Thomas *et al.*, 1999), and represent a whole new group of innate immune molecules awaiting investigation (see reviews by Meylears *et al.*, 2002; Harbige, 2003). The protective role of the mucus secretions of the giant snail, *Achatina fulica*, a gastropod like *Biomphalaria*, is also enhanced by the presence of an antimicrobial factor and an agglutinin present in this secretion (Mitra *et al.*, 1988; Ehara *et al.*, 2002). Interestingly, the giant snail mucus contains a glycoprotein that generates H_2O_2, so killing bacteria (Ehara *et al.*, 2002).

B. The gut barrier

The vector gut, which forms a physical barrier protecting the underlying tissues from digestive enzymes and infectious agents, is also the first point of contact between the parasite ingested with the blood meal and the vector epithelial surfaces. The midgut, in particular, is vulnerable to invasion since this is where the blood meal is stored during the digestive process (Romoser, 1996) and where the parasite has an opportunity to attach to and invade the midgut epithelium. In the midgut there are multiple factors encountered by both the blood meal and the ingested parasites. These include the non-cellular, chitinous peritrophic membrane that surrounds the food and forms a barrier that must be crossed in order for parasites to penetrate the midgut epithelium and enter the haemocoel (Figs 1 and 2). The importance of the peritrophic membrane in the infectious process is evident in *Ae. aegypti* mosquitoes invaded with *Plasmodium gallinaceum*, since infectivity is reduced when the thickness of the peritrophic membrane is increased (Billingsley & Rudin, 1992). The peritrophic membrane may be exploited by parasites during the earliest stages of infection by creating a barrier to the rapid diffusion of digestive enzymes immediately following a blood meal (reviewed by Sacks & Kamhawi, 2001). *P. gallinaceum* ookinetes, *Leishmania* and microfilariae of *Brugia malayi* may all penetrate the peritrophic membrane by means of chitinase which itself is therefore a possible candidate as a transmission-blocking target (Shao *et al.*, 2001).

Other factors/processes, apart from the source, pH and physiological balance of the blood meal and the nutritional and hormonal status of the host (Nguu *et al.*, 1996; Billingsley & Sinden, 1997; Feder *et al.*, 1997) are involved in the developmental success or failure of the parasite in the gut (Fig. 2; reviewed by Nigam *et al.*, 1996; Sinden *et al.*, 1996; see also Hao *et al.*, 2003). These can include:

1. action of digestive enzymes, such as serine proteases (Yan *et al.*, 2001);
2. factors such as specific peptides stimulating parasite differentiation (Fraidenraich *et al.*, 1993);
3. cytotoxic/stimulatory lectin molecules in the lumen and wall of the gut (Maudlin & Welburn, 1987; Mello *et al.*, 1996; Welburn & Maudlin, 1999);
4. glycoprotein receptors on the surface of the midgut for parasite attachment (Wilkins & Billingsley, 2001);
5. presence of bacterial symbionts (Dale & Welburn, 2001);
6. upregulation of immune system genes and antimicrobial peptides (AMPs) (Lehane *et al.*, 2003; Dimopoulos *et al.*, 1998; Hao *et al.*, 2001);
7. nitric oxide killing of parasites (Luckhart *et al.*, 1998; Whitten *et al.*, 2001);
8. role of vector PpO system (Nigam *et al.*, 1997).

In addition, there are many parasite-associated molecules that will determine the outcome of infection in the gut, such as the nature of the surface coat of the insect-stage parasite (summarized in Fig. 3; e.g. Roditi & Liniger, 2002), and the ability of the parasite to counter any reactive oxygen or nitrogen molecules secreted into the gut lumen. Only recently has the importance of researching the insect stage of the parasite been more widely recognized and the role of the gut barrier has received some additional attention (e.g. Dimopoulos *et al.*, 1998; Wilkins & Billingsley, 2001, Hao *et al.*, 2001). Some of the above factors are considered in more detail, so that the true complexity of the interaction of the parasite with the vector gut may be appreciated.

1. Role of digestive enzymes. Regarding the possible role of digestive enzymes in the gut, it has been reported that the transformation of African trypanosomes from their mammalian bloodstream form to procyclics in the tsetse fly midgut is dependent upon a heat-labile factor that is inhibited by a protease inhibitor, a soybean trypsin inhibitor (Imbuga *et al.*, 1992). In addition, Yan *et al.* (2001) have characterized two serine protease genes, Gsp1 and Gsp2, encoding proteins with chymotrysin- and trypsin-like activities and which may be involved in the susceptibility of tsetse flies to trypanosomes. The implication is that trypsin-like enzymes are responsible for refractoriness to infection in tsetse. However, as pointed out by Welburn & Maudlin (1999), feeding trypsin, trypsin inhibitor or anti-trypsin antibody to tsetse flies had no effect on rates of infection and there is no difference in levels of trypsin in susceptible and refractory insects. Similar contradictory results for the influence of proteases on

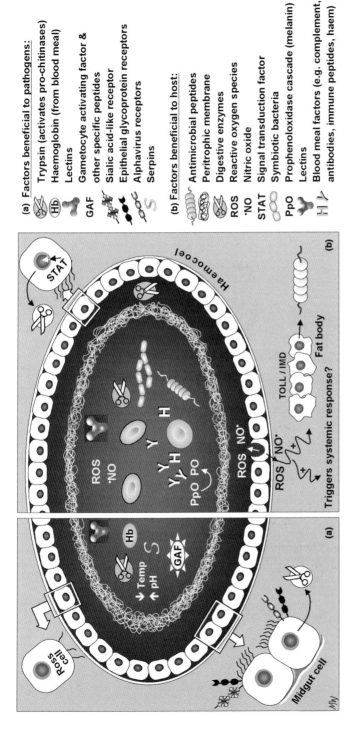

Fig. 2. Host factors and processes involved in the success or failure of pathogens in the gut. (a) Factors beneficial to pathogens. (b) Factors beneficial for the host immune response.

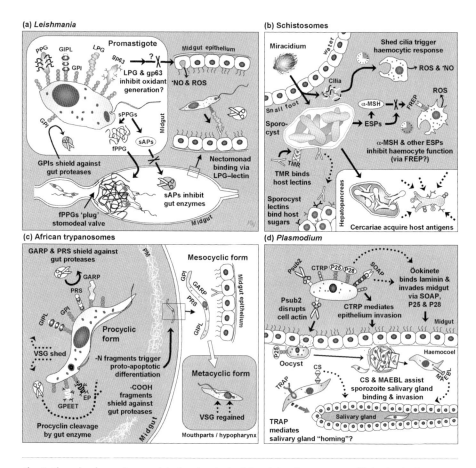

Fig. 3. The role of parasite-associated molecules in determining the outcome of infection in the arthropod vector. Individual panels describe factors in the following models: (a) *Leishmania*–sandfly; (b) schistosome–snail; (c) African trypanosome–tsetse fly; (d) *Plasmodium*–mosquito (for reviews see Landfear & Ignatushchenko, 2001; Sacks & Kamhawi, 2001; Yoshino *et al*., 2001; Cunningham, 2002; Han & Barillas-Mury, 2002; Roditi & Liniger, 2002). Abbreviations: α-MSH, α-melanocyte-stimulating hormone; CS, circumsporozoite protein; CTRP, circumsporozoite and TRAP-related protein; EP, Glu-Pro repeat procyclin; ESPs, excretory–secretory products; fPPG, filamentous PPG; FREP, fibrinogen-related domain protein (host receptor); GARP, glutamate- and alanine-rich protein; GIPL, glycosylinositol phospholipid; GPEET, Gly-Pro-Glu-Glu-Thr repeat procyclin; GPI, glycosylphosphatidylinositol (free or as an anchor); gp63, metalloproteinase (leishmanolysin); LPG, lipophosphoglycan; MAEBL, apical membrane antigen/erythrocyte binding-like protein; •NO, nitric oxide; PM, peritrophic membrane; PPG, proteophosphoglycan; PRS, protease-resistant surface molecule (*Trypanosoma congolense* only); Psub2, subtilisin-like protease; ROS, reactive oxygen species; sAPs, secreted acid phosphatases; SOAP, secreted ookinete adhesive protein; sPPGs, secreted PPGs; TMR, tegumental membrane receptor; TRAP, thrombospondin-related anonymous protein; VSG, variant surface glycoprotein.

Plasmodium infectivity in mosquitoes have been reported (Shahabuddin *et al*., 1998). Both *T. cruzi* and *T. rangeli*, however, appear to be insensitive to the digestive proteases in the vector midgut (Garcia & Azambuja, 1991).

2. Peptides stimulating development. There are reports of non-enzymic factors stimulating parasite development in the gut. Thus, Fraidenraich *et al.* (1993) and Garcia *et al.* (1995) described a peptide from the hindgut of the South American reduviid, *Triatoma infestans*, which activated adenyl cyclase in *T. cruzi* epimastigote membranes so that they differentiated *in vitro* into metacyclic trypomastigotes. This 10 kDa peptide had the same amino acid sequence as chicken α^{D}-globin, and since the insects had been fed on chickens it was concluded that this peptide was derived from the latter. In contrast, in the closely related reduviid, *R. prolixus*, a lytic factor against specific strains of *T. cruzi* has been reported (Azambuja *et al.*, 1989). Thus, host-derived factors must also be taken into account during the development of parasites in haematophagous insects. Another example is provided by the detection of a mosquito exflagellation factor in the midgut of *Ae. aegypti* and the anal secretions of *Anopheles stephensi* that can induce *Plasmodium* gametogenesis (reviewed by Shahabuddin *et al.*, 1998). The factor has now been identified as xanthurenic acid (also known as gametocyte activating factor) but, as pointed out by Shahabuddin *et al.* (1998), there appears to be an interplay between xanthurenic acid, pH of the midgut and temperature for optimal gametogenesis. Again, however, there is also an input derived from the host for gametocyte activation (Arai *et al.*, 2001). The hard tick, *Boophilus microplus*, also has a unique peptide, active against both bacteria and fungi, and which is a fragment of bovine α-haemoglobin from the blood meal produced by enzymic processing in the gut lumen (Fogaça *et al.*, 1999). A role for host-derived factors has also been described in the tick *Ixodes pacificus*, which ingests a borreliacidal protein factor from lizard blood with activity against midgut *Borrelia* spirochaetes (Lane & Quistad, 1998).

3. Gut lectins. Many workers have reported the possible role of lectins in the gut of vectors in determining the outcome of parasite development in this organ (Grubhoffer *et al.*, 1997). One of the first papers was published by Pereira *et al.* (1981) who, working with *R. prolixus*, described lectins in the crop, midgut and haemolymph of this vector with specificity for N-acetylmannosamine, N-acetylgalactosamine and D-galactose residues, respectively. It was suggested that these lectins may have been involved in the differentiation and division of *T. cruzi* in the *Rhodnius* gut. Subsequent work by Ratcliffe *et al.* (1996) also detected a lectin in the *Rhodnius* gut which was not inhibited with any of the residues used above, although there was a positive correlation between the agglutination of the *T. cruzi* strain used and the infectivity of the insects (Mello *et al.*, 1996). In contrast, Maudlin & Welburn (1987) and Welburn *et al.* (1989) reported a midgut lectin in tsetse flies that was inhibited by D-galactosamine and which was involved in the development of African trypanosomes in these insects. Biochemical purification of the tsetse fly lectin has been undertaken and it suggested that there are at least two midgut lectins of 26 and 29 kDa which upon feeding the inhibitory sugar produced 100 % infection rates (Maudlin & Welburn, 1987; Grubhoffer *et al.*, 1997).

The possible role of lectins in midgut interactions of other parasites with their hosts include *Leishmania major* with its sandfly host (Volf *et al.*, 1998, 2001; Sacks & Kamhawi, 2001), *Plasmodium* spp. with mosquitoes (Shahabuddin *et al.*, 1998; Chen & Billingsley, 1999) and the nematode *Brugia pahangi* with the mosquito *Ae. aegypti* (Ham *et al.*, 1991).

4. Midgut glycoprotein receptors. Although lectins may be important determinants of parasite infectivity in the vector (see above), there are other host-associated molecules in the midgut to which the parasite must bind to complete its life cycle. Our knowledge of these molecules is strictly limited (Roditi & Liniger, 2002). It has been shown with *Plasmodium* that ookinetes can develop into oocysts and sporozoites after injection into the haemocoel of *Drosophila*, but that the ookinetes and the sporozoites cannot penetrate the gut epithelium or salivary glands, respectively (Schneider & Shahabuddin, 2000). As discussed by Roditi & Liniger (2002), this indicates that host factors are actively involved in guiding the parasites through the vector tissues. Wilkins & Billingsley (2001) have also analysed the complex mixture of oligosaccharides on the midgut epithelium of *An. stephensi* and found a high proportion of *N*-linked GlcNAc- and GalNAc-terminal oligosaccharides. Similar carbohydrates have been shown by Zieler *et al.* (1999) to be involved in the interaction of the *Plasmodium* ookinete with the midgut epithelium. The ookinetes of *P. gallinaceum* have also been reported to invade a specific cell type, termed 'Ross cells', in *Ae. aegypti* (Shahabuddin & Pimenta, 1998; Han *et al.*, 2000), so characterization of the profile of carbohydrates on the surface of these cells may be worthwhile. Finally, once the midgut epithelial cell has been penetrated by the ookinete, the parasite comes to rest between the basal plasma membrane and the basal lamina. The latter becomes disrupted so that the developing oocyst eventually comes to lie within the haemocoel by a process similar to wound repair (reviewed by Gare *et al.*, 2003). Again, molecular interactions occur between the parasite and the host since a 215 kDa surface moiety of *Plasmodium berghei* ookinetes binds to both laminin and collagen IV components of the basal lamina (Arrighi & Hurd, 2002). Other examples of the role of molecular interactions in the invasion of the midgut by parasites comes from studies of the interaction of *T. rangeli* with the midgut of *R. prolixus* in which lectins not only blocked carbohydrates exposed on the midgut, but also inhibited the attachment of the parasites (Oliveira & De Souza, 1997). The entry of arboviruses also appears to be mediated by receptor–ligand interactions involving glycoproteins, since entry is blocked by monoclonal antibodies (Ludwig *et al.*, 1991).

5. Presence of bacterial symbionts. An additional complication to understanding the role of carbohydrate moieties in vector–parasite interactions in the insect midgut is the presence of endosymbionts in the gut of many vectors. These have been described in many haematophagous arthropod vectors, such as reduviid bugs (Beard *et al.*, 2002; Eichler & Schaub, 2002) and tsetse flies (Dale & Welburn, 2001) in which they are thought to provide essential nutrients such as vitamin B. They are mainly found in those insects relying solely on blood throughout their life cycle. These include sucking lice, bed bugs, reduviid bugs, fleas and tsetse flies, but not mosquitoes, sandflies or blackflies since in these only the adult females need blood for maturation of the eggs (Romoser, 1996). In tsetse flies, Dale & Welburn (2001) showed that the vectorial capacity of the insects was significantly reduced by the specific artificial removal of the endosymbiont *Sodalis glossinidius* by feeding with the antibiotic streptozotocin. Apparently, *Sodalis* interferes with the efficiency of the tsetse fly immune system by producing GlcNAc, from bacterial chitinase activity on the peritrophic membrane, and this sugar inhibits the trypanocidal activity of the tsetse midgut lectin (Maudlin & Welburn, 1987).

6. Upregulation of immune genes. There is now considerable evidence that the midgut is an immune-reactive organ that is capable of detecting and responding to the presence of parasites, such as *Plasmodium*, by upregulation of several immune genes (Fig. 4; Dimopoulos *et al.*, 1997, 1998; Luckhart *et al.*, 1998; Barillas-Mury *et al.*, 2000; Dimopoulos, 2003). Rapid progress is being made in understanding vector–parasite interactions at all stages of invasion using powerful molecular techniques, with much work concentrating on mosquitoes infected by *Plasmodium* (reviewed by Barillas-Mury *et al.*, 2000; Dimopoulos *et al.*, 2001, 2002; Lowenberger, 2001; Dimopoulos, 2003). No doubt this progress will be accelerated with the publication of the *Anopheles gambiae* mosquito (Holt *et al.*, 2002) and *Plasmodium* spp. (Carlton *et al.*, 2002; Gardner *et al.*, 2002) genome sequences. In addition, these methods are now being applied to other systems, including sandfly–*Leishmania* (reviewed by Sacks & Kamhawi, 2001), *Biomphalaria glabrata* snail–*Schistosoma* (Zhang *et al.*, 2001) and tsetse fly–African trypanosome (Hao *et al.*, 2001; Boulanger *et al.*, 2002a; Lehane *et al.*, 2003) interactions. Molecular studies are also under way on the midguts of the stable fly, *Stomoxys calcitrans* (Boulanger *et al.*, 2002b), the reduviid bugs *Triatoma infestans* (Kollien & Billingsley, 2002) and *R. prolixus* (M. M. A. Whitten, G. A. Schaub, F. Sun, A. J. Nappi & N. A. Ratcliffe, unpublished) and the soft tick *Ornithodoros moubata* (Nakajima *et al.*, 2002), following blood- or infected meals.

In *An. gambiae* mosquitoes infected with *Plasmodium*, several AMPs have been isolated, including one defensin and two cecropins (reviewed by Dimopoulos *et al.*, 2001; see also Eggleston *et al.*, 2000), and four defensin and four cecropin genes have been reported (Christophides *et al.*, 2002). In addition, a novel AMP, gambicin, has been reported with some anti-*Plasmodium* activity (Vizioli *et al.*, 2001). In *Ae. aegypti* mosquitoes, three isoforms, A, B and C, of insect defensins have been detected as well

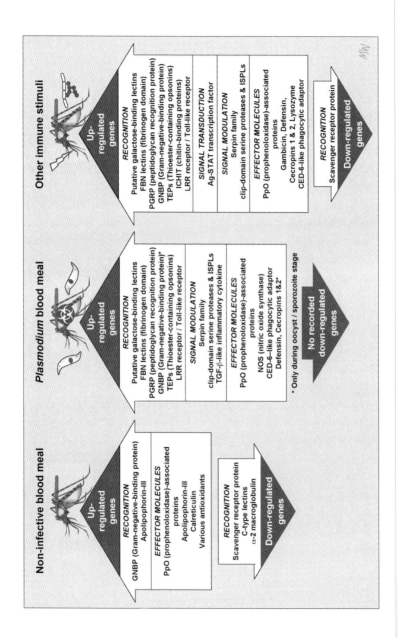

Fig. 4. Changes to *Anopheles* immune-related gene expression following a blood meal, infection with *Plasmodium* spp. parasites or other immune stimuli (e.g. bacteria). Based on Luckhart *et al.* (1998); Muller *et al.* (1999); Oduol *et al.* (2000); Dimopoulos *et al.* (2001, 2002); Ribeiro (2003). Abbreviations: CED-6, *Caenorhabditis elegans* death gene; clip, repeated N-terminal cysteine residue motif; ISPLs, immune-related serine protease-like sequences; LRR, leucine-rich repeat receptor; STAT, signal transducer and activator transcription; TGF-β, transforming growth factor β.

as two cecropins, A and B (Table 2, reviewed by Lowenberger, 2001; see also Cho *et al.*, 1997; Gao *et al.*, 1999, Gao & Fallon, 2000).

Several important points have emerged from these studies as admirably reported by Barillas-Mury *et al.* (1996, 1999, 2000), Richman *et al.* (1997), Dimopoulos *et al.* (1997, 1998, 2000, 2001, 2002), Han *et al.* (1999), Lowenberger *et al.* (1999), Lowenberger (2001), Christophides *et al.* (2002) and Dimopoulos (2003) and these are summarized as follows.

(a) First, following infection with bacteria or parasites many different genes are upregulated and/or immune effector molecules are transcribed, and not just those associated with AMPs. In *An. gambiae* infected with *P. berghei*, for example, mRNA levels increase in the midgut and abdominal wall after 24 h for putative Gram-negative-binding (AgGNBP), chitin-binding (ICHIT) and immune-related serine-protease-like (ISPL5) proteins, as well as for an infection-responsive galactose-binding lectin (IGALE20), nitric oxide synthase (NOS) and defensin (Fig. 4; Barillas-Mury *et al.*, 2000). It may be difficult to interpret the net effect of genes that can be both transiently up- and downregulated following infection (e.g. Dimopoulos *et al.*, 2002). The problem with the gut is to differentiate between genes that are upregulated in response to a natural 11- to 40-fold increase in bacterial numbers following a blood meal (Shahabuddin *et al.*, 1998) and those that are involved in defence against eukaryotic parasites such as *Plasmodium*. The AMPs, for example, are likely to be important in controlling bacterial pathogens in the gut. The dynamics of the production of the AMPs following an infective blood meal may be instructive. Thus, initially, 20–30 h following parasite ingestion when the ookinetes begin to invade the midgut epithelium, several immune genes are upregulated, including those for defensin (Shahabuddin *et al.*, 1998) which is the major AMP in adult mosquitoes (Lowenberger *et al.*, 1995). The presence of asexual parasites alone in the blood meal is not sufficient to trigger the appearance of defensin and removal of the gut bacteria with antibiotics fails to impede the antiparasite immune response (Dimopoulos *et al.*, 1997).

(b) Second, the induction of mosquito immunity is not confined to the mid-gut, but also occurs systemically, even before *Plasmodium* invades through the basal lamina of the midgut epithelium. Thus, signalling may be occurring between immune-competent tissues during infection. The tissue involved in this systemic response is mainly the fat body (Dimopoulos *et al.*, 2001), although studies on blood cells that are known to produce AMPs in other insect species (Cociancich *et al.*, 1994) are mainly limited to haemocyte-like

Table 2. Distribution of AMPs in vector arthropods

Vector	AMP	Location	Trigger	Major roles	Reference*
Mosquito *Aedes aegypti*	Defensin	Haemolymph (synthesized in fat body), cell line cultures	Injected *Br. pahangi* microfilariae	Disrupts development of parasites	1–5
	Defensins A–C		Bacteria in haemocoel	Disrupts oocyst growth, antibacterial	
	Lysozyme	Cultured cells	Bacteria	Antibacterial (G+)	6
	Gambicin	Salivary gland	NR	Antimicrobial	7
	Attacin	NR	NR	Antibacterial	8
Mosquito *Anopheles darlingi*	Lysozyme	Salivary gland (constitutive)	NR	Antibacterial (G+)	9
Mosquito *Anopheles gambiae*	Defensins A–D	Midgut, abdomen	NR	Antibacterial (G+), antiplasmodial	10–12
	Cecropins 1–4	NR (Constitutive)	NR	Antibacterial (G+ and G−)	10
	Lysozyme		NR	Antibacterial (G+)	13
	Gambicin	Anterior midgut, thorax, haemocyte-like cell lines	Bacterial challenge (not by normal gut flora)	Antimicrobial (G+ and G−), antifungal, antiplasmodial (ookinete)	14
Mosquito *Anopheles* spp.	Cecropins	NR	NR	Disrupts development of *Plasmodium*	15
Mosquito *Culex pipiens*	Defensin Gambicin Cecropin A Cecropin B	Haemolymph, fat body, midgut	Bacteria injection / None	Antibacterial and antiparasitic	16
Blackfly *Simulium damnosum*	Lysozyme-like factor 4 kDa cecropin-like factor	Haemolymph	Bacteria, LPS, *Onchocerca* microfilariae (or endosymbionts?)	Antibacterial (G+), Antiparasitic, antibacterial (G−)	17
Sandfly *Lutzomyia longipalpis*	4 and 33 kDa factors	Haemolymph	Bacteria injection	Antibacterial	18

Table 2. cont.

Vector	AMP	Location	Trigger	Major roles	Reference*
Tsetse *Glossina morsitans*	Diptericin	Gut (constitutive)	None	Controls symbiotic gut bacteria	19
	Defensin Attacin Cecropin	Synthesis in fat body, released into haemolymph	Trypanosomes in gut, bacterial infection	Confines trypanosomes to gut?	20
Soft tick (argasid) *Ornithodoros moubata*	Defensins A–D	Midgut (constitutive)	Blood meal, bacteria injection	Antibacterial	21
	Lysozyme	Gut	NR	Antibacterial (G+)	22
Hard ticks (ixodid) *Dermacentor variabilis, Boophilus microplus* and *Rhipicephalus appendiculatus*	Defensin	Haemocyte, released into plasma	*Borrelia burgdorferi* challenge	Antibacterial, most effective with lysozyme	23, 24
	α-Haemoglobin fragment	Gut – bovine blood component (passes into haemolymph?)	Not relevant	Antibacterial (G+) and antifungal	25
	Unknown factor	Haemolymph	Bacteria in haemocoel	Antibacterial, *not* antiprotozoan	26
Reduviid bug *Triatoma infestans*	Lysozyme	Gut	Blood meal	Antibacterial (G+)	27
Reduviid bug *Rhodnius prolixus*	Novel defensins	NR	NR	NR	28, 29
	Defensin A	Synthesis in fat body and gut; released into haemolymph	Bacteria in haemocoel	Antibacterial	30
	Hemiptericin	NR	NR	NR	31
	7 kDa factor	Haemolymph	Bacteria injection	Antibacterial	32

G+, Gram-positive; G–, Gram-negative; NR, not reported.

*References: 1, Chalk *et al.* (1994); 2, Dimopoulos *et al.* (1997); 3, Lowenberger *et al.* (1995); 4, Gao *et al.* (1999); 5, V. Kokoza and others (unpublished), reported in Shin *et al.* (2003b); 6, Gao & Fallon (2000); 7, Valenzuela *et al.* (2002); 8, reviewed by Paskewitz & Christensen (1996); 9, Moreira-Ferro *et al.* (1998); 10, Christophides *et al.* (2002); 11, Richman *et al.* (1997); 12, Dimopolous *et al.* (1998); 13, Kang *et al.* (1996); 14, Vizioli *et al.* (2001); 15, Gwadz *et al.* (1989); 16, Bartholomay *et al.* (2003); 17, Kläger *et al.* (2002); 18, Nimmo *et al.* (1997); 19, Hao *et al.* (2001); 20, Boulanger *et al.* (2002a); 21, Nakajima *et al.* (2001, 2002); 22, Kopacek *et al.* (1999); 23, Ceraul *et al.* (2002); 24, Johns *et al.* (2001); 25, Fogaça *et al.* (1999); 26, Watt *et al.* (2001); 27, Kollien *et al.* (2003); 28, Hoffmann & Hetru (1992); 29, Chernysh *et al.* (1996); 30, Lopez *et al.* (2003); 31, Cociancich *et al.* (1994); 32, Azambuja *et al.* (1986).

cell lines (Vizioli *et al.*, 2001). Signalling pathways are probably triggered in the fat body by nitric oxide (˙NO) diffusing readily from the gut (Barillas-Mury *et al.*, 2000).

(c) Third, during development of *Plasmodium* in the mosquito, the majority of the parasites are killed and it has been calculated that of the tens of thousands of gametocytes ingested in the blood meal only 50–100 öokinetes are produced, and of these fewer than five survive to develop into oocysts (Sinden & Billingsley, 2001). What evidence do we have to indicate a role for innate immune factors in the reduction in these parasite numbers? Several mosquito molecules have the ability to kill malarial parasites and other pathogens. These include defensins, cecropins, gambicin, lysozyme and nitric oxide, as well as components of the PpO cascade. Gwadz *et al.* (1989) and Shahabuddin *et al.* (1998) have tested the possible role of defensin, cecropin, magainin and other AMPs in killing *P. gallinaceum* oocysts by injecting these AMPs into *Ae. aegypti* mosquitoes. They showed that injection on day 3 post-infection significantly reduced oocyst density while injection at day 5 produced abnormal oocysts. In addition, incubation of sporozoites with defensin severely damaged the parasites. Shahabuddin *et al.* (1998) concluded that öokinetes and early oocysts were resistant to defensin while late oocysts and sporozoites were susceptible due to developmental changes in the plasma membrane of the parasites. One weakness of these experiments is that the AMPs used were not from mosquitoes, but from other insect species and although all these defensins have a high amino acid homology their modes of action are not identical (Shahabuddin *et al.*, 1998). The results of these experiments may therefore not accurately reflect the role of the mosquito defensin. Indeed, the work of Lowenberger *et al.* (1999) with bacteria-inoculated *Ae. aegypti* (presumably with enhanced levels of defensin) infected with *P. gallinaceum* failed to demonstrate a reduction in numbers of oocysts but, as pointed out by the authors, the situation is complicated by differences in strains of mosquitoes and parasites used in these experiments. Regarding gambicin, it is induced by malaria infection but not by endogenous bacteria after feeding and has been shown to have a modest öokinete killing property (Vizioli *et al.*, 2001). Also potentially important in the killing of the parasites are nitric oxide and the components of the PpO cascade. These are dealt with separately below.

Apart from mosquitoes, detailed work on midgut immune responses of other arthropod vectors includes that with the tsetse fly, *Glossina morsitans*, to African trypanosomes (Hao *et al.*, 2001; Yan *et al.*, 2001; Boulanger *et al.*, 2002a; Lehane *et al.*, 2003), of the stable fly, *Stomoxys calcitrans*, to African trypanosomes (Munks *et al.*, 2001; Boulanger *et al.*, 2002b) and of the soft tick, *O. moubata* towards bacteria (Nakajima *et al.*, 2001, 2002).

At least four different AMPs have been detected in tsetse flies infected with bacteria or trypanosomes and these include a cecropin, an attacin, a defensin and a diptericin. At least one of these, diptericin, is expressed constitutively in the gut and may control symbiotic bacteria (Hao *et al.*, 2001). What is interesting, however, is the systemic appearance of cecropin, attacin and defensin in the haemolymph as mature peptides or as mRNA in the fat body following gut infection *per os* with trypanosomes (Boulanger *et al.*, 2002a). Normally, trypanosomes do not invade the tsetse fly haemolymph so the presence of the AMPs may confine the parasites to the gut. In addition, the transcription of attacin and defensin in the fat body is induced more by procyclic parasites than by bloodstream forms and so the tsetse immune response is pathogen-specific (Hao *et al.*, 2001). More recently, Lehane *et al.* (2003) reported the sequencing of 21 427 expressed sequence tags from the midgut of adult *Glossina* that included 68 unique genes with putative immune-related functions. These were macroarrayed and their transcriptional profiles investigated following bacterial or *T. brucei brucei* challenge. With both types of infection many genes were downregulated, but those upregulated in response to trypanosome challenge included a number of Toll and Imd genes as well as genes involved in antioxidative stress. The Toll and Imd gene products are involved in signalling pathways leading to immune peptide production (see Section C2b below), while the antioxidant gene products may protect the midgut against haem molecules liberated from the blood meal (see Section B7 below) as well as against the reactive oxygen molecules generated by the immune response. Lehane *et al.* (2003) also detected nine putative serpins in the tsetse midgut and postulated that such large numbers may reflect the need to inactivate the complement and coagulation cascades in the blood meal.

The stable fly, *Stomoxys calcitrans* (which does not vector trypanosomes despite feeding on many of the same hosts as tsetse) has been used as a model to study the possible role of midgut AMPs as determinants of infection (Munks *et al.*, 2001; Boulanger *et al.*, 2002b). At least three AMPs have been found that are specific to adult anterior midgut tissues and which may act synergistically. One of these AMPs, termed stomoxyn, is unique to *Stomoxys* and has broad-spectrum activity against bacteria and fungi. It also kills bloodstream but not procyclic forms of *T. brucei brucei* so that it must act quickly against the former before the latter stage appears (Boulanger *et al.*, 2002b). Boulanger *et al.* (2002b) suggested that the presence of stomoxyn in the anterior midgut can occur at high concentrations and probably explains why *Stomoxys* is not a vector of trypanosomes. *Stomoxys*, unlike the tsetse fly, *Glossina*, feeds on rotting fruit as well as vertebrate blood, so that the threat of microbial challenge in the gut will be higher than in tsetse and necessitate the production of a unique AMP such as stomoxyn. The other two *Stomoxys* AMPs are the defensins, Smd1 and Smd2. These too are expressed exclusively in the midgut both constitutively and by up-regulation of the Smd genes following a blood meal which it is believed they protect from microbial attack (Munks *et al.*, 2001). Unlike tsetse flies, mosquitoes and *Drosophila*, the *Stomoxys* AMPs are only upregulated in the midgut following immune challenge and not systemically in the fat body or other tissues (Munks *et al.*, 2001).

Similarly, in the soft tick, *O. moubata*, four defensins, namely A, B, C and D, have been detected that occur constitutively in the midgut and that are mainly expressed in this organ following a blood meal (Nakajima *et al.*, 2001, 2002).

7. Nitric oxide killing of parasites.
With notable exceptions (Dikkeboom *et al.*, 1988; Adema *et al.*, 1994; Dimopoulos *et al.*, 1998; Luckhart *et al.*, 1998; Luckhart & Rosenberg, 1999; Crampton & Luckhart, 2001; Hahn *et al.*, 2000, 2001a, b; Han *et al.*, 2000; Luckhart & Li, 2001; Whitten *et al.*, 2001; M. M. A. Whitten, G. A. Schaub, F. Sun, A. J. Nappi & N. A. Ratcliffe, unpublished), very few studies have looked at the role of free radicals such as superoxide (O_2^-) and $\cdot$NO in invertebrate vector immunity. This is surprising considering that toxic reactive oxygen and nitrogen molecules play vital roles in vertebrate immunity and antiparasitic defence (reviewed extensively by, e.g., Wientjes & Segal, 1995; Stafford *et al.*, 2002). In the gut of vectors, only the NOS gene in mosquitoes (Luckhart *et al.*, 1998; Dimopoulos *et al.*, 1998) and in the reduviid bug, *Rhodnius* (M. M. A. Whitten, G. A. Schaub, F. Sun, A. J. Nappi & N. A. Ratcliffe, unpublished) has been investigated with most workers reporting on the interaction of free radicals with parasites following invasion of the haemolymph (see Section C2c below). In the mosquito, *An. gambiae*, Dimopoulos *et al.* (1998) identified a NOS gene that is transcriptionally activated by both bacteria and *Plasmodium* parasites and is particularly active in midgut cells. Luckhart *et al.* (1998) discovered that the mosquito, *An. stephensi*, limits *Plasmodium* development with synthesis of $\cdot$NO produced by inducible NOS (AsNOS) in the midgut wall. The NOS gene was probably initially induced by bacterial growth in the blood meal while later increases after 1–3 days correlated with öokinete invasion and development of the oocyst. In addition, feeding mosquitoes the NOS substrate, L-arginine, produced a 28 % reduction in infection, while provision of the NOS inhibitor, L-NAME, significantly increased oocyst numbers in the midgut wall. The induction of NOS expression was not confined to the midgut, but also occurred in the carcass and haemolymph following an infective blood meal. In contrast to the vertebrate NOS with four isoforms encoded by different genes, in insects there appears to be a single NOS gene type which may be a 'primitive composite' having combined features of all the mammalian NOS isoforms (Luckhart & Rosenberg, 1999; Luckhart & Li, 2001). In vertebrates, NOS expression is driven by a suite of cytokines, including TGF-β and, significantly, Crampton & Luckhart (2001) have discovered a

mosquito homologue of TGF-β, called As60A, which is induced in the midgut and carcass of *An. stephensi* in response to *Plasmodium* infection. This As60A has a Rel-binding site that is characteristic of insect immune genes and may well be an inflammatory cytokine controlling AsNOS. In addition, Luckhart *et al.* (2003) have shown that during feeding *An. stephensi* ingests mammalian latent TGF-β that is rapidly activated in the mosquito midgut, possibly by haem and ˙NO, and can alter the development of *Plasmodium falciparum*.

In *Rhodnius*, Whitten and others (M. M. A. Whitten, G. A. Schaub, F. Sun, A. J. Nappi & N. A. Ratcliffe, unpublished) have shown that NOS gene expression is upregulated in the crop, midgut and fat body following *in vivo* challenge with *T. cruzi*, *T. rangeli* and bacterial endotoxin. Elevations in NOS activity in the crop wall occur during the response to infection by both species of trypanosome. However, increased ˙NO is produced in the midgut only in response to *T. cruzi* and not *T. rangeli*, and could be a reason why the latter parasite successfully crosses the midgut barrier, while *T. cruzi* is confined to the alimentary canal. High concentrations of ˙NO are also generated in the rectum following trypanosome infection (M. M. A. Whitten, G. A. Schaub, F. Sun, A. J. Nappi & N. A. Ratcliffe, unpublished), particularly with *T. rangeli*, which unlike *T. cruzi*, does not require colonization of the rectum to complete its life cycle.

Also of great relevance is the recent finding that in *Drosophila* the inhibition of NOS increased larval sensitivity to Gram-negative bacteria and abrogated induction of the AMP, diptericin (Foley & O'Farrell, 2003). NOS activity was shown to be essential for a robust immune response. ˙NO alone is sufficient to trigger immunity in the absence of infection and probably mediates an early step of signal transduction.

Finally, mention must be made of problems that vectors and parasites have in dealing with free haem produced from the digestion of the blood meal in the vector midgut. Haem contains iron and has the potential to catalyse the formation of activated oxygen radicals that cause oxidative damage to tissues and cells (Oliveira *et al.*, 2000). In *Plasmodium* feeding on haemoglobin within vertebrate erythrocytes, haem is sequestered into a dark-brown pigment called haemozoin. Haemozoin has now been shown to be produced in the *Rhodnius* midgut and in the helminth, *S. mansoni*, thus reducing radical damage (Oliveira *et al.*, 2002) both in the vector gut and parasite tissues. In blood-feeding parasites, such as *Plasmodium* and *Schistosoma*, haem conversion to haemozoin is not complete so the potential for oxidative damage is countered by a switch to anaerobic metabolism and dramatic increases in levels of antioxidants such as superoxide dismutase, glutathione reductase and peroxidase, etc. (Oliveira & Oliveira, 2002). This clearly illustrates further complications of the parasite–vector interrelationship.

8. Role of vector PpO system. The PpO system in arthropods is known to be involved in the hardening and darkening of the exoskeleton following moulting as well as in the immune defence reactions towards invading parasites (e.g. see references in Ratcliffe, 1991; Söderhäll & Cerenius, 1998). Most research on the role of PpO in arthropod immunity has focused on the haemolymph stage of the parasite during which the haemocyte and the plasma components of PpO interact. The role of the haemolymph PpO in the vector immune reactions is dealt with in more detail below (see Sections C1b and C2a below). Attention, however, was drawn to a possible role for the PpO system in the vector midgut during immune reactions to parasites by the observation that in a refractory mosquito strain, L3-5, of *An. gambiae*, the *Plasmodium* ökinetes became surrounded by a melanized capsule while no such response was observed in a susceptible strain (Collins *et al.*, 1986). Subsequent work showed the presence of phenoloxidase (PO) in the basal membrane and cellular basal membrane of the midgut of infected refractory mosquitoes but no such activity in the infected susceptible strain (Paskewitz *et al.*, 1988). However, since melanization appears to be confined to the apex of the ökinete, facing the haemolymph (Dimopoulos *et al.*, 2001), it is likely that the PpO components are haemolymph- rather than midgut-derived and that the activation of the PpO system is associated with damage to the midgut wall caused by parasite invasion. There is, however, some indication that midgut PpO may participate in the vector's immune response to parasites. Thus, intracellular melanization of different species of nematode filarial larvae in mosquito guts has often been reported (e.g. Nayar *et al.*, 1989). In addition, and more convincing, is the observation that midgut homogenates of both refractory and susceptible strains of tsetse flies, *Glossina palpalis palpalis* and *G. morsitans morsitans*, have an active PpO system and that this has higher levels of activation in refractory flies (Y. Nigam & N. A. Ratcliffe, unpublished).

C. The haemolymph stage

Many parasites, including *Leishmania* spp., *T. cruzi* and the African trypanosomes, remain in the gut to complete their development and are transmitted to their vertebrate host during feeding. Even though they are apparently confined to the vector gut, the African trypanosomes can stimulate a systemic immune response (see Section B6 above and Boulanger *et al.*, 2002a). Other important parasites, such as *Plasmodium* spp., *T. rangeli*, nematode microfilariae, schistosomes and arboviruses all invade the vector haemolymph at some stage in their life cycles (Fig. 1). Once within the haemolymph, they are then exposed to a whole range of cellular and humoral immune defences. These can be divided into:

    1. The cellular reactions to parasites including:

        (a) phagocytosis by invertebrate leucocytes, termed haemocytes (e.g. Hillyer *et al.*, 2003);

(b) encapsulation which may involve haemocytes and/or humoral plasma factors (e.g. Beerntsen *et al.*, 2000) including components of the PpO cascade (Johnson *et al.*, 2003).

2. Humoral factors interacting with parasites such as:
(a) the PpO system (e.g. Johnson *et al.*, 2003);
(b) pattern recognition receptors (PRRs, e.g. Dimopoulos, 2003), including lectins, non-self recognition and intracellular signalling pathways;
(c) killing factors such as AMPs (e.g. Lowenberger, 2001), reactive oxygen species (ROS) and reactive nitrogen intermediates, including ˙NO (e.g. Whitten *et al.*, 2001).

1. Cellular reactions to parasites. These reactions are mediated rapidly by the blood cells (haemocytes) freely circulating in the haemolymph. Little is known about the possible involvement of fixed or sessile cells or what proportion of the total haemocyte population is composed of these cells in vectors, although in non-vectors they may constitute a major component of the total haemocytes (Ratcliffe *et al.*, 1985). There have been a number of studies of invertebrate vector haemocytes to include mosquitoes (Andreadis & Hall, 1976; Drif & Brehélin, 1983; Kaaya & Ratcliffe, 1982; Kaaya *et al.*, 1986; Hernández *et al.*, 1999; Hillyer & Christensen, 2002), six species of South American hemipteran bugs (Jones, 1965; Barracco *et al.*, 1987; Azambuja *et al.*, 1991; Hypša & Grubhoffer, 1997), tsetse flies (East *et al.*, 1980; Kaaya & Ratcliffe, 1982; Molyneux *et al.*, 1986), sandflies (Brazil *et al.*, 1996), blackflies (Luckhart *et al.*, 1992; Hagen & Kläger, 2001), ticks (Kühn & Haug, 1994; Inoue *et al.*, 2001) and the pulmonate snail *Biomphalaria* spp. (Lie *et al.*, 1987; Barracco *et al.*, 1993; Matricon-Gondran & Letocart, 1999). It is difficult to make widely agreed generalizations about the haemocytes of such a diverse group of invertebrates, although the publications by Price & Ratcliffe (1974), Ratcliffe & Rowley (1981), Kaaya & Ratcliffe (1982) and Azambuja *et al.* (1991) may be helpful as they compare a range of vectors. Most invertebrate vectors appear to have phagocytic cells that are described either as amoebocytes in snails or as plasmatocytes or granular cells in arthropods (the term 'granulocyte' should be avoided as granulocytes are a common leucocyte type in vertebrates and there is little evidence of true homology). Some arthropod vectors, such as *Anopheles albimanus* (Hernández *et al.*, 1999), apparently have both granular cells and plasmatocytes (as well as prohaemocytes) while others, including *Ae. aegypti* (Hillyer & Christensen, 2002), are reported only to have the granular cell type (as well as oenocytoids, adipohaemocytes and thrombocytoids) [see Ratcliffe & Price (1974) for a discussion of vector haemocyte terminology]. In addition, the granular cells, or a larger cell type termed oenocytoids in arthropods, often have PpO activity. Most important, however, is the fact that the granular cells and/or the plasmatocytes may also produce cytokines (e.g. Trenczek & Faye, 1988; Anggraeni & Ratcliffe, 1991; Lavine & Strand, 2002) and be responsible for non-self recognition in association with PRRs (see Section C2b below).

(a) Phagocytosis is the way that animals remove small numbers of micro-organisms from the circulation by ingestion by single leucocytes while larger numbers of microbes, as well as metazoan parasites, are sequestered in modified capsules/nodules usually formed from layers of leucocytes (Ratcliffe, 1993). These latter structures often become melanized to form 'black bodies' attached to various organs of the body (see Section 1b below). What evidence do we have to indicate a role for phagocytosis in vector immunity? The phagocytosis of *Plasmodium* sporozoites by mosquito haemocytes has been reported *in vivo*. This occurred in *An. albimanus* after intra-haemocoelic injection (Hernández-Martinez *et al.*, 2002) and in *Ae. aegypti* following natural infections (Hillyer *et al.*, 2003). It is not known whether the ingested parasites were killed or whether they survived to use the haemocyte as a haven, as is the case for the vertebrate stages of many protozoan parasites (Bogdan & Röllonghoff, 1999). It is known, however, that only 10–25 % of sporozoites released from the mosquito oocyst invade the salivary glands (Sinden & Billingsley, 2001) so that the cellular defences could be involved in this reduction. In the case of *T. rangeli*, the parasites are taken up by the *R. prolixus* haemocytes and multiply therein before escaping and invading the salivary glands (Tobie, 1970). Arboviruses have also been reported within the haemocytes of ticks where they undergo enormous replication to form 'virus factories' (Kleiboeker *et al.*, 1998). With many of these parasites, it is unclear if entry to the haemocyte is mediated by phagocytosis or by active invasion by the parasite in a parasitophorous vacuole, as with *T. cruzi* entering mammalian macrophages under the control of the TGF-β signalling pathway (Bogdan & Röllinghoff, 1999).

Our knowledge of phagocytosis at the molecular level in invertebrate vectors is limited due to the small number of haemocytes in the circulation of many vector species, especially mosquitoes, sandflies and ticks. For example, in the mosquito *Ae. aegypti*, Christensen *et al.* (1989) estimated that at most there are 2000 haemocytes present. It is therefore often very difficult to study the phagocytosis process in vector haemocytes *in vitro* to identify the many complex components and their interactions. Scientists have therefore turned to the use of cell lines to study many aspects of *Drosophila* and mosquito immunity. Caution must, however, be exercised in the interpretation of results from these cell lines. Thus, it is known that phagocytosis and cellular encapsulation involve the co-operation between different haemocyte types (Anggraeni & Ratcliffe, 1991; Pech & Strand, 1996; Lavine & Strand, 2002) that must be mediated by the exchange of signals. Such cell type–cell type communication will be absent in cell cultures composed of just one cell type. Also, as pointed out by Boman

(2000), cell lines are artefacts and the adhesion of a bacterium to a host receptor results in signal transduction between different cell types and often between different organs. All of these components may thus have to act in concert to exert the appropriate immune response. Work with cell lines has, however, provided information on phagocytosis in vector species. Thus, using a haemocyte-like cell line and dsRNA knockout experiments, a thioester-containing protein from *An. gambiae* (aTEPI) has been detected that promotes the phagocytosis of Gram-negative (but not Gram-positive) bacteria (Levashina *et al.*, 2001). These authors also showed that *Anopheles* TEP-I has significant sequence homology to mammalian complement factor C3. Most importantly, they detected TEP-I in mosquito haemocytes by immunostaining, thus indicating the value of the use of the mosquito cell line. A homologue of the TEPIV gene of *Drosophila* is also upregulated in *Glossina* infected with trypanosomes (Lehane *et al.*, 2003). In addition, a cell line from *Aedes albopictus* has been used to detect a c-Jun amino-terminal kinase (JNK)-like protein which upon inhibition prevents the endocytosis and phagocytosis of West Nile virus (Mizutani *et al.*, 2003a, b).

(b) Encapsulation reactions to parasites and the role of PpO. Encapsulation has been described in reduviid bugs, such as *R. prolixus*, against trypanosomes (Tobie, 1970; Takle, 1988; Azambuja *et al.*, 1999), in mosquitoes in response to *Plasmodium* ökinetes, oocysts and filarial worms (Christensen, 1986; Collins *et al.*, 1986; Nigam *et al.*, 1996; Vernick, 1997; Beerntsen *et al.*, 2000), in the tick *Dermacentor variabilis* injected with bacteria (Ceraul *et al.*, 2002) and in *Bi. glabrata* snails to sporocysts of the trematode worms *S. mansoni*, *Echinostoma paraensei* and *Echinostoma lindoense* (Loker & Bayne, 1986; Lie *et al.*, 1987; Yoshino *et al.*, 2001). The encapsulation reaction, or nodule formation as it is often termed in response to large numbers of micro-organisms, described in these vectors often differs considerably.

In *R. prolixus*, following invasion of *T. rangeli* from the midgut into the haemocoel, the parasites rapidly multiply and only in later stages of infection do parasites and haemocytes clump together to form nodules (Takle, 1988). Mello *et al.* (1996) have shown that the haemolymph of *R. prolixus* contains a galactoside-binding lectin that enhances the formation of haemocyte–trypanosome clumps *in vitro*. In addition, there is very little evidence for melanization of these structures, despite reports of the ability of reduviid bugs to form melanized nodules in response to non-parasite, biological materials (Alvarenga *et al.*, 1990). Possibly, the trypanosomes suppress the PpO system that generates melanin (see Section C2a below) by releasing surface components to modulate the activation of the PpO via reduction of the protease activity of the fat body (Gomes *et al.*, 1999, 2003). The *T. rangeli* also multiply in the haemocytes, although it is unclear as to how they gain entry into these cells. It is important to emphasize that in reduviid bugs the extent of the immune response and subsequent host infection depend on the strain/species of both the parasite and host utilized (e.g. Mello *et al.*, 1996; Hecker *et al.*, 1990; Machado *et al.*, 2001; Whitten *et al.*, 2001; Carvalho-Moreira *et al.*, 2003).

The importance of the parasite and vector strain/species in determining the vector immune response is also seen in other interactions, including *Plasmodium* and filarial worm–mosquito systems (see also Section B8 above). Thus, in mosquitoes, a refractory, genetically selected strain of *An. gambiae*, L3-5, encapsulates New World and Asian strains of *P. falciparum*, but fails to efficiently encapsulate *P. falciparum* strains from Africa. Since *An. gambiae* is limited in range to Africa, then sympatric parasites have apparently evolved mechanisms to avoid the encapsulation response (Vernick, 1997). In addition, in the case of *An. gambiae* resistant to the primate parasite, *Plasmodium cynomolgi*, three quantitative trait loci (QTL) termed *Pen1*, *Pen2* and *Pen3* on chromosome 2 control melanotic encapsulation (Zheng *et al.*, 1997). The sequence of *Pen1* has revealed 48 genes potentially involved in the refractory mechanism (Thomasova *et al.*, 2002). In *An. albimanus*, refractory and permissive mosquitoes exist for filarial worms such as *Br. pahangi*, and Beerntsen *et al.* (2000) have discussed the genetics of mosquito vector competence. Thus, the construction of a genetic linkage map for *Ae. aegypti* has allowed the identification of chromosomal regions of genes influencing susceptibility to *P. gallinaceum*, *Dirofilaria immitis* and *Brugia* spp. The f^m gene on chromosome 1 controls susceptibility to *Brugia* spp., while the f^t gene on the same chromosome controls *D. immitis* susceptibility. The *pls* allele on chromosome 2 controls susceptibility to *P. gallinaceum* and another gene(s) in the same region modulates the susceptibility of *Ae. aegypti* to yellow fever virus (reviewed by Beerntsen *et al.*, 2000).

In mosquitoes, the actual encapsulation response towards *Plasmodium* öökinetes and oocysts, and to filarial worms has been described many times (e.g. reviewed by Christensen, 1986; Paskewitz *et al.*, 1988; Nayar *et al.*, 1995; Nigam *et al.*, 1996; Paskewitz & Christensen, 1996; Vernick, 1997; Beerntsen *et al.*, 2000) and differences are apparent between the responses mounted towards these parasites. Thus, with *Plasmodium* öökinetes and oocysts in refractory mosquitoes, once the parasite has penetrated the midgut epithelium and lies between the midgut basal membrane and the basal lamina, as early as 16 h after a blood meal, then granules of melanin begin to be deposited to form a melanotic capsule around the parasite. Haemocytes and cellular debris are only rarely observed in the vicinity of the capsule (Paskewitz *et al.*, 1988) so that the components of the PpO cascade, which are responsible for melanin production (Fig. 5), are usually regarded as originating from the haemolymph plasma (see Section C2a below). This contrasts with the melanotic encapsulation of filarial worms in

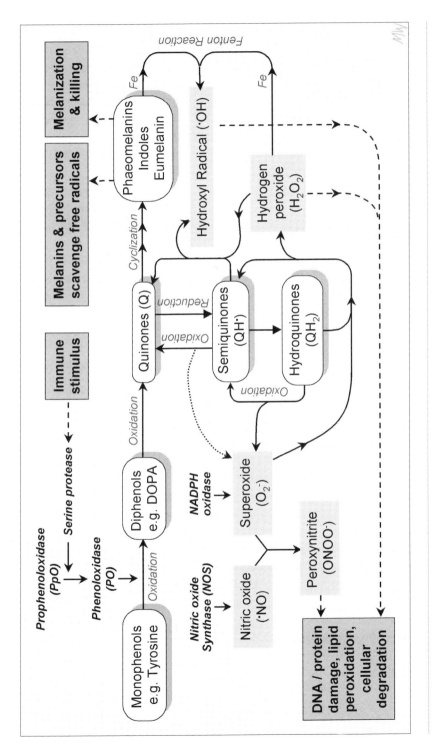

Fig. 5. Simplified outline of the PpO cascade and the interactions of quinone intermediates with reactive oxygen species.

which haemocytes appear to contribute to capsule formation, although the extent of haemocyte involvement seems to depend on the location of the parasite lodgement (reviewed by Christensen, 1986). It may also be significant for parasite cellular encapsulation that PpO activation can generate a peroxidase with cell adhesion properties (Johansson *et al.*, 1995). There seem to be several mechanisms by which encapsulated parasites are killed (Dimopoulos, 2003) and these may or may not involve the toxic molecules generated during activation of the PpO system (see Section C2a below). In addition, surface molecules on the parasites are likely to initiate the melanotic encapsulation response (Beerntsen *et al.*, 2000).

What exactly is the role of the PpO system in the encapsulation reaction of mosquitoes towards parasites? Superficially, melanin appears to form the structural basis of the mosquito capsule but when the encapsulation reaction is initiated *in vitro* in the presence of the PpO inhibitor diethyldithiocarbamate, transparent capsular material is still deposited on the surface of the larval filariae (microfilariae) (Chen & Laurence, 1987). The transparent material appears to consist of a protein–carbohydrate complex (see Vernick, 1997) that forms as a result of the cross-linking of amino groups of proteins by quinones generated from activation of PO (see Section C2a below and Fig. 5; Christensen & Nappi, 1988) by components of the parasite. Liu *et al.* (1997) also showed that *in vitro* the melanization of *D. immitis* microfilariae by the mosquito *Armigeres subalbatus* is a two-step process involving encasement by a transparent capsule followed by melanization of this capsular material. They suggested that PO forms part of the transparent capsule, which is subsequently melanized by means of interaction of PO with a tyrosine substrate also present in the capsule wall. Use of a double subgenomic Sindbis virus to knock out the copper-binding region of *Ar. subalbatus* PO gene resulted in almost complete inhibition of the melanotic encapsulation reaction towards *D. immitis* (Shiao *et al.*, 2001). Also relevant are the lectin-binding properties reported for the melanotic capsules of *Anopheles quadrimaculatus* around *Br. malayi*, indicating the presence of several glycoproteins (Nayar *et al.*, 1995). Since melanotic encapsulation has been reported to occur mainly in several mosquito species with high levels of haemagglutinins (Nayar & Knight, 1997) these molecules may bind to the carbohydrates in the wall of the microfilariae to activate the PpO system and produce the capsular material. Indeed, there is good evidence from non-vector insects that purified or recombinant haemolymph lectins can activate the PpO system from the same insect species (see Section C2a; Chen *et al.*, 1995; Yu *et al.*, 1999). Finally, the reason why blackflies, such as *Simulium ornatum*, are unable to melanize and encapsulate parasites and other pathogens is unknown, especially since their haemolymph has both PpO and serine protease activity (Hagen *et al.*, 1994, 1995, 1997).

Natural populations of *Bi. glabrata* snails are highly polymorphic with regard to their resistance to parasitic trematode worms such as *S. mansoni* (reviewed by Ataev & Coustau, 1999). Inbred laboratory strains of snails with variable resistance to *S. mansoni* and *Echinostoma caproni* have now been produced to study the role of host immunity in the resistance process (Lie *et al.*, 1987; Richards *et al.*, 1992; Langland & Morand, 1998). The resistance of adult snails to parasites appears to be controlled by a single gene locus (Richards *et al.*, 1992). Host resistance in *Bi. glabrata* is mainly characterized by haemocytic encapsulation followed by cytotoxic killing of the parasites. These processes probably involve soluble haemolymph proteins, some of which may act as non-self recognition molecules (see Section C2b below; Loker & Bayne, 2001; Yoshino *et al.*, 2001). Lie *et al.* (1987) have identified two types of encapsulation response (i.e. fast or slow) depending on the strain of *Bi. glabrata* infected with *S. mansoni* or *E. lindoense*, and also depending on the host species (Ataev & Coustau, 1999). In the fast reaction, thick capsules composed of six or more layers of amoebocytes are formed around the parasite sporocysts. In highly resistant snails, sporocysts may be encapsulated within several hours of host invasion. Subsequently, the amoebocytes phagocytose the damaged parasite integument which succumbs to cytotoxic factors from the plasma and the amoebocyte surface membrane (see Section C2c below; Bayne *et al.*, 1980; Lie *et al.*, 1987; Hahn *et al.*, 2000). In the slow reaction, *S. mansoni* sporocysts are rapidly surrounded by several layers of flattened amoebocytes. Once the parasite is completely surrounded, however, no additional cells attach, so that chemotactic attraction seems to be weaker than with the fast reacting snails (Lie *et al.*, 1987). Although the *Bi. glabrata* haemocytes have both serine protease and PO activities (Bahgat *et al.*, 2002), no mention has been made of the deposition of melanin in these capsules. What is particularly remarkable is the fact that in susceptible snails the host seems unaware of the existence of the worm so that the parasite may avoid detection in some way (Adema & Loker, 1997). Finally, the encapsulation reaction by the snail haemocytes depends on their ability to adhere to and spread on the outside of the parasite. Factors that may influence haemocyte adhesion include the excretory–secretory products from the parasite that inhibit *Bi. glabrata* haemocyte-spreading (Degaffe & Loker, 1998), a 36 kDa *Bi. glabrata* haemocyte polypeptide with sequence homology to the human adhesion molecules called selectins (Duclermortier *et al.*, 1999) and a β-integrin subunit that has been cloned from a *Bi. glabrata* cell line (Davids *et al.*, 1999).

2. Humoral factors interacting with parasites. These include components of the PpO cascade, PRRs, AMPs, and ROS and •NO. Many of these are undoubtedly produced and released by the haemocytes although these cells may not be apparent at the site of their activity. In addition, these humoral factors act in concert for the immune system to identify and kill the invading micro-organism/parasite.

(a) The PpO system. The role of PpO during encapsulation of parasites has already been described above (see Section C1b), although few details of the PpO cascade and its activation have been given. PpOs are the inactive forms of POs that are copper-containing enzymes found in numerous micro-organisms, plants and animals (Ashida & Yamazaki, 1990). For vector insects, the site of origin of the PpO complex has been reported to be both the haemocytes and the plasma (Nigam *et al.*, 1996), while PpO is also associated with the insect vector gut (Nayar *et al.*, 1989; see Section B8 above). In the tick *O. moubata*, however, PO activity is mainly detected in the plasma, although few details are available of PO regulation or function (Kadota *et al.*, 2002). The PpO gene products in *An. gambiae* lack predicted signal peptides, indicating that the PpOs are released by rupture of the haemocytes rather than by secretion from these cells (Christophides *et al.*, 2002).

PpO is activated by serine proteases to form PO which then catalyses the oxidation of monophenols, such as tyrosine, to diphenols, including DOPA (dihydroxy-phenolalanine). DOPA is then oxidized by PO to form reactive quinones and these cross-link non-enzymically to form the melanin in a melanotic capsule (Fig. 5; Paskewitz & Christensen, 1996; Söderhäll & Cerenius, 1998; Ashida & Brey, 1997; Beerntsen *et al.*, 2000). This is a much simplified outline of the PpO cascade since other enzymic pathways exist such as the decarboxylation of DOPA by DOPA decarboxylase and are detailed in Nappi & Vass (1993) and Paskewitz & Christensen (1996). The complexity of the PpO system is not surprising considering the presence of nine PpO genes in *An. gambiae*, ten or more non-digestive serine proteases with clip domains (repeated cysteine residues) and three serpins involved in regulation of the mosquito PpO cascade (Dimopoulos *et al.*, 2001; Christophides *et al.*, 2002). Important points to emphasize are:

(i) There appears to be no change in PpO gene transcription following immune challenge of *An. gambiae* or *An. stephensi* with *Plasmodium* spp. so that PpOs are constantly present in the haemolymph and are regulated at the post-transcriptional level (Cui *et al.*, 2000; Dimopoulos, 2003).

(ii) mRNAs for other enzymes associated with the PpO cascade are, however, upregulated following immune challenge. These include mRNAs for phenyl-alanine hydroxylase (PAH) which hydroxylates phenylalanine to produce tyrosine, the substrate for PO (Johnson *et al.*, 2003). Thus, following immune challenge in *Ae. aegypti* with *D. immitis*, the PAH transcript was shown to be upregulated in haemocytes. In addition, a PpO-activating serine protease (AgSp14D1) with a clip domain that closely resembles the PpO-activating enzymes previously described from insects (Jiang *et al.*, 1998; Lee *et al.*, 1998;

Satoh *et al.*, 1999) shows enhanced transcription following infection of *An. gambiae* with *Plasmodium* sp. (Chun *et al.*, 2000; see also Paskewitz *et al.*, 1999; Ribeiro, 2003).

(iii) The PpO system functions in non-self recognition of pathogens and parasites. The role of the PpO system in non-self recognition is generally accepted, although the actual recognition process is mediated by PRRs (see Section C2b below) that initiate PpO activation (Söderhäll & Thörnqvist, 1997). The binding of PRRs to so-called pathogen-associated molecular patterns (PAMPs) on pathogen and parasite surfaces [such as lipopoly-saccharides (LPS), peptidoglycans and β-1,3-glucans] activates proteolytic cascades involving serine proteases and serpins. These result not only in the activation of PpO but also in intracellular signalling pathways leading to AMP production (Fig. 5; Söderhäll & Cerenius, 1998; Dimopoulos, 2003). The activation of PpO by purified PRRs, i.e. haemolymph lectins in this case, has been described in non-vector insects such as the cockroach, *Blaberus discoidalis*, and the tobacco hornworm, *Manduca sexta* (Chen *et al.*, 1995; Yu *et al.*, 1999).

(iv) The activation of PpO leads to the production of killing factors. Some of the precursors of melanin, such as quinones, semiquinones, quinone methides and trihydroxyphenyls, generated during PpO activation are toxic to micro-organisms and parasites (Söderhäll *et al.*, 1996; Nappi & Vass, 1993). The activation of PpO with the production of melanotic capsules should not be considered in isolation since quinones can undergo redox-cycling and are potential sources of ROS, such as O_2^-, H_2O_2 and hydroxyl ($^\bullet OH$) radicals (Fig. 5; reviewed by Nappi & Vass, 1993, 1998a; Nappi *et al.*, 1995). Thus, foreign organisms sequestered in melanotic capsules would be killed by ROS that are confined in their activity to a pigmented cage which would protect the host from radical damage. In addition, ROS are capable of interacting with $^\bullet NO$ to generate highly toxic peroxynitrite ($ONOO^-$) molecules and all these molecules are generated by *Drosophila* haemocytes during melanotic encapsulation (Nappi & Vass, 1998b; Nappi *et al.*, 2000). Recently, Whitten *et al.* (2001), working on the immune system of *R. prolixus* in response to *T. rangeli*, found that the activation kinetics of PpO, ROS and $^\bullet NO$ were synchronized at identical stages of infectivity and demonstrated the potential effectiveness of interaction between the reactive products from these pathways (see Section C2a below). Finally, whatever the exact mechanism involved, recent work on high- and low-density populations of non-vector insects has shown that a definite correlation exists between increased 'gregarization' and enhanced resistance to pathogens as reflected in higher melanization and PpO enzyme activity in the haemo-lymph, midgut and cuticle (e.g. Wilson *et al.*, 2001). Correlations have also been reported between levels of haemolymph PO activity and susceptibility of tsetse flies, *Glossina* spp., to infection with *Trypanosoma brucei rhodesiense* (Nigam *et al.*, 1997) and the necessity of PpO activation for the production of antiviral activity (Ourth & Renis, 1993).

(b) PRRs, non-self recognition and intracellular signalling pathways. From work on the host defences of *Drosophila*, much progress has been made in understanding factors involved in the recognition of invading micro-organisms and subsequent activation of insect innate immunity. Numerous papers and reviews have been published in this area recently since many of the key mechanisms involved in innate immunity are conserved from insects to mammals (e.g. Khush & Lemaitre, 2000; Lowenberger, 2001; Boutros *et al.*, 2002; Gottar *et al.*, 2002; Hoffmann & Reichhart, 2002; Ligoxygakis *et al.*, 2002; Rämet *et al.*, 2002; Hultmark, 2003; Tzou *et al.*, 2002; Weber *et al.*, 2003a; Kurata, 2004). Progress has been accelerated by the availability of the *Drosophila* genomic sequence and of specific mutants for key players in the signalling pathways. Information gleaned about *Drosophila* innate immunity also proved invaluable when attention turned to examining the immune reactions of mosquitoes (e.g. Richman & Kafatos, 1995; Barillas-Mury *et al.*, 2000; Dimopoulos *et al.*, 2001, 2002; Shin *et al.*, 2002, 2003a; Dimopoulos, 2003; McGuinness *et al.*, 2003).

The innate immune system recognizes molecules present on the surfaces of invading pathogens such as LPS on Gram-negative bacteria, peptidoglycans on Gram-positive bacteria and β-1,3-glucans on both bacteria and fungi. These factors have been collectively termed PAMPs (Medzhitov & Janeway, 2002) and include combinations of sugars, proteins and nucleic acid components. Studies have now shown that molecular patterns may also be involved in parasite recognition (McGuinness *et al.*, 2003). There has been some discussion regarding the use of the term 'molecular patterns' since molecules of the microbe or parasite may interact with host receptors without the need for a high degree of multimerization (Beutler, 2003). Also important is the fact that the components of the Gram-positive cell wall, i.e. teichoic acids and peptidoglycans, differ significantly from one pathogen species to another, as illustrated with *Streptococcus pneumoniae* and *Staphylococcus aureus*, so that there is probably a diversity in the host-recognition components (Weber *et al.*, 2003b). Thus, teichoic acids of these two pathogens have different backbones and may have variable D-alanine substitutions and lipid linkages that can affect their immunostimulation properties (reviewed by Weber *et al.*, 2003b). The variability in structure of the peptidoglycan component of Gram-positive bacteria is even greater than for teichoic acid with the key inflammatory element being the glycan backbone (Weber *et al.*, 2003b).

Receptors binding PAMPs are the PRRs and these may be humoral and free in the blood circulation or associated with the effector cell surface or interior. Humoral PRRs in vectors include lectins in insects (Chen & Billingsley, 1999; Lavine & Strand, 2002; Lehane *et al.*, 2003), ticks (Grubhoffer & Jindrák, 1998) and the mollusc *Bi. glabrata* (Horák & van der Knaap, 1997). Other PRRs in insects are Gram-negative bacteria-binding proteins (GNBPs) (Lee *et al.*, 1996), peptidoglycan recognition proteins (PGRPs) (Yoshida *et al.*, 1996), β-glucan-binding proteins (β-GBPs, Ochiai & Ashida, 1988; Söderhäll *et al.*, 1988) and a phagocytosis-promoting, thioester-containing protein (TEP-I) (Levashina *et al.*, 2001; see Section C1a above). In addition, fibrinogen-related domain proteins (FREPs) are found in the haemolymph of the snail *Bi. glabrata* (Adema *et al.*, 1997) and in the mosquito *An. gambiae* (Christophides *et al.*, 2002), and are hypothesized to be lectins involved in non-self recognition. Other possible PRRs have been reported, including haemolin in *Drosophila* (Bettencourt *et al.*, 1997) as well as a calreticulin-like protein and an apolipophorin-III from the waxmoth *Galleria mellonella* (Choi *et al.*, 2002; Whitten *et al.*, 2004). Some of these molecules, including calreticulin (Asgari & Schmidt, 2003), the PGRPs (Lavine & Strand, 2002) and lectins (Yoshino *et al.*, 2001) are not only present free in the circulation, but may also be associated with the haemocyte surface (see also Ogden *et al.*, 2001).

PRRs on the haemocyte surface are less well characterized and include the Toll family of receptors of which there are 8 gene members in *Drosophila* (Imler & Hoffmann, 2000) and 13 in *Anopheles* (Christophides *et al.*, 2002). Toll-like members are also present in other vector species including the tsetse fly *G. palpalis palpalis* (Luo & Zheng, 2000). Additional cell-surface receptors may include the scavenger receptors (SRs), of which there are at least three family members in both *Drosophila* (Pearson *et al.*, 1995) and in *Anopheles* (Christophides *et al.*, 2002). SRs in vertebrates are known to bind to a range of ligands, including low-density lipoproteins (LDLs) and PAMPs such as LPS and peptidoglycans (Sankala *et al.*, 2002). SRs are also involved in the ingestion of *P. falciparum*-infected erythrocytes by macrophages and in the killing of *S. mansoni* via host LDLs (Sankala *et al.*, 2002). In addition, there is evidence for the presence of LDL-like receptors on the haemocytes of *Biomphalaria* that may serve as receptors for soluble *Schistosoma* components (Yoshino *et al.*, 2003). In *Drosophila*, one type of SR, termed dSR-CI, has also been shown to recognize both Gram-positive and Gram-negative bacteria, so that their role in non-self recognition appears to be conserved from insects to humans (Rämet *et al.*, 2001). A third group of possible PRRs include the integrins which are cell receptors that often bind to the extracellular matrix proteins fibronectin and collagen by means of the Arg-Gly-Asp (RGD) sequence. Mammalian integrins have also been reported to bind bacterial wall LPS and, although *Drosophila* cells migrate via integrin receptors (Martin-Bermudo *et al.*, 1999), our knowledge of their role in vector immunity is limited (reviewed by Yoshino *et al.*, 2001;

Lavine & Strand, 2002). The haemocytes of the noctuid moth *Pseudoplusia includens* have been shown to express integrin subunits and may function in recognition of both microbial and abiotic targets (Lavine & Strand, 2001). A fragment of a haemocyte β_1 integrin receptor has also been cloned from *Biomphalaria* haemocytes and may be involved in cell–cell interactions during the encapsulation of parasites (Yoshino *et al.*, 2001).

A major problem is to understand how all these PAMPs and PRRs come together to mediate the recognition of invading micro-organisms and parasites. Despite our vast knowledge of the *Drosophila* genome, Tzou *et al.* (2002) stated that the mechanism for detection of infectious microbes is largely unknown. It is known that in *An. gambiae* several of the seven PGRP genes and one GNBP gene are upregulated upon infection with bacteria or malaria and this results in activation of intracellular signalling pathways (Dimopoulos, 2003). In *Drosophila*, these signalling pathways include the Toll/Dif, Imd/Relish and JAK/STAT pathways (Fig. 6) which upon stimulation lead to immune enhancement that usually culminates in the induction of genes for the AMPs and activation of the cellular defences (e.g. Hoffmann & Reichart, 2002; Tzou *et al.*, 2002; Hultmark, 2003; Kurata, 2004). The Toll and Imd pathways both involve NF-κB-like transcription factors, including three Rel proteins, namely Dorsal, DIF and Relish (Fig. 6; Khush & Lemaitre, 2000). It is generally stated that the Toll pathway is largely activated by fungal infection and Gram-positive bacteria while the Imd (immune deficiency) pathway is mainly activated by Gram-negative bacteria. Less is known about the JAK/STAT pathway, although it is involved in developmental processes and regulation of cellular immunity (Zeidler *et al.*, 2000). Hultmark (2003) also emphasized that fungi, for example, can also induce cecropin, with antifungal activity, via a Relish-dependent pathway so that generalizations about activation of the signalling pathways by whole classes of micro-organisms should be made with care. In addition, *Drosophila* mutants that activate the Toll pathway result in overreactive haemocytes that form melanotic masses similar to cellular responses to parasites (Carton & Nappi, 2001). It is therefore possible that the main role of the Toll pathway is involvement in cellular reactivity (Hultmark, 2003).

The PGRPs from Gram-positive bacteria, of which there are 13 family members in *Drosophila* and 7 in *Anopheles* (Hultmark, 2003; Kurata, 2004), can be grouped in *Drosophila* into two classes: short PGRPs (PGRP-S) and long PGRPs (PGRP-L). PGRP-S is humoral and activates the Toll pathway in response to Gram-positive *Micrococcus luteus*, while PGRP-L is probably cell-associated and activates Relish in response not only to peptidoglycan, but also Gram-negative bacterial LPS (Hultmark, 2003; Kurata, 2004). The PGRPs then probably distinguish between different types of invading bacteria and activate appropriate immune reactions (Kurata, 2004). In

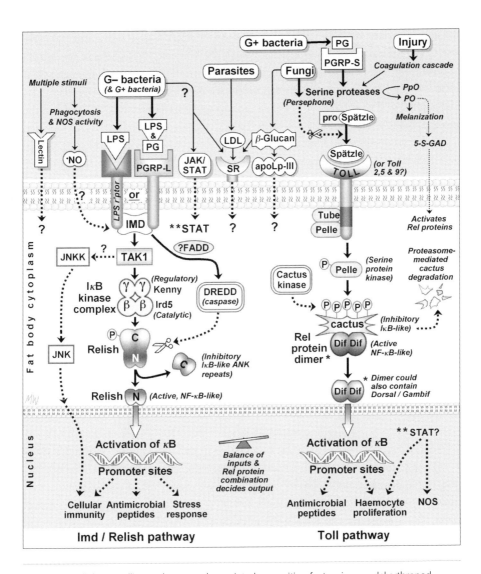

Fig. 6. Intracellular signalling pathways and associated recognition factors in a model arthropod vector. Upon immune stimulation, the Toll, Imd/Relish and JAK/STAT pathways lead to immune enhancement that usually culminates in the induction of genes for AMPs and activation of cellular defences. The Toll and Imd pathways both involve NF-κB-like transcription factors, including the Rel proteins, Dorsal, DIF and Relish. The role of several factors, including lectins, SRs and the JNK pathway, are not presently well defined. Abbreviations not in main text: ANK, ankyrin; DIF, dorsal-related immunity factor; DREDD, death-related ced-3/Nedd2-like protein; FADD, Fas-associated death domain protein; G–, Gram-negative; G+, Gram-positive; IκB, inhibitory protein κB; IMD, immune deficiency; JAK, Janus kinase; JNK, c-Jun amino-terminal kinase; JNKK, JNK kinase; PG, peptidoglycan; 5-S-GAD, N-β-alanyl-5-S-glutathionyl-3,4-dihydroxyphenylalanine; STAT, signal transducer and activator transcription; TAK1, TGF-β-activated kinase 1.

mosquitoes, however, knowledge of the pathways regulating the immune response is limited, although four Toll-like receptors (TLRs), namely AgToll, AgToll6, AgTrex and AgToll9, have been cloned and characterized from *An. gambiae*, but at present little is known about their functions (Luna *et al.*, 2002). Shin *et al.* (2002, 2003a) have also cloned and characterized the gene homologous to *Drosophila* Relish from the mosquito *Ae. aegypti* and have shown that three Relish transcripts can be induced by bacterial injection, but not by blood feeding (see also Barillas-Mury *et al.*, 1996). In Relish-mediated, immune-deficient, transgenic *Aedes*, microbial challenge results in extreme susceptibility to infection by Gram-negative bacteria due to reduced levels of AMP gene expression (Shin *et al.*, 2003a). There is also evidence of the STAT pathway in *An. gambiae* which may regulate inducible NOS (iNOS) (Barillas-Mury *et al.*, 1999, 2000).

The above illustrates the complexity of immune recognition and reactivity in invertebrates invaded by parasites and pathogens. There appear to be multiple putative PRRs that interact with an array of PAMPs to trigger a diversity of signalling pathways. This diversity probably results from the enormous range of microbial and macrobial organisms that are ever-threatening in the environment. Even two Gram-positive bacteria, such as *Streptococcus pneumoniae* and *Staphylococcus aureus*, have significant differences in their cell-wall compositions and this variation in PAMPs may impose the need for diversity in the PRRs (see above and Weber *et al.*, 2003b). We do know from work with *Drosophila* that Toll receptors are involved in the recognition pathways but, in contrast to mammalian TLRs, Toll does not apparently act directly as a PAMP sensor, but is activated by the Spatzle ligand. The latter itself is activated following proteolytic cleavage by serine proteases following infection, but Spatzle does not appear to function directly as a PRR (Hultmark, 2003; Kurata, 2004) so that both Toll and Spatzle lie on a pathway downstream from the actual recognition event (Janeway & Medzhitov, 1999). In addition, serine proteases are involved in activation of PpO and *An. gambiae* contains the clip-domain serine proteases, AgSP14D1, AgSP14D2 and AgSP14A with sequences characteristic of PpO-activating enzymes (Dimopoulos, 2003). With the current emphasis on the Toll receptor system, the following observations also need integrating into the chain of recognition events:

(i) There are frequent accounts of the elevation of haemolymph lectins following parasitization of invertebrate vectors. For example, in the vector snail *Bi. glabrata*, following exposure to the trematode worm *E. paraensei*, titres of haemolymph lectin increase to 8–16 times those of unexposed control snails. In addition, infection increased the heterogeneity of plasma lectins and provoked the appearance of unique specificities not present in uninfected controls (Loker *et al.*, 1994). A similar reaction was recorded in blackflies

(*Simulium ornatum*) inoculated with the filarial nematode *Onchocerca lienalis*, with significantly elevated levels of haemolymph lectin detected in comparison with the controls (Ham *et al.*, 1988a). Most interesting was also the change in the specificity of the haemolymph lectin from D(+)-mannose-binding by 5 days post-infection to *N*-acetyl-D-galactosamine by 7 days (Smail & Ham, 1988; Ham *et al.*, 1988b). Significantly, this switch in lectin specificity was correlated with changes in the surface glycans of *Onchocerca lienalis* with the exposed carbohydrates correlating with the haemolymph lectin specificity. Ham *et al.* (1995) suggested that the blackfly lectins may function in parasite recognition and immune stimulation. In snail–trematode interactions, similar modulations of snail haemolymph lectins and surface carbohydrates of parasites have often been recorded and believed to be related to recognition processes by circulating haemocytes and/or avoidance mechanisms by parasites (reviewed by Horák & van der Knaap, 1997; Yoshino *et al.*, 2001; Castillo & Yoshino, 2002).

(ii) Haemolymph lectins of non-vector insects and molluscs have been purified and shown to act as opsonins for a range of test particles, including fungal spores and erythrocytes (Renwrantz & Stahmer, 1983; Pendland *et al.*, 1988; Drif & Brehélin, 1989; Richards & Ratcliffe, 1990). In some of these studies, opsonization was inhibited by simple sugars and the phagocytosis experiments were conducted in the absence of plasma factors, i.e. without other humoral PRRs.

(iii) The haemolymph of several non-vector insects contains multiple lectins, with different carbohydrate specificities (e.g. Kubo *et al.*, 1993; Chen *et al.*, 1993; Wilson *et al.*, 1999) which in the cockroach, *Blaberus discoidalis*, function selectively in the absence of plasma to opsonize Gram-positive and Gram-negative bacteria, and yeast. In addition, one of these molecules (BDL1), which is a C-type lectin with mannose/glucose specificity, also activates the PpO cascade, probably through activation of the plasma serine proteases (Chen *et al.*, 1995).

(iv) Insect apolipophorin-III (apoLp-III), a lipid transport apolipoprotein present in the haemolymph, has recently been shown to play a crucial role in innate immunity. apoLp-III acts as a PRR, binding to bacterial LPS and lipoteichoic acid as well as fungal β-1,3-glucan (Dettloff & Wiesner, 1999; Halwani *et al.*, 2000; Whitten *et al.*, 2004). It also enhances phagocytosis (Wiesner *et al.*, 1997), increases cellular encapsulation and modulates cell attachment and spreading (Whitten *et al.*, 2004).

(v) Other relevant observations, mentioned previously above, include the possible role of SRs and integrins as possible PRRs. Interestingly, the mammalian apolipoprotein recognition motif for SRs is an amphipathic α-helix (Williams *et al.*, 2000) and this motif is a highly conserved structural feature of insect apoLp-III.

Clearly, the Toll complex is not the complete answer to recognition and signalling by the vector immune system following microbial and parasitic invasion. The correct approach, therefore, must be that described by Underhill (2003) for understanding the Toll-like receptors (TLRs) of the mammalian immune system. Thus, full activation of the immune response involves a networking system in which receptors recognize different ligands independently to activate signalling cascades that interact within the cell. Any microbe may be recognized by multiple PRRs and this is regarded as necessary to prevent pathogens from avoiding recognition (Underhill, 2003). It is already known that different classes of mammalian TLRs, such as TLR1, 2 and 6, form heterodimers, each component contributing to a different aspect of recognition (Beutler *et al.*, 2003). In invertebrate vectors, therefore, the same invading organism may be recognized, for example, by the Toll receptors, the haemolymph or membrane-associated lectins, the integrins and/or the SRs. It is possible that one or more of these PRRs may recruit other proteins to form a recognition/signalling complex as postulated for the TLR4 signalling assembly of mammals (Underhill, 2003). The co-operation of a C-type lectin (dectin-1) with TLR2 in eliciting immune response to zymosan has already been reported in mammals (Gantner *et al.*, 2003). Likewise, the two LPS receptors, CD11/CD18 integrins and CD14 (the latter transferring the ligand to TLR2 in LPS recognition) share a common signalling pathway into the cell (Ingalls *et al.*, 1998). We can only speculate about the details of such co-operation during immune recognition in invertebrates given our present knowledge.

(c) Killing factors such as AMPs, ROS and nitrogen intermediates such as •NO. Many parasites spend one or more phases of their life cycle in the haemocoel of their vector hosts. Examples of these parasites in arthropods include *Plasmodium* spp., *T. rangeli*, filarial worms such as *Wuchereria bancrofti*, *Br. pahangi* and *Onchocerca* spp., probably *Borrelia burgdorferi* and numerous arboviruses, while the trematode worms *S. mansoni* and *E. paraensei* invade the haemocoel of the snail *Bi. glabrata*. In the haemolymph, these parasites are exposed to a range of toxic molecules, including the AMPs, low-molecular-mass, possibly non-peptidergic antimicrobial compounds, ROS, •NO, products of PpO activation (see Sections B8, C1b and C2a above) and transferrins. It is therefore not perhaps surprising that over 80 % of *Plasmodium* sporozoites are rapidly removed from the haemolymph before invasion of the salivary glands (Dimopoulos, 2003). Details of the vector AMPs are given in Section B6 (above) as they are often assumed to form important components of the arthropod gut barrier to infection. The distribution of AMPs among vector arthropods is summarized in Table 2. In mosquitoes and tsetse flies, exposure of the insect to parasites in the midgut produces upregulation of immune genes to produce a systemic response in the haemolymph and other tissues (see Section B6 above). In many arthropod vectors, the fat body, haemocytes, midgut and salivary glands are involved in synthesizing the AMPs (e.g. Dimopoulos *et al.*, 2001; Blandin *et al.*, 2002). In gastropods, such as *Biomphalaria*, AMPs have not so far been reported, with these vectors relying on ROS and ˙NO for their protection (see below). There is conflicting evidence about the importance of the arthropod AMPs in defence against *Plasmodium* and filarial worms. Experiments supporting a possible role for these molecules in killing of *Plasmodium* are described above (see Section B6). Some of the experiments conducted were suboptimal and used, for example, AMPs from non-vector insects (e.g. Gwadz *et al.*, 1989). More recently, new powerful tools based on reverse genetics with RNA interference (RNAi) and transgenic techniques have been developed for testing the antiparasite activity of AMPs in immune-deficient mosquitoes (Blandin *et al.*, 2002; Shin *et al.*, 2003b). In these studies, Blandin *et al.* (2002) reported that in *An. gambiae*, knockout of the defensin gene had no effect on *Plasmodium* development, while V. Kokoza and others (unpublished, reported in Shin *et al.*, 2003b) found that overexpressing the defensin gene in *Ae. aegypti* produced 65–70 % inhibition of *P. gallinaceum* oocyst growth. Also contradictory are the results of experiments by Lowenberger *et al.* (1996, 1999) working with *Ae. aegypti* and *An. gambiae* compared with Bartholomay *et al.* (2003) using *Culex pipiens pipiens*. Both groups pre-injected bacteria to induce immune-stimulation of AMPs and then challenged with *Plasmodium* or filarial worms. Immune-stimulation reduced the incidence of both *Br. pahangi* and *P. gallinaceum* in *Ae. aegypti* and of *P. berghei* in *An. gambiae* while, in contrast, *W. bancrofti* developed normally in bacteria-inoculated *C. pipiens pipiens*. Such contradictions may result from differential efficiency of the immune systems of the various mosquito species and/or differential resistance or avoidance mechanisms of the parasites used (Bartholomay *et al.*, 2003).

Other interesting antimicrobial compounds have recently been described from non-vector insects and actively kill bacteria and fungi, but have yet to be tested against parasites. These include two cationic peptides isolated from infected blowflies (*Calliphora vicinia*), referred to as alloferons (Chernysh *et al.*, 2002), and a group of low-molecular-mass (less than 1 kDa) constitutively expressed compounds isolated from different insect species. The latter are particularly interesting as they represent an entirely separate but neglected group of antimicrobial compounds and include β-alanyl-tyrosine and 5-*S*-GAD from Dipterans (Leem *et al.*, 1996; Chernysh *et al.*, 2002). 5-*S*-GAD is a DOPA derivative that activates the Rel transcription factor, as well as having broad-spectrum antibacterial activity brought about by reactions with bacterial terminal oxidases that generate H_2O_2 (Leem *et al.*, 1996; Natori *et al.*, 1999).

The final group of killing factors are the ROS and $^{\bullet}$NO which are also important components of the gut barrier (see Section B7 above) and are generated by PpO activation in capsules (see Section C1b above). They are briefly mentioned again here as they undoubtedly have important roles in killing parasites in the haemocoel and because considerable details of their functions in the haemolymph of molluscan hosts are available.

In *Drosophila*, during the melanotic encapsulation of the eggs of the intrahaemocoelic parasitoid *Leptopilina boulardi* (Nappi *et al.*, 1995, 2000; Nappi & Vass, 1998a, b; Vass & Nappi, 1998, 2000, 2001), cytotoxic molecules such as O_2^- and H_2O_2 have been detected in immune-reactive plasmatocytes involved in the encapsulation response. The precise nature of the killing process, however, has not been fully elucidated since *Drosophila* strains with elevated levels of O_2^- or H_2O_2 failed to kill encapsulated parasitoids (Nappi *et al.*, 1995). The enhanced production of $^{\bullet}$NO in immune-reactive *Drosophila*, however, was also recorded during the destruction of the parasitoid eggs (Nappi *et al.*, 2000). Thus, $^{\bullet}$NO, O_2^- and H_2O_2 probably interact during cytotoxicity not only together to produce highly toxic $ONOO^-$ and $^{\bullet}$OH radicals (see details in Miles *et al.*, 1996), but also with components of the PpO system. $^{\bullet}$NO and ROS also probably act as a signalling molecule during insect immunity, activating NF-κB and other pathways (Nappi *et al.*, 2000). Recently, the importance of $^{\bullet}$NO as a signalling molecule in the innate immune response of *Drosophila* towards Gram-negative bacteria via the Imd pathway has been described (Foley & O'Farrell, 2003). In addition, the role of SRs in $^{\bullet}$NO regulation has been reported (Matsuno *et al.*, 1997).

In vector arthropods, much less is known about ROS and $^{\bullet}$NO activities in the haemolymph and the roles of these molecules in host defence. Luckhart *et al.* (1998) reported that the induction of NOS expression by an infective blood meal in the mosquito *An. stephensi* was not confined to the midgut, but also occurred in the carcass and haemolymph (see Section B7 above). Tahar *et al.* (2002) also showed that in *An. gambiae* upregulation of immune-related genes depended upon the species of the parasite infecting the mosquito. Thus, in the model *P. berghei–An. gambiae* system, gametocytes and asexual stages triggered NOS expression in midguts. In contrast, in the natural *P. falciparum–An. gambiae* system this did not occur, although in carcasses NOS expression was increased. Absence of or low NOS expression was interpreted as possible evidence of repression by *P. falciparum* (Tahar *et al.*, 2002). This serves as a timely warning regarding the interpretation of results obtained from model parasite–host associations. Evidence for the cytotoxic roles for ROS and ˙NO towards *Plasmodium* in the haemolymph of mosquitoes is limited to work under way at the Kafatos laboratory. This group has shown that in a refractory (R) strain of *An. gambiae* to malaria, insects have elevated levels of constitutively expressed redox-related genes compared with the susceptible (S) strain (Kafatos, 2001).

Other potential cytotoxic factors in mosquitoes have also been identified and include peroxidase (Zhao *et al.*, 2001), as well as factors involved in iron metabolism such as transferrins and iron-regulatory proteins (IRPs) (Yoshiga *et al.*, 1997; Beerntsen *et al.*, 2000; Dimopoulos *et al.*, 2001; Fallon & Sun, 2001; Lowenberger 2001; Zhang *et al.*, 2002). Transferrin is transcriptionally upregulated in *Ae. aegypti* and *Ar. subalbatus*, melanotically encapsulating the filarial worms *D. immitis* and *Br. pahangi*, respectively (Beerntsen *et al.*, 2000). Yoshiga *et al.* (1997) hypothesize that transferrin acts like vertebrate lactoferrin and sequesters iron from invading organisms to prevent infection. Two mosquito IRPs have been cloned from *Ae. aegypti* and *An. gambiae* and probably control the translational regulation of transferrin as well as interacting with ˙NO (Dimopoulos *et al.*, 2001; Zhang *et al.*, 2002).

ROS or ˙NO radicals in the haemolymph may also be important killing factors in other arthropod vectors, including blackflies (Hagen & Kläger, 2001), tsetse flies (Lehane *et al.*, 2003), ticks (Pereira *et al.*, 2001) and the reduviid bug *R. prolixus* (Whitten *et al.*, 2001). In the tsetse fly *G. morsitans morsitans*, 18 putative antioxidant genes have been identified in an expressed sequence tag project, thus indicating the importance of tolerance of oxidative stress in this vector (Lehane *et al.*, 2003). The only detailed study of the role of ROS and ˙NO in the haemolymph stage of a parasite involves *R. prolixus* infected with different strains of *T. rangeli* (Whitten *et al.*, 2001). With the *T. rangeli* H14 strain, which fails to multiply in the haemocoel and invade the salivary glands, the parasites stimulated high levels of O_2^- and PpO as well as increased nitrites and nitrates, indicative of ˙NO, 24 h after infection. In contrast, the *T. rangeli* Choachi strain, which multiplies rapidly in the haemolymph and invades the salivary glands, stimulated significantly lower levels of these radicals. Injection of NADPH oxidase or NOS inhibitors caused significantly higher *Rhodnius* mortalities following infection with either strain. Thus abrogation of these radicals severely compromised the vector immune response.

Finally, it has been known for many years that the haemocytes of various snail species, including the vector *Bi. glabrata* for the trematode parasite *S. mansoni*, produce both ROS and ˙NO to kill invading sporocysts (e.g. Dikkeboom *et al.*, 1988; Adema *et al.*, 1991, 1994; Connors & Yoshino, 1990; Connors *et al.*, 1995; LoVerde, 1998; Hahn *et al.*, 2000, 2001a, b). The sporocysts are usually encapsulated by the haemocytes and then killed by destruction of the tegument (Loker *et al.*, 1982). Utilizing *in vitro* cytotoxicity killing assays and haemocytes from two (or more sometimes) different strains of *Biomphalaria*, namely the M-line and 13-16-R1, which are susceptible and resistant, respectively, to the PR-1 strain of *S. mansoni*, it has been shown that both $^{\bullet}$NO and H_2O_2 are involved in sporocyst killing by the resistant strain (Hahn *et al.*, 2001a, b). Neither toxic $ONOO^-$ nor $^{\bullet}OH$ radicals seemed to be involved (Hahn *et al.*, 2001b). In addition, the injection of human interleukin-1 into both susceptible (M-line) and resistant (13-16-R1) snails increased levels of both phagocytosis and O_2^- production at 24 h exposure to schistosomes, so that the ROS modulation, as with vertebrates, is probably mediated by cytokines (Connors *et al.*, 1995). Also interesting is the report of the possible role of lectin-like receptors on the *Biomphalaria* haemocytes which may be involved in both parasite recognition and killing by stimulating ROS production (Hahn *et al.*, 2000).

D. The salivary glands

Many parasites and pathogens make their way from the haemolymph or the gut to the salivary glands to complete their life cycles (Fig. 1). Thus, *Plasmodium* spp., *T. rangeli*, African trypanosomes such as *Trypanosoma brucei gambiense* and *T. brucei rhodesiense*, as well as numerous arthropod-borne bacteria and viruses invade the salivary glands which are then utilized for their transmission to the vertebrate host during feeding by the vector (for ticks see Munderloh & Kurtti, 1995). Much research on the parasite–salivary gland interaction has concentrated on the attachment of the parasites to the salivary glands and the role of parasite surface/salivary gland molecules in this process (e.g. Sidjanski *et al.*, 1997; Beier, 1998; Barreau *et al.*, 1999; Sultan, 1999; Wengelnik *et al.*, 1999; Menard, 2000; Ghosh *et al.*, 2001; Basseri *et al.*, 2002; Kariu *et al.*, 2002; Matuschewski *et al.*, 2002; Tewari *et al.*, 2002). One of the few reviews of the parasite–salivary gland interaction is given by James (2003). One reason for interest in the attachment of parasites to the salivary glands is that the molecules involved in this process are seen as potential targets for the production of transmission-blocking vaccines and transgenic mosquitoes (Beier, 1998; Brennan *et al.*, 2000; James, 2003). Studies using lectins and monoclonal antibodies have been made to probe the mosquito salivary gland surface and showed that the distal–lateral and/or medial lobes are particularly interactive with these probes and are also targets for parasite invasion (Perrone *et al.*, 1986; Pimenta *et al.*, 1994; Andrews *et al.*, 1997; Barreau *et al.*, 1995, 1999). The actual attachment and invasion of the salivary glands is believed to be mediated by receptor–ligand interactions (e.g. Sidjanski *et al.*, 1997). The specificity of this interaction is essential to ensure that the parasite interacts with the salivary gland as opposed to other organs in the body. Possible parasite ligands for recognition of the salivary gland include the circumsporozoite protein (CSP), thrombospondin-related adhesive protein (TRAP) and apical membrane antigen/erythrocyte-binding-like protein (MAEBL) (Fig. 3) (e.g. Sidjanski *et al.*, 1997; Sultan, 1999; Wengelnik *et al.*, 1999; Tewari *et al.*, 2002). Other molecules involved occur on the surface of the salivary glands and include a salivary gland and midgut peptide 1 (SM1) which is composed of 12 amino acids and binds specifically to the midgut and distal lobe of the salivary gland (Ghosh *et al.*, 2001). Other studies with the salivary glands of *Glossina* (Okolo *et al.*, 1990) and *Rhodnius* (Basseri *et al.*, 2002) have also shown the importance of carbohydrate–ligand interactions in attachment and infectivity of these organs by trypanosomes. Once within the salivary glands the parasites are exposed to an array of factors (Arca *et al.*, 1999; Haddow *et al.*, 2002; Sant Anna *et al.*, 2002), many involved in feeding (e.g. Nussenzveig *et al.*, 1995), but some with potential antiparasitic properties (Valenzuela *et al.*, 2002). For example, defensin, β-GBPs and NOS gene markers are all activated in the salivary glands of mosquitoes at the time of sporozoite invasion (Dimopoulos *et al.*, 1998), while the production of H_2O_2 by mosquito salivary homogenates may be significant in exerting pressure on the parasites to develop antioxidant defence for survival in the vertebrate host (Ribeiro, 1996). Increased ˙NO activity is also seen in *Rhodnius* salivary gland homogenates infected with *T. rangeli* (M. M. A. Whitten, G. A. Schaub, F. Sun, A. J. Nappi & N. A. Ratcliffe, unpublished). Finally, there is also evidence for the presence of the PpO activating system in the salivary glands of *Ae. aegypti* with three serine protease sequences detected, one of which is very similar to the *Bombyx* PpO-activating enzyme (Valenzuela *et al.*, 2002).

FINAL COMMENTS

The above review indicates the complexity of the vector immune system and indicates areas of the parasite–vector interaction that require attention. Rapid progress has been made in understanding the mosquito innate immune system due to the application of molecular techniques, genetic manipulation of the vector and the utilization of the knowledge available from work on *Drosophila*. Even so, the more we learn about the vector immune system the more questions seem to be created. There are many aspects of the parasite–vector interaction that require urgent attention and have been grossly neglected, including:

1. the gut barrier, which is extremely complex, but poorly understood;
2. the salivary glands in which the roles of many of the multiple proteins produced are unknown;
3. the possible role of eicosanoids and endocrine secretions in vector immunity has barely been addressed (see, however, Garcia *et al.*, 2004);
4. the role of vector AMPs and ROS/˙NO in cell signalling (see Salzet, 2002) and parasite killing;
5. the influence of charge in the parasite–vector interaction (Takle & Lackie, 1987; Richman & Kafatos, 1995; Schmidt *et al.*, 1998).

Finally, the above review only represents half of the story as it fails to address, in any detail, the response of the parasite in terms of avoidance of the vector immune response. This response includes the switching of parasite surface components about which our knowledge is rapidly expanding (Fig. 3) (e.g. Schaub *et al.*, 1989; Ferguson *et al.*, 1994; Sacks & Kamhawi, 2001).

ACKNOWLEDGEMENTS

We gratefully acknowledge the National Science Foundation, USA (grant ref. IBN-0095421) for financial support for M. M. A. W., and Dr A. J. Nappi, Department of Animal Health and Biomedical Sciences, University of Wisconsin-Madison, whose encouragement and help allowed the authors to continue their research during the production of this review.

REFERENCES

Adema, C. M. & Loker, E. S. (1997). Specificity and immunobiology of larval digenean–snail interactions. In Advances in Trematode Biology, pp. 209–222. Edited by B. Fried & T. K. Graczyk. Boca Raton, FL: CRC Press.

Adema, C. M., van der Knaap, W. P. W. & Sminia, T. (1991). Molluscan hemocyte-mediated cytotoxicity: the role of reactive oxygen intermediates. *Rev Aquat Sci* **4**, 201–223.

Adema, C. M., van Deutekom-Mulder, E. C., van der Knaap, W. P. W. & Sminia, T. (1994). Schistosomicidal activities of *Lymnaea stagnalis* haemocytes: the role of oxygen radicals. *Parasitology* **109**, 479–485.

Adema, C. M., Hertel, L. A., Miller, R. D. & Loker, E. S. (1997). A family of fibrinogen-related proteins that precipitates parasite-derived molecules is produced by an invertebrate after infection. *Proc Natl Acad Sci U S A* **94**, 8691–8696.

Alvarenga, N. J., Bronfen, E., Alvarenga, R. J. & Barracco, M. A. (1990). Triatomine's hemocytes and granuloma formation around biological and non-biological material. *Mem Inst Oswaldo Cruz* **85**, 377–379.

Andreadis, T. G. & Hall, D. W. (1976). *Neoaplectana carpocaspsae*: encapsulation in *Aedes aegypti* and changes in host hemocytes and hemolymph proteins. *Exp Parasitol* **39**, 252–261.

Andrews, L., Laughinghouse, A. & Sina, B. J. (1997). Lecting binding characterstics of male and female salivary gland proteins of *Anopheles gambiae*: identification and characterization of female specific glycoproteins. *Insect Biochem Mol Biol* **27**, 159–166.

Anggraeni, T. & Ratcliffe, N. A. (1991). Studies on cell–cell co-operation during phago-cytosis by purified haemocyte populations of the wax moth, *Galleria mellonella*. *J Insect Physiol* **37**, 453–460.

Arai, M., Billker, O., Morris, H. R., Panico, M., Delcroix, M., Dixon, D., Ley, S. V. & Sinden, R. E. (2001). Both mosquito-derived xanthurenic acid and a host blood-derived factor regulate gametogenesis of *Plasmodium* in the midgut of the mosquito. *Mol Biochem Parasitol* **116**, 17–24.

Arca, B., Lombardo, F., de Lara Capurro, M., della Torre, A., Dimopoulos, G., James, A. A. & Coluzzi, M. (1999). Trapping cDNAs encoding secreted proteins from the salivary glands of the malaria vector *Anopheles gambiae*. *Proc Natl Acad Sci U S A* **96**, 1516–1521.

Arrighi, R. B. G. & Hurd, H. (2002). The role of *Plasmodium berhei* öokinete proteins in binding to basal lamina components and transformation into oocysts. *Int J Parasitol* **32**, 91–98.

Asano, T. & Ashida, M. (2001). Cuticular pro-phenoloxidase of the silkworm, *Bombyx mori*. *J Biol Chem* **276**, 11100–11112.

Asgari, S. & Schmidt, O. (2003). Is cell surface calreticulin involved in phagocytosis by insect hemocytes? *J Insect Physiol* **49**, 545–550.

Ashida, M. & Brey, P. T. (1997). Recent advances in research on the insect prophenoloxidase cascade. In *Molecular Mechanisms of Immune Responses in Insects*, pp. 135–172. Edited by P. T. Brey & D. Hultmark. London: Chapman & Hall.

Ashida, M. & Yamazaki, H. I. (1990). Biochemistry of the phenoloxidase system in insects: with special reference to its activation. In *Moulting and Metamorphosis*, pp. 239–265. Edited by E. Ohnishi & H. Ishizaki). Tokyo/Berlin: Japan Science Society Press/Springer.

Ataev, G. L. & Coustau, C. (1999). Cellular response to *Echinostoma caproni* infection in *Biomphalaria glabrata* strains selected for susceptibility/resistance. *Dev Comp Immunol* **23**, 187–198.

Azambuja, P., Freitas, C. C. & Garcia, E. S. (1986). Evidence and partial characterization of an inducible antibacterial factor in the haemolymph of *Rhodnius prolixus*. *J Insect Physiol* **32**, 807–812.

Azambuja, P., Mello, C. B., D'Escoffier, L. N. & Garcia, E. S. (1989). *In vitro* cytotoxicity of *Rhodnius prolixus* haemolytic factor and mellitin towards different trypanosomatids. *Braz J Med Biol Res* **22**, 597–599.

Azambuja, P., Garcia, E. S. & Ratcliffe, N. A. (1991). Aspects of classification of hemiptera hemocytes from six triatomine species. *Mem Inst Oswaldo Cruz* **86**, 1–10.

Azambuja, P., Feder, D., Mello, C. B., Gomes, S. A. O. & Garcia, E. S. (1999). Immunity in *Rhodnius prolixus*: trypanosomatid–vector interactions. *Mem Inst Oswaldo Cruz Suppl 1* **94**, 219–222.

Bahgat, M., Doenhoff, M., Kirschfink, M. & Ruppel, A. (2002). Serine protease and phenoloxidase activities in hemocytes of *Biomphalaria glabrata* snails with varying susceptibility to infection with the parasite *Schistosoma mansoni*. *Parasitol Res* **88**, 489–494.

Barillas-Mury, C., Charlesworth, A., Gross, I., Richman, A., Hoffmann, J. A. & Kafatos, F. C. (1996). Immune factor Gambi fl, a new rel family member from the human malaria vector, *Anopheles gambiae*. *EMBO J* **15**, 4691–4701.

Barillas-Mury, C., Han, Y. S., Seeley, D. & Kafatos, F. C. (1999). *Anopheles gambiae* Ag-STAT, a new insect member of the STAT family, is activated in response to bacterial infection. *EMBO J* **18**, 959–967.

Barillas-Mury, C., Wizel, B. & Han, Y. S. (2000). Mosquito immune responses and malaria transmission: lessons from insect model systems and implications for vertebrate innate immunity and vaccine development. *Insect Biochem Mol Biol* **30**, 429–442.

Barracco, M. A., De Olivera, R. & Schlemper, B., Jr (1987). The hemocytes of *Panstrongylus megistus* (Hemiptera: Reduviidae). *Mem Inst Oswaldo Cruz* **82**, 431–438.

Barracco, M. A., Steil, A. A. & Gargioni, R. (1993). Morphological characterization of the hemocyte of the pulmonate snail *Biomphalaria tenagophila*. *Mem Inst Oswaldo Cruz* **88**, 73–83.

Barreau, C., Touray, M., Pimenta, P. F., Miller, L. H. & Vernick, K. D. (1995). *Plasmodium gallinaeceum*: sporozoite invasion of *Aedes aegypti* salivary glands is inhibited by anti-gland antibodies and by lectins. *Exp Parasitol* **81**, 332–343.

Barreau, C., Conrad, J., Fischer, E., Lujan, H. D. & Vernick, K. D. (1999). Identification of surface molecules on salivary glands of the mosquito, *Aedes aegypti*, by a panel of monoclonal antibodies. *Insect Biochem Mol Biol* **29**, 515–526.

Bartholomay, L. C., Farid, H. A., Ramzy, R. M. & Christensen, B. M. (2003). *Culex pipiens pipiens*: characterization of immune peptides and the influence of immune activation on development of *Wuchereria bancrofti*. *Mol Biochem Parasitol* **130**, 43–50.

Basseri, H. R., Tew, I. F. & Ratcliffe, N. A. (2002). Identification and distribution of carbohydrate moieties on the salivary glands of *Rhodnius prolixus* and their possible involvement in attachment/invasion by *Trypanosoma rangeli*. *Exp Parasitol* **100**, 226–234.

Bayne, C. J., Buckley, P. M. & De Wan, P. C. (1980). *Schistosoma mansoni* cytotoxicity of hemocytes from susceptible snail hosts for sporocysts in plasma from resistant *Biomphalaria glabrata*. *Exp Parasitol* **50**, 409–416.

Beard, C. B., Cordon-Rosales, C. & Durvasula, R. V. (2002). Bacterial symbionts of the triatominae and their potential use in control of Chagas disease transmission. *Annu Rev Entomol* **47**, 123–141.

Beerntsen, B. T., James, A. A. & Christensen, B. M. (2000). Genetics of mosquito vector competence. *Microbiol Mol Biol Rev* **64**, 115–137.

Beier, J. C. (1998). Malaria parasite development in mosquitoes. *Annu Rev Entomol* **43**, 519–543.

Bettencourt, R., Lanz-Mendoza, H., Roxstrom-Lindquist, K. & Faye, I. (1997). Cell adhesion properties of hemolin, an insect immune protein of the Ig superfamily. *Eur J Biochem* **250**, 630–637.

Beutler, B. (2003). Not 'molecular patterns' but molecules. *Immunity* **19**, 155–158.

Beutler, B., Hoebe, K., Du, X. & Ulevitch, R. J. (2003). How we detect microbes and respond to them: the Toll-like receptors and their transducers. *J Leukocyte Biol* **74**, 479–485.

Billingsley, P. F. & Rudin, W. (1992). The role of the mosquito peritrophic membrane in bloodmeal digestion and infectivity of *Plasmodium* species. *J Parasitol* **78**, 430–440.

Billingsley, P. F. & Sinden, R. E. (1997). Determinants of malaria–mosquito specificity. *Parasitol Today* **13**, 297–301.

Blandin, S., Moita, L. F., Köcher, T., Wilm, M., Kafatos, F. C. & Levashina, E. A. (2002). Reverse genetics in the mosquito *Anopheles gambiae*: targeted disruption of the *Defensin* gene. *EMBO Rep* **3**, 852–856.

Bogdan, C. & Röllinghoff, M. (1999). How do protozoan parasites survive inside macrophages? *Parasitol Today* **15**, 22–28.

Boman, H. G. (2000). Innate immunity and the normal microflora. *Immunol Rev* **173**, 5–16.

Boulanger, N., Brun, R., Ehret-Sabatier, L., Kunz, C. & Bulet, P. (2002a). Immuno-peptides in the defense reactions of *Glossina morsitans* to bacterial and *Trypanosoma brucei brucei* infections. *Insect Biochem Mol Biol* **32**, 369–375.

Boulanger, N., Munks, R. J. L., Hamilton, J. V., Vovelle, F., Brun, R., Lehane, M. J. & Bulet, P. (2002b). Epithelial innate immunity. A novel antimicrobial peptide with antiparasitic activity in the blood sucking insect *Stomoxys calcitrans*. *J Biol Chem* **277**, 49921–49926.

Boutros, M., Agaisse, H. & Perrimon, N. (2002). Sequential activation of signalling pathways during innate immune responses in *Drosphilia*. *Dev Cell* **3**, 1–20.

Brazil, R. P., Brazil, B. G. & Alves, J. C. M. (1996). The hemocytes of *Lutzomyia longipalpis* (Diptera: Psychodidae). *Mem Inst Oswaldo Cruz Suppl* **Nov. 1996**, Abstract 121.

Brennan, J. D., Kent, M., Dhar, R., Fujioka, H. A. & Kumar, N. (2000). *Anopheles gambiae* salivary gland proteins as putative targets for blocking transmission of malaria parasites. *Proc Natl Acad Sci U S A* **97**, 13859–13864.

Capron, A., Capron, M. & Riveau, G. (2002). Vaccine development against schistosomiasis from concepts to clinical trials. *Br Med Bull* **62**, 139–148.

Carlton, J. M., Angiuoli, S. V., Suh, B. B. & 41 other authors (2002). Genome sequence and comparative analysis of the model rodent malaria parasite *Plasmodium yoelii yoelii*. *Nature* **419**, 512–519.

Carton, Y. & Nappi, A. J. (2001). Immunogenetic aspects of the cellular immune response of *Drosophila* against parasitoids. *Immunogenetics*. **52**, 157–164.

Carvalho-Moreira, C. J., Spata, M. C. D., Coura, J. R., Garcia, E. S., Azambuja, P., Gonzalez, M. S. & Mello, C. B. (2003). *In vivo* and *in vitro* metacyclogenesis tests of two strains of *Trypanosoma cruzi* in the triatomine vectors *Triatoma pseudomaculata* and *Rhodnius neglectus*: short/long-term and comparative study. *Exp Parasitol* **103**, 102–111.

Castillo, M. G. & Yoshino, T. P. (2002). Carbohydrate inhibition of *Biomphalaria glabrata* embryonic (Bge) cell adhesion to primary sporocysts of *Schistosoma mansoni*. *Parasitology* **125**, 513–525.

Ceraul, S. M., Sonenshine, D. E. & Hynes, W. L. (2002). Resistance of the tick *Dermacentor variabilis* (Acari: Ixodidae) following challenge with the bacterium *Escherichia coli* (Enterobacteriales: Enterobacteriaceae). *J Med Entomol* **39**, 376–383.

Chalk, R., Townson, H., Natori, S., Desmond, H. & Ham, P. J. (1994). Purification of an insect defensin from the mosquito, *Aedes aegypti*. *Insect Biochem Mol Biol* **24**, 403–410.

Chen, C. & Billingsley, P. F. (1999). Detection and characterization of a mannan-binding lectin from the mosquito, *Anopheles stephensi* (liston). *Eur J Biochem* **263**, 360–366.

Chen, C., Ratcliffe, N. A. & Rowley, A. F. (1993). Detection, isolation and characterization of multiple lectins from the hemolymph of the cockroach, *Blaberus discoidalis*. *Biochem J* **294**, 181–190.

Chen, C., Durrant, H. J., Newton, R. P. & Ratcliffe, N. A. (1995). A study of novel lectins and their involvement in the activation of the prophenoloxidase system in *Blaberus discoidalis*. *Biochem J* **310**, 23–31.

Chen, C. C. & Laurence, B. R. (1987). Selection of a strain of *Anopheles quadrimaculatus* with high filarial encapsulation rates. *J Parasitol* **73**, 418–419.

Chernysh, S., Cociancich, S., Briand, J. P., Hetru, C. & Bulet, P. (1996). The inducible antibacterial peptides of the hemipteran insect *Palomena prasina*: identification of a unique family of proline-rich peptides and of a novel insect defensin. *J Insect Physiol* **42**, 81–89.

Chernysh, S., Kim, S. I., Bekker, G., Pleskach, V. A., Filatova, N. A., Anikin, V. B., Platonov, V. G. & Bulet, P. (2002). Antiviral and antitumor peptides from insects. *Proc Natl Acad Sci U S A* **99**, 12628–12632.

Cho, W.-L., Fu, T.-F., Chiou, J.-Y. & Chen, C.-C. (1997). Molecular characterization of a defensin gene from the mosquito *Aedes aegypti*. *Insect Biochem Mol Biol* **27**, 351–358.

Choi, J. Y., Whitten, M. M. A., Cho, M. Y., Lee, K. Y., Kim, M. S., Ratcliffe, N. A. & Lee, B. L. (2002). Calreticulin enriched as an early-stage encapsulation protein in wax moth *Galleria mellonella* larvae. *Dev Comp Immunol* **26**, 335–343.

Christensen, B. M. (1986). Immune mechanisms and mosquito–filarial worm relationships. *Symp Zool Soc Lond* **56**, 145–160.

Christensen, B. M. & Nappi, A. J. (1988). Immune responses of arthropods. *ISI Atlas Sci Anim Plant Sci* **S**, 15–19.

Christensen, B. M., Huff, B. M., Miranpuri, G. S., Harris, K. L. & Christensen, L. A. (1989). Hemocyte population changes during the immune response of *Aedes aegypti* to inoculated microfilariae of *Dirofilaria immitis*. *J Parasitol*, **75**, 119–123.

Christophides, G. K., Zdobnov, E., Barillas-Mury, C. & 32 other authors (2002). Immunity-related genes and gene families in *Anopheles gambiae*. *Science* **298**, 159–165.

Chun, J., McMaster, J., Han, Y.-S., Schwartz, A. & Paskewitz, S. M. (2000). Two-dimensional gel analysis of haemolymph proteins from *Plasmodium*-melanizing and -non-melanizing strains of *Anopheles gambiae*. *Insect Mol Biol* **9**, 39–45.

Cociancich, S., Bulet, P., Hetru, C. & Hoffman, J. A. (1994). The inducible antibacterial peptides of insects. *Parasitol Today* **10**, 132–139.

Collins, F. H. & Paskewitz, S. M. (1995). Malaria: current and future prospects for control. *Annu Rev Entomol* **40**, 195–219.

Collins, F. H., Sakai, R. K., Vernick, K. D., Paskewitz, S., Seeley, D. C., Miller, L. H., Collins, W. E., Campbell, C. C. & Gwadz, R. W. (1986). Genetic selection of a *Plasmodium*-refractory strain of the malaria vector, *Anopheles gambiae*. *Science* **234**, 607–610.

Connors, V. A. & Yoshino, T. P. (1990). *In vitro* effect of larval *Schistosoma mansoni* excretory–secretory products on phagocytosis-stimulated superoxide production in hemocytes from *Biomphalaria glabrata*. *J Parasitol* **76**, 895–902.

Connors, V. A., de Buron, I. & Granath, W. O., Jr (1995). *Schistosoma mansoni*: interleukin-1 increases phagocytosis and superoxide production by hemocytes and decreases output of cercariae in schistosome-susceptible *Biomphalaria glabrata*. *Exp Parasitol* **80**, 139–148.

Crampton, A. & Luckhart, S. (2001). The role of As60A, a TGF-β homolog, in *Anopheles stephensi* innate immunity and defense against *Plasmodium* infection. *Infect Genet Evol* **1**, 131–141.

Cui, L., Luckhart, S. & Rosenberg, R. (2000). Molecular characterization of a pro-phenoloxidase cDNA from the malaria mosquito *Anopheles stephensi*. *Insect Mol Biol* **9**, 127–137.

Cunningham, A. C. (2002). Parasitic adaptive mechanisms in infection by *Leishmania*. *Exp Mol Pathol* **72**, 132–141.

Dale, C. & Welburn, S. C. (2001). The endosymbionts of tsetse flies: manipulating host–parasite interactions. *Int J Parasitol* **31**, 628–631.

Davids, B. J., Xiao-Jun, W. & Yoshino, T. P. (1999). Cloning of a β integrin subunit cDNA from an embryonic cell line derived from the freshwater mollusc, *Biomphalaria glabrata*. *Gene* **228**, 213–223.

Degaffe, G. & Loker, E. S. (1998). Susceptibility of *Biomphalaria glabrata* to infection with *Echinostoma paraensei*: correlation with the effect of parasite secretory–excretory products on host hemocyte spreading. *J Invertebr Pathol* **71**, 64–72.

Dettloff, M. & Wiesner, A. (1999). Immune stimulation by lipid-bound apolipophorin III. In *Techniques in Insect Immunology*, pp. 243–251. Edited by A. Wiesner, G. B. Dunphy, V. J. Marmaras, I. Morishima, M. Sugumaran & M. Yamakawa. Fair Haven: SOS Publications.

Dikkeboom, R., van der Knaap, W. P. W., van den Bovenkamp, W., Tijnagel, J. M. G. H. & Bayne, C. J. (1988). The production of toxic oxygen metabolites by hemocytes of different snail species. *Dev Comp Immunol* **12**, 509–520.

Dimopoulos, G. (2003). Insect immunity and its implication in mosquito–malaria interactions. *Cell Microbiol* **5**, 3–14.

Dimopoulos, G., Richman, A., Müller, H.-M. & Kafatos, F. C. (1997). Molecular immune responses of the mosquito *Anopheles gambiae* to bacteria and malaria parasites. *Proc Natl Acad Sci U S A* **94**, 11508–11513.

Dimopoulos, G., Seeley, D., Wolf, A. & Kafatos, F. C. (1998). Malaria infection of the mosquito *Anopheles gambiae* activates immune-responsive genes during critical transition stages of the parasite life cycle. *EMBO J* **17**, 6115–6123.

Dimopoulos, G., Casavant, T. L., Chang, S. & 9 other authors (2000). *Anopheles gambiae* pilot gene discovery project: identification of mosquito innate immunity genes from expressed sequence tags generated from immune-competent cell lines. *Proc Natl Acad Sci U S A* **97**, 6619–6624.

Dimopoulos, G., Müller, H.-M., Levashina, E. A. & Kafatos, F. C. (2001). Innate immune defense against malaria infection in the mosquito. *Curr Opin Immunol* **13**, 79–88.

Dimopoulos, G., Christophides, G. K., Meister, S., Schultz, J., White, K. P., Barillas-Mury, C. & Kafatos, F. C. (2002). Genome expression analysis of *Anopheles gambiae*: responses to injury, bacterial challenge, and malaria infection. *Proc Natl Acad Sci U S A* **99**, 8814–8819.

Drif, L. & Brehélin, M. (1983). The circulating hemocytes of *Culex pipiens* and *Aedes aegypti*: cytology, histochemistry, hemograms and functions. *Dev Comp Immunol* **7**, 687–690.

Drif, L. & Brehélin, M. (1989). Agglutinin mediated immune recognition in *Locusta migratoria* (Insecta). *J Insect Physiol* **35**, 729–736.

Duclermortier, P., Lardans, V., Serra, E., Trottein, F. & Dissous, C. (1999). *Biomphalaria glabrata* embryonic cells express a protein with a domain homologous to the lectin domain of mammalian selectins. *Parasitol Res* **85**, 481–486.

East, J., Molyneux, D. H. & Hillen, N. (1980). Hemocytes of *Glossina*. *Ann Trop Med Parasitol* **74**, 471–474.

Eggleston, P., Lu, W. & Zhao, Y. (2000). Genomic organization and immune regulation of the defensin gene from the mosquito, *Anopheles gambiae*. *Insect Mol Biol* **9**, 481–490.

Ehara, T., Kitajima, S., Kanzawa, N., Tamiya, T. & Tsuchiya, T. (2002). Antimicrobial action of achacin is mediated by L-amino acid oxidase activity. *FEBS Lett* **531**, 509–512.

Eichler, S. & Schaub, G. A. (2002). Development of symbionts in triatomine bugs and the effects of infections with trypanosomatids. *Exp Parasitol* **100**, 17–27.

Fallon, A. M. & Sun, D. (2001). Exploration of mosquito immunity using cells in culture. *Insect Biochem Mol Biol* **31**, 263–278.

Feder, D., Mello, C. B., Garcia, E. S. & Azambuja, P. (1997). Immune responses in *Rhodnius prolixus*: influence of nutrition and ecdysone. *J Insect Physiol* **43**, 513–519.

Ferguson, M. A. J., Brimacombe, J. S., Cottaz, S. & 10 other authors (1994). Glycosyl-phosphatidylinositol molecules of the parasite and the host. *Parasitology* **108**, S45–S54.

Fogaça, A. C., da Silva, P. I., Jr, Miranda, M. T. M., Bianchi, A. G., Miranda, A., Ribolla, P. E. M. & Daffre, S. (1999). Antimicrobial activity of a bovine hemoglobin fragment in the tick *Boophilus microplus*. *J Biol Chem* **274**, 25330–25334.

Foley, E. & O'Farrell, P. H. (2003). Nitric oxide contributes to induction of innate immune responses to gram-negative bacteria in *Drosophila*. *Genes Dev* **17**, 115–125.

Fraidenraich, D., Pena, C., Isola, E. L. & 7 other authors (1993). Stimulation of *Trypanosoma cruzi* adenyl cyclase by an α^D-globin fragment from *Triatoma* hindgut: effect on differentiation of epimastigote to trypomastigote forms. *Proc Natl Acad Sci U S A* **90**, 10140–10144.

Gantner, B. N., Simmons, R. M., Canavera, S. J., Akira, S. & Underhill, D. M. (2003). Collaborative induction of inflammatory responses by dectin-1 and Toll-like receptor 2. *J Exp Med* **197**, 1107–1119.

Gao, Y. & Fallon, A. M. (2000). Immune activation upregulates lysozyme gene expression in *Aedes aegypti* mosquito cell culture. *Insect Mol Biol* **9**, 553–558.

Gao, Y., Hernandez, V. P. & Fallon, A. M. (1999). Immunity proteins from mosquito cell lines include three defensin A isoforms from *Aedes aegypti* and a defensin D from *Aedes albopictus*. *Insect Mol Biol* **8**, 311–318.

Garcia, E. S. & Azambuja, P. (1991). Development and interactions of *Trypanosoma cruzi* within the insect vector. *Parasitol Today* **7**, 240–244.

Garcia, E. S., Gonzalez, M. S., Azambuja, P., Baralle, F. E., Frainderaich, D., Torres, H. N. & Flawia, M. M. (1995). Induction of *Trypanosoma cruzi* metacyclogenesis in the hematophagous insect vector by hemoglobin and peptides carrying alpha-D-globin sequences. *Exp Parasitol* **81**, 255–261.

Garcia, E. S., Machado, E. M. M. & Azambuja, P. (2004). Effects of eicosanoid biosynthesis inhibitors on the prophenoloxidase-activating system and micro-aggregation reactions in the hemolymph of *Rhodnius prolixus* infected with *Trypanosoma rangeli*. *J Insect Physiol* (in press).

Gardner, M.J., Hall, N., Fung, E. & 42 other authors (2002). Genome sequence of the human malaria parasite *Plasmodium falciparum*. *Nature* **419**, 498–511.

Gare, D. C., Piertney, S. B. & Billingsley, P. F. (2003). *Anopheles gambiae* collagen IV genes: cloning, phylogeny and midgut expression associates with blood feeding and *Plasmodium* infection. *Int J Parasitol* **33**, 681–690.

Ghosh, A. K., Ribolla, P. E. M. & Jacobs-Lorena, M. (2001). Targeting *Plasmodium* ligands on mosquito salivary glands and midgut with a phage display peptide library. *Proc Natl Acad Sci U S A* **98**, 13278–13281.

Ghosh, A. K., Moreira, L. A. & Jacobs-Lorena, M. (2002). *Plasmodium*–mosquito interactions, phage display libraries and transgenic mosquitoes impaired for malaria transmission. *Insect Biochem Mol Biol* **32**, 1325–1331.

Gomes, S. A. O., Feder, D., Thomas, N. E. S., Garcia, E. S. & Azambuja, P. (1999). *Rhodnius prolixus* infected with *Trypanosoma rangeli*: *in vivo* and *in vitro* experiments. *J Invertebr Pathol* **73**, 289–293.

Gomes, S. A. O., Feder, D., Garcia, E. S. & Azambuja, P. (2003). Suppression of the prophenoloxidase system in *Rhodnius prolixus* orally infected with *Trypanosoma rangeli*. *J Insect Physiol* **49**, 829–837.

Gottar, M., Gobert, V., Michel, T., Belvin, M., Duyk, G., Hoffmann, J. A., Ferrandon, D. & Royet, J. (2002). The *Drosophila* immune response against Gram-negative bacteria is mediated by a peptidoglycan recognition protein. *Nature* **416**, 640–643.

Gratz, N. G. (1999). Emerging and resurging vector-borne diseases. *Annu Rev Entomol* **44**, 51–75.

Gross, J., Muller, C., Vilcinskas, A. & Hilker, M. (1998). Antimicrobial activity of exocrine glandular secretions, hemolymph, and larval regurgitate of the mustard leaf beetle *Phaedon cochleariae*. *J Invertebr Pathol* **72**, 296–303.

Grubhoffer, L. & Jindrák, L. (1998). Lectins and tick–pathogen interactions: a minireview. *Folia Parasitol* **45**, 9–13.

Grubhoffer, L., Hypša, V. & Volf, P. (1997). Lectins (hemagglutinins) in the gut of the important disease vectors. *Parasite* **4**, 203–216.

Gwadz, R. W., Kaslow, D., Lee, J.-Y., Maloy, L., Zasloff, M. & Miller, L. H. (1989). Effects of magainins and cecropins on the sporogonic development of the malaria parasites in mosquitoes. *Infect Immun* **57**, 2628–2633.

Haddow, J. D., Poulis, B., Haines, L. R., Gooding, R. H., Aksoy, S., Pearson, T. W. (2002). Identification of major soluble salivary gland proteins in teneral G*lossina morsitans morsitans*. *Insect Biochem Mol Biol* **32**, 1045–1053.

Hagen, H. E. & Kläger, S. L. (2001). Integrin-like RGD-dependent cell adhesion mechanism is involved in the rapid killing of *Onchocerca* microfilariae during early infection of *Simulium damnosum* s.l. *Parasitology* **122**, 433–438.

Hagen, H. E., Grunewald, J. & Ham, P. J. (1994). Induction of the prophenoloxidase-activating system of *Simulium* (Diptera: Simuliidae) following *Onchocerca* (Nematoda: Filarioidea) infection. *Parasitology* **109**, 649–655.

Hagen, H. E., Kläger, S. L., Chan, V., Sakanari, J. A., McKerrow, J. H. & Ham, P. J. (1995). *Simulium damnosum* s.l. identification of inducible serine proteases following an *Onchocerca* infection by differential display reverse transcription PCR. *Exp Parasitol* **81**, 249–254.

Hagen, H. E., Kläger, S. L., McKerrow, J. H. & Ham, P. J. (1997). *Simulium damnosum* s.l. isolation and identification of prophenoloxidase following an infection with *Onchocerca* spp. using targeted differential display. *Exp Parasitol* **86**, 213–218.

Hahn, U. K., Bender, R. C. & Bayne, C. J. (2000). Production of reactive oxygen species by hemocytes of *Biomphalaria glabrata*: carbohydrate-specific stimulation. *Dev Comp Immunol* **24**, 531–541.

Hahn, U. K., Bender, R. C. & Bayne, C. J. (2001a). Involvement of nitric oxide in killing of *Schistosoma mansoni* sporocysts by hemocytes from resistant *Biomphalaria glabrata*. *J Parasitol* **87**, 778–785.

Hahn, U. K., Bender, R. C. & Bayne, C. J. (2001b). Killing of *Schistosoma mansoni* sporocysts by hemocytes from resistant *Biomphalaria glabrata*: role of reactive oxygen species. *J Parasitol* **87**, 292–299.

Halwani, A. E., Niven, D. F. & Dunphy, G. B. (2000). Apolipophorin-III and the interactions of lipoteichoic acids with the immediate immune responses of *Galleria mellonella*. *J Invertebr Pathol* **76**, 233–241.

Ham, P. J., Zulu, M. B. & Zahedi, M. B. (1988a). *In vitro* haemagglutination and attenuation of microfilarial motility by haemolymph from individual black-flies (*Simulium ornatum*) infected with *Onchocerca lienalis*. *Med Vet Entomol* **2**, 7–18.

Ham, P. J., Smail, A. J. & Groeger, B. K. (1988b). Surface carbohydrate changes on *Onchocerea lienalis* larvae as they develop from microfilariae to the infective third-stage in *Simulium ornatum*. *J Helminthol* **62**, 195–205.

Ham, P. J., Phiri, J. S. & Nolan, G. P. (1991). Effect of N-acetyl-D-glucosamine on the migration of *Brugia pahangi* microfilariae into the haemocoel of *Aedes aegypti*. *Med Vet Entomol* **5**, 485–493.

Ham, P. J., Hagen, H. E., Baxter, A. J. & Grunewald, J. (1995). Mechanisms of resistance to *Onchocerca* infection in blackflies. *Parasitol Today* **11**, 63–67.

Han, Y. S. & Barillas-Mury, C. (2002). Implications of time bomb model of ookinete invasion of midgut cells. *Insect Biochem Mol Biol* **32**, 1311–1316.

Han, Y. S., Chun, J., Schwartz, A., Nelson, S. & Paskewitz, S. M. (1999). Induction of mosquito hemolymph proteins in response to immune challenge and wounding. *Dev Comp Immunol* **23**, 553–562.

Han, Y. S., Thompson, J., Kafatos, F. C. & Barillas-Mury, C. (2000). Molecular interactions between *Anopheles stephensi* midgut cells and *Plasmodium berghei*: the time bomb theory of öokinete invasion of mosquitoes. *EMBO J* **19**, 6030–6040.

Hao, Z., Kasumba, I., Lehane, M. J., Gibson, W. C., Kwon, J. & Aksoy, S. (2001). Tsetse immune responses and trypanosome transmission: implications for the development of tsetse-based strategies to reduce trypanosomiasis. *Proc Natl Acad Sci U S A* **98**, 12648–12653.

Hao, Z., Kasumba, I., Aksoy, S. (2003). Proventriculus (cardia) plays a crucial role in immunity in tsetse fly (Diptera: Glossinidiae). *Insect Biochem Mol Biol* **33**, 1155–1164.

Harbige, L. S. (2003). Fatty acids, the immune response, and autoimmunity: a question of n-6 essentiality and the balance between n-6 and n-3. *Lipids* **38**, 323–341.

Hecker, H., Schwarzenbach, M. & Rudin, W. (1990). Development and interactions of *Trypanosoma rangeli* in and with the reduviid bug *Rhodnius prolixus*. *Parasitol Res* **76**, 311–318.

Hemingway, J. & Ranson, H. (2000). Insecticide resistance in insect vectors of human disease. *Annu Rev Entomol* **45**, 371–391.

Hernández, S., Lanz, H., Rodriguez, M. H., Torres, J. A., Martinez-Paloma, A. & Tsutsumi, V. (1999). Morphological and cytochemical characterization of female *Anopheles albimanus* (Diptera: Culicidae) hemocytes. *J Med Entomol* **36**, 426–434.

Hernández-Martinez, S., Lanz, H., Rodriguez, M. H., González-Ceron, L. & Tsutsumi, V. (2002). Cellular-mediated reactions to foreign organisms inoculated into the hemocoel of *Anopheles albimanus* (Diptera: Culicidae). *J Med Entomol* **39**, 61–69.

Hillyer, J. F. & Christensen, B. M. (2002). Characterization of hemocytes from the yellow fever mosquito, *Aedes aegypti*. *Histochem Cell Biol* **117**, 431–440.

Hillyer, J. F., Schmidt, S. L. & Christensen, B. M. (2003). Rapid phagocytosis and melanization of bacteria and *Plasmodium* sporozoites by hemocytes of the mosquito *Aedes aegypti*. *J Parasitol* **89**, 62–69.

Hoffmann, J. A. & Hetru, C. (1992). Insect defensins – inducible antibacterial peptides. *Immunol Today* **13**, 411–415.

Hoffmann, J. A. & Reichhart, J. M. (2002). *Drosophila* innate immunity: an evolutionary perspective. *Nat Immunol* **3**, 121–126.

Holt, R. A., Subramanian, G. M., Halpern, A. & 120 other authors (2002). The genome sequence of the malaria mosquito *Anopheles gambiae*. *Science* **298**, 129–149.

Horák, P. & van der Knaap, W. P. W. (1997). Lectins in snail–trematode immune interactions: a review. *Folia Parasitol* **44**, 161–172.

Hultmark, D. (2003). *Drosophila* immunity: paths and patterns. *Curr Opin Immunol* **15**, 12–19.

Hurd, H. (2003). Manipulation of medically important insect vectors by their parasites. *Annu Rev Entomol* **48**, 141–161.

Hypša, V. & Grubhoffer, L. (1997). Two hemocyte populations in *Triatoma infestans*: ultrastructural and lectin-binding characterization. *Folia Parasitol* **44**, 62–70.

Imbuga, M. O., Osir, E. O. & Labongo, V. L. (1992). Inhibitory effect of *Trypanosoma brucei brucei* on *Glossina morsitans* midgut trypsin *in vitro*. *Parasitol Res* **78**, 273–276.

Imler, J.-L. & Hoffmann, J. A. (2000). Signaling mechanisms in the antimicrobial host defense of *Drosophila*. *Curr Opin Microbiol* **3**, 16–22.

Ingalls, R. R., Monks, B. G., Savedra, R., Jr, Christ, W. J., Delude, R. L., Medvedev, A. E., Espevik, T. & Golenbock, D. T. (1998). CD11/CD18 and CD14 share a common lipid A signalling pathway. *J Immunol* **161**, 5413–5420.

Inoue, N., Hanada, K., Tsuji, N., Igarashi, I., Nagasawa, H., Mikami, T. & Fujisaki, K. (2001). Characterization of phagocytic hemocytes in *Ornithodoros moubata* (Acari: Ixodidae). *J Med Entomol* **38**, 514–519.

James, A. A. (2003). Blocking malaria parasite invasion of mosquito salivary glands. *J Exp Biol* **206**, 3817–3821.

Janeway, C. A., Jr & Medzhitov, R. (1999). Innate immunity: lipoproteins take their toll on the host. *Curr Biol* **9**, R879–R882.

Jiang, H., Wang, Y. & Kanost, M. R. (1998). Pro-phenol oxidase activating proteinase from an insect, *Manduca sexta*: a bacteria-inducible protein similar to *Drosophila* easter. *Proc Natl Acad Sci U S A* **95**, 12220–12225.

Johansson, M. W., Lind, M. I., Holmblad, T., Thörnqvist, P. O. & Söderhäll, K. (1995). Peroxinectin, a novel cell adhesion protein from crayfish blood. *Biochem Biophys Res Commun* **216**, 1079–1087.

Johns, R., Sonenshine, D. E. & Hynes, W. L. (2001). Identification of a defensin from the hemolymph of the American dog tick, *Dermacentor variabilis*. *Insect Biochem Mol Biol* **31**, 857–865.

Johnson, J. K., Rocheleau, T. A., Hillyer, J. F., Chen, C. C., Li, J. & Christensen, B. M. (2003). A potential role for phenylalanine hydroxylase in mosquito immune responses. *Insect Biochem Mol Biol* **33**, 345–354.

Jones, J. C. (1965). The hemocytes of *Rhodnius prolixus* Stal. *Biol Bull* **129**, 282–294.

Kaaya, G. P. & Ratcliffe, N. A. (1982). Comparative study of hemocytes and associated cells of some medically important dipterans. *J Morphol* **173**, 351–365.

Kaaya, G. P., Ratcliffe, N. A. & Alemu, P. (1986). Cellular and humoral defenses of *Glossina* (Diptera: Glossinidae): reactions against bacteria, trypanosomes and experimental implants. *J Med Entomol* **23**, 30–43.

Kadota, K., Satoh, E., Ochiai, M. & 7 other authors (2002). Existence of phenol oxidase in the argasid tick *Ornithodoros moubata*. *Parasitol Res* **88**, 781–784.

Kafatos, F. C. (2001). *Malaria: the Mosquito's View. EMBL Research Report*, pp. 1–7.

Kang, D. W., Romans, P. & Lee, J. Y. (1996). Analysis of a lysozyme gene from the malaria vector mosquito, *Anopheles gambiae*. *Gene* **174**, 239–244.

Kariu, T., Uda, M., Yano, K. & Chinzei, Y. (2002). MAEBL is essential for malarial sporozoite infection of the mosquito salivary gland. *J Exp Med* **195**, 1317–1323.

Khush, R. S. & Lemaitre, B. (2000). Genes that fight infection: what the *Drosophila* genome says about animal immunity. *Trends Genet* **16**, 442–449.

Kläger, S. L., Watson, A., Achukwi, D., Hultmark, D. & Hagen, H.-E. (2002). Humoral immune response of *Simulium damnosum* s.l. following filarial and bacterial infections. *Parasitology* **125**, 359–366.

Kleiboeker, S. B., Burrage, T. G., Scoles, G. A., Fish, D. & Rock, D. L. (1998). African swine fever virus infection in the argasid host, *Ornithodoros porcinus porcinus*. *J Virol* **72**, 1711–1724.

Kokoza, V., Ahmed, A., Cho, W. L., Jasinskiene, N., James, A. A. & Raikhel, A. S. (2000). Engineering blood meal-activated systemic immunity in the yellow fever mosquito, *Aedes aegypti*. *Proc Natl Acad Sci U S A* **97**, 9144–9149.

Kollien, A. H. & Billingsley, P. F. (2002). Differential display of mRNAs associated with blood feeding in the midgut of the bloodsucking bug, *Triatoma infestans*. *Parasitol Res* **88**, 1026–1033.

Kollien, A. H., Fechner, S., Waniek, P. J. & Schaub, G. A. (2003). Isolation and characterization of a cDNA encoding for a lysozyme from the gut of the reduviid bug *Triatoma infestans*. *Arch Insect Biochem Physiol* **53**, 134–145.

Kopacek, P., Vogt, R., Jindrak, L., Weise, C. & Safarik, I. (1999). Purification and characterization of the lysozyme from the gut of the soft tick *Ornithodoros moubata*. *Insect Biochem Mol Biol* **29**, 989–997.

Kreier, J. P. & Baker, J. R. (1987). *Parasitic Protozoa*. Bury St Edmunds: Allen & Unwin.

Kubo, T., Kawasaki, K. & Natori, S. (1993). Transient appearance and localization of a 26 kDa lectin, a novel member of the *Periplaneta* lectin family, in regenerating cockroach legs. *Dev Biol* **156**, 381.

Kühn, K. H. & Haug, T. (1994). Ultrastructure, cytochemical and immunocytochemical characterization of hemocytes of the hard tick *Ixodes ricinus* (Acari: Chelicerata). *Cell Tissue Res* **277**, 493–504.

Kurata, S. (2004). Recognition of infectious non-self and activation of immune responses by peptidoglycan recognition protein (PGRP)-family members in *Drosophila*. *Dev Comp Immunol* **28**, 89–95.

Landfear, S. M. & Ignatushchenko, M. (2001). The flagellum and flagellar pocket of trypanosomatids. *Mol Biochem Parasitol* **115**, 1–17.

Lane, R. S. & Quistad, G. B. (1998). Borreliacidal factor in the blood of the western fence lizard (*Sceloporus occidentalis*). *J Parasitol* **84**, 29–34.

Langland, J. & Morand, S. (1998). Heritable non-susceptibility in an allopatric host–parasite system: *Biomphalaria glabrata* (Mollusca) *Echinostoma caproni* (Platyhelminth digenea). *J Parasitol* **84**, 739–742.

Lavine, M. D. & Strand, M. R. (2001). α-Integrin subunits expressed in hemocytes of *Pseudoplusia includens*. In *Keystone Symposium on the Genetic Manipulation of Insects*. Silverthorne, CO: Keystone Symposia 63.

Lavine, M. D. & Strand, M. R. (2002). Insect hemocytes and their role in immunity. *Insect Biochem Mol Biol* **32**, 1295–1309.

Lee, S. Y., Cho, M. Y., Hyun, J. H., Lee, K. M., Homma, K., Natori, S., Kawabata, S., Iwanaga, S. & Lee, B. L. (1998). Molecular cloning of cDNA for pro-phenol-oxidase activating factor 1, a serine protease is induced by lipopolysaccharide or 1,3-β-glucan in coleopteran insect, *Holotrichia diomphalia* larvae. *Eur J Biochem* **257**, 615–621.

Lee, W. J., Lee, J. D., Kravchenko, V. V., Ulevitch, R. J. & Brey, P. T. (1996). Purification and molecular cloning of an inducible gram-negative bacteria-binding protein from the silkworm, *Bombyx mori*. *Proc Natl Acad Sci U S A* **93**, 7888–7893.

Leem, J. Y., Nishimura, C., Kurata, S., Shimada, I., Kobayashi, A. & Natori, S. (1996). Purification and characterization of N-beta-alanyl-5-S-glutathionyl-3,4-dihydroxy-phenylalanine, a novel antibacterial substance of *Sarcophaga peregrina* (flesh fly). *J Biol Chem* **271**, 13575–13577.

Lehane, M. J., Wu, D. & Lehane, S. M. (1997). Midgut-specific immune molecules are produced by the blood-sucking insect *Stomoxys calcitrans*. *Proc Natl Acad Sci U S A* **94**, 11502–11507.

Lehane, M. J., Aksoy, S., Gibson, W. & 7 other authors (2003). Adult midgut expressed sequence tags from the tsetse fly *Glossina morsitans morsitans* and expression analysis of putative immune response genes. *Genome Biol* **4**, R63.

Levashina, E. A., Moita, L. F., Blandin, S., Vriend, G., Lagueux, M. & Kafatos, F. C. (2001). Conserved role of a complement-like protein in phagocytosis revealed by dsRNA knockout in cultured cells of the mosquito, *Anopheles gambiae*. *Cell* **104**, 709–718.

Lie, K. J., Jeong, K. H. & Heyneman, D. (1987). Molluscan host reactions to helminthic infections. In *Immune Responses in Parasitic Infections*, pp. 211–270. Edited by E. J. L. Soulsby. Boca Raton, FL: CRC Press.

Ligoxygakis, P., Pelte, N., Hoffmann, J. A. & Reichhart, J.-M. (2002). Activation of *Drosophila* Toll during fungal infection by a blood serine protease. *Science* **297**, 114–116.

Liu, C. T., Hou, R. F., Ashida, M. & Chen, C. C. (1997). Effects of inhibitors of serine protease, phenoloxidase and dopa decarboxylase on the melanization of *Dirofilaria immitis* microfilariae with *Armigeres subalbatus* hemolymph *in vitro*. *Parasitology* **115**, 57–68.

Loker, E. S. & Bayne, C. J. (1986). Immunity to trematode larvae in the snail *Biomphalaria*. *Symp Zool Soc Lond* **56**, 199–220.

Loker, E. S. & Bayne, C. J. (2001). Molecular studies of the molluscan response to digenean infection. In *Phylogenic Perspectives on the Vertebrate Immune System*, pp. 209–222. Edited by G. Beck, M. Sugumaran & E. L. Cooper. New York: Kluwer/Plenum.

Loker, E. S., Bayne, C. J., Buckley, P. & Kruse, K. T. (1982). Ultrastructure of encapsulation of *Schistosoma mansoni* mother sporocysts by hemocytes of juveniles of the 10-R2 strain of *Biomphalaria glabrata*. *J Parasitol* **68**, 84–94.

Loker, E. S., Couch, L. & Hertel, L. A. (1994). Elevated agglutination titers in plasma of *Biomphalaria glabrata* exposed to *Echinostoma paraensei*: characterization and functional relevance of a trematode-induced response. *Parasitology* **108**, 17–26.

Lopez, L., Morales, G., Ursic, R., Wolff, M. & Lowenberger, C. (2003). Isolation and characterization of a novel insect defensin from *Rhodnius prolixus*, a vector of Chagas disease. *Insect Biochem Mol Biol* **33**, 439–447.

Lounibos, L. P. (2002). Invasions by insect vectors of human disease. *Annu Rev Entomol* **47**, 233–266.

LoVerde, P. T. (1998). Do antioxidants play a role in the schistosome host–parasite interactions? *Parasitol Today* **14**, 284–289.

Lowenberger, C. (2001). Innate immune response of *Aedes aegypti*. *Insect Biochem Mol Biol* **31**, 219–229.

Lowenberger, C., Bulet, P., Charlet, M., Hetru, C., Hodgeman, B., Christensen, B. M. & Hoffmann, J. A. (1995). Insect immunity: isolation of three novel inducible antibacterial defensins from the vector mosquito, *Aedes aegypti*. *Insect Biochem Mol Biol* **25**, 867–873.

Lowenberger, C. A., Ferdig, M. T., Bulet, P., Khalili, S., Hoffmann, J. A. & Christensen, B. M. (1996). *Aedes aegypti*: induced antibacterial proteins reduce the establishment and development of *Brugia malayi*. *Exp Parasitol* **83**, 191–201.

Lowenberger, C. A., Kamal, S., Chiles, J., Paskewitz, S., Bulet, P., Hoffmann, J. A. & Christensen, B. M. (1999). Mosquito–*Plasmodium* interactions in response to immune activation of the vector. *Exp Parasitol* **91**, 59–69.

Luckhart, S. & Li, K. (2001). Transcriptional complexity of the *Anopheles stephensi* nitric oxide synthase gene. *Insect Biochem Mol Biol* **31**, 249–256.

Luckhart, S. & Rosenberg, R. (1999). Gene structure and polymorphism of an invertebrate nitric oxide synthase gene. *Gene* **232**, 25–34.

Luckhart, S., Cupp, M. S. & Cupp, E. W. (1992). Morphological and functional classification of the hemocytes of adult female *Simulium vittatum* (Diptera: Simuliidae). *J Med Entomol* **29**, 457–466.

Luckhart, S., Vodovotz, Y., Cui, L. & Rosenberg, R. (1998). The mosquito *Anopheles stephensi* limits malaria parasite development with inducible synthesis of nitric oxide. *Proc Natl Acad Sci U S A* **95**, 5700–5705.

Luckhart, S., Crampton, A. L., Zamora, R. & 7 other authors (2003). Mammalian transforming growth factor $\beta 1$ activated after ingestion by *Anopheles stephensi* modulates mosquito immunity. *Infect Immun* **71**, 3000–3009.

Ludwig, G. V., Israel, B. A., Christensen, B. M., Yuill, T. M. & Schultz, K. T. (1991). Monoclonal antibodies directed against the envelope glycoproteins of La Crosse virus. *Microb Pathog* **11**, 411–421.

Luna, C., Wang, X., Huang, Y., Zhang, J., Zheng, L. (2002). Characterization of four Toll related genes during development and immune responses in *Anopheles gambiae*. *Insect Biochem Mol Biol* **32**, 1171–1179.

Luo, C. & Zheng, L. (2000). Independent evolution of Toll and related genes in insects and mammals. *Immunogenetics* **51**, 92–98.

Machado, P. E., Eger-Mangrich, I., Rosa, G., Koerich, L. B., Grisard, E. C. & Steindel, M. (2001). Differential susceptibility of triatomines of the genus *Rhodnius* to *Trypanosoma rangeli* strains from different geographical origins. *Int J Parasitol* **31**, 632–634.

Martin-Bermudo, M. D., Alvarez-Garcia, I. & Brown, N. H. (1999). Migration of the *Drosophila* primordial midgut cells requires coordination of diverse PS integrin functions. *Development* **126**, 5161–5169.

Matricon-Gondran, M. & Letocart, M. (1999). Internal defenses of the snail *Biomphalaria glabrata*. *J Invertebr Pathol* **74**, 224–234.

Matsuno, R., Aramaki, Y., Arima, H. & Tsuchiya, S. (1997). Scavenger receptors may regulate nitric oxide production from macrophages stimulated by LPS. *Biochem Biophys Res Commun* **237**, 601–605.

Matuschewski, K., Nunes, A. C., Nussenzweig, V. & Menard, R. (2002). *Plasmodium* sporozoite invasion into insect and mammalian cells is directed by the same dual binding system. *EMBO J* **21**, 1597–1606.

Maudlin, I. & Welburn, S. C. (1987). Lectin mediated establishment of midgut infections of *Trypanosoma congolense* and *Trypanosoma brucei* in *Glossina morsitans*. *Tropenmed Parasitol* **38**, 167–170.

McGuinness, D. H., Dehal, P. K. & Pleass, R. J. (2003). Pattern recognition molecules and innate immunity to parasites. *Trends Parasitol* **19**, 312–319.

McNair, A., Dissous, C., Duvaux-Miret, O. & Capron, A. (1993). Cloning and characterization of the gene encoding the 28 kDa glutathion *S*-transferase of *Schistosoma mansoni*. *Gene* **124**, 245–249.

Medzhitov, R. & Janeway, C. A., Jr (2002). Decoding the patterns of self and nonself by the innate immune system. *Science* **296**, 298–300.

Mello, C. B., Azambuja, P., Garcia, E. S. & Ratcliffe, N. A. (1996). Differential *in vitro* and *in vivo* behavior of three strains of *Trypanosoma cruzi* in the gut and hemolymph of *Rhodnius prolixus*. *Exp Parasitol* **82**, 112–121.

Menard, R. (2000). The journey of the malaria sporozoite through its hosts: two parasite proteins lead the way. *Microbes Infect* **2**, 633–642.

Meylears, K., Cerstiaens, A., Vierstraete, E., Baggerman, G., Michiels, C. W., De Loof, A. & Schoofs, L. (2002). Antimicrobial compounds of low molecular mass are constitutively present in insects: characterisations of a β-alanyl-tyrosine. *Curr Pharm Des* **9**, 159–174.

Miles, A. M., Bohle, D. S., Glassbrenner, P. A., Hansert, B., Wink, D. A. & Grisham, M. B.

(1996). Modulation of superoxide-dependent oxidation and hydroxylation reactions by nitric oxide. *J Biol Chem* **271**, 40–47.

Mitra, D., Sarkar, M. & Allen, A. K. (1988). Purification and characterization of an agglutinin from mucus of the snail – *Achatina fulica. Biochimie* **70**, 1821–1829.

Mizutani, T., Kobayashi, M., Eshita, Y. & 9 other authors (2003a). Characterization of JNK-like protein derived from a mosquito cell line C6/36. *Insect Mol Biol* **12**, 61–66.

Mizutani, T., Kobayashi, M., Eshita, Y. & 9 other authors (2003b). Involvement of the JNK-like protein of the *Aedes albopictus* mosquito cell line, C6/36, in phagocytosis, endocytosis and infection of West Nile virus. *Insect Mol Biol* **12**, 491–499.

Molyneux, D. H., Takle, G., Ibrahim, E. A. & Ingram, G. A. (1986). Insect immunity to Trypanosomatidae. *Symp Zool Soc Lond* **56**, 117–144.

Moreira-Ferro, C. K., Daffre, S., James, A. A. & Marinotti, O. (1998). A lysozyme in the salivary glands of the malaria vector *Anopheles darlingi. Insect Mol Biol* **7**, 257–264.

Muller, H. M., Dimopoulos, G., Blass, C. & Kafatos, F. C. (1999). A hemocyte-like cell line established from the malaria vector *Anopheles gambiae* expresses six prophenol-oxidase genes. *J Biol Chem* **274**, 11727–11735.

Munderloh, U. G. & Kurtti, T. J. (1995). Cellular and molecular interrelationships between ticks and prokaryotic tick-borne pathogens. *Annu Rev Entomol* **40**, 221–243.

Munks, R. J. L., Hamilton, J. V., Lehane, S. M. & Lehane, M. J. (2001). Regulation of midgut defensin production in the blood-sucking insect *Stomoxys calcitrans. Insect Mol Biol* **10**, 561–571.

Nakajima, Y., van der Goes van Naters-Yasui, A., Taylor, D., Yamakawa, M. (2001). Two isoforms of a member of the arthropod defensin family from the soft tick, *Ornithodoros moubata* (Acari: Argasidae). *Insect Biochem Mol Biol* **31**, 747–751.

Nakajima, Y., van der Goes van Naters-Yasui, A., Taylor, D. & Yamakawa, M. (2002). Antibacterial peptide defensin is involved in midgut immunity of the soft tick, *Ornithodoros moubata. Insect Mol Biol* **11**, 611–618.

Nappi, A. J. & Vass, E. (1993). Melanogenesis and the generation of cytotoxic molecules during insect cellular immune reactions. *Pigment Cell Res* **6**, 117–126.

Nappi, A. J. & Vass, E. (1998a). Hydrogen peroxide production in immune-reactive *Drosophila melanogaster. J Parasitol* **84**, 1150–1157.

Nappi, A. J. & Vass, E. (1998b). Hydroxyl radical formation resulting from the interaction of nitric oxide and hydrogen peroxide. *Biochim Biophys Acta* **1380**, 55–63.

Nappi, A. J., Vass, E., Frey, F. & Carton, Y. (1995). Superoxide anion generation in *Drosophila* during melanotic encapsulation of parasites. *Eur J Cell Biol* **68**, 450–456.

Nappi, A. J., Vass, E., Frey, F. & Carton, Y. (2000). Nitric oxide involvement in *Drosophila* immunity. *Nitric Oxide Biol Chem* **4**, 423–430.

Natori, S., Shiraishi, H., Hori, S. & Kobayashi, A. (1999). The role of *Sarcophaga* defense molecules in immunity and metamorphosis. *Dev Comp Immunol* **23**, 317–328.

Nayar, J. K. & Knight, J. W. (1997). Hemagglutinins in mosquitoes and their role in the immune response to *Brugia malayi* (Filarioidea: Nematoda) larvae. *Comp Biochem Physiol* **118A**, 1321–1326.

Nayar, J. K., Knight, J. W. & Vickery, A. C. (1989). Intracellular melanisation in the mosquito *Anopheles quadrimaculatus* Say (Diptera: Culicidae) against the filarial nematodes, *Brugia* sp. (Nematoda: Filarioidea). *J Med Entomol* **26**, 159–166.

Nayar, J. K., Mikarts, L. L., Chikilian, M. L., Knight, J. W. & Bradley, T. J. (1995). Lectin binding to extracellularly melanized microfilariae of *Brugia malayi* from the hemocoel of *Anopheles quadrimaculatus*. *J Invertebr Pathol* **66**, 277–286.

Nguu, E. K., Osir, E. O., Imbuga, M. O. & Olembo, N. K. (1996). The effect of host blood in the *in vitro* transformation of blood-stream trypanosomes by tsetse midgut homogenates. *Med Vet Entomol* **10**, 317–322.

Nigam, Y., de Azambuja, P. & Ratcliffe, N. A. (1996). Insect vector immunity. In *New Directions in Invertebrate Immunology*, pp. 401–442. Edited by K. Söderhäll, S. Iwanaga & G. R. Vasta. New Haven, NJ: SOS Publications.

Nigam, Y., Maudlin, I., Welburn, S. & Ratcliffe, N. A. (1997). Detection of phenoloxidase activity in the hemolymph of tsetse flies, refractory and susceptible to infection with *Trypansoma brucei rhodesiense*. *J Invertebr Pathol* **69**, 279–281.

Nimmo, D. D., Ham, P. J., Ward, R. D. & Maingon, R. (1997). The sandfly *Lutzomyia longipalpis* shows specific humoral responses to bacterial challenge. *Med Vet Entomol* **11**, 324–328.

Nussenzveig, R. H., Bentley, D. L. & Ribeiro, J. M. (1995). Nitric oxide loading of the salivary nitric-oxide-carrying hemoproteins (nitrophorins) in the blood-sucking bug *Rhodnius prolixus*. *J Exp Biol* **198**, 1093–1098.

Ochiai, M. & Ashida, M. (1988). Purification of a β-1,3-glucan recognition protein in the prophenoloxidase activating system from hemolymph of the silkworm, *Bombyx mori*. *J Biol Chem* **263**, 12056–12062.

Oduol, F., Xu, J. N., Niare, O., Natarajan, R. & Vernick, K. D. (2000). Genes identified by an expression screen of the vector mosquito *Anopheles gambiae* display differential molecular immune response to malaria parasites and bacteria. *Proc Natl Acad Sci U S A* **97**, 11397–11402.

Ogden, C. A., deCathelineau, A., Hoffmann, P. R., Bratton, D., Ghebrehiwet, B., Fadok, V. A. & Henson, P. M. (2001). C1q and mannose binding lectin engagement of cell surface calreticulin and CD91 initiates macropinocytosis and uptake of apoptotic cells. *J Exp Med* **194**, 781–796.

Okolo, C. J., Jenni, L., Molyneux, D. H. & Wallbanks, K. R. (1990). Surface carbohydrate differences of *Glossina* salivary glands and infectivity of *Trypanosoma brucei gambiense* to *Glossina*. *Ann Soc Belg Med Trop* **70**, 39–47.

Oliveira, M. A. & De Souza, W. (1997). Interaction of *Trypanosoma rangeli* with midgut cells of *Rhodnius prolixus* treated with lectins. *Mem Inst Oswaldo Cruz Suppl* **Nov. 1997**, Abstract 093.

Oliveira, M. F., Silva, J. R., Dansa-Petretski, M., de Souza, W., Braga, C. M. S., Masuda, H. & Oliveira, P. L. (2000). Haemozoin formation in the midgut of the blood-sucking insect *Rhodnius prolixus*. *FEBS Lett* **477**, 95–98.

Oliveira, M. F., Timm, B. L., Machado, E. A. & 10 other authors (2002). On the pro-oxidant effects of haemozoin. *FEBS Lett* **512**, 139–144.

Oliveira, P. L. & Oliveira, M. F. (2002). Vampires, Pasteur and reactive oxygen species. Is the switch from aerobic to anaerobic metabolism a preventive antioxidant defence in blood-feeding parasites? *FEBS Lett* **525**, 3–6.

Ourth, D. D. & Renis, H. E. (1993). Antiviral melanization reaction of *Heliothis virescens* hemolymph against DNA and RNA viruses *in vitro*. *Comp Biochem Physiol B Biochem Mol Biol* **105**, 719–723.

Parola, P. & Raoult, D. (2001). Ticks and tickborne bacterial diseases in humans: an emerging infectious threat. *Clin Infect Dis* **32**, 897–928.

Paskewitz, S. M. & Christensen, B. M. (1996). Immune responses of vectors. In *The Biology of Disease Vectors*, pp. 371–392. Edited by B. J. Beaty & W. C. Marquardt. Boulder, CO: University Press of Colorado

Paskewitz, S. M., Brown, M. R., Lea, A. O. & Collins, F. H. (1988). Ultrastructure of the encapsulation of *Plasmodium cynomolgi* (B strain) on the midgut of a refractory strain of *Anopheles gambiae*. *J Parasitol* **74**, 432–439.

Paskewitz, S. M., Reese-Stardy, S. & Gorman, M. J. (1999). An easter-like serine protease from *Anopheles gambiae* exhibits changes in transcript levels following immune challenge. *Insect Mol Biol* **8**, 329–338.

Pearson, A., Lux, A. & Krieger, M. (1995). Expression cloning of dSR-CI, a class C macrophage-specific scavenger receptor from *Drosophila melanogaster*. *Proc Natl Acad Sci U S A* **92**, 4056–4060.

Pech, L. L. & Strand, M. R. (1996). Granular cells are required for encapsulation of foreign targets by insect haemocytes. *J Cell Sci* **109**, 2053–2060.

Pendland, J. C., Heath, M. A. & Boucias, D. G. (1988). Function of a galactose-binding lectin from *Spodoptera exigua* larval haemolymph: opsonisation of blastopores from entomopathogenous hypomycetes. *J Insect Physiol* **34**, 533–540.

Pereira, L. S., Oliveira, P. L., Barja-Fidalgo, C. & Daffre, S. (2001). Production of reactive oxygen species by hemocytes from the cattle tick *Boophilus microplus*. *Exp Parasitol* **99**, 66–72.

Pereira, M. E. A., Andrade, A. F. B. & Ribeiro, J. M. C. (1981). Lectins of distinct specificity in *Rhodnius prolixus* interact selectively with *Trypanosoma cruzi*. *Science* **211**, 597–600.

Perrone, J. B., DeMaio, J. & Spielman, A. (1986). Regions of mosquito salivary glands distinguished by surface lectin-binding characteristics. *Insect Biochem* **16**, 313–318.

Pimenta, P. F., Touray, M. & Miller, L. (1994). The journey of malaria sporozoites in the mosquito salivary gland. *J Eukaryot Microbiol* **41**, 608–624.

Price, C. D. & Ratcliffe, N. A. (1974). A reappraisal of insect haemocyte classification by the examination of blood from fifteen insect orders. *Z Zellforsch Mikrosk Anat* **147**, 313–324.

Rämet, M., Pearson, A., Manfruelli, P., Li, X., Koziel, H., Göbel, V., Chung, E., Krieger, M. & Ezekowitz, R. A. B. (2001). *Drosophila* scavenger receptor CI is a pattern recognition receptor for bacteria. *Immunity* **15**, 1027–1038.

Rämet, M., Manfruelli, P., Pearson, A., Mathey-Prevot, B. & Ezekowitz, R. A. B. (2002). Functional genomic analysis of phagocytosis and identification of a *Drosophila* receptor for *E. coli*. *Nature* **416**, 644–648.

Ratcliffe, N. A. (1991). The prophenoloxidase system and its role in arthropod immunity. In *Phylogenesis of Immune Functions*, pp. 45–71. Edited by G. W. Warr & N. Cohen. Boca Raton, FL: CRC Press.

Ratcliffe, N. A. (1993). Cellular defense responses of insects: unresolved problems. In *Parasites and Pathogens of Insects*, pp. 267–304. Edited by N. E. Beckage, S. N. Thompson & B. A. Federici. London: Academic Press.

Ratcliffe, N. A. & Rowley, A. F. (editors) (1981). *Invertebrate Blood Cells*, Vols 1 and 2. London: Academic Press.

Ratcliffe, N. A., Rowley, A. F., Fitzgerald, S. W. & Rhodes, C. P. (1985). Invertebrate immunity: basic concepts and recent advances. *Int Rev Cytol* **97**, 183–350.

Ratcliffe, N. A., Nigam, Y., Mello, C. B., Garcia, E. S. & Azambuja, P. (1996). *Trypanosoma cruzi* and erythrocyte agglutinins: a comparative study of occurrence and properties in the gut and hemolymph of *Rhodnius prolixus*. *Exp Parasitol* **83**, 83–93.

Renwrantz, L. & Stahmer, A. (1983). Opsonizing properties of an isolated hemolymph agglutinin and demonstration of lectin-like recognition molecules at the surface of hemocytes from *Mytilus edulis*. *J Comp Physiol* **149B**, 535–546.

Ribeiro, J. M. C. (1996). NAD(P)H-dependent production of oxygen reactive species by the salivary glands of the mosquito *Anopheles albimanus*. *Insect Biochem Mol Biol* **26**, 715–720.

Ribeiro, J. M. C. (2003). A catalogue of *Anopheles gambiae* transcripts significantly more or less expressed following a blood meal. *Insect Biochem Mol Biol* **33**, 865–882.

Richards, E. H. & Ratcliffe, N. A. (1990). Direct binding and lectin-mediated binding of erythrocytes to haemocytes of the insect, *Extatosomoa tiaratum*. *Dev Comp Immunol* **14**, 269–281.

Richards, C. S., Knight, M. & Lewis, F. A. (1992). Genetics of *Biomphalaria glabrata* and its effect on the outcome of *Schistosoma mansoni* infection. *Parasitol Today* **8**, 171–174.

Richman, A. & Kafatos, F. C. (1995). Immunity to eukaryotic parasites in vector insects. *Curr Opin Immunol* **8**, 14–19.

Richman, A. M., Dimopoulos, G., Seeley, D. & Kafatos, F. C. (1997). *Plasmodium* activates the innate immune response of *Anopheles gambiae* mosquitoes. *EMBO J* **16**, 6114–6119.

Riveau, G., Poulain-Godefroy, P., Dupré, L., Remoué, F., Mielcarek, N., Locht, C. & Capron, A. (1998). Glutathione *S*-transferases of 28 kDa as major vaccine candidates against schistosomiasis. *Mem Inst Oswaldo Cruz* **93**, 87–94.

Roditi, I. & Liniger, M. (2002). Dressed for success: the surface coats of insect-borne protozoan parasites. *Trends Microbiol* **10**, 128–134.

Romoser, W. F. (1996). The vector alimentary system. In *The Biology of Disease Vectors*, pp. 298–317. Edited by B. J. Beaty & W. C. Marquardt. Boulder, CO: University Press of Colorado.

Sacks, D. & Kamhawi, S. (2001). Molecular aspects of parasite–vector and vector–host interactions in leishmaniasis. *Annu Rev Microbiol* **55**, 453–483.

Salzet, M. (2002). Antimicrobial peptides are signalling molecules. *Trends Immunol* **23**, 283–284.

Sankala, M., Brännström, A., Schulthess, T., Bergmann, U., Morgunova, E., Engel, J., Tryggvason, K. & Pikkarainen, T. (2002). Characterization of recombinant soluble macrophage scavenger receptor MARCO. *J Biol Chem* **277**, 33378–33385.

Sant Anna, M. R. V., Araugo, J. G. V. C., Pereira, M. H., Pesquero, J. L., Diotaiuti, L., Lehane, S. M. & Lehane, M. J. (2002). Molecular cloning and sequencing of salivary gland-specific cDNAs of the blood-sucking bug *Triatoma brasiliensis* (Hemiptera: Reduviidae). *Insect Mol Biol* **11**, 585–593.

Satoh, D., Horii, A., Ochiai, M. & Ashida, M. (1999). Prophenoloxidase-activating enzyme of the silkworm, *Bombyx mori*. Purification, characterization and cDNA cloning. *J Biol Chem* **274**, 7441–7453.

Schaub, G. A., Grunfelder, C. G., Zimmerman, D. & Peters, W. (1989). Binding of lectin-gold conjugates by two *Trypanosoma cruzi* strains in ampullae and rectum of *Triatoma infestans*. *Acta Trop* **46**, 291–301.

Schmidt, J., Kleffmann, T. & Schaub, G. A. (1998). Hydrophobic attachment of *Trypanosoma cruzi* to a superficial layer of the rectal cuticle in the bug *Triatoma infestans*. *Parasitol Res* **84**, 527–536.

Schneider, D. & Shahabuddin, M. (2000). Malaria parasite development in a *Drosophila* model. *Science* **288**, 2376–2379.

Shahabuddin, M. & Pimenta, P. F. (1998). *Plasmodium gallinaceum* preferentially invades vesicular ATPase-expressing cells in *Aedes aegypti* midgut. *Proc Natl Acad Sci U S A* **95**, 3385–3389.

Shahabuddin, M., Fields, I., Bulet, P., Hoffmann, J. A. & Miller, L. H. (1998). *Plasmodium gallinaceum*: differential killing of some mosquito stages of the parasite by insect defensin. *Exp Parasitol* **89**, 103–112.

Shao, L., Devenport, M. & Jacobs-Lorena, M. (2001). The peritrophic matrix of hematophagous insects. *Arch Insect Biochem Physiol* **47**, 119–125.

Shiao, S. H., Higgs, S., Adelman, Z., Christensen, B. M., Liu, S. H. & Chen, C. C. (2001). Effect of prophenoloxidase expression knockout on the melanization of microfilariae in the mosquito *Armigeres subalbatus*. *Insect Mol Biol* **10**, 315–321.

Shin, S. W., Kokoza, V. A., Ahmed, A. & Raikhel, A. S. (2002). Characterization of three alternatively spliced isoforms of the Rel/NF-κB transcription factor relish from the mosquito *Aedes aegypti*. *Proc Natl Acad Sci U S A* **99**, 9978–9983.

Shin, S. W., Kokoza, V., Lobkov, I. & Raikhel, A. S. (2003a). Relish-mediated immune deficiency in the transgenic mosquito *Aedes aegypti*. *Proc Natl Acad Sci U S A* **100**, 2616–2621.

Shin, S. W., Kokoza, V. A. & Raikhel, A. S. (2003b). Transgenesis and reverse genetics of mosquito innate immunity. *J Exp Biol* **206**, 3835–3843.

Sidjanski, S. P., Vanderberg, J. P. & Sinnis, P. (1997). *Anopheles stephensi* salivary glands bear receptors for regions of the circumsporozoite protein of *Plasmodium falciparum*. *Mol Biochem Parasitol* **90**, 33–41.

Sinden, R. E. & Billingsley, P. F. (2001). *Plasmodium* invasion of mosquito cells: hawk or dove? *Trends Parasitol* **17**, 209–212.

Sinden, R. E., Butcher, G. A., Billker, O. & Fleck, S. L. (1996). Regulation of infectivity of *Plasmodium* to the mosquito vector. *Adv Parasitol* **38**, 53–117.

Smail, A. J. & Ham, P. J. (1988). *Onchocerca* induced haemolymph lectins in blackflies: confirmation by sugar inhibition of erythrocyte agglutination. *Trop Med Parasitol* **39**, 82–83.

Snow, K. (2000). Could malaria return to Britain? *Biologist* **47**, 176–180.

Söderhäll, K. & Cerenius, L. (1998). Role of the prophenoloxidase-activating system in invertebrate immunity. *Curr Opin Immunol* **10**, 23–28.

Söderhäll, K. & Thörnqvist, P.-O. (1997). Crustacean immunity – a short review. In *Fish Vaccinology*, pp. 45–51. Edited by R. Gudding, A. Lillehaug, P. J. Midtlyng & F. Brown. Basel: Karger.

Söderhäll, K., Rögener, W., Söderhäll, I., Newton, R. P. & Ratcliffe, N. A. (1988). The properties and purification of a *Blaberus craniifer* plasma protein which enhances the activation of haemocyte prophenoloxidase by a β-1,3-glucan. *Insect Biochem* **18**, 323–330.

Söderhäll, K., Cerenius, L. & Johansson, M. W. (1996). The prophenoloxidase activating system in invertebrates. In *New Directions in Invertebrate Immunology*, pp. 229–253. Edited by K. Söderhäll, S. Iwanaga & G. Vasta. New Jersey: SOS Publications.

Stafford, J. L., Galvez, F., Goss, G. G. & Belosevic, M. (2002). Induction of nitric oxide and respiratory burst response in activated goldfish macrophages requires potassium channel activity. *Dev Comp Immunol* **26**, 445–459.

Sultan, A. A. (1999). Molecular mechanisms of malaria sporozoite motility and invasion of host cells. *Int Microbiol* **2**, 155–160.

Tahar, R., Boudin, C., Thiery, I. & Bourgouin, C. (2002). Immune response of *Anopheles gambiae* to the early sporogonic stages of the human malaria parasite *Plasmodium falciparum*. *EMBO J* **21**, 6673–6680.

Takle, G. B. (1988). Studies on the cellular immune responses of insects toward the insect pathogen *Trypanosoma rangeli*. *J Invertebr Pathol* **51**, 64–72.

Takle, G. B. & Lackie, A. M. (1987). Investigation of the possible role of hemocyte and parasite surface charge in the clearance of *Trypanosoma rangeli* from the insect hemocoel. *J Invertebr Pathol* **50**, 336–338.

Tewari, R., Spaccapelo, R., Bisoni, F., Holder, A. A. & Crisanti, A. (2002). Function of region I and II adhesive motifs of *Plasmodium falciparum* circumsporozoite protein in sporozoite motility and infectivity. *J Biol Chem* **277**, 47613–47618.

Thomas, S., Andrews, A., Hay, P. & Bourgoise, S. (1999). The anti-microbial activity of maggot secretions: results of a preliminary study. *J Tissue Viability* **9**, 127–132.

Thomasova, D., Ton, L. Q., Coploy, R. R. & 7 other authors (2002). Comparative genomic analysis in the region of a major *Plasmodium*-refractoriness locus of *Anopheles gambiae*. *Proc Natl Acad Sci U S A* **99**, 8179–8184.

Tobie, E. J. (1970). Observations on the development of *Trypanosoma rangeli* in the hemocoel of *Rhodnius prolixus*. *J Invertebr Pathol* **15**, 118–125.

Trenczek, T. & Faye, I. (1988). Synthesis of immune proteins in primary cultures of fat body from *Hyalophora cecropia*. *Insect Biochem* **18**, 299–312.

Tzou, P., De Gregorio, E. & Lemaitre, B. (2002). How *Drosophila* combats microbial infection: a model to study innate immunity and host–pathogen interactions. *Curr Opin Microbiol* **5**, 102–110.

Underhill, D. M. (2003). Toll-like receptors: networking for success. *Eur J Immunol* **33**, 1767–1775.

US Centers for Disease Control and Prevention (1999). Recommendations for the use of Lyme disease vaccine: recommendations of the Advisory Committee on Immunization Practices (ACIP). *MMWR* **48 (RR-7)**, 1–25.

Valenzuela, J. G., Pham, V. M., Garfield, M. K., Francischetti, I. M. & Ribeiro, J. M. C. (2002). Toward a description of the sialome of the adult female mosquito *Aedes aegypti*. *Insect Biochem Mol Biol* **32**, 1101–1122.

Vass, E. & Nappi, A. J. (1998). Prolonged oviposition decreases the ability of the parasitoid *Leptopilina boulardi* to suppress the cellular immune response of its host *Drosophila melanogaster*. *Exp Parasitol* **89**, 86–91.

Vass, E. & Nappi, A. J. (2000). Developmental and immunological aspects of *Drosophila*–parasitoid relationships. *J Parasitol* **86**, 1259–1270.

Vass, E. & Nappi, A. J. (2001). Fruit fly immunity. *Bioscience* **51**, 529–535.

Vernick, K. D. (1997). Mechanisms of immunity and refractoriness in insect vectors of eukaryotic parasites. In *Molecular Mechanisms of Immune Responses in Insects*, pp. 261–309. Edited by P. T. Brey & D. Hultmark. London: Chapman & Hall.

Vizioli, J., Bulet, P., Hoffmann, J. A., Kafatos, F. C., Müller, H.-M. & Dimopoulos, G. (2001). Gambicin: a novel immune responsive antimicrobial peptide from the malaria vector *Anopheles gambiae*. *Proc Natl Acad Sci U S A* **98**, 12630–12635.

Volf, P., Kiewegova, A. & Svobodova, M. (1998). Sandfly midgut lectin: effect of galacto-samine on *Leishmania major* infections. *Med Vet Entomol* **12**, 151–154.

Volf, P., Svobodová, M. & Dvoráková, E. (2001). Bloodmeal digestion and *Leishmania major* infections in *Phlebotomus duboscqi*: effect of carbohydrates inhibiting midgut lectin activity. *Med Vet Entomol* **15**, 281–286.

Watt, D. M., Walker, A. R., Lamza, K. A. & Ambrose, N. C. (2001). Tick *Theileria* interactions in response to immune activation of the vector. *Exp Parasitol* **97**, 89–94.

Weber, J. R., Moreillon, P. & Tuomanen, E. I. (2003a). Innate sensors for gram-positive bacteria. *Curr Opin Immunol* **15**, 408–415.

Weber, J. R., Freyer, D., Alexander, C., Schröder, N. W. J., Reiss, A., Küster, C., Pfeil, D., Tuomanen, E. I. & Schumann, R. R. (2003b). Recognition of pneumococcal peptidoglycan: an expanded, pivotal role for LPS binding protein. *Immunity* **19**, 269–279.

Welburn, S. C. & Maudlin, I. (1999). Tsetse–trypanosome interactions: rites of passage. *Parasitol Today* **15**, 399–403.

Welburn, S. C., Maudlin, I. & Ellis, D. S. (1989). Rate of trypanosome killing by lectins in midguts of different species and strains of *Glossina*. *Med Vet Entomol* **3**, 77–82.

Wengelnik, K., Spaccapelo, R., Naitza, S., Robson, K. J., Janse, C. J., Bistoni, F., Waters, A. P. & Crisanti, A. (1999). The A-domain and the thrombospondin-related motif of *Plasmodium falciparum* TRAP are implicated in the invasion process of mosquito salivary glands. *EMBO J* **18**, 5195–5204.

Whitten, M. M. A., Mello, C. B., Gomes, S. A. O., Nigam, Y., Azambuja, P., Garcia, E. S. & Ratcliffe, N. A. (2001). Role of superoxide and reactive nitrogen intermediates in *Rhodnius prolixus* (Reduviidae)/*Trypanosoma rangeli* interactions. *Exp Parasitol* **98**, 44–57.

Whitten, M. M. A., Tew, I. F., Lee, B. L. & Ratcliffe, N. A. (2004). A novel role for an insect apolipoprotein (apolipophorin-III) in β-1,3-glucan pattern recognition and cellular encapsulation reactions. *J Immunol* (in press).

Wientjes, F. B. & Segal, A. W. (1995). NADPH oxidase and the respiratory burst. *Cell Biol* **6**, 357–365.

Wiesner, A., Losen, S., Kopacek, P., Weise, C. & Götz, P. (1997). Isolated apolipophorin III from *Galleria mellonella* stimulates the immune reactions of this insect. *J Insect Physiol* **43**, 383–391.

Wilkins, S. & Billingsley, P. F. (2001). Partial characterization of oligosaccharides expressed on midgut microvillar glycoproteins of the mosquito, *Anopheles stephensi* Liston. *Insect Biochem Mol Biol* **31**, 937–948.

Williams, D. L., de la Llera-Moya, M., Thuahnai, S. T., Lund-Katz, S., Connelly, M. A., Azhar, S., Anantharamaiah, G. M. & Phillips, M.C. (2000). Binding and cross-linking studies show that scavenger receptor B1 interacts with multiple sites in apolipoprotein A-I and identify the class A amphipathic alpha-helix as a recognition motif. *J Biol Chem* **275**, 18897–18904.

Wilson, K., Cotter, S. C., Reeson, A. F. & Pell, J. K. (2001). Melanism and disease resistance in insects. *Ecol Lett* **4**, 637–649.

Wilson, R., Chen, C. & Ratcliffe, N. A. (1999). Innate immunity in insects: the role of multiple, endogenous serum lectins in the recognition of foreign invaders in the cockroach, *Blaberus discoidalis*. *J Immunol* **162**, 1590–1596.

Wittmann, E. J. & Baylis, M. (2000). Climate change: effects on *Culicoides*-transmitted viruses and implications for the UK. *Vet J* **160**, 107–117.

Yan, J., Cheng, Q., Li, C.-B. & Aksoy, S. (2001). Molecular characterization of two serine proteases expressed in gut tissue of the African trypansome vector, *Glossina morsitans morsitans*. *Insect Mol Biol* **10**, 47–56.

Yoshida, H., Kinoshita, K., Ashida, M. (1996). Purification of a peptidoglycan recognition protein from hemolymph of the silkworm, *Bombyx mori*. *J Biol Chem* **271**, 13854–13860.

Yoshiga, T., Hernandez, V. P., Fallon, A. M. & Law, J. H. (1997). Mosquito transferrin, an acute-phase protein that is up-regulated upon infection. *Proc Natl Acad Sci U S A* **94**, 12337–12342.

Yoshino, T. P., Boyle, J. P. & Humphries, J. E. (2001). Receptor–ligand interactions and cellular signalling at the host–parasite interface. *Parasitology* **123**, S143–S157.

Yoshino, T. P., Dinguirard, N. & Johnston, L. (2003). LDL-binding receptors on circulating haemocytes of the freshwater snail *Biomphalaria glabrata*. In *Ninth International Congress of ISDCI, St Andrews, Scotland, Programme and Abstracts*, p. 55. Edited by E. A. Dyrynda.

Yu, X.-Q., Gan, H. & Kanost, M. R. (1999). Immulectin, an inducible C-type lectin from an insect, *Manduca sexta*, stimulates activation of plasma prophenol oxidase. *Insect Biochem Mol Biol* **29**, 585–597.

Zeidler, M. P., Bach, E. A. & Perrimon, N. (2000). The roles of the *Drosophila* JAK/STAT pathway. *Oncogene* **19**, 2598–2606.

Zhang, D., Dimopoulos, G., Wolf, A., Minana, B., Kafatos, F. C. & Winzerling, J. J. (2002). Cloning and molecular characterization of two mosquito iron regulatory proteins. *Insect Biochem Mol Biol* **32**, 579–589.

Zhang, S.-M., Léonard, P. M., Adema, C. M. & Loker, E. S. (2001). Parasite-responsive IgSF members in the snail *Biomphalaria glabrata*: characterization of novel genes with tandemly arranged IgSF domains and a fibrinogen domain. *Immunogenetics* **53**, 684–694.

Zhao, X., Smartt, C. T., Li, J. & Christensen, B. M. (2001). *Aedes aegypti* peroxidase gene characterization and developmental expression. *Insect Biochem Mol Biol* **31**, 481–490.

Zheng, L., Cornel, A. J., Wang, R., Erfle, H., Voss, H., Ansorge, W., Kafatos, F. C. & Collins, F. H. (1997). Quantitative trait loci for refractoriness of *Anopheles gambiae* to *Plasmodium cynomolgi* B. *Science* **276**, 425–428.

Zieler, H., Garon, C. F., Fischer, E. R. & Shahabuddin, M. (1998). Adhesion of *Plasmodium gallinaceum* to the *Aedes aegypti* midgut: sites of parasite attachment and morphological changes in the ookinete. *J Euk Microbiol* **45**, 512–520.

Zieler, H., Nawrocki, J. P. & Shahabuddin, M. (1999). *Plasmodium gallinaceum* ookinetes adhere specifically to the midgut epithelium of *Aedes aegypti* by interaction with a carbohydrate ligand. *J Exp Biol* **202**, 485–495.

Transmission of plant viruses by nematodes

Stuart A. MacFarlane and David J. Robinson

Scottish Crop Research Institute, Invergowrie, Dundee DD2 5DA, UK

INTRODUCTION

Plant viruses use a variety of mechanisms to enable them to spread away from an initial focus of infection into the wider environment. Many viruses are able to infect the developing seed and so are disseminated to new areas when the seed is shed. In addition, viruses may be taken up, for example by insect pests and fungal pathogens as they feed on or multiply within the plant, and subsequently can be transmitted to further uninfected plants by the vector organism. It is becoming clear that rather than being a non-specific, merely passive process, transmission of plant viruses by vector organisms is a highly specific process, in which particular virus proteins are adapted to allow interaction with the vector.

Two groups of viruses, the tobraviruses and the nepoviruses, use plant-parasitic nematodes as their transmission vector (Taylor & Brown, 1997; MacFarlane *et al.*, 2002). Nematodes are present in almost all environments both on land and in the sea and have many different lifestyles, including parasitism on animals and plants. Over 20 000 species of nematodes are known, of which perhaps 10 % are plant parasites. Considering their diversity it is perhaps surprising that, again, only two groups of nematodes, the trichodorid nematodes and the longidorid nematodes, are responsible for transmitting viruses. Plant-parasitic nematodes in their own right have a considerable impact on agriculture. Worldwide, yield losses attributed to nematode attack have been estimated at 5–12 % per year, although if nematode numbers are allowed to build up by a failure of control practices then yield loss can be far higher (Barker & Koenning, 1998). The characteristics of the diseases caused by tobraviruses

SGM symposium 63: Microbe–vector interactions in vector-borne diseases.
Editors S. H. Gillespie, G. L. Smith & A. Osbourn. Cambridge University Press. ISBN 0 521 84312 X ©SGM 2004

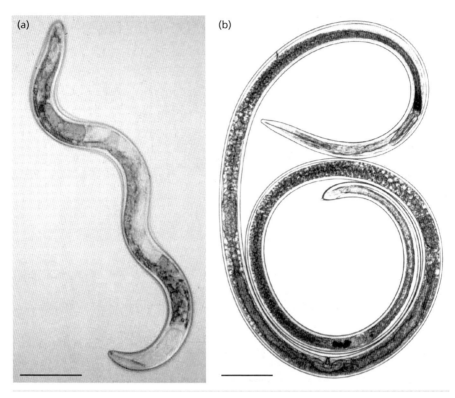

Fig. 1. Micrographs of (a) *P. pachydermus* and (b) *X. index*. Bars, 100 μm.

and nepoviruses are greatly influenced by the biology of the nematode vectors and strategies for disease control are primarily designed to target the nematode.

TRICHODORIDS AND TOBRAVIRUSES

Nematodes of the genera *Trichodorus* and *Paratrichodorus* (Fig. 1a) (family Trichodoridae), commonly referred to as trichodorids, are the vectors of all three tobraviruses, *Tobacco rattle virus* (TRV), *Pea early-browning virus* (PEBV) and *Pepper ringspot virus* (PepRSV). However, only 14 of the approximately 75 species in these two genera have been shown to be virus vectors (Table 1) and each of these 14 species transmits a different selection of tobravirus strains (Ploeg *et al.*, 1992a, b). Trichodorids occur worldwide, but are almost entirely restricted to freely draining, usually sandy soils and rarely occur in clay or silt soils. They feed on the roots of many plant species where they aggregate around the zone of root elongation, 1–3 mm behind the apical meristem (Pitcher, 1967). If the number of feeding nematodes is sufficiently large, root growth may be retarded, leading to the classic 'stubby root' symptom, and this may be accompanied by chlorosis and wilting of the foliage.

Table 1. Tobraviruses, their distribution, the main diseases they cause and their trichodorid vectors

Tobravirus species	Geographical distribution	Main diseases	Trichodorid vector species
TRV	Europe, North America, Japan, New Zealand	Potato spraing, tobacco rattle, gladiolus notched leaf, hyacinth malaria, aster ringspot, sugar beet yellow blotch	*Trichodorus cylindricus, T. primitivus, T. similis, T. sparsus, T. viruliferus, Paratrichodorus allius, P. anemones, P. hispanus, P. minor, P. nanus, P. pachydermus, P. porosus, P. teres, P. tunisiensis*
PEBV	Europe, North Africa	Pea early-browning, broad bean yellow band	*T. primitivus, T. viruliferus, P. anemones, P. pachydermus, P. teres*
PepRSV	South America	Ringspot in pepper, yellow band in tomato and artichoke	*P. minor*

TRV has one of the widest known host ranges of any plant virus and causes disease in many crop species (Table 1; Harrison & Robinson, 1986; MacFarlane, 1999). It originally got its name from the rattling sounds made when the wind blows through the dry, necrotic leaves of infected tobacco plants. Economically, it is most important as the cause of a form of spraing disease of potatoes, similar to, but distinct from that caused by *Potato mop-top virus*. Consignments of potatoes in which as few as 2–3 % of the tubers show the disfiguring symptoms of spraing are likely to be rejected by processors or supermarkets. PEBV and PepRSV have more restricted geographical distributions and host ranges than TRV and both cause diseases of local importance (Table 1). Tobraviruses also infect many wild and weed species, and may be transmitted in infected seed of some crop species as well as several weed species (Cooper & Harrison, 1973).

LONGIDORIDS AND NEPOVIRUSES

A few of the approximately 350 species of longidorid nematode in the genera *Xiphinema* (Fig. 1b), *Longidorus* and *Paralongidorus* (family Longidoridae), transmit viruses from the genus *Nepovirus* (Table 2) (Taylor & Brown, 1997). However, only 12 of about 40 nepoviruses have been demonstrated to be transmitted by these nematodes and it is possible that some nepoviruses have other transmission mechanisms. Longidorids are less restricted by soil type than trichodorids, but nevertheless are most frequently found in lighter soils, particularly sandy loams and loams. *Longidorus* species (needle nematodes) feed exclusively at, or just behind the root tips, but some *Xiphinema* species (dagger nematodes) will feed elsewhere on the root system, provided the roots are young and actively growing. Root-tip feeding by both genera leads to characteristic galling and root systems may be severely damaged or even completely destroyed, with inevitable consequences for the aerial parts of the plant (Griffin & Epstein, 1964).

Table 2. Nematode-transmitted nepoviruses, their distribution, the main diseases they cause and their longidorid vectors

Nepovirus species	Geographical distribution	Main diseases	Longidorid vector species
Arabis mosaic virus (ArMV)	Europe, Japan, New Zealand	Yellow dwarf of raspberry, mosaic and yellow crinkle of strawberry, nettlehead of hops	*Xiphinema diversicaudatum*
Artichoke Italian latent virus (AILV)	Southern Italy, Greece, Bulgaria	Chlorotic mottle of chicory	*Longidorus apulus, L. fasciatus*
Beet ringspot virus (BRSV)	Europe	Beet ringspot, potato pseudo-aucuba	*L. elongatus*
Cherry rasp leaf virus (CRLV)*	Western North America	Cherry rasp leaf, flat apple	*X. americanum, X. californicum, X. rivesi*
Grapevine fanleaf virus (GFLV)	Worldwide	Fanleaf and yellow mosaic of grapevine	*X. index, X. italiae*
Mulberry ringspot virus (MRSV)	Japan	Mulberry ringspot	*L. martini*
Peach rosette mosaic virus (PRMV)	North America	Peach rosette mosaic, degeneration of grapevine	*L. diadecturus, X. americanum, X. rivesi*
Raspberry ringspot virus (RpRSV)	Europe, Turkey	Raspberry ringspot and leafcurl, strawberry ringspot, redcurrant spoonleaf	*L. elongatus, L. macrosoma, Paralongidorus maximus*
Strawberry latent ringspot virus (SLRSV)*	Europe	Mottle and decline of strawberry and raspberry	*X. diversicaudatum*
Tobacco ringspot virus (TRSV)	North America	Tobacco ringspot, soybean bud blight, blueberry necrotic ringspot, ringspot of cucurbits	*X. americanum, X. californicum, X. intermedium, X. rivesi, X. tarjanense*
Tomato black ring virus (TBRV)	Europe	Ringspot of raspberry and strawberry, potato bouquet, lettuce ringspot	*L. attenuatus*
Tomato ringspot virus (ToRSV)	North America, Japan, former USSR	Stem pitting of peach and cherry, yellow bud mosaic of *Prunus* spp., apple union necrosis and decline, yellow vein and decline of grapevine	*X. americanum, X. bricolensis, X. californicum, X. intermedium, X. rivesi, X. tarjanense*

*Taxonomic proposals currently under consideration would remove CRLV and SLRSV from the genus *Nepovirus* and place them in the new genera *Cheravirus* and *Sadwavirus*, respectively.

Nepoviruses are responsible for causing disease in a wide range of crop plants (Table 2), particularly in long-lived, perennial crops such as fruit trees or grapevines. An exception is *Tobacco ringspot virus*, which causes bud blight in soybeans as well as diseases in other annual crops such as tobacco and cucurbits. Many nepoviruses persist by being seed-transmitted or through vegetative propagation from infected sources.

Fig. 2. Aerial view of patchy field damage caused to raspberry plants by RpRSV.

CHARACTERISTICS OF DISEASES CAUSED BY NEMATODE-TRANSMITTED VIRUSES

Unlike insect-transmitted viruses, tobraviruses and nepoviruses cannot be spread over long distances by their vectors and, although they have other means by which they can move from site to site, their distribution is often limited. In contrast, once established at a site they may be very persistent and difficult to eradicate; the soil itself acts as a reservoir of infection between crops.

The nematodes move only short distances, in the order of centimetres, laterally in the soil and hence the distributions of the viruses they transmit tend to be patchy within a particular field (Fig. 2).

Movement of soil during cultivation is a significant factor in local spread and the patches of infection are often elongated in the direction of cultivation. Soil movement may also be an important means of spread to new sites. This may be either natural soil movement, for example by wind blow, or the result of human activities, for example on the wheels of agricultural machinery or together with planting material.

The other principal means by which nematode-borne viruses move from site to site is in propagules of their plant hosts. Many are transmitted in the seed produced on infected plants and infected seed may be spread by natural means or, in the case of crop species, by man. Movement of infected vegetative propagating material of perennial crop species, including tubers, bulbs and corms, is also an important means of spread in some instances. In general, nematode-transmitted viruses either have very wide plant host ranges, including wild plants and weed species, or their hosts are long-lived perennial species, such as fruit trees. Thus, the cycle of transmission from vector to plant and back to vector is unlikely to be broken by lack of availability of susceptible plant hosts.

CONTROL MEASURES

Measures that have been used in an attempt to control diseases caused by nematode-borne viruses can be conveniently divided into four categories: avoidance of introduction, chemical methods, cultural methods and plant resistance.

Avoidance of introduction

Because nematode-borne viruses are so difficult to eliminate from a site, it is sensible to take precautions not to introduce them into new sites. Particularly in the case of perennial crops, it is important to start by planting healthy material. In the UK, there are voluntary certification schemes for planting material of raspberries, strawberries and *Ribes* (blackcurrants and gooseberries), although there is no legal requirement for growers to plant certified stock. In the England and Wales schemes (but not in the Scottish schemes), stock for certification can be grown only on land that has previously been tested for being free from soil-borne viruses. In the case of potatoes, the identification of cultivars that can be infected with TRV without showing obvious symptoms, and that can then be sources for acquisition of the virus by nematodes (Xenophontos *et al.*, 1998), suggests that seed tubers of these cultivars may be a means of spread of the virus to new sites. However, in the absence of any certification scheme or testing regime, there is no way for the ware grower to determine the health status of his seed.

Nematodes, with or without the virus they transmit, can only too easily be introduced to new sites along with soil. It is therefore sensible to take measures to minimize the movement of soil from infected sites. Some virus-vector nematodes, e.g. *Xiphinema americanum* in Europe, are controlled by plant quarantine regulations.

Chemical methods

Chemicals that can be applied to the soil to control nematodes fall into two classes. Fumigants, such as D-D (dichloropropane-dichloropropene), methylbromide and dazomet, kill nematodes and weed seeds, but are phytotoxic. Indeed, they are also toxic to mammals and their safe application is a specialist task. Non-fumigants include oxime-carbamates, such as aldicarb and oxamyl, as well as organophosphates and methylcarbamates. They do not kill the nematodes, but disrupt their nervous system and prevent them feeding. However, the effect is reversible and the chemicals are effective for only a single season at most. Moreover, the effect of both classes of compound is limited to the upper layers of the soil and nematodes are likely to survive at depths to which the chemicals do not penetrate. Nevertheless, reductions of 80–90 % in nematode populations have been achieved and this is sufficient to decrease disease incidence in annual crops to very low levels (Thomason & McKenry, 1975). The effects of nematicide application on crop plant development in the field are illustrated in Fig. 3.

Indeed, aldicarb is widely used to control spraing disease caused by TRV in potato (Figs 4 and 5). However, like most chemical nematicides and fumigants, aldicarb is environmentally hazardous and its use is banned in several countries, including Bulgaria and Sweden, and is severely restricted in many others, including Belgium, Canada, Germany, Israel, Italy, Norway and Switzerland. Moreover, aldicarb will be totally banned throughout the European Union from 2007. In perennial crops, such as raspberries, nematicides cannot be used. Control of nepoviruses has been attempted by a combination of healthy planting stock, soil sterilization and weed control, but has rarely been successful (Hewitt *et al.*, 1962; Pitcher, 1975). The alternative is to avoid planting on affected sites.

Cultural methods

Nematode activity is inhibited at low soil temperatures and this is probably why, in the Netherlands, delaying bulb planting from September to December decreases infection by TRV. Conversely, the practice of planting crops under polythene, designed to conserve warmth and moisture at the early stages of plant growth, may incidentally provide ideal conditions for nematode activity and consequently virus transmission at a critical developmental stage of the crop.

Crop rotation is unlikely to be beneficial, because the viruses and their vectors have very wide host ranges. Even in the case of *Xiphinema index*, which reproduces only on grapevine and fig, and is the vector of *Grapevine fanleaf virus* (GFLV), at least five grapevine-free years are required to eliminate vector and virus, because the nematodes can survive on root fragments remaining in the soil (Taylor & Brown, 1997). Trichodorids will reproduce on barley, but it is a very poor host for TRV. Thus, growing barley for 3 or more years can eliminate the virus from the nematode population, although reintroduction of the virus to a vector population with known ability to transmit remains a danger. Use of non-host crops in these ways needs to be coupled with efficient weed control, because weeds can provide alternative hosts for vector, virus or both. In contrast, rigorous weed control in a potato crop has been shown to increase the incidence of TRV infection, presumably because the vector nematodes were compelled to feed on the potatoes, a less preferred host (Cooper & Harrison, 1973).

Fig. 3. Effect of nematicide application on the control of disease caused by ArMV in strawberry. Treated plants are on the left and untreated plants are on the right.

Fig. 4. Delayed emergence of potato (cv. Nadine) caused by TRV infection (left) compared to uninfected plants (right).

Fig. 5. Spraing disease of potato (cv. Santé) tuber caused by infection with TRV.

Plant resistance

For many nematode-borne viruses, the optimum method of control is likely to be the deployment of virus-resistant cultivars, but because the diseases are of only local importance, the development of such cultivars has rarely been a high priority in breeding programmes. Natural sources of resistance to GFLV have been deployed with limited success (Walker *et al.*, 1994). In raspberries, genes for immunity to *Arabis mosaic virus* (ArMV), *Raspberry ringspot virus* (RpRSV) and *Tomato black ring virus* (TBRV) have been identified (Jennings, 1964), but cultivars containing these genes have not proved useful in practice because of the rapid emergence of resistance-breaking virus variants from field populations. Although potato cultivars resistant to or tolerant of TRV infection are available, spraing-susceptible cultivars are frequently grown even at sites where the virus is known to be present, because the choice of cultivar is largely market-driven; the requirements of processors and supermarkets for produce with particular characteristics severely limits the grower's choice of cultivar.

Tobacco plants resistant to TRV have been produced by inserting the coat protein (CP) gene into the plant genome (Van Dun & Bol, 1988; Angenent *et al.*, 1990). These plants were found to be resistant to infection by mechanical leaf inoculation with the homologous TRV isolate or isolates that were serologically closely related. However, they were not resistant to infection by inoculation with viral RNA or by a serologically distinct isolate; this form of resistance would therefore not be effective against the wide variety of TRV serotypes that are known to occur in nature. Furthermore, it has been demonstrated that this type of CP-mediated resistance was not effective when the virus was transmitted by vector nematodes (Ploeg *et al.*, 1993).

The only other approach to pathogen-derived transgenic resistance to TRV that has been tested used a construct derived from the polymerase read-through gene (Bem *et al.*, 2000). In glasshouse tests, tobacco plants transformed with this construct were protected against heterologous isolates of TRV and this approach may provide a practically useful source of resistance.

Transgenes based on various parts of the genomes of several nepoviruses have been shown to confer resistance to the homologous virus in model plants (*Nicotiana* spp.) (e.g. Pacot-Hiriart *et al.*, 1999; Spielmann *et al.*, 2000; Sun *et al.*, 2001). Some of these constructs have also been used to transform crop plants, mainly grapevines, but no agriculturally useful resistance has yet been described.

Little attention has been paid to possible resistance to the vectors. Resistance to *X. index* feeding occurs in some *Vitis* species, including some hybrids used as rootstocks (Coiro *et al.*, 1990), and shows promise for the control of GFLV.

INVESTIGATION OF THE TRANSMISSION MECHANISM

Working with nematodes

The study of virus transmission by nematodes is made difficult by several factors, the most obvious of which is that the nematodes transmit virus to the roots of plants which in the natural situation are hidden from view in the soil. The nematodes themselves cannot be propagated in pure culture and for most experiments have to be isolated by washing and filtration of field soil. Even after this procedure the virus vector nematodes are obtained as a mixture with many other non-vector soil nematodes and may require individual picking by hand under a microscope before they can be used. To study transmission efficiency and specificity it is necessary to identify natural populations of nematodes that are free of virus (non-viruliferous). Virus-infected plants are then grown in soil containing known numbers of these nematodes to allow them to acquire the virus. Subsequently, the nematodes are collected by washing them from the soil and are transferred to fresh, sterile soil containing healthy plants. Lastly, the plants are taken from the soil, their roots are washed to remove the nematodes and the presence of (transmitted) virus within the roots is assayed. Even when the transmission assay is modified, using very small seedlings grown in sterilized sand, the entire procedure can take 3 months to complete.

Nevertheless, despite these difficulties, several detailed studies of virus-vector nematodes have been carried out. The feeding of these nematodes *in situ* on plant roots has been observed using an underground laboratory (Pitcher, 1967), although more frequently these experiments have used plants grown *in vitro* on the surface of agar plates (Wyss & Zunke, 1985; Karanastasi, 2001; Karanastasi *et al.*, 2001). Studies are also in progress to understand the biochemistry and molecular biology of virus-vector nematodes and include the preparation of antibodies against molecules on the outer cuticle of *Xiphinema* and to proteins secreted by the gland cells, which are thought to play a central role in the feeding process (Chen *et al.*, 2003). In addition, by the use of transmission tests with recombinant viruses whose full sequences are known, it has been possible to determine which of the viral proteins encoded by tobraviruses and nepoviruses are required for successful transmission by nematodes.

Sites of virus retention in the nematode

Trichodorid and longidorid nematodes are both ectoparasites that remain outside the roots that they feed on and use an extendable stylet to pierce cells generally located at, or just behind the root tip (Fig. 6). Although both juvenile and adult nematodes can transmit viruses, the ability to transmit is lost during the moult that accompanies the transition between these stages. In addition, virus is not passed from adult females into their eggs (Taylor & Raski, 1964). This indicates that virus is not carried within the

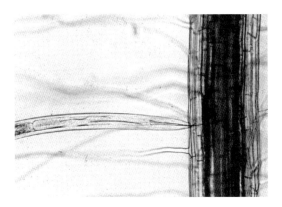

Fig. 6. *X. index* feeding on a root of a fig plant. Bar, 100 µm.

tissues of the nematode's body, but is associated with internal cuticular surfaces of the feeding apparatus that are shed or altered during the moult.

In longidorids, the feeding apparatus consists of a hollow, retractable tube (the odontostyle) that is connected to the anterior end of the oesophagus (Taylor & Brown, 1997). After piercing the plant cell wall, the cell contents pass through the central cavity of the odontostyle and into the oesophagus. When longidorids have been allowed to feed on the roots of plants infected with a virus that they transmit, virus particles are found specifically associated with surfaces within the feeding apparatus, but the location of these surfaces varies in different nematode species. In most *Longidorus* vectors, virus particles are adsorbed to the inner surface of the odontostyle. In *Xiphinema* species, virus particles attach to the lining throughout the oesophagus, but not to the odontostyle (Taylor & Robertson, 1969, 1970a).

Trichodorids possess a solid, curved tooth or stylet (the onchiostyle), which is used to penetrate the root cell wall (Fig. 7). Subsequently, the onchiostyle is withdrawn from the cell wall and a feeding tube is formed, probably from secretions of the oesophageal glands, which extends from the nematode into the root cell and acts as a suction tube for the ingestion of the cell contents. This feeding tube is left behind when the nematode ceases feeding and moves away. Tobravirus particles become associated with the cuticular lining throughout the oesophagus, but not with the onchiostyle (Taylor & Robertson, 1970b). Virus particles appear to be extremely stable when adsorbed to the nematode oesophageal cuticle. It was reported that *Paratrichodorus pachydermus* nematodes could transmit virus even after soil samples containing them had been stored for 3 years (van Hoof, 1970).

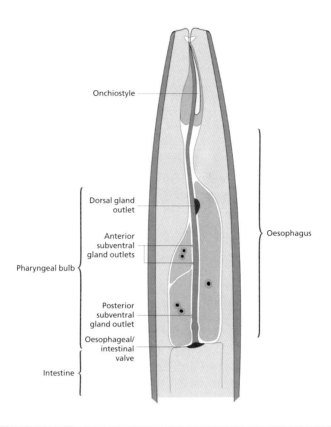

Fig. 7. Diagram of a longitudinal section of the head of a trichodorid nematode.

A more recent study used electron microscopy to determine the pattern of distribution of defined TRV and PEBV isolates in the oesophagus following uptake by feeding trichodorids in both compatible combinations, where transmission of the particular virus isolate by the nematode species did occur, and non-compatible combinations, where transmission of the virus isolate by the nematode did not occur (Karanastasi *et al.*, 1999; Karanastasi, 2001). For the transmitting combinations, two patterns of virus particle retention were noted. For *P. pachydermus* with TRV isolates PaY4 or PpK20 and *Paratrichodorus hispanus* with TRV-PhP5, virus particles were attached to the lining of all parts of the oesophagus. Alternatively, for *Paratrichodorus anemones* with TRV-PaY4 and *Trichodorus primitivus* with TRV-TpO38, virus particles were located only in the middle and posterior parts.

For the non-transmitting combination of PEBV-TpA56 with *P. pachydermus*, virus particles were observed only in the posterior part of the pharyngeal bulb, close to the junction of the oesophagus with the intestine. A similar distribution was seen for the combinations of PEBV-TpA56 with *P. anemones*, PEBV-TpA56 with *T. primitivus* and TRV-TpO1 with *T. primitivus*. The last two of these combinations were found to be competent for transmission in previous work but did not transmit in this series of experiments. These observations suggested that the location of the virus particles in relation to the outlets of the three secretory glands in the pharyngeal bulb determines whether the virus is transmitted or not. Secretions from two glands (the anterior subventral gland and the dorsal gland), whose outlets are located in the anterior part of the pharyngeal bulb, are suggested to move forwards and to be released into the plant during feeding. Virus particles that are located anterior of the outlets of these glands could be flushed into the plant along with the secretions. In contrast, the third gland, the posterior subventral gland, probably functions during ingestion and its secretions are moved backwards into the intestine. Thus, virus particles located in the posterior part of the pharyngeal bulb are pushed into the intestine rather than forwards into the plant. For transmission to take place, virus particles must bind to the surface of the nematode oesophagus forward of the anterior ventrosublateral gland outlet. In contrast, binding in the posterior part of the pharyngeal bulb does not result in transmission during nematode feeding. Thus, retention of virus particles by the nematode is not sufficient to ensure successful transmission, and directed release of retained particles is a crucial step in the process.

This accounts for the observation that although a large proportion of nematodes may acquire virus during feeding, only a small number of them may transmit the virus. For example, six of eight *P. pachydermus* nematodes examined by electron microscopy after feeding on infected plants were found to contain retained virus particles. However, when transferred to bait plants, only 2 of 50 nematodes from the same population transmitted the virus.

For longidorids too, there is evidence that uptake and even retention of virus by the nematode does not guarantee transmission. For example, TRV was found, by indicator tests, in the body of *Xiphinema diversicaudatum*, a vector of nepoviruses, but was not transmitted by this nematode during feeding (van Hoof, 1967). In a recent study of GFLV and ArMV, RT-PCR was used to examine the presence of virus in nematodes fed on plants infected with each of the viruses (Belin *et al.*, 2001). ArMV was not detected in extracts of *X. index*, whereas GFLV was readily detected in these nematodes, suggesting that ArMV is perhaps not transmitted by these nematodes because it is not retained following feeding. Because *Xiphinema* spp. nematodes are able to retain and transmit nepoviruses over longer periods than are *Longidorus* spp. nematodes, it was suggested that these two groups of nematodes might differ in their transmission mechanisms. Thus, it was proposed that for *Xiphinema*, specificity of transmission is determined by specific adsorption of virus particles at sites of retention. However, for

Longidorus, it was proposed that retention is non-specific, but differences in dissociation of virus particles determine whether or not transmission occurs. Some support for this notion came from electron microscopy studies of *Longidorus macrosoma* in which particles of both the Scottish and English strains of the nepovirus RpRSV were found to be adsorbed to the inner surface of the odontostyle even though these nematodes only transmitted the English strain (Trudgill & Brown, 1978). Similarly, in an earlier study it was found that although two different nepoviruses, RpRSV and TBRV, were retained in the mouthparts of *Longidorus elongatus*, only one was naturally transmitted during feeding (Taylor & Robertson, 1969).

Whatever the sites of retention of virus particles, transmission must involve dissociation of particles from these sites. This is probably brought about by changes in pH or ionic conditions, or by enzymes in the glandular secretions that accompany the initiation of feeding. Only a fraction of the retained particles are released at any one time and the vectors can transmit serially to several plants.

It is also interesting to note that the host plant can influence virus transmission, as experiments have shown that *X. index* can acquire GFLV from roots of both grapevine and *Chenopodium quinoa*, but can transmit the virus only to grapevine. Apparently, *C. quinoa* lacks a 'factor' that is necessary for release of virus from the nematode, or the root cells of *C. quinoa* do not support virus replication after delivery by nematodes (Trudgill & Brown, 1980).

Nematode feeding behaviour

The paradox of plant virus transmission by nematodes is that the virus can only multiply when introduced into a living cell, but that the cell contents are consumed during nematode feeding. There have been several studies of trichodorid feeding behaviour, all of which concluded that even partial withdrawal of cell contents by the nematode resulted in cell death. However, in a recent study (Karanastasi *et al.*, 2001), observation of *P. anemones* nematodes feeding on *Nicotiana tabacum* roots in a growth chamber using video-enhanced interference light microscopy produced different results.

Initially the nematode moved around an area of root apparently searching for an actively growing region. The nematode then 'explored' the surface of the chosen root by repeatedly rubbing its 'lips' against the root epidermis, a process which normally lasted between 2 and 4 h. Next, the nematode began a series of 10–15 probes using its stylet usually without piercing the cell wall. Subsequently, a further number of cells (usually four) was pierced by repeated thrusts of the onchiostyle. However, the nematode then departed within a few seconds without causing any substantial or irreversible damage to these cells. This phase of 'unsuccessful' feeding usually lasted for between 15 and 30 s and was followed by the true feeding phase.

Here the nematode fed continuously for between 1 to 2 h, moving from cell to cell, although during the first 30 min of this period up to half of the cells were not completely penetrated and the nematode abandoned these cells before any cell contents were withdrawn. During the next 30 min the feeding was much more complete, the nematode consuming usually 20 to 25 cells. In the last stage, incomplete feeding occurred in up to 10 % of targeted cells, sometimes following disturbance by other nematodes. Of cells that were attacked during this phase, 10–25 % were observed to remain alive, as shown by the continuation of cytoplasmic streaming, and possibly these cells are the source of infection during successful virus transmissions. It is also relevant to note that cells immediately adjacent to those fed upon by nematodes were not affected by the process and remained alive.

During actual feeding, after penetration of the cell wall, the nematode uses pharyngeal gland secretions to form a flexible feeding tube 2–3 μm long around the stylet tip. The tube is attached to the cell wall, may become longer as the nematode feeds, and remains attached to the cell wall after feeding and departure of the nematode. Once the tube is formed, the nematode continues to thrust its stylet, but at a slower rate, and during this time a gelatinous plug is formed on the inner surface of the cell wall that acts as a one-way valve allowing secretions to pass into the cell from the nematode but preventing cell contents leaking out of the cell. Salivation usually lasts for less than 1 min and triggers a massive disorganization of the cell architecture whereby the cytoplasm migrates from all areas of the cell to the feeding site. Often the nucleus also migrates to the feeding site where it swells and ruptures. Occasionally this process is interrupted and the nematode departs before the nucleus becomes disorganized. In these few cases the cell may survive, providing an opportunity for virus introduced during salivation to initiate an infection.

Once the cell contents have been sufficiently aggregated, the nematode thrusts its stylet deeper to penetrate the plug and widen the aperture of the feeding tube. The cytoplasm and collapsed nucleus are then ingested through the feeding tube into the nematode during a period of between 10 and 90 s. If the cell being fed upon was itself infected with virus it must be at this stage that virus particles are adsorbed onto the oesophageal surface to begin the transmission cycle. It is not known what the minimum number of virus particles necessary to initiate an infection by nematode transmission is. However, as the ingested plant cell contents are not fully liquefied and are unlikely to flow freely within the nematode oesophagus before passing into the intestine, it is likely that, to ensure sufficient virus particles adsorb to the oesophageal surface, many thousands would have to be ingested.

ANALYSIS OF VIRUS PROTEINS INVOLVED IN NEMATODE TRANSMISSION

Transmission of tobraviruses and nepoviruses by nematodes requires a specific interaction between the virus and its vector. The specificity can exist at a wider level when we consider why it is, for example, that tobraviruses are transmitted only by trichodorids, but not by longidorids or any other group of plant-parasitic nematodes. At a narrower level we can ask why it is that one isolate of TRV (PpK20) can be transmitted by a particular species of trichodorid, *P. pachydermus*, whereas another TRV isolate (PaY4) is transmitted by both *P. pachydermus* and *P. anemones*. The specificity in the interaction between virus and vector could occur at the level of virus retention in the nematode mouthparts or of release into the plant during feeding, and there is evidence for both. However, little is known about the chemical nature of the sites of virus retention in the nematode. In contrast, for both nepoviruses and tobraviruses, experiments carried out using cloned viruses containing defined mutations have revealed that several viral proteins may be important determinants of transmissibility (MacFarlane, 2003).

Nepoviruses and tobraviruses do not appear to share any structural characteristics that may explain their transmission properties, as nepoviruses form spherical virus particles, whereas tobraviruses form rod-shaped particles. In early studies it was found that there was often a correlation between the specific serotype of the transmitted virus and the particular species of vector nematode. This was interpreted as evidence that the virus CP, of both tobraviruses and nepoviruses, was an important transmission determinant.

Both nepoviruses and tobraviruses have genomes divided into two RNA molecules (Mayo & Robinson, 1996; MacFarlane, 1999). Further experiments made use of pseudorecombinant virus isolates in which the two viral RNAs were separated and mixed with each of the viral RNAs from a different isolate. The behaviour of the pseudorecombinants could be compared to those of the parental isolates, revealing whether RNA1 or RNA2 (or both) determined the transmission characteristics. Results from these studies showed that for both nepoviruses and tobraviruses the smaller genomic RNA (RNA2) determined both the serological and transmission properties of the virus.

More recently, sequencing of the genomes of many nepovirus and tobravirus isolates has revealed that RNA1 of these viruses encodes proteins involved in RNA replication and, for tobraviruses, viral cell-to-cell movement. RNA2 of nepoviruses encodes the CP as well as proteins involved in RNA2 replication and virus movement. For the tobraviruses, RNA2 varies considerably between isolates in size and gene content,

Nepoviruses: GFLV

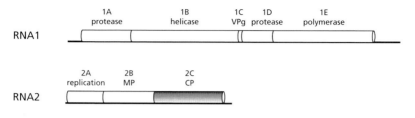

Fig. 8. Genome organization of GFLV. MP, Virus movement protein; CP, coat protein. The shaded section denotes virus protein demonstrated to have a role in nematode transmission.

but always encodes at least the CP. To determine more specifically which of the different viral genes encoded by RNA2 of nepoviruses and tobraviruses are required for nematode transmission, several groups have constructed and systematically mutated infectious cDNA clones of these viruses.

Nepovirus transmission determinant

The molecular details of nepovirus transmission have concentrated on GFLV (Fig. 8), which is transmitted by *X. index*, and ArMV, which is transmitted by *X. diversicaudatum*. These viruses have similar genome organizations and RNA2 encodes a polyprotein that is cleaved into three functional proteins: 2A (28 kDa), required for replication of RNA2; 2B (38 kDa), a cell-to-cell movement protein that localizes to tubules that allow passage of virus between cells; and 2C (56 kDa), the virus CP (Ritzenthaler *et al.*, 1995; Gaire *et al.*, 1999). When recombinant GFLV RNA2 carrying various portions of ArMV RNA2 was examined, it was found that the ArMV 2C (CP) could not encapsidate GFLV RNAs and so could not be tested for transmission (Belin *et al.*, 1999). Nevertheless, replacing either or both of the GFLV 2A and 2B proteins with those of ArMV did not affect the ability of the virus either to infect plants systemically or to be transmitted by *X. index* (Belin *et al.*, 2001). Thus, there is no evidence to link the proteins from RNA1 or the 2A and 2B proteins with transmission, and currently it is thought that the CP is the only virus protein involved in transmission by nematodes.

Tobravirus transmission determinants

The tobravirus RNA2 frequently undergoes deletion of internal sequences and recombination of the 3′ terminal region with the analogous part of RNA1. These mutations occur particularly when the virus is propagated by repeated mechanical passaging rather than by nematode transmission, although recombination probably does occur in the field even with transmission by nematodes (Vassilakos *et al.*, 2001). Indeed, transmission by nematodes has been proposed to act as a bottleneck in the

Tobraviruses: TRV

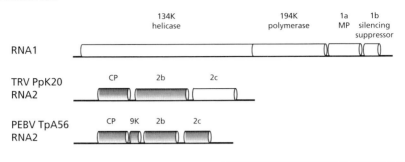

Fig. 9. Tobravirus genome organization. MP, Virus movement protein; CP, coat protein. The shaded sections denote virus proteins demonstrated to have a role in nematode transmission.

tobravirus life cycle that maintains the fitness of the virus population by screening out defective virus genomes (Hernández *et al.*, 1996). One consequence of the instability of RNA2 is that many of the tobravirus genome sequences, obtained from isolates that have been maintained for long periods in the glasshouse, are incomplete and lack one or more viral genes that we now know to be involved in transmission. For example, the one available sequence of PepRSV RNA2 encodes only the CP and we would not expect this isolate to be transmissible by nematodes (MacFarlane, 1999).

By ensuring that the viruses were analysed as soon as possible after isolation from field nematodes, it has been possible to sequence and make infectious clones of four transmissible tobravirus isolates (MacFarlane *et al.*, 1996, 1999; Hernández *et al.*, 1997; Vassilakos *et al.*, 2001). Analysis of these cloned isolates, TRV PpK20, TRV PaY4, TRV TpO1 and PEBV TpA56, has allowed us identify which viral genes are involved in transmission and how the specificity of the vector nematode might be determined.

The RNA2 of TRV PpK20 and TRV PaY4 encodes three genes; the CP gene, the 2b gene and the 2c gene. TRV TpO1 and PEBV TpA56 RNA2 encodes the same three genes as well as a small ORF potentially encoding a 9 kDa protein located between the CP and 2b genes (Fig. 9). Western blot analysis has confirmed that the CP, 2b and 2c proteins are expressed in plants, whereas there is no direct evidence that the 9 kDa protein is expressed (Schmitt *et al.*, 1998; Visser & Bol, 1999).

The transmission determinants of three of the four cloned viruses have been identified. For PEBV TpA56, which is transmitted by *T. primitivus*, mutation of the genes encoding the CP, 2b and 2c proteins all affect transmission to some degree. Interestingly, whereas the 2b gene was absolutely required for transmission, deletion mutation of the 2c gene significantly reduced but did not abolish transmission (MacFarlane *et al.*, 1996;

Schmitt *et al.*, 1998). In contrast, although the CP and 2b genes of TRV PpK20 and TRV PaY4 were required for transmission, deletion of the 2c gene had no effect on transmission (Hernández *et al.*, 1997; Vassilakos *et al.*, 2001). A possible explanation for the differential requirement for the 2c protein in transmission might be that this protein is involved in the interaction with *Trichodorus* spp. nematodes (the vector of PEBV TpA56), but not with *Paratrichodorus* spp. nematodes (the vectors of TRV PpK20 and TRV PaY4).

The tobravirus 2b protein is not an integral component of the virus particle, and does not co-purify and is not co-immunoprecipitated with virions. However, models for the mechanism of tobravirus transmission assume that the 2b protein must interact with virus particles and perhaps form a bridge between the virion and molecules on the nematode oesophageal surface. In this way, the tobravirus 2b protein is thought to function in a similar manner to the potyvirus helper component protein (HC-Pro), which is essential for transmission of these viruses by aphids. The C-terminal domain of the tobravirus CP, which is between 20 and 30 amino acids in length, is mostly unstructured and extends from the outer surface of the virus particle where it is the major antigenic determinant (Brierley *et al.*, 1993; Legorburu *et al.*, 1996; Blanch *et al.*, 2001). Deletion of all or part of this domain prevents nematode transmission, making this part of the CP a prime candidate for interaction with the 2b protein and/or nematode oesophageal cuticle. A weak interaction between the C-terminal domain of the TRV PpK20 CP and the 2b protein was detected using the yeast two-hybrid system and, more recently, immunogold localization (IGL) experiments have shown that in plants the TRV and PEBV 2b proteins associate with virus particles (Visser & Bol, 1999; Vellios *et al.*, 2002).

Investigation of the transmission behaviours of TRV PpK20, which is transmitted by *P. pachydermus*, and TRV PaY4, which is transmitted by *P. pachydermus* and *P. anemones*, have implicated the 2b protein in determining vector specificity. For example, the 2b protein encoded by wild-type TRV PaY4 was able to complement transmission of mutant TRV PaY4 lacking a functional 2b gene (Vassilakos *et al.*, 2001). In contrast, although they are both transmitted by *P. pachydermus*, the 2b protein encoded by wild-type TRV PpK20 was not able to complement transmission of the TRV PaY4 2b mutant by this nematode. These results correlate with those of an IGL study of a hybrid TRV in which the 2b gene of TRV PpK20 was replaced with that of TRV PaY4. Here, the TRV PaY4 2b protein did not associate with TRV PpK20 virus particles. Indeed Western blotting revealed that, for the hybrid virus, the 2b protein did not accumulate in plants, suggesting that a failure to interact with the virus CP led to rapid degradation of the 2b protein (Vellios *et al.*, 2002). In a subsequent study, replacement of the CP C-terminal domain of the TRV PpK20 hybrid virus with the same region from the TRV PaY4 CP did not rescue accumulation of the TRV PaY4 2b protein and association with virions (R. Holeva and S. MacFarlane, unpublished). Our current hypothesis is that vector specificity is determined primarily by the 2b protein, but that there is also specificity in the interaction between the CP and the 2b protein. In addition, the C-terminal domain may not be the only part of the CP that interacts with the 2b protein.

The tools and techniques are now in place to enable us to understand at the molecular level the structure and surface characteristics of the viral proteins that mediate interaction with the nematode vector. To reach a similar level of understanding of the nematodes themselves will require a significant investment of effort and ingenuity that is essential if we are to fully appreciate the subtleties and complexities of the transmission process.

ACKNOWLEDGEMENTS

We would like to thank our colleagues at SCRI for helpful discussions and for providing us with images for this paper. SCRI is grant-aided by the Scottish Executive Environment and Rural Affairs Department.

REFERENCES

Angenent, G. C., Van den Ouweland, J. M. V. & Bol, J. F. (1990). Susceptibility to virus infection of transgenic tobacco plants expressing structural and nonstructural genes of tobacco rattle virus. *Virology* **175**, 191–198.

Barker, K. R. & Koenning, S. R. (1998). Development of sustainable systems for nematode management. *Annu Rev Phytopathol* **36**, 165–205.

Belin, C., Schmitt, C., Gaire, F., Walter, B., Demangeat, G. & Pinck, L. (1999). The nine C-terminal residues of the Grapevine fanleaf nepovirus movement protein are critical for systemic virus spread. *J Gen Virol* **80**, 1347–1356.

Belin, C., Schmitt, C., Demangeat, G., Komar, V., Pinck, L. & Fuchs, M. (2001). Involvement of RNA2-encoded proteins in the specific transmission of Grapevine fanleaf virus by its nematode vector *Xiphinema index*. *Virology* **291**, 161–171.

Bem, F., Robinson, D. J., Barker, H., Reavy, B. & Plastira, V. (2000). Viral protection in tobacco plants transformed with the 59K polymerase read-through gene of tobacco rattle tobravirus. In *Abstracts of the EMBO Workshop, Plant Virus Invasion and Host Defence, Kolymbari, Greece*, p. 58.

Blanch, E. W., Robinson, D. J., Hecht, L. & Barron, L. D. (2001). A comparison of the solution structures of tobacco rattle and tobacco mosaic viruses from Raman optical activity. *J Gen Virol* **82**, 1499–1502.

Brierley, K. M., Goodman, B. A. & Mayo, M. A. (1993). A mobile element on a virus particle surface identified by nuclear magnetic resonance spectroscopy. *Biochem J* **293**, 657–659.

Chen, Q., Curtis, R. H., Lamberti, F., Moens, M., Brown, D. J. F. & Jones, J. T. (2003). Production and characterisation of monoclonal antibodies to antigens from *Xiphinema index*. *Nematology* **5**, 349–366.

Coiro, M. I., Taylor, C. E., Borgo, M. & Lamberti, F. (1990). Resistance of grapevine rootstocks to *Xiphinema index. Nematologia mediterranea* **18**, 119–121.

Cooper, J. I. & Harrison, B. D. (1973). The role of weed hosts and the distribution and activity of vector nematodes in the ecology of tobacco rattle virus. *Ann Appl Biol* **73**, 53–66.

Gaire, F., Schmitt, C., Stussi-Garaud, C., Pinck, L. & Ritzenthaler, C. (1999). Protein 2A of Grapevine fanleaf nepovirus is implicated in RNA2 replication and colocalises to the replication site. *Virology* **264**, 25–36.

Griffin, G. D. & Epstein, A. H. (1964). Association of the dagger nematode, *Xiphinema americanum*, with stunting and winter kill of ornamental spruce. *Phytopathology* **54**, 177–180.

Harrison, B. D. & Robinson, D. J. (1986). Tobraviruses. In *The Plant Viruses*, Vol. 2, pp. 339–369. Edited by M. H. V. van Regenmortel & H. Fraenkel-Conrat. New York: Plenum.

Hernández, C., Carette, J. E., Brown, D. J. F. & Bol, J. F. (1996). Serial passage of tobacco rattle virus under different selection conditions results in deletion of structural and non-structural genes in RNA2. *J Virol* **70**, 4933–4940.

Hernández, C., Visser, P. B., Brown, D. J. F. & Bol, J. F. (1997). Transmission of tobacco rattle virus isolate PpK20 by its nematode vector requires one of the two non-structural genes in the viral RNA2. *J Gen Virol* **78**, 465–467.

Hewitt, W. B., Goheen, A. C., Raski, D. J. & Gooding, G. W. (1962). Studies on virus diseases of the grapevine in California. *Vitis* **3**, 57–83.

Jennings, D. L. (1964). Studies on the inheritance in the red raspberry of immunities from three nematode-borne viruses. *Genetica* **35**, 152–164.

Karanastasi, E. (2001). *Acquisition, retention and transmission of tobravirus particles by nematodes of the family Trichodoridae, and intervention of the trichodorid feeding behaviour and pharyngeal ultrastructure.* PhD thesis, University of Dundee.

Karanastasi, E., Roberts, I. M., MacFarlane, S. A. & Brown, D. J. F. (1999). The location of specifically retained tobacco rattle virus particles in the oesophageal tract of the vector *Paratrichodorus anemones. Helminthologia* **36**, 62–63.

Karanastasi, E., Wyss, U. & Brown, D. (2001). Observations on the feeding behaviour of *Paratrichodorus anemones* in relation to tobravirus transmission. *Meded Rijksuniv Gent Fak Landbouwkd Toegep Biol Wet* **66**, 599–608.

Legorburu, F. J., Robinson, D. J. & Torrance, L. (1996). Features on the surface of the tobacco rattle tobravirus particle that are antigenic and sensitive to proteolytic digestion. *J Gen Virol* **77**, 855–859.

MacFarlane, S. A. (1999). The molecular biology of the tobraviruses. *J Gen Virol* **80**, 2799–2807.

MacFarlane, S. A. (2003). Molecular determinants of the transmission of plant viruses by nematodes. *Mol Plant Pathol* **4**, 211–215.

MacFarlane, S. A., Wallis, C. V. & Brown, D. J. F. (1996). Multiple genes involved in the nematode transmission of pea early browning virus. *Virology* **219**, 417–422.

MacFarlane, S. A., Vassilakos, N. & Brown, D. J. F. (1999). Similarities in the genome organization of tobacco rattle virus and pea early-browning virus isolates that are transmitted by the same vector nematode. *J Gen Virol* **80**, 273–276.

MacFarlane, S. A., Neilson, R. & Brown, D. J. F. (2002). Nematodes. *Adv Bot Res* **36**, 169–198.

Mayo, M. A. & Robinson, D. J. (1996). Nepoviruses: molecular biology and replication. In *The Plant Viruses, Vol. 5, Polyhedral Virions and Bipartite RNA Genomes*, pp. 139–185. Edited by B. D. Harrison & A. F. Murant. New York: Plenum.

Pacot-Hiriart, C., le Gall, O., Candresse, T., Delbos, R. P. & Dunez, J. (1999). Transgenic tobaccos transformed with a gene encoding a truncated form of the coat protein of tomato black ring nepovirus are resistant to viral infection. *Plant Cell Rep* **19**, 203–209.

Pitcher, R. S. (1967). The host–parasite relations and ecology of *Trichodorus viruliferous* on apple roots, as observed from an underground laboratory. *Nematologica* **13**, 547–557.

Pitcher, R. S. (1975). Chemical and cultural control of nettlehead and related virus diseases of hop. In *Nematode Vectors of Plant Viruses*, pp. 447–448. Edited by F. Lamberti, C. E. Taylor & J. W. Seinhorst. London: Plenum.

Ploeg, A. T., Brown, D. J. F. & Robinson, D. J. (1992a). Acquisition and subsequent transmission of tobacco rattle virus isolates by *Paratrichodorus* and *Trichodorus* nematode species. *Neth J Plant Pathol* **98**, 291–300.

Ploeg, A. T., Brown, D. J. F. & Robinson, D. J. (1992b). The association between species of *Trichodorus* and *Paratrichodorus* vector nematodes and serotypes of tobacco rattle tobravirus. *Ann Appl Biol* **121**, 619–630.

Ploeg, A. T., Robinson, D. J. & Brown, D. J. F. (1993). RNA2 of tobacco rattle virus encodes the determinants of transmissibility by trichodorid vector nematodes. *J Gen Virol* **74**, 1463–1466.

Ritzenthaler, C., Schmitt, A. C., Michler, P., Stussi-Garaud, C. & Pinck, L. (1995). Grapevine fanleaf nepovirus P38 putative movement protein is located on tubules *in vivo*. *Mol Plant–Microbe Interact* **8**, 379–387.

Schmitt, C., Mueller, A.-M., Mooney, A., Brown, D. J. F. & MacFarlane, S. A. (1998). Immunological detection and mutational analysis of the RNA2-encoded nematode transmission proteins of pea early-browning virus. *J Gen Virol* **79**, 1281–1288.

Spielmann, A., Krastanova, S., Douet-Orhant, V. & Gugerli, P. (2000). Analysis of transgenic grapevine (*Vitis rupestris*) and *Nicotiana benthamiana* plants expressing an *Arabis mosaic virus* coat protein gene. *Plant Sci* **156**, 235–244.

Sun, F., Xiang, Y. & Sanfaçon, H. (2001). Homology-dependent resistance to tomato ringspot nepovirus in plants transformed with the VPg-protease coding region. *Can J Plant Pathol* **23**, 292–299.

Taylor, C. E. & Brown, D. J. F. (1997). *Nematode Vectors of Plant Viruses*. Wallingford: CABI.

Taylor, C. E. & Raski, D. J. (1964). On the transmission of Grapevine fanleaf virus by *Xiphinema index*. *Nematologica* **10**, 489–495.

Taylor, C. E. & Robertson, W. M. (1969). The location of raspberry ringspot and tomato black ring viruses in the nematode vector, *Longidorus elongatus*. *Ann Appl Biol* **64**, 233–237.

Taylor, C. E. & Robertson, W. M. (1970a). Sites of virus retention in the alimentary tract of the nematode vectors, *Xiphinema diversicaudatum* (Micol.) and *X. index* (Thorne and Allen). *Ann Appl Biol* **66**, 375–380.

Taylor, C. E. & Robertson, W. M. (1970b). Location of tobacco rattle virus in the nematode vector *Trichodorus pachydermus* Seinhorst. *J Gen Virol* **6**, 179–182.

Thomason, I. J. & McKenry, M. (1975). Chemical control of nematode vectors of plant viruses. In *Nematode Vectors of Plant Viruses,* pp. 447–448. Edited by F. Lamberti, C. E. Taylor & J. W. Seinhorst. London: Plenum.

Trudgill, D. L. & Brown, D. J. F. (1978). Ingestion, retention and transmission of two strains of raspberry ringspot virus by *Longidorus macrosoma*. *J Nematol* **10**, 85–89.

Trudgill, D. L. & Brown, D. J. F. (1980). Effect of the bait plant on transmission of viruses by *Longidorus* and *Xiphinema* species. In *Annual Report of the Scottish Horticultural Research Institute for 1979*, p. 120. Dundee.

van Dun, C. M. P. & Bol, J. F. (1988). Transgenic tobacco plants accumulating tobacco rattle virus coat protein resist infection with tobacco rattle virus and pea early-browning virus. *Virology* **167**, 649–652.

van Hoof, H. A. (1967). Het mechanisme van de virusoverbrenging door nematoden. *Neth J Plant Pathol* **73**, 193–194.

van Hoof, H. A. (1970). Onderzoek van Virussen die Samenhangen me de Grond Aardappel-rattelvirus. In *Annual Report of the Institute for Plant Protection for 1969*, pp. 91–92. Wageningen, The Netherlands.

Vassilakos, N., Vellios, E. K., Brown, E. C., Brown, D. J. F. & MacFarlane, S. A. (2001). Tobravirus 2b protein acts *in trans* to facilitate transmission by nematodes. *Virology* **279**, 478–487.

Vellios, E., Duncan, G., Brown, D. & MacFarlane, S. (2002). Immunogold localization of tobravirus 2b nematode transmission helper protein associated with virus particles. *Virology* **300**, 118–124.

Visser, P. B. & Bol, J. F. (1999). Nonstructural proteins of *Tobacco rattle virus* which have a role in nematode-transmission: expression pattern and interaction with viral coat protein. *J Gen Virol* **80**, 3272–3280.

Walker, M. A., Wolpert, J. A. & Weber, E. (1994). Viticultural characteristics of VR hybrid rootstocks in a vineyard site infected with grapevine fanleaf virus. *Vitis* **33**, 19–23.

Wyss, U. & Zunke, U. (1985). *In vitro* feeding of root parasitic nematodes. In *Plant Nematology Laboratory Handbook,* pp. 147–162. Edited by B. M. Zuckerman, W. F. Mai & M. B. Harrison. Amhurst, MA: University of Massachusetts Agricultural Experimental Station.

Xenophontos, S., Robinson, D. J., Dale, M. F. B. & Brown, D. J. F. (1998). Evidence for persistent, symptomless infection of some potato cultivars with tobacco rattle virus. *Potato Res* **41**, 255–265.

Wolbachia host–symbiont interactions

Mark J. Taylor

Filariasis Research Laboratory, Molecular and Biochemical Parasitology, Liverpool School of Tropical Medicine, Pembroke Place, Liverpool L3 5QA, UK

INTRODUCTION

Wolbachia are obligate intracellular bacteria that infect a wide range of invertebrate hosts (O'Neill *et al.*, 1997). As a genus they are most closely related to the arthropod-transmitted bacteria *Ehrlichia* within the Rickettsiaceae (Fig. 1a). *Wolbachia* are widespread throughout insects and also infect crustaceans, mites and filarial nematodes. The extent of their distribution is still not well understood but they are already among the most abundant and widespread bacterial symbionts so far described. *Wolbachia* have evolved a diversity of associations with their hosts, ranging from being pathogens and reproductive parasites in arthropods to mutualists in filarial nematodes. A common target of *Wolbachia* is the host's reproductive system, which is manipulated to enhance the spread of the symbiont throughout populations. In arthropods, the bacteria have developed a bewildering repertoire of strategies with which to manipulate host reproduction. They can affect the fertility of hosts by controlling compatibility between infected males and uninfected females (cytoplasmic incompatibility), alter sex ratios by killing males or by inducing asexual production of female offspring (parthenogenesis) and even turn males into females (feminization). Not all associations, however, are weighted in favour of the bacteria. In a species of parasitic wasp which feeds on *Drosophila*, egg production is dependent on *Wolbachia* (Dedeine *et al.*, 2001), and in filarial nematodes the symbionts are strict mutualists, essential for larval and embryo development and the survival of adult worms (Taylor, 2002). The diversity of association and phenotype in *Wolbachia* provides an excellent opportunity to study the nature, mechanisms and evolution of these different host–symbiont interactions.

SGM symposium 63: Microbe–vector interactions in vector-borne diseases.
Editors S. H. Gillespie, G. L. Smith & A. Osbourn. Cambridge University Press. ISBN 0 521 84312 X ©SGM 2004

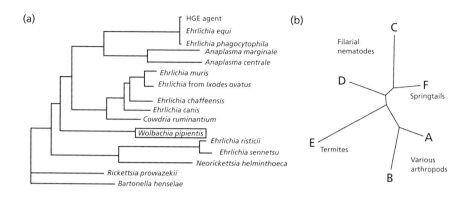

Fig. 1. (a) Phylogeny of *Wolbachia* within the rickettsiae. Phylogenetic tree based on the citrate synthase gene (*gltA*) and adapted from Fenollar *et al.* (2003). (b) Phylogenetic relationship of the six 'supergroups' of *Wolbachia* based on the *ftsZ* gene and adapted from Lo *et al.* (2002).

In this chapter I will summarize the types of *Wolbachia*–host associations, including examples that influence vectors of human disease, provide an overview of the association of *Wolbachia* with filarial nematodes, the subject of my own research, and discuss the applied aspects of using *Wolbachia* as a tool and target for disease and vector control. For a more in-depth account of the associations of *Wolbachia* and arthropods I refer the reader to the book '*Influential Passengers – Inherited Microorganisms and Arthropod Reproduction*' (O'Neill *et al.*, 1997) and other more recent reviews (Bourtzis & Braig, 1999; Stouthamer *et al.*, 1999; Bandi *et al.*, 2001; Charlat *et al.*, 2003) and websites (http://www.wolbachia.sols.uq.edu.au/index.html, http://filariasis.net).

WOLBACHIA–HOST ASSOCIATIONS IN ARTHROPODS

Cytoplasmic incompatibility (CI)

CI is a mechanism unique to *Wolbachia* (Hoffman & Turelli, 1997). CI results in embryonic lethality in mating between infected males and uninfected females or when mating pairs are infected with incompatible strains. It occurs frequently in many diverse groups of insects and mites and isopods, including vectors of human disease such as the *Culex* mosquitoes, in which it was first discovered (Yen & Barr, 1971). CI can be either unidirectional, in which incompatibility arises from mating between infected males and uninfected females, or bidirectional, where incompatibility results from crosses between individuals infected with different strains of *Wolbachia*. The degree of CI induced is dependent upon host factors, bacterial density and environmental factors. The outcome of CI is a reproductive advantage for infected females, which can mate successfully with all males, whereas uninfected females can only successfully mate with uninfected males. As only females transmit the symbionts, this promotes the spread of the infection through populations. It is this feature that has led to the proposed use of *Wolbachia* as a means to drive the infection or transgenes into vectors of disease (discussed below).

The mechanisms that cause CI are currently being hotly pursued. The bacteria are believed to somehow modify infected male sperm. This modification prevents successful fertilization unless it occurs with females infected with the same bacterial strain, which is able to 'rescue' the modified sperm. In parasitic wasps CI is a consequence of a delay in nuclear envelope breakdown resulting in asynchrony of the male and female pronuclei and the loss of paternal chromosomes at the first meiosis. These observations suggest that the modification and rescue mechanism may be due to the bacteria manipulating cell cycle regulatory proteins (Tram & Sullivan, 2002).

Feminization

The ability of *Wolbachia* to change genetic males into functional females appears to be restricted to terrestrial crustaceans and has been mostly studied in woodlice (Rigaud, 1997). In infected populations *Wolbachia* controls sex determination, as all infected females are genetically male. It is thought the plasticity of crustacean sex determination, in which males and females are easily manipulated to change sex, favours the expression of this phenotype in these hosts and its apparent absence from other arthropods. Although the mechanism of feminization has yet to be defined it is thought possibly to be due to the influence of *Wolbachia* on the androgenic hormones that control male differentiation.

Parthenogenesis

Parthenogenesis, the development of individuals from unfertilized eggs, is common throughout different orders of insects and other arthropods. *Wolbachia*-induced parthenogenesis has been described in a number of the Hymenoptera and has been most well characterized in parasitoid wasps (Stouthamer, 1997). In these species, females normally produce haploid males from unfertilized eggs and diploid females from fertilized eggs. *Wolbachia* infection, which occurs in females only, results in the production of females from both unfertilized and fertilized eggs. The generation of diploid females occurs by the process of gamete duplication, owing to the failure of chromosomal segregation. Consequently, infected females produce twice as many daughters as uninfected females, again facilitating the spread of *Wolbachia*.

Male killing

Another way of modifying the sex ratio in favour of females and their symbionts is simply to kill males, a mechanism which has been exploited by a number of bacterial species. Male-killing *Wolbachia* have been described in butterflies, ladybirds and *Drosophila* and may be the predominant cause of this phenomenon in insects (Hurst & Jiggins, 2000). The killing occurs early in male embryogenesis, although the mechanism remains unknown. The adaptive significance of male killing can be reflected in benefits to the surviving female progeny, who face reduced competition and obtain a free meal by consuming their dead brothers (Hurst & Jiggins, 2000).

Pathogenic effects

The idea of a vertically transmitted symbiont being pathogenic is counter to our understanding of the evolution of virulence. That has nevertheless been ignored by a strain of *Wolbachia* that infects *Drosophila*, which leads to tissue degeneration and death of adult flies (Min & Benzer, 1997). The so-called 'popcorn' strain (on account of the appearance of heavily infected cells) proliferates in the brain, muscles, retina and oocytes of adults but not the larval stages. How or why this strain has developed a virulence phenotype is unknown. Werren (1997) suggests two possible explanations. Either the strain is a mutant which has lost control of replication, or it is a natural variant that has been selected for higher virulence. In either case the strain could be important in dissecting the host and bacterial mechanisms which regulate bacterial growth. The virulence phenotype would also have obvious potential applications in the control of pests or vectors of disease.

Other effects on arthropod hosts

Whilst most associations of *Wolbachia* contribute little if any benefit to their arthropod hosts, some cases of partial or complete effects on host fitness and physiology exist. The most convincing evidence comes from a study on a parasitic wasp, *Asobara tabida*. In this species, removal of *Wolbachia* with antibiotics resulted in the unexpected blocking of oogenesis. No other deleterious effects were observed on male or female wasps and the effect was shown to be density dependent (Dedeine *et al.*, 2001). A remarkably similar event has also been observed in *Drosophila* with a functional mutation in *Sxl*, the master regulator of sex determination and a gene essential for oogenesis (Starr & Cline, 2002). In mutant flies infected with *Wolbachia*, oogenesis was reinstated, although infection of other mutants failed to reverse these defects, suggesting that the bacteria interact directly with the *Sxl* protein. *Wolbachia* super-infection in *Aedes albopictus* results in both CI and benefits to female longevity and rates of egg laying and hatching in infected females, which would be predicted to promote the spread of infection (Dobson *et al.*, 2002a). *Wolbachia* appears not to stimulate or modulate insect antibacterial immune responses (Bourtzis *et al.*, 2000) although the mechanisms by which it achieves immunological anonymity are unknown.

The association between *Wolbachia* and filarial nematodes nevertheless provides the clearest example of 'classical' mutualism and may be the only example within the rickettsiae (Taylor, 2002).

WOLBACHIA–HOST ASSOCIATIONS IN NEMATODES

Filarial nematodes are an important group of parasitic worms, some of which cause major human diseases throughout the tropics. The parasites have a complex life cycle involving transmission via a blood-feeding arthropod vector, which may itself be infected with its own *Wolbachia* symbionts. Phylogenetic studies, however, show that nematode and vector *Wolbachia* form distinct 'supergroups' and the source of symbiosis probably occurred independently (Bandi *et al.*, 1998; Lo *et al.*, 2002; Fig. 1b). Filarial nematodes infect a wide variety of vertebrate hosts although, inevitably, most attention has focused on disease-causing species in humans. Although bacterial symbionts are found in other nematode species (Forst & Clarke, 2002), *Wolbachia* so far appears to be confined to the filarial nematodes. The bacteria are found in the majority of species of filarial nematodes studied to date, including the major pathogenic species that infect humans (*Wuchereria bancrofti* and *Brugia malayi*, the causes of lymphatic filariasis, and *Onchocerca volvulus*, which causes onchocerciasis or 'river blindness') and animals (*Dirofilaria immitis*, the cause of dog heartworm). The few notable exceptions include a rodent filaria, *Acanthocheilonema viteae*, the human parasites *Loa loa* and *Mansonella perstans*, a deer parasite, *Onchocerca flexuosa*, and a horse parasite, *Setaria equina* (Taylor & Hoerauf, 1999; Chirgwin *et al.*, 2002; McGarry *et al.*, 2003; Büttner *et al.*, 2003; Grobusch *et al.*, 2003). These species have been particularly useful in defining the influence of bacteria on infected species. In particular, antibiotic treatment has no effect on *A. viteae* (Hoerauf *et al.*, 1999), supporting the notion that the target of antibiotic-mediated effects is *Wolbachia*. Although infected nematode species show a congruent phylogeny with their symbionts, at present the phylogeny of filariae appears unstable and it is not possible to determine whether aposymbiotic species have either 'lost' their symbionts or diverged from the lineage prior to the acquisition of *Wolbachia* (Casiraghi *et al.*, 2001). In the case of *O. flexuosa*, which is the only aposymbiotic species within the genus *Onchocerca*, it is likely that the symbiosis has been 'lost', as all other species in this genus are infected. Intriguingly, this species has a very short adult worm lifespan of 1–2 years compared with 10 years or more for the other members of the genus, which would be consistent with a role for *Wolbachia* in the survival of the long-lived adult worms.

In those species that host *Wolbachia*, infection is ubiquitous throughout all individual worms, developmental stages and isolates from different geographical areas, as would be predicted for a strictly mutualistic association. In male and female worms the bacteria colonize the hypodermal cord cells (Fig. 2), large syncytial cells that run

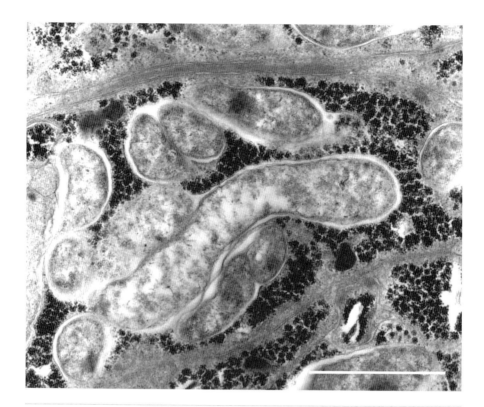

Fig. 2. Ultrastructure of *Wolbachia* in the hypodermal cells of *Onchocerca ochengi*. Bar, 1 μm.

the length of the worm and are important areas of metabolic activity. In adult female worms the symbionts are also found in the reproductive tissues, eggs and embryonic stages as a result of their vertical mode of transmission.

Evidence to support a mutualistic association comes from experiments with antibiotics to clear the *Wolbachia* infection. Antibiotics shown to be effective against nematode *Wolbachia* include the tetracyclines and rifampicin, with other antibiotics having little or no effect (Taylor, 2002; Hoerauf *et al.*, 2000; Townson *et al.*, 2000; Volkmann *et al.*, 2003). Antibiotic treatment results in dramatic inhibition or blocking of larval and embryonic development with longer-term effects on the fertility and viability of adult worms (Taylor & Hoerauf, 1999). It has so far proved impossible to 'cure' nematodes of their symbionts without causing sterility or death of the worm. The dependency of nematodes on *Wolbachia* is consistent with their long and congruent evolution (Bandi *et al.*, 1998; Casiraghi *et al.*, 2001), a genome two-thirds the size of *Wolbachia* from *Drosophila* (Sun *et al.*, 2001) and the evolutionary rate and lack of positive selection of the major outer-membrane protein (Bazzocchi *et al.*, 2000; Jiggins *et al.*, 2002).

The function(s) provided by the symbionts remains unknown, but the multiple effects of antibiotic depletion on the nematode suggest that the worms have become dependent on the bacteria for a diverse range of biological processes. Analysis of the population dynamics of *Wolbachia* in *B. malayi* gives an insight into the role of the bacteria in different life cycle stages of the worm (McGarry *et al.*, 2004). Microfilariae (L1) and the L2–L3 stages, which develop in the mosquito vector, have the lowest numbers of bacteria and the lowest ratios of bacterial cells to nematode cells. An unexpected finding was that within a week of infecting the mammalian host the bacteria rapidly grew to colonize the developing worms. This rapid rate of growth continued during the L4 stage and peaked just prior to the moult into adult worms. At this point male and female worms contain similar numbers of bacteria and these levels are maintained or slightly increased in adult males. In females the number of bacteria increases with age, predominantly because of the growth of bacteria in the reproductive tissues and embryonic stages. This rapid growth of bacteria in developing larvae in the mammalian host helps to explain why L3–L4 stages are particularly susceptible to the effects of antibiotic treatment compared with adults. The inhibition of larval and embryonic development is probably due to the bacteriostatic effect of tetracyclines, which would be most effective against the stages in which the bacteria grow most rapidly. The bacteriostatic effects of tetracycline and the low rate of growth of bacteria in adult worms probably dictate the requirement for long-term courses of antibiotics to affect adult worm viability (Langworthy *et al.*, 2000; Chirgwin *et al.*, 2003). Overall, the dynamics of bacterial infection throughout the nematode life cycle support a major role for *Wolbachia* in the development of larval and embryonic stages and the survival of adult worms in the mammalian host. In microfilariae and the developmental stages within the mosquito the symbiont may be commensal or have subtle or redundant activity. The dependency of the nematodes on *Wolbachia* has been exploited in a new approach to the treatment of filarial diseases using antibiotics (Taylor & Hoerauf, 2001).

WOLBACHIA AS A TOOL AND TARGET FOR DISEASE AND VECTOR CONTROL

Clinical trials have already established the effectiveness of doxycycline treatment in human onchocerciasis and lymphatic filariasis (Hoerauf *et al.*, 2000, 2001, 2003a, b).

A 6-week course of treatment results in the long-term depletion of *Wolbachia* and profound effects on worm fertility and reductions in microfilariae. So far macrofilaricidal activity has not been observed in these trials, although current trials of longer treatments and follow-up show very encouraging results, and emphasize that adult worm viability may only be affected after a long time. Trials are also under way to determine the shortest course of treatment to be effective at removing the bacteria, and also to determine the effect of *Wolbachia* depletion on disease pathogenesis. In addition to their role as symbionts, *Wolbachia* have also been implicated as a major cause of inflammatory responses induced by filarial worms (Taylor *et al.*, 2000; Brattig *et al.*, 2000; Saint Andre *et al.*, 2002), the stimulation of immune responses associated with disease presentation (Saint Andre *et al.*, 2002; Punkosdy *et al.*, 2003) and inflammatory adverse reactions to anti-filarial treatment (Taylor *et al.*, 2000; Cross *et al.*, 2001; Keiser *et al.*, 2002). The hope is that antibiotic depletion of bacteria will not only lead to the elimination of the parasite, but also prevent the onset of inflammatory-mediated disease (Taylor, 2003).

A variety of approaches have been suggested for the use of *Wolbachia* as a tool to control vectors of disease (Sinkins *et al.*, 1997; Bourtzis & Braig, 1999; Townson, 2002). *Wolbachia* have been reported from many vector species, including the mosquitoes *Culex* spp. and *Aedes* spp., sandflies and tsetse flies, but so far they have not been detected in anophelines (Cheng *et al.*, 2000; Ono *et al.*, 2001; Ruang-Areerate *et al.*, 2003). They may be used directly to modify sex ratios through CI in a similar manner to sterile insect techniques, use CI as a driving mechanism for transgenic vectors or be used as a paratransgenic expression system. Proof of principle experiments on the use of CI in field populations of *Culex pipiens*, a vector of lymphatic filariasis, were successfully carried out in the 1960s prior to the knowledge of bacterial symbionts' aetiology (Laven, 1967). However, a number of serious problems restrict the application of this method to the scale required to control these globally distributed vectors of disease (Sinkins *et al.*, 1997). CI could also be used to drive the spread of genetically modified insects or bacteria into natural populations (Curtis & Sinkins, 1998; Dobson, 2003). The use of the pathogenic 'popcorn' strain has been proposed as a novel approach to reducing the lifespan of disease vectors (Bourtzis & Braig, 1999; Brownstein *et al.*, 2003). The advantages of this late-stage pathogenic activity would be particularly appropriate to mosquito–pathogen dynamics, where reducing adult lifespan could prevent the transmission of disease but allow egg laying (which occurs mostly in the younger adults) and so prevent selection pressure on the vectors to become resistant (Bourtzis & Braig, 1999). Ultimately, characterization of the genes for CI and virulence could be used to transform the vectors themselves. In fact this has already occurred naturally in a beetle in which *Wolbachia* genes have been transferred directly into the host's genome (Kondo *et al.*, 2002). Further studies into the extent of lateral gene transfer in other *Wolbachia*–host systems and to determine whether expression and function of these genes occurs are eagerly awaited. Although there is good theoretical and some encouraging experimental support for these approaches, there is still a considerable amount of work to be done before the feasibility and practicality of applying *Wolbachia* and/or their adaptations to vector control strategies is realized.

FUTURE APPROACHES TO UNDERSTANDING *WOLBACHIA*–HOST INTERACTIONS

The presence of mutualistic, commensal, parasitic and pathogenic associations within the same group of bacteria presents a unique opportunity to unravel the evolutionary biology of these associations and identify mechanisms which are of importance for the expression of each phenotype. Owing to the fastidious nature of their intracellular lifestyle, the molecular genetics and biochemistry of *Wolbachia* have been difficult to study. This is beginning to change with the complete genome sequencing of *Wolbachia* from a variety of hosts (Slatko *et al.*, 1999) and the possibility of growing and maintaining different strains of the bacteria in cell culture (Dobson *et al.*, 2002b). Some models are particularly well suited to proteomic and genomic approaches to dissect the molecular machinery and communication between symbiont and host. In both *Wolbachia*/*Drosophila* and *Wolbachia*/*B. malayi* systems complete genome sequences of both bacteria and host have been or soon will be determined.

The range of associations and plasticity of phenotype displayed by *Wolbachia* encompass all of the major symbiotic interactions. In the majority of cases and in a number of bizarre, brutal and wonderful ways, the bacteria induce changes in their hosts, which promote their transmission to new hosts and populations. Their abundance and diverse range of taxonomic hosts are evidence of the success of their lifestyle and may well be only a fraction of the extent of their repertoire. Understanding the nature and extent of this repertoire is an exciting and important target for future research.

ACKNOWLEDGEMENTS

I would like to thank the Wellcome Trust for Senior Fellowship support.

REFERENCES

Bandi, C., Anderson, T. J., Genchi, C. & Blaxter, M. L. (1998). Phylogeny of *Wolbachia* in filarial nematodes. *Proc R Soc Lond B Biol Sci* **265**, 2407–2413.

Bandi, C., Dunn, A. M., Hurst, G. D. & Rigaud, T. (2001). Inherited microorganisms, sex-specific virulence and reproductive parasitism. *Trends Parasitol* **17**, 88–94.

Bazzocchi, C., Jamnongluk, W., O'Neill, S. L., Anderson, T. J., Genchi, C. & Bandi, C. (2000). *wsp* gene sequences from the *Wolbachia* of filarial nematodes. *Curr Microbiol* **41**, 96–100.

Bourtzis, K. & Braig, H. R. (1999). The many faces of *Wolbachia*. In *Rickettsiae and Rickettsial Diseases at the Turn of the Third Millenium*, pp. 199–219. Edited by D. Raoult & P. Brouqui. Paris: Elsevier.

Bourtzis, K., Pettigrew, M. M. & O'Neill, S. L. (2000). *Wolbachia* neither induces nor suppresses transcripts encoding antimicrobial peptides. *Insect Mol Biol* **9**, 635–639.

Brattig, N. W., Rathjens, U., Ernst, M., Geisinger, F., Renz, A. & Tischendorf, F. W. (2000). Lipopolysaccharide-like molecules derived from *Wolbachia* endobacteria of the filaria *Onchocerca volvulus* are candidate mediators in the sequence of inflammatory and antiinflammatory responses of human monocytes. *Microbes Infect* **2**, 1147–1157.

Brownstein, J. S., Hett, E. & O'Neill, S. L. (2003). The potential of virulent *Wolbachia* to modulate disease transmission by insects. *J Invertebr Pathol* **84**, 24–29.

Büttner, D. W., Wanji, S., Bazzocchi, C., Bain, O. & Fischer, P. (2003). Obligatory symbiotic *Wolbachia* endobacteria are absent from *Loa loa*. *Filaria J* **2**, 10.

Casiraghi, M., Anderson, T. J., Bandi, C., Bazzocchi, C. & Genchi, C. (2001). A phylogenetic analysis of filarial nematodes: comparison with the phylogeny of *Wolbachia* endosymbionts. *Parasitology* **122**, 93–103.

Charlat, S., Hurst, G. D. & Mercot, H. (2003). Evolutionary consequences of *Wolbachia* infections. *Trends Genet* **19**, 217–223.

Cheng, Q., Ruel, T. D., Zhou, W., Moloo, S. K., Majiwa, P., O'Neill, S. L. & Aksoy, S. (2000). Tissue distribution and prevalence of *Wolbachia* infections in tsetse flies, *Glossina* spp. *Med Vet Entomol* **14**, 44–50.

Chirgwin, S. R., Porthouse, K. H., Nowling, J. M. & Klei, T. R. (2002). The filarial endosymbiont *Wolbachia* sp. is absent from *Setaria equina*. *J Parasitol* **88**, 1248–1250.

Chirgwin, S. R., Nowling, J. M., Coleman, S. U. & Klei, T. R. (2003). *Brugia pahangi* and *Wolbachia*: the kinetics of bacteria elimination, worm viability, and host responses following tetracycline treatment. *Exp Parasitol* **103**, 16–26.

Cross, H. F., Haarbrink, M., Egerton, G., Yazdanbakhsh, M. & Taylor, M. J. (2001). Severe reactions to filarial chemotherapy and release of *Wolbachia* endosymbionts into blood. *Lancet* **358**, 1873–1875.

Curtis, C. F. & Sinkins, S. P. (1998). *Wolbachia* as a possible means of driving genes into populations. *Parasitology* **116**, S111–115.

Dedeine, F., Vavre, F., Fleury, F., Loppin, B., Hochberg, M. E. & Bouletreau, M. (2001). Removing symbiotic *Wolbachia* bacteria specifically inhibits oogenesis in a parasitic wasp. *Proc Natl Acad Sci U S A* **98**, 6247–6252.

Dobson, S. L. (2003). Reversing *Wolbachia*-based population replacement. *Trends Parasitol* **19**, 128–133.

Dobson, S. L., Marsland, E. J. & Rattanadechakul, W. (2002a). Mutualistic *Wolbachia* infection in *Aedes albopictus*: accelerating cytoplasmic drive. *Genetics* **160**, 1087–1094.

Dobson, S. L., Marsland, E. J., Veneti, Z., Bourtzis, K. & O'Neill, S. L. (2002b). Characterization of *Wolbachia* host cell range via the in vitro establishment of infections. *Appl Environ Microbiol* **68**, 656–660.

Fenollar, F., La Scola, B., Inokuma, H., Dumler, J. S., Taylor, M. J. & Raoult, D. (2003). Culture and phenotypic characterization of a *Wolbachia pipientis* isolate. *J Clin Microbiol* **41**, 5434–5441.

Forst, S. & Clarke, D. (2002). Bacteria-nematode symbiosis. In *Entomopathogenic Nematology*, pp. 57–77. Edited by R. Gaugler. Wallingford, UK: CABI.

Grobusch, M. P., Kombila, M., Autenrieth, I., Mehlhorn, H. & Kremsner, P. G. (2003). No evidence of *Wolbachia* endosymbiosis with *Loa loa* and *Mansonella perstans*. *Parasitol Res* **90**, 405–408.

Hoerauf, A., Nissen-Pahle, K., Schmetz, C. & 7 other authors (1999). Tetracycline therapy targets intracellular bacteria in the filarial nematode *Litomosoides sigmodontis* and results in filarial infertility. *J Clin Invest* **103**, 11–18.

Hoerauf, A., Volkmann, L., Hamelmann, C., Adjei, O., Autenrieth, I. B., Fleischer, B. & Büttner, D. W. (2000). Endosymbiotic bacteria in worms as targets for a novel chemotherapy in filariasis. *Lancet* **355**, 1242–1243.

Hoerauf, A., Mand, S., Adjei, O., Fleischer, B. & Büttner, D. W. (2001). Depletion of *Wolbachia* endobacteria in *Onchocerca volvulus* by doxycycline and microfilaridermia after ivermectin treatment. *Lancet* **357**, 1415–1416.

Hoerauf, A., Mand, S., Volkmann, L., Büttner, M., Marfo-Debrekyei, Y., Taylor, M., Adjei, O. & Büttner, D. W. (2003a). Doxycycline in the treatment of human onchocerciasis: kinetics of *Wolbachia* endobacteria reduction and of inhibition of embryogenesis in female *Onchocerca* worms. *Microbes Infect* **5**, 261–273.

Hoerauf, A., Mand, S., Fischer, K., Kruppa, T., Marfo-Debrekyei, Y., Debrah, A. Y., Pfarr, K. M., Adjei, O. & Büttner, D. W. (2003b). Doxycycline as a novel strategy against bancroftian filariasis-depletion of *Wolbachia* endosymbionts from *Wuchereria bancrofti* and stop of microfilaria production. *Med Microbiol Immunol* **192**, 211–216.

Hoffman, A. A. & Turelli, M. (1997). Cytoplasmic incompatibility in insects. In *Influential Passengers: Inherited Microorganisms and Arthropod Reproduction*, pp. 102–122. Edited by S. L. O'Neill, A. A. Hoffmann & J. H. Werren. Oxford: Oxford University Press.

Hurst, G. D. & Jiggins, F. M. (2000). Male-killing bacteria in insects: mechanisms, incidence, and implications. *Emerg Infect Dis* **6**, 329–336.

Jiggins, F. M., Hurst, G. D. & Yang, Z. (2002). Host-symbiont conflicts: positive selection on an outer membrane protein of parasitic but not mutualistic Rickettsiaceae. *Mol Biol Evol* **19**, 1341–1349.

Keiser, P. B., Reynolds, S. M., Awadzi, K., Ottesen, E. A., Taylor, M. J. & Nutman, T. B. (2002). Bacterial endosymbionts of *Onchocerca volvulus* in the pathogenesis of posttreatment reactions. *J Infect Dis* **185**, 805–811.

Kondo, N., Nikoh, N., Ijichi, N., Shimada, M. & Fukatsu, T. (2002). Genome fragment of *Wolbachia* endosymbiont transferred to X chromosome of host insect. *Proc Natl Acad Sci U S A* **99**, 14280–14285.

Langworthy, N. G., Renz, A., Mackenstedt, U., Henkle-Duhrsen, K., de Bronsvoort, M. B., Tanya, V. N., Donnelly, M. J. & Trees, A. J. (2000). Macrofilaricidal activity of tetracycline against the filarial nematode *Onchocerca ochengi*: elimination of *Wolbachia* precedes worm death and suggests a dependent relationship. *Proc R Soc Lond B Biol Sci* **267**, 1063–1069.

Laven, H. (1967). Eradication of *Culex pipiens fatigans* through cytoplasmic incompatibility. *Nature* **216**, 383–384.

Lo, N., Casiraghi, M., Salati, S., Bazzocchi, C. & Bandi, C. (2002). How many *Wolbachia* supergroups exist? *Mol Biol Evol* **19**, 341–346.

McGarry, H. F., Pfarr, K., Egerton, G. & 8 other authors (2003). Evidence against *Wolbachia* symbiosis in *Loa loa*. *Filaria J* **2**, 9.

McGarry, H. F., Egerton, G. & Taylor, M. J. (2004). Population dynamics of *Wolbachia* bacterial endosymbionts of *Brugia malayi*. *Mol Biochem Parasitol* (in press).

Min, K. T. & Benzer, S. (1997). *Wolbachia*, normally a symbiont of *Drosophila*, can be virulent, causing degeneration and early death. *Proc Natl Acad Sci U S A* **94**, 10792–10796.

O'Neill, S. L., Hoffmann, A. A. & Werren, J. H. (1997). *Influential Passengers: Inherited Microorganisms and Arthropod Reproduction*. Oxford: Oxford University Press.

Ono, M., Braig, H. R., Munstermann, L. E., Ferro, C. & O'Neill, S. L. (2001). *Wolbachia* infections of phlebotomine sand flies (Diptera: Psychodidae). *J Med Entomol* **38**, 237–241.

Punkosdy, G. A., Addiss, D. G. & Lammie, P. J. (2003). Characterization of antibody responses to *Wolbachia* surface protein in humans with lymphatic filariasis. *Infect Immun* **71**, 5104–5114.

Rigaud, T. (1997). Inherited microorganisms and sex determination of arthropod hosts. In *Influential Passengers: Inherited Microorganisms and Arthropod Reproduction*, pp. 81–101. Edited by S. L. O'Neill, A. A. Hoffmann & J. H. Werren. Oxford: Oxford University Press.

Ruang-Areerate, T., Kittayapong, P., Baimai, V. & O'Neill, S. L. (2003). Molecular phylogeny of *Wolbachia* endosymbionts in Southeast Asian mosquitoes (Diptera: Culicidae) based on wsp gene sequences. *J Med Entomol* **40**, 1–5.

Saint Andre, A., Blackwell, N. M., Hall, L. R. & 9 other authors (2002). The role of endosymbiotic *Wolbachia* bacteria in the pathogenesis of river blindness. *Science* **295**, 1892–1895.

Sinkins, S. P., Curtis, C. F. & O'Neill, S. L. (1997). The potential application of inherited symbiont systems to pest control. In *Influential Passengers: Inherited Microorganisms and Arthropod Reproduction*, pp. 155–175. Edited by S. L. O'Neill, A. A. Hoffmann & J. H. Werren. Oxford: Oxford University Press.

Slatko, B. E., O'Neill, S. L., Scott, A. L., Werren, J. L. & Blaxter, M. L. (1999). The *Wolbachia* Genome Consortium. *Microb Comp Genomics* **4**, 161–165.

Starr, D. J. & Cline, T. W. (2002). A host parasite interaction rescues *Drosophila* oogenesis defects. *Nature* **418**, 76–79.

Stouthamer, R. (1997). *Wolbachia*-induced parthenogenesis. In *Influential Passengers: Inherited Microorganisms and Arthropod Reproduction*, pp. 102–122. Edited by S. L. O'Neill, A. A. Hoffmann & J. H. Werren. Oxford: Oxford University Press.

Stouthamer, R., Breeuwer, J. A. & Hurst, G. D. (1999). *Wolbachia pipientis*: microbial manipulator of arthropod reproduction. *Annu Rev Microbiol* **53**, 71–102.

Sun, L. V., Foster, J. M., Tzertzinis, G., Ono, M., Bandi, C., Slatko, B. E. & O'Neill, S. L. (2001). Determination of *Wolbachia* genome size by pulsed-field gel electrophoresis. *J Bacteriol* **183**, 2219–2225.

Taylor, M. J. (2002). *Wolbachia* bacterial endosymbionts. In *The Filaria*, pp. 143–153. Edited by T. R. Klei & T. V. Rajan. Boston: Kluwer.

Taylor, M. J. (2003). *Wolbachia* in the inflammatory pathogenesis of human filariasis. *Ann N Y Acad Sci* **990**, 444–449.

Taylor, M. J. & Hoerauf, A. (1999). *Wolbachia* bacteria of filarial nematodes. *Parasitol Today* **15**, 437–442.

Taylor, M. J. & Hoerauf, A. (2001). A new approach to the treatment of filariasis. *Curr Opin Infect Dis* **14**, 727–731.

Taylor, M. J., Cross, H. F. & Bilo, K. (2000). Inflammatory responses induced by the filarial nematode *Brugia malayi* are mediated by lipopolysaccharide-like activity from endosymbiotic *Wolbachia* bacteria. *J Exp Med* **191**, 1429–1436.

Townson, H. (2002). *Wolbachia* as a potential tool for suppressing filarial transmission. *Ann Trop Med Parasitol* **96** (Suppl. 2), S117–S127.

Townson, S., Hutton, D., Siemienska, J., Hollick, L., Scanlon, T., Tagboto, S. K. & Taylor, M. J. (2000). Antibiotics and *Wolbachia* in filarial nematodes: antifilarial activity of rifampicin, oxytetracycline and chloramphenicol against *Onchocerca gutturosa*, *Onchocerca lienalis* and *Brugia pahangi*. *Ann Trop Med Parasitol* **94**, 801–816.

Tram, U. & Sullivan, W. (2002). Role of delayed nuclear envelope breakdown and mitosis in *Wolbachia*-induced cytoplasmic incompatibility. *Science* **296**, 1124–1126.

Volkmann, L., Fischer, K., Taylor, M. J. & Hoerauf, A. (2003). Antibiotic therapy in murine filariasis (*Litomosoides sigmodontis*): comparative effects of doxycycline and rifampicin on *Wolbachia* and filarial viability. *Trop Med Int Health* **8**, 392–401.

Werren, J. H. (1997). *Wolbachia* run amok. *Proc Natl Acad Sci U S A* **94**, 11154–11155.

Yen, J. H. & Barr, A. R. (1971). New hypothesis of the cause of cytoplasmic incompatibility in *Culex pipiens* L. *Nature* **232**, 657–658.

Pathogenic strategies of *Anaplasma phagocytophilum*, a unique bacterium that colonizes neutrophils

Jason A. Carlyon and Erol Fikrig

Section of Rheumatology, Department of Internal Medicine, Yale University School of Medicine, The Anylan Center for Medical Research and Education, New Haven, CT 06520-8031, USA

INTRODUCTION

Human granulocytic ehrlichiosis (HGE) is an emerging zoonosis caused by *Anaplasma phagocytophilum*, an obligate intracellular bacterium with a unique tropism for neutrophils (Dumler *et al.*, 2001; Chen *et al.*, 1994). Since documentation of the first human case in 1994, HGE has been increasingly recognized in the United States (US) and Europe. Cases of HGE occur in areas endemic for ticks of the *Ixodes persulcatus* complex, which are vectors for the disease. In the US, these areas comprise the Northeast, the upper Midwest and northern California (Dumler & Bakken, 1998; Richter *et al.*, 1996; Telford *et al.*, 1996; Pancholi *et al.*, 1995; Chen *et al.*, 1994). Cases of or serological evidence for HGE have been documented in the following European countries – Slovenia, the Netherlands, Sweden, Norway, Spain, Croatia, Poland (Blanco & Oteo, 2002), the Czech Republic (Hulinska *et al.*, 2002), Belgium (Guillaume *et al.*, 2002), Greece (Daniel *et al.*, 2002), Germany (Fingerle *et al.*, 1999), Bulgaria (Christova & Dumler, 1999), Denmark (Lebech *et al.*, 1998) and Switzerland (Pusterla *et al.*, 1998). The annual incidence of HGE in these areas ranges from 2·3 to 16·1 cases per 100 000 persons (McQuiston *et al.*, 1999). In one study, however, 71 of 475 Wisconsin resident serum samples contained antibodies against the HGE agent, which suggests that the annual incidence in this area may be considerably higher (Bakken *et al.*, 1998). In nature, *A. phagocytophilum* cycles between its tick vector and mammalian hosts, the primary of which are the white-footed mouse, *Peromyscus leucopus*, and the white-tailed deer, *Odocoileus virginianus*. Humans are accidental hosts (Dumler & Bakken, 1998; Ogden *et al.*, 1998).

SGM symposium 63: Microbe–vector interactions in vector-borne diseases.
Editors S. H. Gillespie, G. L. Smith & A. Osbourn. Cambridge University Press. ISBN 0 521 84312 X ©SGM 2004

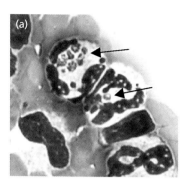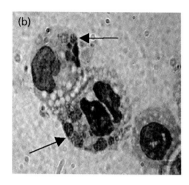

Fig. 1. Detection of morulae. Peripheral blood cells from an *A. phagocytophilum*-infected C3H/HeJ mouse (a) and infected rHL-60 cells (b) were subjected to Giemsa staining. The morulae, or intravacuolar colonies of *A. phagocytophilum*, stain darker than the host neutrophils' nuclei and are indicated by arrows. Images provided courtesy of Dr Venetta Thomas, Yale University School of Medicine.

Clinical symptoms associated with HGE include non-specific manifestations such as fever, chills, headache, malaise and myalgia. More distinguishing complications consist of leukopenia, thrombocytopenia, and elevated levels of C-reactive protein and hepatic transaminases. The cardinal manifestation of HGE, however, is the infection of neutrophils and bone marrow progenitors (Bakken *et al.*, 1994; Chen *et al.*, 1994). *A. phagocytophilum*-infected monocytes are infrequently observed and one report detected the bacterium in a macrophage (Walker & Dumler, 1997). Monocytes are also susceptible to infection *in vitro* and infected monocytes have been observed in experimentally infected monkeys (Foley *et al.*, 1999) and sheep (Klein *et al.*, 1997; Foggie, 1951). Though usually self-limiting, severe manifestations can result and include prolonged fever, shock, seizures, pneumonitis, acute renal failure, haemorrhages,rhabdomyolysis, opportunistic infections and death (Aguero-Rosenfeld *et al.*, 1996; Shea *et al.*, 1996; Wong & Grady, 1996; Hardalo *et al.*, 1995; Bakken *et al.*, 1994). Individuals most susceptible to severe and/or fatal cases are those with pre-existing immune dysfunction, such as chronic lymphocytic leukaemia, or those that have undergone radiation therapy or organ transplantation. The drug of choice for treating HGE is a tetracycline antibiotic, usually doxycycline (Dumler & Bakken, 1998; Ogden *et al.*, 1998; Bakken *et al.*, 1996).

A. phagocytophilum is a small coccobacillary Gram-negative bacterium, typically 0·2–2 µm in diameter, that replicates within the confines of a host cell vacuole to form a microcolony called a morula (Chen *et al.*, 1994; Rikihisa, 1991). The morulae, the term of which is Latin for 'mulberry', appear as round inclusions 2–7 µm in diameter and can be detected in infected cells by Romanowsky stains, such as Wright or Giemsa stains. The microcolonies are often stippled in appearance and stain darker than the host cell's nucleus (Bakken, 1998; Chen *et al.*, 1994; Rikihisa, 1991) (Fig. 1).

Until recently, *A. phagocytophilum* was originally referred to as 'the agent of HGE'. It is, however, morphologically indistinguishable from and is cross-reactive with antisera to *Ehrlichia phagocytophila* and *Ehrlichia equi*, the causative agents of tick-borne fever in sheep and equine granulocytic ehrlichiosis in horses, respectively (Goodman *et al.*, 1996; Bakken *et al.*, 1994; Chen *et al.*, 1994). Like the HGE agent, *E. phagocytophila* and *E. equi* target granulocytes within their mammalian hosts. They possess only minor variations in their 16S rRNA gene sequences, and their GroESL amino acid sequences are 100 % identical. Because their genetic sequences are more closely related to *Anaplasma marginale* and *Ehrlichia platys* than to other *Ehrlichiae*, these three bacteria were collectively reclassified as *A. phagocytophilum* (Dumler *et al.*, 2001). In view of this new delineation and because no other *Anaplasma* species is known to cause human infection, we recently recommended the term 'human anaplasmosis' (HA) be applied to describe *A. phagocytophilum* infection in humans (Carlyon & Fikrig, 2003). The term HGE would still apply to infections caused by *Ehrlichia ewingii*, which has been identified in granulocytes of afflicted patients (Buller *et al.*, 1999).

The trademark manifestation of *A. phagocytophilum* infection is the colonization of neutrophils. Neutrophils represent the first line of defence against invading pathogens and constitute approximately 75 % of the total circulating leukocytes. Patients suffering from neutrophil deficiency or dysfunction experience recurrent, life-threatening infections, which evidences the importance of these cells to fighting microbial infection. Neutrophils phagocytose bacteria and fungi and kill them using an armamentarium of oxidative-dependent and -independent killing mechanisms (Burg & Pillinger, 2001). Moreover, they have high rates of apoptosis and very short half-lives in peripheral blood (Akgul *et al.*, 2001; Savill *et al.*, 1989). Therefore, the ability of *A. phagocytophilum* to establish infection within the hostile intracellular environment of the short-lived neutrophil presents a striking paradox.

Since documentation of the first human case, *A. phagocytophilum* has been the subject of intense study. The human promyelocytic leukaemia cell line HL-60 is susceptible to *A. phagocytophilum* infection (Goodman *et al.*, 1996), which allows for *in vitro* studies of the organism. A murine model of HA has been developed that parallels haematologic parameters and histopathological lesions similar to those observed for human disease (Hodzic *et al.*, 1998). Use of these laboratory models has begun to delineate the organism's invasion and intracellular survival mechanisms. This review covers aspects regarding the tropism *A. phagocytophilum* has for neutrophils, as well as the different strategies it employs for host adaptation, evasion of the humoral immune response, circumventing and subverting neutrophil killing mechanisms, delaying neutrophil apoptosis, and exploiting the neutrophil chemokine response to efficiently propagate infection.

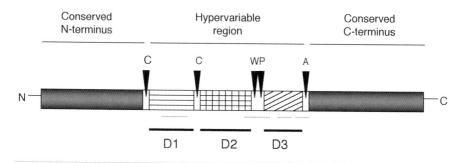

Fig. 2. Representative schematic of an *A. phagocytophilum* P44 paralogue. Each P44 protein comprises conserved N- and C-terminal regions that flank a central HVR. Five absolutely conserved amino acids, C, C, W, P and A (denoted by arrowheads), subdivide the HVR into three domains (marked D1, D2 and D3 and distinguished by different fill patterns of the corresponding bars). Four dashed lines indicate regions of the HVR that display high surface exposure probability.

A. PHAGOCYTOPHILUM CARRIES A PARALOGOUS GENE FAMILY BELIEVED TO BE INVOLVED IN ANTIGENIC VARIATION

The maintenance of *A. phagocytophilum* in nature is dependent on its ability to efficiently cycle between ticks and mammals. It follows that the organism would benefit from strategies by which it can evade the humoral immune response to persist within its mammalian host. This would consequentially enhance its opportunity for transmission back to its tick vector. Indeed, infection can persist in mice, ruminants, and in some instances, humans (Dumler & Bakken, 1996; Telford *et al.*, 1996). Even when *A. phagocytophilum* morulae are no longer microscopically detectable in laboratory mice, their blood is still infective to naïve mice (Telford *et al.*, 1996). Additionally, the bacterium interacts with a huge diversity of cell types as it passes between arthropods and mammals and must therefore rapidly adapt to these different environments. *A. phagocytophilum* possesses a series of polymorphic genes known as the *p44* gene family (Ijdo *et al.*, 1998; Murphy *et al.*, 1998; Zhi *et al.*, 1998) that is thought to play a role in antigenic variation and host adaptation.

Members of this paralogous gene family encode immunodominant outer-membrane proteins that range in size from 42 to 49 kDa. The proteins share a common structure consisting of three major parts – a central hypervariable region (HVR) with 94 amino acid residues flanked by N-terminal and C-terminal conserved regions comprising 58 and 65 amino acids, respectively (Ijdo *et al.*, 1998; Murphy *et al.*, 1998; Zhi *et al.*, 1998) (Fig. 2). Amino acid similarity among the HVRs ranges from 19·9 to 32·8 % (Zhi *et al.*, 1999). The HVR is denoted at each end and partitioned into three additional domains by five absolutely conserved amino acid residues (Barbet *et al.*, 2003; Lin *et al.*, 2002). The hypervariable domains exhibit strong predictability of being hydrophilic and surface-exposed and thus likely provide a source for surface phenotype diversity.

Moreover, one of the domains is flanked by two conserved cystine residues, which could potentially form a disulfide bond and expose the domain as a hydrophilic loop on the bacterial surface (Lin *et al.*, 2002).

The *A. phagocytophilum* P44 proteins are orthologous (62 % similarity and 53 % identity) to those encoded by the *A. marginale msp2* gene family (Ijdo *et al.*, 1998). *A. marginale*, which is closely related to *A. phagocytophilum*, is another tick-borne pathogen that infects bovine erythrocytes and can remain in its ruminant host for the animal's lifetime (Palmer *et al.*, 2000). In cattle persistently infected with *A. marginale*, there are cyclic peaks of bacteraemia every 2–6 weeks containing different immuno-protective MSP2 variants (French *et al.*, 1999; Eid *et al.*, 1996; Kieser *et al.*, 1990). These variants carry epitopes recognized by antibodies that appear subsequent but not prior to the respective bacteraemic peaks, which indicates that antigenic variation of MSP2 proteins is responsible for the bacterium's persistence (Palmer *et al.*, 2000; Eid *et al.*, 1996). Sequence variation of the MSP2 proteins is achieved by segmental gene conversion of a single polycistronic expression site by different pseudogenes of the *msp2* family which are truncated and cannot otherwise encode full-length proteins (Brayton *et al.*, 2002; Barbet *et al.*, 2000; Palmer *et al.*, 1994). Like the *p44* genes, *msp2* pseudogenes consist of two conserved regions flanking a central HVR (Brayton *et al.*, 2001). Because of the similarity between the gene families and because both pathogens can manifest as persistent infections, the *p44* genes are believed to play a similar role in HA to that of the *msp2* genes during bovine anaplasmosis.

At least 18 *p44* paralogues are carried by *A. phagocytophilum* (Barbet *et al.*, 2003; Zhi *et al.*, 1999). Unlike the *msp2* genes, which are grouped all together, the *p44* genes are scattered throughout the genome. Zhi *et al.* (1999, 2002b) found that certain genes exist independently, while others are clustered in tandemly arranged groups of two or three. The single genes and the first genes of each cluster carry AUG start codons and ribosome-binding sites, while the remaining genes of each cluster initiate with TCT (Zhi *et al.*, 1999, 2002b). It is unknown whether TCT represents an alternative start codon for *A. phagocytophilum*.

P44 variability is regulated at the transcriptional level and occurs irrespective of immune pressure. At least 43 different *p44* transcripts have been documented thus far, some of which exhibit host-specific expression patterns (Barbet *et al.*, 2003; Ijdo *et al.*, 2002; Lin *et al.*, 2002; Zhi *et al.*, 1999, 2002b). For instance, *p44-18* (as denoted by Zhi *et al.*, 1999, 2002b) is the predominant transcript expressed during murine and equine infection. Its expression in the tick, however, is nearly non-existent following transmission from infected horses. Its expression is at least partially regulated by temperature as mRNA levels in HL-60 cells cultivated at 37 °C, but not 24 °C, increase by over 100-fold. These data suggest that P44-18 may play a role in establishing mammalian infection. Conversely, *p44-1* and *-2* are barely detectable during mammalian infection, but constitute the majority of *p44* mRNAs expressed in infected, fed ticks. While a diverse milieu of *p44* transcripts are expressed in the tick midgut, those of *p44-1* and *-2* are exclusive in the salivary glands, which indicates that the encoded proteins may be involved in bacterial transmission (Zhi *et al.*, 2002b). Ijdc *et al.* (2002) observed that no *p44* genes are expressed in ticks until they partake of a blood meal. Specific and reversible changes in P44 expression profiles were also observed when *A. phagocytophilum* was transferred from HL-60 cells to ISE6 tick cells and subsequently back to HL-60 cells (Barbet *et al.*, 2003; Jauron *et al.*, 2001).

Expression of *p44* genes during human infection is quite diverse as the number of transcripts detected in HA patient blood ranges from a predominantly single allele to as many as 17. Among the different copies detected, the five aforementioned amino acids are absolutely maintained, which indicates that *p44* sequence variation is not random, but restricted (Barbet *et al.*, 2003; Caspersen *et al.*, 2002; Lin *et al.*, 2002). Furthermore, the variant sequences are maintained in nature as determined by a comparison of the Southern hybridization profiles of *A. phagocytophilum* isolates obtained from HA patients 5 years apart (Lin *et al.*, 2002).

Though several intact *p44* genes are carried by *A. phagocytophilum*, only a single expression locus with a predicted promoter sequence has been identified (Barbet *et al.*, 2003), which lends credence to the speculation that variation of *p44* transcripts is achieved in a manner similar to the gene conversion mechanism of *A. marginale*. An alternative mechanism, however, has been shown to occur for *p44-18*, a pseudogene that lies downstream, out of frame, from *p44-1*, which carries an AUG start codon. Expression of *p44-18* is achieved by a unique splicing event that combines the 5′ end of *p44-1* with *p44-18* and excises intervening mRNA as a circular intermediate. Support for this mechanism is provided by the fact that the splicing event is also observed if a plasmid carrying the *p44-1-18* sequence is transformed into *Escherichia coli* (Zhi *et al.*, 2002a).

Over the past few years, a great deal of information regarding the *p44* gene family has been amassed. Current evidence suggests the likelihood that they undergo antigenic variation in a manner similar to that of *A. marginale* and serve *A. phagocytophilum* by enhancing its ability to persist within its mammalian host. The environment- and/or host-specific expression of *p44-1*, *-2* and *-18* indicates that certain genes play significant roles in different stages of the bacterium's zoonotic life cycle. Future investigations should delineate the roles of additional *p44* family members and define their expression mechanisms.

A. PHAGOCYTOPHILUM USES PSGL-1 OF HUMAN NEUTROPHILS AND HL-60 CELLS AS A LIGAND FOR CELLULAR ENTRY

A. phagocytophilum primarily colonizes neutrophils, which suggests that it recognizes specific molecular determinants on the neutrophil surface. *A. phagocytophilum*'s tropism for human neutrophils is explained, at least in part, by the expression of P-selectin glycoprotein ligand-1 (PSGL-1) (Herron *et al.*, 2000) and sialylated and α1,3-fucosylated glycans (Goodman *et al.*, 1999) on neutrophil surfaces. PSGL-1 is a mucin, or a glycoprotein with multiple Ser/Thr-linked oligosaccharides (O-glycans) and repeating peptide motifs. It is present on granulocytes, monocytes and subsets of lymphocytes and plays a key role in the early steps of inflammation as it facilitates the tethering of leukocytes to the endothelium of inflamed tissues. This is accomplished by stereospecific interactions of PSGL-1 with P-selectin expressed on activated platelets and endothelial cells, E-selectin expressed on cytokine activated endothelial cells, and L-selectin expressed on other leukocytes (McEver, 2001; McEver & Cummings, 1997; Kansas, 1996). PSGL-1 undergoes post-translational modifications such as α2,3-sialylation and α1,3-fucosylation of the core 2 O-glycans decorating its entirety. Formation of the α2,3-sialylated, α1,3-fucosylated tetrasaccharide capping structure, sialyl Lewis x (sLe^x; NeuAcα2,3Galβ1,4[Fucα1,3]GlcNAc), at the N terminus of PSGL-1 is critical for proper selectin binding. The sLe^x moiety is, in fact, crucial for proper interaction of all selectin ligands with their corresponding selectins (McEver, 2001, 2002; Lowe, 1997; McEver & Cummings, 1997; Kansas, 1996). In addition to sLe^x, the N terminus of PSGL-1 binds P-selectin via cooperative interactions with tyrosine sulfate residues, as well as primary peptide determinants flanking the sLe^x-capped core 2 O-glycan (Leppanen *et al.*, 1999, 2000, 2002; Somers *et al.*, 2000).

The efforts that identified PSGL-1 as a receptor for *A. phagocytophilum* were spear-headed by Goodman and colleagues, who initiated their studies by hypothesizing that sLe^x might be important for *A. phagocytophilum* binding because of its rich expression on HL-60 cells and those bone marrow and peripheral blood cells susceptible to infection (Herron *et al.*, 2000; Goodman *et al.*, 1999). Treating these different cell types with each of four different anti-sLe^x monoclonal antibodies prevented *A. phagocyto-philum* infection, but had no effect on bacterial binding. Proper binding of sLe^x to its selectin counter-receptor depends on the presence of its terminal α2,3-linked sialic acid. Treatment of HL-60 cells with neuraminidase, which eliminates sialic acid residues, abolished both *A. phagocytophilum* binding and invasion (Goodman *et al.*, 1999). These studies indicated that sLe^x expression and sialylation, which may or may not occur in the context of sLe^x, are necessary for *A. phagocytophilum* adhesion. Moreover, the importance of sLe^x to *A. phagocytophilum* binding suggested that the bacterium's receptor might be a selectin ligand.

HL-60-A2 is a clonal cell line that lacks expression of α1,3-fucosyltransferase-VII (Fuc-TVII), which mediates the terminal step in the biosynthesis of α1,3-fucosylated glycans that decorate selectin ligands of neutrophils (Goodman *et al.*, 1999). This modification is crucial for sLe^x synthesis and for PSGL-1 and other selectin ligands to bind their respective selectins (McEver, 2002; Lowe, 1997; Kansas, 1996). HL-60-A2 cells are highly resistant to *A. phagocytophilum* infection. Reconstituting α1,3-fucosyltransferase activity in these cells restores functional sLe^x expression and, consequentially, renders them susceptible to *A. phagocytophilum* invasion. This suggested that the *A. phagocytophilum* receptor was an α1,3-fucosylated cell surface glycoprotein (Goodman *et al.*, 1999).

With this information in hand, Herron and colleagues focused their attention on the predominant selectin ligand expressed on neutrophils, PSGL-1. Pretreatment of human neutrophils or HL-60 cells with monoclonal antibodies against the PSGL-1 N terminus prevented *A. phagocytophilum* binding, while antibodies directed against either a proximal region of PSGL-1 or other sLe^x-modified or related mucin-modified proteins had no effect. The B lymphoblastoid cell line BJAB, which is non-permissive to *A. phagocytophilum* infection, lacks PSGL-1. It also fails to express Fuc-TVII and therefore cannot properly express sLe^x. Coexpression of PSGL-1 and Fuc-TVII, but neither protein alone, in BJAB cells rendered them susceptible to *A. phagocytophilum* infection. These data were interpreted to indicate that the organism binds specifically to α1,3-fucosylated human PSGL-1, presumably near the N terminus, much in the same fashion as P-selectin (Herron *et al.*, 2000).

The selectin mimicry strategy of *A. phagocytophilum*, however, is more complex than originally predicted. Like P-selectin, *A. phagocytophilum* requires both the PSGL-1 N terminus and α2,3-sialylated, α1,3-fucosylated glycans for effective adhesion to human neutrophils. Unlike P-selectin, it is unnecessary that each determinant lies on the same ligand, as *A. phagocytophilum* binds to microspheres coupled independently with sLe^x-free PSGL-1 glycopeptide and peptide-free sLe^x (Yago *et al.*, 2003). Also unlike P-selectin, *A. phagocytophilum* binds to PSGL-1 glycopeptides lacking tyrosine sulfation or a specific core-2 orientation of sLe^x on the O-glycan. Furthermore, P-selectin binding to PSGL-1 necessitates that sLe^x lies exclusively on O-glycans. In contrast, *A. phagocytophilum* binds to Chinese hamster ovary cells transfected to express PSGL-1 in cooperation with sLe^x on both *N*- and O-glycans. P-selectin interaction with PSGL-1 also requires Ca^{2+} (McEver, 2002), while *A. phagocytophilum* binding is Ca^{2+}-independent (Goodman *et al.*, 1999). Thus *A. phagocytophilum* binds cooperatively to a non-sulfated N-terminal peptide in human PSGL-1 and to sLe^x that may lie on PSGL-1 or other glycoproteins.

Whereas *A. phagocytophilum* binding to human neutrophils requires expression of PSGL-1 and α1,3-fucosyltransferases that construct sLe^x, binding to murine neutrophils requires expression of 1,3-fucosyltransferases, but not PSGL-1. *A. phagocytophilum* binding to Fuc-TIV$^{-/-}$/Fuc-TVII$^{-/-}$ neutrophils *in vitro* and infection of Fuc-TIV$^{-/-}$/Fuc-TVII$^{-/-}$ mice is nearly abolished relative to that of wild-type mice. In contrast, infection of PSGL-1$^{-/-}$ mice is highly similar to that of wild-type mice. *A. phagocytophilum* binding to PSGL-1$^{-/-}$ neutrophils *in vitro* is modestly reduced (Carlyon *et al.*, 2003). This differential binding can be explained, at least in part, by the fact that *A. phagocytophilum* recognizes a short amino acid sequence found in the N terminus of human, but not murine, PSGL-1 (Yago *et al.*, 2003). It therefore appears that *A. phagocytophilum* utilizes at least two adhesins, which bind cooperatively to at least two ligands on both human and murine neutrophils. One adhesin might bind to α1,3-fucosylated and possibly α2,3-sialylated glycans found on both human and murine neutrophils. A second adhesin would recognize the amino acid sequence found in human PSGL-1 and may also recognize a structurally similar region on a different protein on murine neutrophils. Alternatively, a third adhesin might bind to another ligand found on murine, but not human, neutrophils. Studies by Park and colleagues suggest that a P44 paralogue may facilitate binding to human PSGL-1 (Park *et al.*, 2003). *A. phagocytophilum*'s use of PSGL-1 and α1,3-fucosylated, α2,3-sialylated glycans as a receptor-mediated pathway for cellular adhesion explains not only its tropism for neutrophils but also one means by which it evades destruction by the neutrophil phagocytic pathway.

A. PHAGOCYTOPHILUM RESIDES WITHIN A VACUOLE THAT EVADES LYSOSOMAL FUSION

After docking to its target surface receptor(s), *A. phagocytophilum* colonizes its host neutrophil, thereby indicating that it evades, withstands, or subverts phagolysosome formation. Studies of several intracellular pathogens have revealed an array of such strategies. For instance, *Mycobacterium tuberculosis* (Clemens & Horwitz, 1996), *Chlamydia psittaci* (Eissenberg & Wyrick, 1981), *Legionella pneumophila* (Bozue & Johnson, 1996), *Salmonella typhimurium* (Rathman *et al.*, 1997) and *Ehrlichia risticii* (Wells & Rikihisa, 1988) reside in developmentally arrested phagosomes, as they do not fuse with lysosomes nor do they become acidified. *Francisella tularensis* (Fortier *et al.*, 1995) and *Coxiella burnetii* (Heinzen *et al.*, 1996), however, are able to survive within the acidic environment of the phagolysosome. Other organisms, such as *Listeria monocytogenes* (Theriot, 1995), *Rickettsia* spp. (Winkler, 1990) and *Trypanosoma cruzi* (Andrews *et al.*, 1990) escape from the phagosome prior to lysosomal fusion. Still others, such as *Chlamydia trachomatis*, reside in vacuoles that remain independent from the endocytic pathway yet acquire markers of exocytotic vesicles (Hackstadt *et al.*, 1996; Scidmore *et al.*, 1996).

Phagosomes undergo a series of maturation events that culminate in lysosomal fusion (Gagnon *et al.*, 2002; Garin *et al.*, 2001). Phagosomal maturation is a dynamic process that results in loss of certain components from the phagosome as they are recycled back to the plasma membrane, followed by the acquisition of endosomal and lysosomal markers, and finally lumen acidification (Desjardins, 2003; Pitt *et al.*, 1992a, b). By screening for colocalization of these markers with an intracellular pathogen's inclusion, it is possible to elucidate the endosomal stage within which the organism resides. Lysosome-associated membrane protein 1 and CD63 are glycoproteins found in the membranes of late endosomes and lysosomes (Chen *et al.*, 1986). In *A. phagocytophilum*-infected HL-60 cells, neither protein associates with vacuoles containing the organism, but both associate with other inclusions, particularly with those in close proximity (Mott *et al.*, 1999; Webster *et al.*, 1998). Myeloperoxidase, another lysosomal marker, exhibits a highly similar localization pattern (Mott *et al.*, 1999). Mature phago-lysosome lumens have low pH and can therefore be labelled with the compounds 3-(2,4-dinitroanilino)-3′-amino-N-methyldipropylamine and acridine orange (Anderson *et al.*, 1984). Both markers accrue in vacuoles of *A. phagocytophilum*-infected HL-60 cells, including those adjacent to vacuoles that contain the bacteria. Yet, neither compound accumulates within the same compartment as the bacterium (Mott *et al.*, 1999; Webster *et al.*, 1998). It is not surprising, then, that *A. phagocytophilum*-containing inclusions also do not label with antibody directed against vacuole-type H^+-ATPase (Mott *et al.*, 1999), which is involved in acidifying vesicular compartments such as lysosomes, endosomes and the Golgi apparatus (Lukacs *et al.*, 1990). *A. phagocytophilum*-infected ovine neutrophils retain the ability to phagocytose latex beads and *Candida albicans* and incorporate each into phagolysosomes. However, *A. phagocytophilum*-containing phagosomes do not incorporate latex beads or *C. albicans* nor do they accumulate phagolysosome markers (Gokce *et al.*, 1999). In all these studies, lysosomes were observed accumulating in close proximity around the periphery of the *A. phagocytophilum*-containing inclusion. Synaptic vesicle fusion is GTP-dependent (Vitale *et al.*, 1995), which offers speculation that the organism may block a GTPase activity present on the phagosomal membrane. Indeed, *A. phagocytophilum* inclusions do not colocalize with the small GTPase, Rab5, which is required for endosomal membrane trafficking (Mott *et al.*, 1999; Vitale *et al.*, 1995). These studies collectively demonstrate that *A. phagocytophilum* does not globally inhibit phagosome–lysosome fusion or lysosomal function, but instead specifically blocks lysosomal fusion with the protective vacuole in which it resides. This inhibition is dependent on bacterial protein synthesis, as oxytetracycline treatment results in maturation of the *A. phagocyto-philum*-containing vacuoles into phagolysosomes (Gokce *et al.*, 1999).

In addition to preventing luminal acidification and lysosomal fusion, *A. phago-cytophilum* also subverts an endocytic pathway to create its intracellular haven. In infected HL-60 cells, 43 % of *A. phagocytophilum*-containing vacuoles incorporate BSA–gold following 60 min incubation at 37 °C, which strongly suggests these compartments to be part of an endocytic pathway. Incubation at 13 °C, a temperature known to inhibit early–late endosomal fusion, reduces the percentage of vacuoles colocalizing with BSA–gold and bacteria to 13 % (Webster *et al.*, 1998). Moreover, some of the intracellular inclusions that contain *A. phagocytophilum* and BSA–gold also label with an antibody that recognizes mannose-6-phosphate receptor (MPR), a late endosomal marker (Webster *et al.*, 1998; Geuze *et al.*, 1984). Using immunofluorescence microscopy, Webster and colleagues observed that 68 % of *A. phagocytophilum* inclusions were MPR-positive (Webster *et al.*, 1998). A study by Mott *et al.* (1999), however, suggested that MPR does not colocalize with *A. phagocytophilum*-containing vacuoles. The inclusions also do not label for transferrin receptor Rab5 or early endosomal antigen I (Mott *et al.*, 1999; Webster *et al.*, 1998), each of which is a marker for early endosomes (Mu *et al.*, 1995; Vitale *et al.*, 1995; Geuze *et al.*, 1984). Furthermore, they do not label for the general endocytic markers clathrin heavy chain, α-adaptin or annexin I, II, IV or VI (Mott *et al.*, 1999).

The *A. phagocytophilum* compartment fails to label for several endocytic markers yet retains the ability to accumulate BSA–gold, which indicates that the organism modifies the normal endocytic development of its inclusion. Indeed, 25 % of *A. phagocytophilum* inclusions within infected HL-60 cells label with markers for major histocompatibility complex (MHC) class I and class II molecules (Mott *et al.*, 1999). These molecules, which are present on the surface and in the cytoplasmic membrane of leukocytes, are distributed differently in the endosomal–lysosomal pathway. MHC class I molecules are directed from the Golgi apparatus to the plasma membrane and are excluded during endocytosis (Neefjes *et al.*, 1990). MHC class II molecules are trafficked from the Golgi network to class II-containing endosomes, which are subsequently transported to the plasma membrane (Amigorena *et al.*, 1994). Thus the acquisition or slow removal of both MHC class I and class II molecules on *A. phagocytophilum* inclusions is due to the organism's alteration of normal membrane molecule sorting and maturation along the endosomal–lysosomal pathway.

Lastly, the exocytic pathway is not involved in formation of the protective vacuole in which *A. phagocytophilum* resides. Neither β-coatomer protein nor C6-NBD-ceramide, both of which are markers for the Golgi apparatus and Golgi-derived vesicles, colocalize with the inclusions (Mott *et al.*, 1999; Duden *et al.*, 1991; Lipsky & Pagano, 1983). Furthermore, inhibiting antegrade transport from the Golgi network with brefeldin A does not affect intracellular *A. phagocytophilum* growth (Mott *et al.*, 1999; Fujiwara *et al.*, 1988). Taken together, these studies demonstrate that *A. phagocytophilum* resides in an inclusion that does not fuse with lysosomes or Golgi-derived vesicles. This phagosome is unique from those occupied by other micro-organisms in that it displays characteristics of neither early nor late endosomes yet retains MHC class I and II molecules as well as the ability to accumulate BSA–gold. Therefore, this strategy functions as an effective means of molecular disguise that protects *A. phagocytophilum* from the normal intracellular killing pathways of its host cell.

A. PHAGOCYTOPHILUM INHIBITS NEUTROPHIL O_2^- PRODUCTION

One of the primary means by which neutrophils destroy phagocytosed bacteria is through the production of toxic oxygen intermediates derived from superoxide anion (O_2^-), including hydrogen peroxide and hypochlorous acid (Leusen *et al.*, 1996; Wientjes & Segal, 1995; Clark *et al.*, 1987). Furthermore, the pH change resulting from the O_2^- influx triggers a potassium-dependent release of proteases into the phagosome (Reeves *et al.*, 2002). Collectively, these products kill and degrade the ingested micro-organisms. O_2^- is produced by NADPH oxidase, a tightly controlled, rapidly activatable enzymic complex (Leusen *et al.*, 1996; Wientjes & Segal, 1995; Clark *et al.*, 1987). In resting neutrophils, the oxidase is inactive and its unassembled components are localized in different parts of the cell (Leusen *et al.*, 1996; Clark *et al.*, 1987). The membrane-bound components are cytochrome b_{558}, which is comprised of gp91^{phox} and p22^{phox} (Parkos *et al.*, 1987), and Rap1a (Babior, 1999). The additional subunits, which are present in the cytosol, are p40^{phox}, p47^{phox}, p67^{phox} and Rac2 (Fig. 3a) (Leusen *et al.*, 1996; Wientjes & Segal, 1995; Abo *et al.*, 1991; Knaus *et al.*, 1991; Volpp *et al.*, 1988). NADPH oxidase can be activated by either receptor-dependent means, which include fMLP, certain bacteria and Fc receptor cross-linking, or by receptor-independent mechanisms, such as the protein kinase C activator, phorbol myristate acetate (PMA) (Babior, 1999). Rac2 essentially serves as the molecular switch for initiating the respiratory burst (Abo *et al.*, 1991; Knaus *et al.*, 1991). Upon stimulation, Rac2 switches from an inactive, GDP-bound status to its active, GTP-bound form, dissociates from its Rho guanine-nucleotide-dissociation inhibitor, and migrates to join cytochrome b_{558} (Abo *et al.*, 1994; Quinn *et al.*, 1993). Concurrent with Rac2 migration, p47^{phox} becomes heavily phosphorylated and it, along with p40^{phox}, p67^{phox} and Rap1a, colocalize with p22^{phox} and gp91^{phox} to form the active complex (Leusen *et al.*, 1996; Gabig *et al.*, 1995; Volpp *et al.*, 1988). Once assembled, NADPH oxidase accepts electrons from NADPH at the cytosolic side of the membrane and donates them to oxygen at the other side, thereby generating O_2^- either outside the cell or within the phagosome (Fig. 3b) (Leusen *et al.*, 1996; Wientjes & Segal, 1995). Rac2 and p67^{phox} mediate the electron transfer reactions through gp91^{phox} (Diebold & Bokoch, 2001).

A. phagocytophilum's intra-neutrophil survival indicates that it employs strategies for combating the oxygen-dependent killing mechanisms of its host cell. Indeed,

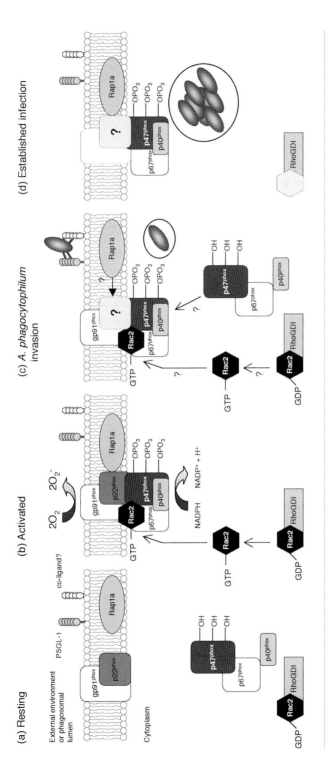

Fig. 3. *A. phagocytophilum* inhibits neutrophil NADPH oxidase. (a) In a resting neutrophil, the components of NADPH oxidase are unassembled. (b) Upon activation, Rac2 dissociates from its RhoGDI inhibitor, switches from a GDP- to GTP-bound status and migrates to the site of oxidase assembly. Concurrently, p47^{phox} becomes heavily phosphorylated and it, along with p67^{phox} and p40^{phox}, associates at the site of complex formation. At the expense of NADPH, the assembled enzyme converts 2 mol oxygen to O$_2^-$, which is formed outside the cell or within the phagosomal lumen. (c) One-half to a few hours following the addition of *A. phagocytophilum*, neutrophils lose the ability to generate O$_2^-$. Partial degradation of p22^{phox} has been reported to occur within this time frame. Yet, results thus far are inconclusive. It is currently unknown whether *A. phagocytophilum* inhibits other steps preceding NADPH oxidase activity (denoted by ?). It is also unknown whether *A. phagocytophilum* binds a non-PSGL-1 ligand(s) that may be involved in this process. (d) Neutrophils or HL-60 cells with established *A. phagocytophilum* infection (24 and ≥48 h, respectively) demonstrate losses of *rac2* and *gp91*^{phox} transcripts and corresponding losses of protein. As both Rac2 and gp91^{phox} are integral to NADPH oxidase activity, their loss clearly contributes to lack of O$_2^-$ production by cells with established infection.

PMA-induced O_2^- production by HL-60 cells (Carlyon *et al.*, 2002; Wang *et al.*, 2002; Banerjee *et al.*, 2000), IFN-γ-stimulated HL-60 cells, and HL-60 cells differentiated along the neutrophil lineage with all-trans-retinoic acid (rHL-60) (Banerjee *et al.*, 2000) is severely inhibited following *A. phagocytophilum* infection. Infected human neutrophils also exhibit repressed O_2^- production, whether activated by Fc receptor cross-linking (Wang *et al.*, 2002), fMLP, *E. coli* or PMA (Choi & Dumler, 2003; Mott & Rikihisa, 2000). These inhibitory effects also translate to *in vivo* infection since neutrophils from *A. phagocytophilum*-infected mice, as well as HA patients, exhibit reduced NADPH oxidase activity (Carlyon *et al.*, 2003; Wang *et al.*, 2002). Although not directly proven, this likely contributes to the increased susceptibility to opportunistic infections associated with severe cases of HA.

The strategies utilized by *A. phagocytophilum* for inhibiting NADPH oxidase are complex and multi-factorial, targeting the enzyme at the levels of expression and activation. *A. phagocytophilum* specifically inhibits expression of *gp91*[phox] and *rac2*, while the mRNA levels of the other NADPH oxidase components are unaffected (Carlyon *et al.*, 2002; Banerjee *et al.*, 2000). Loss of *gp91*[phox] transcript has been demonstrated for infected HL-60 and rHL-60 cells and also for murine infection (Banerjee *et al.*, 2000). Inhibition of *rac2* expression has been demonstrated *in vitro* for infected HL-60 and rHL-60 cells and human neutrophils. This occurs rapidly, as neutrophil *rac2* transcript levels decline 50-fold 24 h after *A. phagocytophilum* infection (Carlyon *et al.*, 2002). Loss of *rac2* and *gp91*[phox] mRNA translates into severe deficiencies of both proteins (Fig. 3d) (Carlyon *et al.*, 2002; Banerjee *et al.*, 2000). An HL-60 cell line stably transfected to express both *rac1* and *gp91*[phox] from the CMV immediate early promoter retained the ability to produce O_2^- 5 days post-*A. phagocytophilum* infection, thereby directly linking the bacterium's inhibition of *rac2* and *gp91*[phox] transcription to loss of NADPH oxidase function. Furthermore, *A. phagocytophilum* intracellular survival was severely inhibited in these cells (Carlyon *et al.*, 2002). Thus repression of *gp91*[phox] and *rac2* transcription is essential to bacterial proliferation.

Transcriptional repression of NADPH oxidase components, however, represents only one of apparently multiple strategies *A. phagocytophilum* uses to inhibit O_2^- production. Neutrophils emerge from the bone marrow with a full armamentarium of each of the pre-formed NADPH oxidase components. Furthermore, they are terminally differentiated. Therefore, altering *rac2* and *gp91*[phox] transcription is likely most relevant to infection of neutrophil progenitors in the bone marrow. *A. phagocytophilum* infection of neutrophil precursors in bone marrow has been demonstrated *in vitro* (Klein *et al.*, 1997), as well as for HA patients (2001; Trofe *et al.*, 2001) and experimentally infected mice (Hodzic *et al.*, 2001) and monkeys (Foley *et al.*, 1999). It is tempting to speculate that during the initial phase of infection, *A. phagocytophilum* infects peripherally circulating neutrophils and is carried to the bone marrow, where it egresses to invade neutrophil precursors. As the infected cells differentiate, *A. phagocytophilum* exerts its repressive effects on *rac2* and *gp91^{phox}* transcription, thereby leading to the emergence of NADPH-oxidase-deficient neutrophils.

The question remains, however, as to how *A. phagocytophilum* deals with the pre-formed NADPH oxidase components. Two independent studies using human neutrophils demonstrated that O_2^- production is inhibited following a few hours (Choi & Dumler, 2003; Wang *et al.*, 2002) or 30 min (Mott & Rikihisa, 2000) incubation with *A. phagocytophilum*. Thus, prior to inhibiting transcription of *rac2* and *gp91^{phox}* once inside the host cell, *A. phagocytophilum* initially inhibits NADPH oxidase activity during invasion (Fig. 3c). This rapid inhibition is dose- and contact-dependent and likely results from interaction of an *A. phagocytophilum* carbohydrate component and a neutrophil surface protein. Host, but not bacterial, protein synthesis is required, which hints that a neutrophil protein with a short half-life may be involved in blocking NADPH oxidase activation (Mott & Rikihisa, 2000). This notion is partially supported by a study suggesting that neutrophil NADPH oxidase activity stimulated by fMLP or leukotriene B$_4$, but not by PMA, is regulated by a short-lived protein (Stringer & Edwards, 1995).

A study by Mott *et al.* (2002) reported minor to moderately reduced p22^{phox} protein levels in neutrophils and HL-60 cells following 30 min and 7 days, respectively, incubation with *A. phagocytophilum*. Levels of gp91^{phox}, p40^{phox}, p47^{phox} and p67^{phox} were unchanged in this study. In addition, *A. phagocytophilum* had no effect on the PMA-induced phosphorylation of p47^{phox}. The authors also attempted to demonstrate that *A. phagocytophilum* does not prevent PMA-induced migration of p40^{phox}, p47^{phox} and p67^{phox} to the neutrophil cell membrane (Mott *et al.*, 2002). However, both p47^{phox} and p40^{phox} were readily detected in membrane fractions in the absence of PMA stimulation. As it is well-established that both subunits remain cytosolic until activation (Babior, 1999; Leusen *et al.*, 1996; Wientjes & Segal, 1995), this raises concern regarding the integrity of the fractions. The authors concluded that partial degradation of p22^{phox} was solely responsible for the ≥95 % decrease in NADPH oxidase activity. Evidence for such a mechanism has been demonstrated for PLB-985 cells incubated with succinyl acetone, an inhibitor of haem biosynthesis. These cells exhibit considerable reduction in the expression of p22^{phox} and the mature 91 kDa form of gp91^{phox}, but not its 61 kDa precursor, which results in improper formation of the cytochrome b_{558} heterodimer and a profound reduction in NADPH oxidase activity (Yu *et al.*, 1997). However, short-term incubation with *A. phagocytophilum* does not affect mature gp91^{phox} levels. Also, the low degree to which p22^{phox} levels may be reduced in *A. phagocytophilum*-incubated cells does not approach that observed for succinyl acetone-incubated cells. As only a minor fraction of each NADPH oxidase component is utilized for O_2^- generation (Leusen *et al.*, 1996), there should be enough p22^{phox} present to allow proper enzymic function. Furthermore, if p22^{phox} degradation were the key factor in oxidase inhibition, then HL-60 cells transfected to express Rac and gp91^{phox}, but not p22^{phox}, should have been unable to produce O_2^- following several days of *A. phagocytophilum* infection (Carlyon *et al.*, 2002). Thus, while partial reduction of p22^{phox} following incubation with *A. phagocytophilum* could possibly contribute to rapid loss of NADPH oxidase activity, this phenomenon and the severity to which it occurs needs to be confirmed.

Another means by which *A. phagocytophilum* may potentially protect itself from oxidative damage is by the O_2^- neutralizing enzyme superoxide dismutase (SOD). SODs are metalloenzymes that catalyse the rapid dismutation of O_2^- to H_2O_2 and O_2. There are three types of SODs, which are distinguishable by the prosthetic metals at their active sites: manganese (MnSOD), copper or zinc (Cu/ZnSOD) or iron (FeSOD). FeSODs and MnSODs protect bacteria from reactive oxygen species (ROS) generated from bacterial metabolism, while Cu/ZnSODs are believed to play a role in defence against external ROS (Lynch & Kuramitsu, 2000). The genomic location and transcriptional expression of a *sodB* homologue, which encodes an FeSOD, has been reported for *A. phagocytophilum* (Ohashi *et al.*, 2002). However, its functional contribution remains to be determined.

A. PHAGOCYTOPHILUM HINDERS NEUTROPHIL APOPTOSIS

Peripherally circulating neutrophils have very brief half-lives of only 6–12 h (Akgul *et al.*, 2001; Homburg & Roos, 1996). However, their lifespan can be extended *in vitro* via cytokine treatment or *in vivo* during inflammation (Moulding *et al.*, 1998; Lee *et al.*, 1993; Brach *et al.*, 1992; Colotta *et al.*, 1992). Given their short-lived nature, neutrophils would seem to be unsuitable host cells for intracellular pathogens. Yet, *A. phagocytophilum* establishes residence and regenerates itself within their confines. Considering *A. phagocytophilum*'s slow growth rate, prolonging the lifespan of its host neutrophil would benefit both its survival and pathogenesis. Indeed, this bacterium appears to delay neutrophil apoptosis, both *in vitro* and *in vivo* (Scaife *et al.*, 2003; Yoshiie *et al.*, 2000). Following 24 h *in vitro* cultivation, nearly 100 % of human neutrophils are apoptotic. In contrast, >50 % of neutrophils incubated with *A. phagocytophilum* remain non-apoptotic for up to 48 h and 10–20 % remain so at 96 h, as determined by cellular morphology. Such results were obtained for purified neutrophils, as well as for neutrophils in the presence of monocytes and lymphocytes. Thus the presence of other leukocytes cannot override the antiapoptotic effect of *A. phagocytophilum*. Furthermore, untreated neutrophils display considerably higher proportions of cytoplasmic histone-associated DNA fragments as compared to those incubated with *A. phagocytophilum*. This effect is time-dependent, as the T_{50} for neutrophils to become apoptotic is $12 \cdot 2 \pm 2 \cdot 5$ h versus $45 \cdot 0 \pm 9 \cdot 8$ h for *A. phagocytophilum*-infected cells. It is also dose-dependent and requires *A. phagocytophilum* surface proteins, but not carbohydrates. Oxytetracycline treatment does not affect *A. phagocytophilum*'s influence on apoptosis, which indicates that bacterial protein synthesis is not involved (Yoshiie *et al.*, 2000). Treatment with monodansylcadaverine, a transglutaminase inhibitor that hinders receptor-mediated endocytosis (Levitzki *et al.*, 1980), blocks the antiapoptotic effect, which suggests that this event is likely preceded by bacterial internalization. Little is known regarding the host cellular mechanisms involved in the *A. phagocytophilum*-induced apoptosis delay except that neither protein kinase A nor NF-κB activation is involved. Addition of blocking antibody to IL-1β or YVAD-CMK, each of which inhibits caspase-1 and accelerates apoptosis (Watson *et al.*, 1998; Thornberry *et al.*, 1992), has no effect on *A. phagocytophilum*'s antiapoptotic effect. However, the eukaryotic protein synthesis inhibitor cycloheximide, which blocks IL-1β and caspase-1 up-regulation (Watson *et al.*, 1998), accelerates neutrophil apoptosis, regardless of *A. phagocytophilum* infection (Yoshiie *et al.*, 2000).

Results obtained with *A. phagocytophilum*-infected ovine neutrophils *ex vivo* are similar to those observed for human neutrophils infected *in vitro*. *A. phagocytophilum*-infected neutrophils isolated from infected sheep on the peak day of bacteraemia (day 2) demonstrate apoptotic rates of 30 ± 7 % and 35 ± 2 % for annexin V binding and morphology, respectively, following 22 h cultivation *in vitro*. Corresponding values obtained for uninfected neutrophils are 52 ± 5 % and 78 ± 2 %. Treatment with the transcription inhibitor actinomycin D considerably accelerates apoptosis of control neutrophils, while infected cells isolated on the day of peak bacteraemia are less sensitive. Infected cells isolated after the *A. phagocytophilum* burden had waned (day 4) demonstrate actinomycin D sensitivity comparable to that of uninfected cells (Scaife *et al.*, 2003).

The pronounced effect that *A. phagocytophilum* exhibits on the apoptosis of neutrophils does not extend to HL-60 cells. *A. phagocytophilum*-infected HL-60 cells are considerably more apoptotic than uninfected cells as demonstrated by their arrest in G_1 phase and concurrent reductions in expression of PCNA, pRB, cyclin D3, cyclin E and p21[WAF1/CIP1], each of which are involved in cell cycling (Bedner *et al.*, 1998; Hsieh *et al.*, 1997; Cordon-Cardo, 1995; Ewen *et al.*, 1993; Mercer *et al.*, 1991). Expression of the antiapoptotic factor Bcl-2 is also reduced during infection. *A. phagocytophilum* infection also leads to cleavage of the DNA repair enzyme poly(ADP-ribose) polymerase, with a concordant increase in caspase-3, both of which are linked to cells undergoing apoptosis (DiPietrantonio *et al.*, 2000; Lazebnik *et al.*, 1994; Kaufmann *et al.*, 1993). Moreover, intracellular levels of ceramide, a lipid second messenger involved in caspase-3 activation (DiPietrantonio *et al.*, 1998), are increased in *A. phagocytophilum*-infected HL-60 cells. *A. phagocytophilum*-induced apoptosis delay is therefore a neutrophil-specific process and not a global consequence of infection.

It is clear that *A. phagocytophilum* retards neutrophil apoptosis. However, *A. phagocytophilum* has a long doubling time, which poses the question as to what benefit is afforded the bacterium by briefly extending the life of its host cell. As discussed later, one means by which *A. phagocytophilum* propagates is by stimulating neutrophil IL-8 production to recruit naïve cells to the sites of infection. Extending its host cell life would potentially enable the organism to maximize this and other survival strategies.

A. PHAGOCYTOPHILUM EXPLOITS NEUTROPHIL CHEMOKINE PRODUCTION

Cytokines play significant roles in the antimicrobial response as they help control infection. However, they may also contribute to disease pathology. Several studies have characterized the cytokine response associated with *A. phagocytophilum* infection, the majority of which overwhelmingly indicate a cytokine response weighted toward a Th1 phenotype. Sera from HA patients (Dumler *et al.*, 2000) and experimentally inoculated mice (Akkoyunlu & Fikrig, 2000; Martin *et al.*, 2000), as well as supernatants from experimentally infected bone marrow progenitors, HL-60 and rHL-60 cells (Klein *et al.*, 2000), demonstrate elevated levels of IFN-γ and IL-10. Levels of proinflammatory cytokines TNF-α, IL-1β, IL-4 and IL-6 are not elevated. IFN-γ protects against several different obligate intracellular bacterial infections and likely does so against *A. phagocytophilum*. Such protection carries a consequence though, as IFN-γ is also damaging to host cells (Fresno *et al.*, 1997). Indeed, in HA, clinical illness and tissue pathology are more severe than predicted by bacterial load (Dumler & Bakken, 1998; Walker & Dumler, 1997).

Chemokines are chemotactic cytokines predominantly secreted by inflammatory cells and mediate leukocyte activation and chemotaxis. Expression of these molecules is induced by a variety of stimuli, including proinflammatory cytokines, lipopolysaccharide and other bacterial products (Mahalingam & Karupiah, 1999; Luster, 1998; Adams & Lloyd, 1997). *A. phagocytophilum* infection of bone marrow progenitors, HL-60 cells and rHL-60 cells *in vitro* also results in the elevated expression of chemokines IL-8, monocyte chemotactic protein-1 (MCP-1), macrophage inflammatory protein (MIP)-1α and -β, and RANTES (Akkoyunlu *et al.*, 2001; Klein *et al.*, 2000). Supernatants from HA patient sera and experimentally infected neutrophils display elevated levels of IL-8 (Akkoyunlu *et al.*, 2001). This phenomenon appears to be specifically induced by a heat-stable *A. phagocytophilum* component independent of classic LPS, as neither heat inactivation nor polymyxin B treatment of bacterial preparations inhibits chemokine secretion (Klein *et al.*, 2000).

IL-8 is a powerful neutrophil chemoattractant (Mahalingam & Karupiah, 1999) whose expression is up-regulated during *A. phagocytophilum* infection. The bacterium induces IL-8 secretion from rHL-60 cells in dose- and time-dependent manners. Incubation with a recombinant form of one of the P44 homologues also stimulates IL-8 production (Akkoyunlu *et al.*, 2001). IL-8 enhances neutrophil phagocytosis, as well as O_2^- and granule release (Baggiolini *et al.*, 1994), which raises the question as to why *A. phagocytophilum* specifically induces its production. The answer paradoxically lies in the fact that increased IL-8 levels effectively recruit naïve neutrophils to facilitate bacterial dissemination. Indeed, chemotaxis of human neutrophils is induced by supernatants from *A. phagocytophilum*-infected, but not uninfected, rHL-60 cells in a fashion similar to that by recombinant IL-8. Human neutrophils respond to IL-8 through the receptors CXCR1 and CXCR2 (Mahalingam & Karupiah, 1999). *A. phagocytophilum*-infected rHL-60 cells demonstrate increased surface expression of CXCR2, but not CXCR1 (Akkoyunlu *et al.*, 2001).

While mice have a homologue of human CXCR2, the murine equivalent of IL-8 has yet to be identified. However, murine MIP-2 and KC do bind murine CXCR2 to facilitate neutrophil chemotaxis (Cacalano *et al.*, 1994). Incubation of *A. phagocytophilum* with murine splenocytes (Akkoyunlu *et al.*, 2001) or purified neutrophils (Carlyon *et al.*, 2003) results in significant increases in KC and MIP-2 relative to uninfected controls. CXCR2$^{-/-}$ and wild-type mice pretreated with CXCR2 antiserum each exhibit considerable decreases in bacterial load compared to control background or mice pretreated with control serum, respectively, following *A. phagocytophilum* infection. This indicates that CXCR2-induced neutrophil migration plays an important role in *A. phagocytophilum* dissemination during murine infection. Furthermore, wild-type mice pretreated with CXCR2 antiserum prior to the administration of *A. phagocyto-philum* exhibit significant decreases in neutrophil chemotaxis into the peritoneal cavity as compared to that of infected mice pretreated with control serum (Akkoyunlu *et al.*, 2001). As *A. phagocytophilum* is an obligate intracellular bacterium, its establishment and maintenance of infection are dependent on its transfer to naïve neutrophils or precursor cells. *A. phagocytophilum* facilitates neutrophil IL-8 production as a strategy for attracting neutrophils to sites of infection for further propagation.

CONCLUDING REMARKS

A. phagocytophilum is a highly unique pathogen because of its tropism for neutrophils. This organism employs several strategies, some of them unprecedented, to flourish within the diverse milieus encountered during its zoonotic life cycle, including the hostile intracellular environment of the neutrophil. Differential expression of its P44 paralogues likely facilitates host adaptation and evasion of the humoral immune response. It utilizes as yet to be defined adhesins to bind to PSGL-1 of human neutrophils and subsequently protects itself from oxidative damage by inhibiting NADPH oxidase activity. Once inside, *A. phagocytophilum* manipulates vacuolar trafficking and neutrophil gene expression to create a haven within the confines of its host cell. It promotes IL-8 secretion to recruit naïve neutrophils and propagate infection. Furthermore, *A. phagocytophilum* delays neutrophil apoptosis, which not only partially compensates for its long doubling time, but also likely extends its host cell's life such that it can optimally execute each of the aforementioned strategies. Continued study of *A. phagocytophilum* will promote a more thorough understanding of intracellular pathogens, as well as neutrophil biology.

ACKNOWLEDGEMENTS

We regret that we could not cite several important original papers due to space limitation. We thank Dr Venetta Thomas for providing photographs of A. phagocytophilum-infected cells.

REFERENCES

(2001). Case records of the Massachusetts General Hospital. Weekly clinicopathological exercises. Case 37-2001. A 76-year-old man with fever, dyspnea, pulmonary infiltrates, pleural effusions, and confusion. *N Engl J Med* **345**, 1627–1634.

Abo, A., Pick, E., Hall, A., Totty, N., Teahan, C. G. & Segal, A. W. (1991). Activation of the NADPH oxidase involves the small GTP-binding protein p21rac1. *Nature* **353**, 668–670.

Abo, A., Webb, M. R., Grogan, A. & Segal, A. W. (1994). Activation of NADPH oxidase involves the dissociation of p21rac from its inhibitory GDP/GTP exchange protein (rhoGDI) followed by its translocation to the plasma membrane. *Biochem J* **298**, 585–591.

Adams, D. H. & Lloyd, A. R. (1997). Chemokines: leucocyte recruitment and activation cytokines. *Lancet* **349**, 490–495.

Aguero-Rosenfeld, M. E., Horowitz, H. W., Wormser, G. P., McKenna, D. F., Nowakowski, J., Munoz, J. & Dumler, J. S. (1996). Human granulocytic ehrlichiosis: a case series from a medical center in New York State. *Ann Intern Med* **125**, 904–908.

Akgul, C., Moulding, D. A. & Edwards, S. W. (2001). Molecular control of neutrophil apoptosis. *FEBS Lett* **487**, 318–322.

Akkoyunlu, M. & Fikrig, E. (2000). Gamma interferon dominates the murine cytokine response to the agent of human granulocytic ehrlichiosis and helps to control the degree of early rickettsemia. *Infect Immun* **68**, 1827–1833.

Akkoyunlu, M., Malawista, S. E., Anguita, J. & Fikrig, E. (2001). Exploitation of interleukin-8-induced neutrophil chemotaxis by the agent of human granulocytic ehrlichiosis. *Infect Immun* **69**, 5577–5588.

Amigorena, S., Drake, J. R., Webster, P. & Mellman, I. (1994). Transient accumulation of new class II MHC molecules in a novel endocytic compartment in B lymphocytes. *Nature* **369**, 113–120.

Anderson, R. G., Falck, J. R., Goldstein, J. L. & Brown, M. S. (1984). Visualization of acidic organelles in intact cells by electron microscopy. *Proc Natl Acad Sci U S A* **81**, 4838–4842.

Andrews, N. W., Abrams, C. K., Slatin, S. L. & Griffiths, G. (1990). A T. cruzi-secreted protein immunologically related to the complement component C9: evidence for membrane pore-forming activity at low pH. *Cell* **61**, 1277–1287.

Babior, B. M. (1999). NADPH oxidase: an update. *Blood* **93**, 1464–1476.

Baggiolini, M., Dewald, B. & Moser, B. (1994). Interleukin-8 and related chemotactic cytokines – CXC and CC chemokines. *Adv Immunol* **55**, 97–179.

Bakken, J. S. (1998). The discovery of human granulocytotropic ehrlichiosis. *J Lab Clin Med* **132**, 175–180.

Bakken, J. S., Dumler, J. S., Chen, S. M., Eckman, M. R., Van Etta, L. L. & Walker, D. H. (1994). Human granulocytic ehrlichiosis in the upper Midwest United States. A new species emerging? *JAMA (J Am Med Assoc)* **272**, 212–218.

Bakken, J. S., Krueth, J., Wilson-Nordskog, C., Tilden, R. L., Asanovich, K. & Dumler, J. S. (1996). Clinical and laboratory characteristics of human granulocytic ehrlichiosis. *JAMA (J Am Med Assoc)* **275**, 199–205.

Bakken, J. S., Goellner, P., Van Etten, M. & 8 other authors (1998). Seroprevalence of human granulocytic ehrlichiosis among permanent residents of northwestern Wisconsin. *Clin Infect Dis* **27**, 1491–1496.

Banerjee, R., Anguita, J., Roos, D. & Fikrig, E. (2000). Infection by the agent of human granulocytic ehrlichiosis prevents the respiratory burst by down-regulating gp91phox. *J Immunol* **164**, 3946–3949.

Barbet, A. F., Lundgren, A., Yi, J., Rurangirwa, F. R. & Palmer, G. H. (2000). Antigenic variation of *Anaplasma marginale* by expression of MSP2 mosaics. *Infect Immun* **68**, 6133–6138.

Barbet, A. F., Meeus, P. F., Belanger, M. & 8 other authors (2003). Expression of multiple outer membrane protein sequence variants from a single genomic locus of *Anaplasma phagocytophilum*. *Infect Immun* **71**, 1706–1718.

Bedner, E., Burfeind, P., Hsieh, T. C., Wu, J. M., Aguero-Rosenfeld, M. E., Melamed, M. R., Horowitz, H. W., Wormser, G. P. & Darzynkiewicz, Z. (1998). Cell cycle effects and induction of apoptosis caused by infection of HL-60 cells with human granulocytic ehrlichiosis pathogen measured by flow and laser scanning cytometry. *Cytometry* **33**, 47–55.

Blanco, J. R. & Oteo, J. A. (2002). Human granulocytic ehrlichiosis in Europe. *Clin Microbiol Infect* **8**, 763–772.

Bozue, J. A. & Johnson, W. (1996). Interaction of *Legionella pneumophila* with *Acanthamoeba castellanii*: uptake by coiling phagocytosis and inhibition of phagosome-lysosome fusion. *Infect Immun* **64**, 668–673.

Brach, M. A., deVos, S., Gruss, H. J. & Herrmann, F. (1992). Prolongation of survival of human polymorphonuclear neutrophils by granulocyte-macrophage colony-stimulating factor is caused by inhibition of programmed cell death. *Blood* **80**, 2920–2924.

Brayton, K. A., Knowles, D. P., McGuire, T. C. & Palmer, G. H. (2001). Efficient use of a small genome to generate antigenic diversity in tick-borne ehrlichial pathogens. *Proc Natl Acad Sci U S A* **98**, 4130–4135.

Brayton, K. A., Palmer, G. H., Lundgren, A., Yi, J. & Barbet, A. F. (2002). Antigenic variation of *Anaplasma marginale* msp2 occurs by combinatorial gene conversion. *Mol Microbiol* **43**, 1151–1159.

Buller, R. S., Arens, M., Hmiel, S. P. & 9 other authors (1999). *Ehrlichia ewingii*, a newly recognized agent of human ehrlichiosis. *N Engl J Med* **341**, 148–155.

Burg, N. D. & Pillinger, M. H. (2001). The neutrophil: function and regulation in innate and humoral immunity. *Clin Immunol* **99**, 7–17.

Cacalano, G., Lee, J., Kikly, K., Ryan, A. M., Pitts-Meek, S., Hultgren, B., Wood, W. I. & Moore, M. W. (1994). Neutrophil and B cell expansion in mice that lack the murine IL-8 receptor homolog. *Science* **265**, 682–684.

Carlyon, J. A. & Fikrig, E. (2003). Invasion and survival strategies of *Anaplasma phagocytophilum*. *Cell Microbiol* **5**, 743–754.

Carlyon, J. A., Chan, W. T., Galan, J., Roos, D. & Fikrig, E. (2002). Repression of rac2 mRNA expression by *Anaplasma phagocytophila* is essential to the inhibition of superoxide production and bacterial proliferation. *J Immunol* **169**, 7009–7018.

Carlyon, J. A., Akkoyunlu, M., Xia, L., Yago, T., Wang, T., Cummings, R. D., McEver, R. P. & Fikrig, E. (2003). Murine neutrophils require alpha1,3-fucosylation but not PSGL-1 for productive infection with *Anaplasma phagocytophilum*. *Blood* **102**, 3387–3395.

Caspersen, K., Park, J. H., Patil, S. & Dumler, J. S. (2002). Genetic variability and stability of *Anaplasma phagocytophila* msp2 (p44). *Infect Immun* **70**, 1230–1234.

Chen, J. W., Chen, G. L., D'Souza, M. P., Murphy, T. L. & August, J. T. (1986). Lysosomal membrane glycoproteins: properties of LAMP-1 and LAMP-2. *Biochem Soc Symp* **51**, 97–112.

Chen, S. M., Dumler, J. S., Bakken, J. S. & Walker, D. H. (1994). Identification of a granulocytotropic *Ehrlichia* species as the etiologic agent of human disease. *J Clin Microbiol* **32**, 589–595.

Choi, K. S. & Dumler, J. S. (2003). Early induction and late abrogation of respiratory burst in *A. phagocytophilum*-infected neutrophils. *Ann N Y Acad Sci* **990**, 488–493.

Christova, I. S. & Dumler, J. S. (1999). Human granulocytic ehrlichiosis in Bulgaria. *Am J Trop Med Hyg* **60**, 58–61.

Clark, R. A., Leidal, K. G., Pearson, D. W. & Nauseef, W. M. (1987). NADPH oxidase of human neutrophils. Subcellular localization and characterization of an arachidonate-activatable superoxide-generating system. *J Biol Chem* **262**, 4065–4074.

Clemens, D. L. & Horwitz, M. A. (1996). The *Mycobacterium tuberculosis* phagosome interacts with early endosomes and is accessible to exogenously administered transferrin. *J Exp Med* **184**, 1349–1355.

Colotta, F., Re, F., Polentarutti, N., Sozzani, S. & Mantovani, A. (1992). Modulation of granulocyte survival and programmed cell death by cytokines and bacterial products. *Blood* **80**, 2012–2020.

Cordon-Cardo, C. (1995). Mutations of cell cycle regulators. Biological and clinical implications for human neoplasia. *Am J Pathol* **147**, 545–560.

Daniel, S. A., Manika, K., Arvanitidou, M., Diza, E., Symeonidis, N. & Antoniadis, A. (2002). Serologic evidence of human granulocytic ehrlichiosis, Greece. *Emerg Infect Dis* **8**, 643–644.

Desjardins, M. (2003). ER-mediated phagocytosis: a new membrane for new functions. *Nat Rev Immunol* **3**, 280–291.

Diebold, B. A. & Bokoch, G. M. (2001). Molecular basis for Rac2 regulation of phagocyte NADPH oxidase. *Nat Immunol* **2**, 211–215.

DiPietrantonio, A. M., Hsieh, T. C., Olson, S. C. & Wu, J. M. (1998). Regulation of G1/S transition and induction of apoptosis in HL-60 leukemia cells by fenretinide (4HPR). *Int J Cancer* **78**, 53–61.

DiPietrantonio, A. M., Hsieh, T. C. & Wu, J. M. (2000). Specific processing of poly(ADP-ribose) polymerase, accompanied by activation of caspase-3 and elevation/reduction of ceramide/hydrogen peroxide levels, during induction of apoptosis in host HL-60 cells infected by the human granulocytic ehrlichiosis (HGE) agent. *IUBMB Life* **49**, 49–55.

Duden, R., Griffiths, G., Frank, R., Argos, P. & Kreis, T. E. (1991). Beta-COP, a 110 kd protein associated with non-clathrin-coated vesicles and the Golgi complex, shows homology to beta-adaptin. *Cell* **64**, 649–665.

Dumler, J. S. & Bakken, J. S. (1996). Human granulocytic ehrlichiosis in Wisconsin and Minnesota: a frequent infection with the potential for persistence. *J Infect Dis* **173**, 1027–1030.

Dumler, J. S. & Bakken, J. S. (1998). Human ehrlichioses: newly recognized infections transmitted by ticks. *Annu Rev Med* **49**, 201–213.

Dumler, J. S., Trigiani, E. R., Bakken, J. S., Aguero-Rosenfeld, M. E. & Wormser, G. P. (2000). Serum cytokine responses during acute human granulocytic ehrlichiosis. *Clin Diagn Lab Immunol* **7**, 6–8.

Dumler, J. S., Barbet, A. F., Bekker, C. P., Dasch, G. A., Palmer, G. H., Ray, S. C., Rikihisa, Y. & Rurangirwa, F. R. (2001). Reorganization of genera in the families *Rickettsiaceae* and *Anaplasmataceae* in the order *Rickettsiales*: unification of some species of *Ehrlichia* with *Anaplasma*, *Cowdria* with *Ehrlichia* and *Ehrlichia* with *Neorickettsia*, descriptions of six new species combinations and designation of *Ehrlichia equi* and 'HGE agent' as subjective synonyms of *Ehrlichia phagocytophila*. *Int J Syst Evol Microbiol* **51**, 2145–2165.

Eid, G., French, D. M., Lundgren, A. M., Barbet, A. F., McElwain, T. F. & Palmer, G. H. (1996). Expression of major surface protein 2 antigenic variants during acute *Anaplasma marginale* rickettsemia. *Infect Immun* **64**, 836–841.

Eissenberg, L. G. & Wyrick, P. B. (1981). Inhibition of phagolysosome fusion is localized to *Chlamydia psittaci*-laden vacuoles. *Infect Immun* **32**, 889–896.

Ewen, M. E., Sluss, H. K., Sherr, C. J., Matsushime, H., Kato, J. & Livingston, D. M. (1993). Functional interactions of the retinoblastoma protein with mammalian D-type cyclins. *Cell* **73**, 487–497.

Fingerle, V., Goodman, J. L., Johnson, R. C., Kurtti, T. J., Munderloh, U. G. & Wilske, B. (1999). Epidemiological aspects of human granulocytic Ehrlichiosis in southern Germany. *Wien Klin Wochenschr* **111**, 1000–1004.

Foggie, A. (1951). Studies on the infectious agent of tick-borne fever in sheep. *J Pathol Bacteriol* **53**, 1–15.

Foley, J. E., Lerche, N. W., Dumler, J. S. & Madigan, J. E. (1999). A simian model of human granulocytic ehrlichiosis. *Am J Trop Med Hyg* **60**, 987–993.

Fortier, A. H., Leiby, D. A., Narayanan, R. B., Asafoadjei, E., Crawford, R. M., Nacy, C. A. & Meltzer, M. S. (1995). Growth of *Francisella tularensis* LVS in macrophages: the acidic intracellular compartment provides essential iron required for growth. *Infect Immun* **63**, 1478–1483.

French, D. M., Brown, W. C. & Palmer, G. H. (1999). Emergence of *Anaplasma marginale* antigenic variants during persistent rickettsemia. *Infect Immun* **67**, 5834–5840.

Fresno, M., Kopf, M. & Rivas, L. (1997). Cytokines and infectious diseases. *Immunol Today* **18**, 56–58.

Fujiwara, T., Oda, K., Yokota, S., Takatsuki, A. & Ikehara, Y. (1988). Brefeldin A causes disassembly of the Golgi complex and accumulation of secretory proteins in the endoplasmic reticulum. *J Biol Chem* **263**, 18545–18552.

Gabig, T. G., Crean, C. D., Mantel, P. L. & Rosli, R. (1995). Function of wild-type or mutant Rac2 and Rap1a GTPases in differentiated HL60 cell NADPH oxidase activation. *Blood* **85**, 804–811.

Gagnon, E., Duclos, S., Rondeau, C., Chevet, E., Cameron, P. H., Steele-Mortimer, O., Paiement, J., Bergeron, J. J. & Desjardins, M. (2002). Endoplasmic reticulum-mediated phagocytosis is a mechanism of entry into macrophages. *Cell* **110**, 119–131.

Garin, J., Diez, R., Kieffer, S., Dermine, J. F., Duclos, S., Gagnon, E., Sadoul, R., Rondeau, C. & Desjardins, M. (2001). The phagosome proteome: insight into phagosome functions. *J Cell Biol* **152**, 165–180.

Geuze, H. J., Slot, J. W., Strous, G. J., Peppard, J., von Figura, K., Hasilik, A. & Schwartz, A. L. (1984). Intracellular receptor sorting during endocytosis: comparative immunoelectron microscopy of multiple receptors in rat liver. *Cell* **37**, 195–204.

Gokce, H. I., Ross, G. & Woldehiwet, Z. (1999). Inhibition of phagosome-lysosome fusion in ovine polymorphonuclear leucocytes by *Ehrlichia* (*Cytoecetes*) *phagocytophila*. *J Comp Pathol* **120**, 369–381.

Goodman, J. L., Nelson, C., Vitale, B., Madigan, J. E., Dumler, J. S., Kurtti, T. J. & Munderloh, U. G. (1996). Direct cultivation of the causative agent of human granulocytic ehrlichiosis. *N Engl J Med* **334**, 209–215.

Goodman, J. L., Nelson, C. M., Klein, M. B., Hayes, S. F. & Weston, B. W. (1999). Leukocyte infection by the granulocytic ehrlichiosis agent is linked to expression of a selectin ligand. *J Clin Invest* **103**, 407–412.

Guillaume, B., Heyman, P., Lafontaine, S., Vandenvelde, C., Delmee, M. & Bigaignon, G. (2002). Seroprevalence of human granulocytic ehrlichiosis infection in Belgium. *Eur J Clin Microbiol Infect Dis* **21**, 397–400.

Hackstadt, T., Rockey, D. D., Heinzen, R. A. & Scidmore, M. A. (1996). *Chlamydia trachomatis* interrupts an exocytic pathway to acquire endogenously synthesized sphingomyelin in transit from the Golgi apparatus to the plasma membrane. *EMBO J* **15**, 964–977.

Hardalo, C. J., Quagliarello, V. & Dumler, J. S. (1995). Human granulocytic ehrlichiosis in Connecticut: report of a fatal case. *Clin Infect Dis* **21**, 910–914.

Heinzen, R. A., Scidmore, M. A., Rockey, D. D. & Hackstadt, T. (1996). Differential interaction with endocytic and exocytic pathways distinguish parasitophorous vacuoles of *Coxiella burnetii* and *Chlamydia trachomatis*. *Infect Immun* **64**, 796–809.

Herron, M. J., Nelson, C. M., Larson, J., Snapp, K. R., Kansas, G. S. & Goodman, J. L. (2000). Intracellular parasitism by the human granulocytic ehrlichiosis bacterium through the P-selectin ligand, PSGL-1. *Science* **288**, 1653–1656.

Hodzic, E., Ijdo, J. W., Feng, S. & 7 other authors (1998). Granulocytic ehrlichiosis in the laboratory mouse. *J Infect Dis* **177**, 737–745.

Hodzic, E., Feng, S., Fish, D., Leutenegger, C. M., Freet, K. J. & Barthold, S. W. (2001). Infection of mice with the agent of human granulocytic ehrlichiosis after different routes of inoculation. *J Infect Dis* **183**, 1781–1786.

Homburg, C. H. & Roos, D. (1996). Apoptosis of neutrophils. *Curr Opin Hematol* **3**, 94–99.

Hsieh, T. C., Aguero-Rosenfeld, M. E., Wu, J. M. & 9 other authors (1997). Cellular changes and induction of apoptosis in human promyelocytic HL-60 cells infected with the agent of human granulocytic ehrlichiosis (HGE). *Biochem Biophys Res Commun* **232**, 298–303.

Hulinska, D., Votypka, J., Plch, J., Vlcek, E., Valesova, M., Bojar, M., Hulinsky, V. & Smetana, K. (2002). Molecular and microscopical evidence of *Ehrlichia* spp. and *Borrelia burgdorferi* sensu lato in patients, animals and ticks in the Czech Republic. *New Microbiol* **25**, 437–448.

Ijdo, J. W., Sun, W., Zhang, Y., Magnarelli, L. A. & Fikrig, E. (1998). Cloning of the gene encoding the 44-kilodalton antigen of the agent of human granulocytic ehrlichiosis and characterization of the humoral response. *Infect Immun* **66**, 3264–3269.

IJdo, J., Wu, C., Telford, S. R., 3rd & Fikrig, E. (2002). Differential expression of the p44 gene family in the agent of human granulocytic ehrlichiosis. *Infect Immun* **70**, 5295–5298.

Jauron, S. D., Nelson, C. M., Fingerle, V., Ravyn, M. D., Goodman, J. L., Johnson, R. C., Lobentanzer, R., Wilske, B. & Munderloh, U. G. (2001). Host cell-specific expression of a p44 epitope by the human granulocytic ehrlichiosis agent. *J Infect Dis* **184**, 1445–1450.

Kansas, G. S. (1996). Selectins and their ligands: current concepts and controversies. *Blood* **88**, 3259–3287.

Kaufmann, S. H., Desnoyers, S., Ottaviano, Y., Davidson, N. E. & Poirier, G. G. (1993). Specific proteolytic cleavage of poly(ADP-ribose) polymerase: an early marker of chemotherapy-induced apoptosis. *Cancer Res* **53**, 3976–3985.

Kieser, S. T., Eriks, I. S. & Palmer, G. H. (1990). Cyclic rickettsemia during persistent *Anaplasma marginale* infection of cattle. *Infect Immun* **58**, 1117–1119.

Klein, M. B., Miller, J. S., Nelson, C. M. & Goodman, J. L. (1997). Primary bone marrow progenitors of both granulocytic and monocytic lineages are susceptible to infection with the agent of human granulocytic ehrlichiosis. *J Infect Dis* **176**, 1405–1409.

Klein, M. B., Hu, S., Chao, C. C. & Goodman, J. L. (2000). The agent of human granulocytic ehrlichiosis induces the production of myelosuppressing chemokines without induction of proinflammatory cytokines. *J Infect Dis* **182**, 200–205.

Knaus, U. G., Heyworth, P. G., Evans, T., Curnutte, J. T. & Bokoch, G. M. (1991). Regulation of phagocyte oxygen radical production by the GTP-binding protein Rac 2. *Science* **254**, 1512–1515.

Lazebnik, Y. A., Kaufmann, S. H., Desnoyers, S., Poirier, G. G. & Earnshaw, W. C. (1994). Cleavage of poly(ADP-ribose) polymerase by a proteinase with properties like ICE. *Nature* **371**, 346–347.

Lebech, A. M., Hansen, K., Pancholi, P., Sloan, L. M., Magera, J. M. & Persing, D. H. (1998). Immunoserologic evidence of Human Granulocytic Ehrlichiosis in Danish patients with Lyme neuroborreliosis. *Scand J Infect Dis* **30**, 173–176.

Lee, A., Whyte, M. K. & Haslett, C. (1993). Inhibition of apoptosis and prolongation of neutrophil functional longevity by inflammatory mediators. *J Leukoc Biol* **54**, 283–288.

Leppanen, A., Mehta, P., Ouyang, Y. B. & 7 other authors (1999). A novel glycosulfopeptide binds to P-selectin and inhibits leukocyte adhesion to P-selectin. *J Biol Chem* **274**, 24838–24848.

Leppanen, A., White, S. P., Helin, J., McEver, R. P. & Cummings, R. D. (2000). Binding of glycosulfopeptides to P-selectin requires stereospecific contributions of individual tyrosine sulfate and sugar residues. *J Biol Chem* **275**, 39569–39578.

Leppanen, A., Penttila, L., Renkonen, O., McEver, R. P. & Cummings, R. D. (2002). Glycosulfopeptides with O-glycans containing sialylated and polyfucosylated polylactosamine bind with low affinity to P-selectin. *J Biol Chem* **277**, 39749–39759.

Leusen, J. H., Verhoeven, A. J. & Roos, D. (1996). Interactions between the components of the human NADPH oxidase: intrigues in the phox family. *J Lab Clin Med* **128**, 461–476.

Levitzki, A., Willingham, M. & Pastan, I. (1980). Evidence for participation of transglutaminase in receptor-mediated endocytosis. *Proc Natl Acad Sci U S A* **77**, 2706–2710.

Lin, Q., Zhi, N., Ohashi, N., Horowitz, H. W., Aguero-Rosenfeld, M. E., Raffalli, J., Wormser, G. P. & Rikihisa, Y. (2002). Analysis of sequences and loci of p44 homologs expressed by *Anaplasma phagocytophila* in acutely infected patients. *J Clin Microbiol* **40**, 2981–2988.

Lipsky, N. G. & Pagano, R. E. (1983). Sphingolipid metabolism in cultured fibroblasts: microscopic and biochemical studies employing a fluorescent ceramide analogue. *Proc Natl Acad Sci U S A* **80**, 2608–2612.

Lowe, J. B. (1997). Selectin ligands, leukocyte trafficking, and fucosyltransferase genes. *Kidney Int* **51**, 1418–1426.

Lukacs, G. L., Rotstein, O. D. & Grinstein, S. (1990). Phagosomal acidification is mediated by a vacuolar-type H(+)-ATPase in murine macrophages. *J Biol Chem* **265**, 21099–21107.

Luster, A. D. (1998). Chemokines – chemotactic cytokines that mediate inflammation. *N Engl J Med* **338**, 436–445.

Lynch, M. & Kuramitsu, H. (2000). Expression and role of superoxide dismutases (SOD) in pathogenic bacteria. *Microbes Infect* **2**, 1245–1255.

Mahalingam, S. & Karupiah, G. (1999). Chemokines and chemokine receptors in infectious diseases. *Immunol Cell Biol* **77**, 469–475.

Martin, M. E., Bunnell, J. E. & Dumler, J. S. (2000). Pathology, immunohistology, and cytokine responses in early phases of human granulocytic ehrlichiosis in a murine model. *J Infect Dis* **181**, 374–378.

McEver, R. P. (2001). Adhesive interactions of leukocytes, platelets, and the vessel wall during hemostasis and inflammation. *Thromb Haemostasis* **86**, 746–756.

McEver, R. P. (2002). Selectins: lectins that initiate cell adhesion under flow. *Curr Opin Cell Biol* **14**, 581–586.

McEver, R. P. & Cummings, R. D. (1997). Role of PSGL-1 binding to selectins in leukocyte recruitment. *J Clin Invest* **100**, 485–491.

McQuiston, J. H., Paddock, C. D., Holman, R. C. & Childs, J. E. (1999). The human ehrlichioses in the United States. *Emerg Infect Dis* **5**, 635–642.

Mercer, W. E., Shields, M. T., Lin, D., Appella, E. & Ullrich, S. J. (1991). Growth suppression induced by wild-type p53 protein is accompanied by selective down-regulation of proliferating-cell nuclear antigen expression. *Proc Natl Acad Sci U S A* **88**, 1958–1962.

Mott, J. & Rikihisa, Y. (2000). Human granulocytic ehrlichiosis agent inhibits superoxide anion generation by human neutrophils. *Infect Immun* **68**, 6697–6703.

Mott, J., Barnewall, R. E. & Rikihisa, Y. (1999). Human granulocytic ehrlichiosis agent and *Ehrlichia chaffeensis* reside in different cytoplasmic compartments in HL-60 cells. *Infect Immun* **67**, 1368–1378.

Mott, J., Rikihisa, Y. & Tsunawaki, S. (2002). Effects of *Anaplasma phagocytophila* on NADPH oxidase components in human neutrophils and HL-60 cells. *Infect Immun* **70**, 1359–1366.

Moulding, D. A., Quayle, J. A., Hart, C. A. & Edwards, S. W. (1998). Mcl-1 expression in human neutrophils: regulation by cytokines and correlation with cell survival. *Blood* **92**, 2495–2502.

Mu, F. T., Callaghan, J. M., Steele-Mortimer, O. & 7 other authors (1995). EEA1, an early endosome-associated protein. EEA1 is a conserved alpha-helical peripheral membrane protein flanked by cysteine "fingers" and contains a calmodulin-binding IQ motif. *J Biol Chem* **270**, 13503–13511.

Murphy, C. I., Storey, J. R., Recchia, J., Doros-Richert, L. A., Gingrich-Baker, C., Munroe, K., Bakken, J. S., Coughlin, R. T. & Beltz, G. A. (1998). Major antigenic proteins of the agent of human granulocytic ehrlichiosis are encoded by members of a multigene family. *Infect Immun* **66**, 3711–3718.

Neefjes, J. J., Stollorz, V., Peters, P. J., Geuze, H. J. & Ploegh, H. L. (1990). The biosynthetic pathway of MHC class II but not class I molecules intersects the endocytic route. *Cell* **61**, 171–183.

Ogden, N. H., Woldehiwet, Z. & Hart, C. A. (1998). Granulocytic ehrlichiosis: an emerging or rediscovered tick-borne disease? *J Med Microbiol* **47**, 475–482.

Ohashi, N., Zhi, N., Lin, Q. & Rikihisa, Y. (2002). Characterization and transcriptional analysis of gene clusters for a type IV secretion machinery in human granulocytic and monocytic ehrlichiosis agents. *Infect Immun* **70**, 2128–2138.

Palmer, G. H., Eid, G., Barbet, A. F., McGuire, T. C. & McElwain, T. F. (1994). The immunoprotective *Anaplasma marginale* major surface protein 2 is encoded by a polymorphic multigene family. *Infect Immun* **62**, 3808–3816.

Palmer, G. H., Brown, W. C. & Rurangirwa, F. R. (2000). Antigenic variation in the persistence and transmission of the ehrlichia *Anaplasma marginale*. *Microbes Infect* **2**, 167–176.

Pancholi, P., Kolbert, C. P., Mitchell, P. D., Reed, K. D., Jr, Dumler, J. S., Bakken, J. S., Telford, S. R., 3rd & Persing, D. H. (1995). *Ixodes dammini* as a potential vector of human granulocytic ehrlichiosis. *J Infect Dis* **172**, 1007–1012.

Park, J., Choi, K. S. & Dumler, J. S. (2003). Major surface protein 2 of *Anaplasma phagocytophilum* facilitates adherence to granulocytes. *Infect Immun* **71**, 4018–4025.

Parkos, C. A., Allen, R. A., Cochrane, C. G. & Jesaitis, A. J. (1987). Purified cytochrome b from human granulocyte plasma membrane is comprised of two polypeptides with relative molecular weights of 91,000 and 22,000. *J Clin Invest* **80**, 732–742.

Pitt, A., Mayorga, L. S., Stahl, P. D. & Schwartz, A. L. (1992a). Alterations in the protein composition of maturing phagosomes. *J Clin Invest* **90**, 1978–1983.

Pitt, A., Mayorga, L. S., Schwartz, A. L. & Stahl, P. D. (1992b). Transport of phagosomal components to an endosomal compartment. *J Biol Chem* **267**, 126–132.

Pusterla, N., Weber, R., Wolfensberger, C., Schar, G., Zbinden, R., Fierz, W., Madigan, J. E., Dumler, J. S. & Lutz, H. (1998). Serological evidence of human granulocytic ehrlichiosis in Switzerland. *Eur J Clin Microbiol Infect Dis* **17**, 207–209.

Quinn, M. T., Evans, T., Loetterle, L. R., Jesaitis, A. J. & Bokoch, G. M. (1993). Translocation of Rac correlates with NADPH oxidase activation. Evidence for equimolar translocation of oxidase components. *J Biol Chem* **268**, 20983–20987.

Rathman, M., Barker, L. P. & Falkow, S. (1997). The unique trafficking pattern of *Salmonella typhimurium*-containing phagosomes in murine macrophages is independent of the mechanism of bacterial entry. *Infect Immun* **65**, 1475–1485.

Reeves, E. P., Lu, H., Jacobs, H. L. & 7 other authors (2002). Killing activity of neutrophils is mediated through activation of proteases by K+ flux. *Nature* **416**, 291–297.

Richter, P. J., Jr, Kimsey, R. B., Madigan, J. E., Barlough, J. E., Dumler, J. S. & Brooks, D. L. (1996). *Ixodes pacificus* (Acari: Ixodidae) as a vector of *Ehrlichia equi* (*Rickettsiales: Ehrlichieae*). *J Med Entomol* **33**, 1–5.

Rikihisa, Y. (1991). The tribe Ehrlichieae and ehrlichial diseases. *Clin Microbiol Rev* **4**, 286–308.

Savill, J. S., Wyllie, A. H., Henson, J. E., Walport, M. J., Henson, P. M. & Haslett, C. (1989). Macrophage phagocytosis of aging neutrophils in inflammation. Programmed cell death in the neutrophil leads to its recognition by macrophages. *J Clin Invest* **83**, 865–875.

Scaife, H., Woldehiwet, Z., Hart, C. A. & Edwards, S. W. (2003). *Anaplasma phagocytophilum* reduces neutrophil apoptosis in vivo. *Infect Immun* **71**, 1995–2001.

Scidmore, M. A., Fischer, E. R. & Hackstadt, T. (1996). Sphingolipids and glycoproteins are differentially trafficked to the *Chlamydia trachomatis* inclusion. *J Cell Biol* **134**, 363–374.

Shea, K. W., Calio, A. J., Klein, N. C. & Cunha, B. A. (1996). *Ehrlichia equi* infection associated with rhabdomyolysis. *Clin Infect Dis* **22**, 605.

Somers, W. S., Tang, J., Shaw, G. D. & Camphausen, R. T. (2000). Insights into the molecular basis of leukocyte tethering and rolling revealed by structures of P- and E-selectin bound to SLe(X) and PSGL-1. *Cell* **103**, 467–479.

Stringer, R. E. & Edwards, S. W. (1995). Potentiation of the respiratory burst of human neutrophils by cycloheximide: regulation of reactive oxidant production by a protein(s) with rapid turnover? *Inflamm Res* **44**, 158–163.

Telford, S. R., 3rd, Dawson, J. E., Katavolos, P., Warner, C. K., Kolbert, C. P. & Persing, D. H. (1996). Perpetuation of the agent of human granulocytic ehrlichiosis in a deer tick-rodent cycle. *Proc Natl Acad Sci U S A* **93**, 6209–6214.

Theriot, J. A. (1995). The cell biology of infection by intracellular bacterial pathogens. *Annu Rev Cell Dev Biol* **11**, 213–239.

Thornberry, N. A., Bull, H. G., Calaycay, J. R. and 24 other authors (1992). A novel heterodimeric cysteine protease is required for interleukin-1 beta processing in monocytes. *Nature* **356**, 768–774.

Trofe, J., Reddy, K. S., Stratta, R. J., Flax, S. D., Somerville, K. T., Alloway, R. R., Egidi, M. F., Shokouh-Amiri, M. H. & Gaber, A. O. (2001). Human granulocytic ehrlichiosis in pancreas transplant recipients. *Transplant Infect Dis* **3**, 34–39.

Vitale, G., Alexandrov, K., Ullrich, O. & 7 other authors (1995). The GDP/GTP cycle of Rab5 in the regulation of endocytotic membrane traffic. *Cold Spring Harbor Symp Quant Biol* **60**, 211–220.

Volpp, B. D., Nauseef, W. M. & Clark, R. A. (1988). Two cytosolic neutrophil oxidase components absent in autosomal chronic granulomatous disease. *Science* **242**, 1295–1297.

Walker, D. H. & Dumler, J. S. (1997). Human monocytic and granulocytic ehrlichioses. Discovery and diagnosis of emerging tick-borne infections and the critical role of the pathologist. *Arch Pathol Lab Med* **121**, 785–791.

Wang, T., Malawista, S. E., Pal, U., Grey, M., Meek, J., Akkoyunlu, M., Thomas, V. & Fikrig, E. (2002). Superoxide anion production during *Anaplasma phagocytophila* infection. *J Infect Dis* **186**, 274–280.

Watson, R. W., Rotstein, O. D., Parodo, J., Bitar, R., Marshall, J. C., William, R. & Watson, G. (1998). The IL-1 beta-converting enzyme (caspase-1) inhibits apoptosis of inflammatory neutrophils through activation of IL-1 beta. *J Immunol* **161**, 957–962.

Webster, P., IJdo, J. W., Chicoine, L. M. & Fikrig, E. (1998). The agent of Human Granulocytic Ehrlichiosis resides in an endosomal compartment. *J Clin Invest* **101**, 1932–1941.

Wells, M. Y. & Rikihisa, Y. (1988). Lack of lysosomal fusion with phagosomes containing *Ehrlichia risticii* in P388D1 cells: abrogation of inhibition with oxytetracycline. *Infect Immun* **56**, 3209–3215.

Wientjes, F. B. & Segal, A. W. (1995). NADPH oxidase and the respiratory burst. *Semin Cell Biol* **6**, 357–365.

Winkler, H. H. (1990). Rickettsia species (as organisms). *Annu Rev Microbiol* **44**, 131–153.

Wong, S. & Grady, L. J. (1996). Ehrlichia infection as a cause of severe respiratory distress. *N Engl J Med* **334**, 273.

Yago, T., Leppanen, A., Carlyon, J. A., Akkoyunlu, M., Karmakar, S., Fikrig, E., Cummings, R. D. & McEver, R. P. (2003). Structurally distinct requirements for binding of PSGL-1 and sialyl Lewis x to *Anaplasma phagocytophilum* and P-selectin. *J Biol Chem* **278**, 37987–37997.

Yoshiie, K., Kim, H. Y., Mott, J. & Rikihisa, Y. (2000). Intracellular infection by the human granulocytic ehrlichiosis agent inhibits human neutrophil apoptosis. *Infect Immun* **68**, 1125–1133.

Yu, L., Zhen, L. & Dinauer, M. C. (1997). Biosynthesis of the phagocyte NADPH oxidase cytochrome b558. Role of heme incorporation and heterodimer formation in maturation and stability of gp91phox and p22phox subunits. *J Biol Chem* **272**, 27288–27294.

Zhi, N., Ohashi, N., Rikihisa, Y., Horowitz, H. W., Wormser, G. P. & Hechemy, K. (1998). Cloning and expression of the 44-kilodalton major outer membrane protein gene of the human granulocytic ehrlichiosis agent and application of the recombinant protein to serodiagnosis. *J Clin Microbiol* **36**, 1666–1673.

Zhi, N., Ohashi, N. & Rikihisa, Y. (1999). Multiple p44 genes encoding major outer membrane proteins are expressed in the human granulocytic ehrlichiosis agent. *J Biol Chem* **274**, 17828–17836.

Zhi, N., Ohashi, N. & Rikihisa, Y. (2002a). Activation of a p44 pseudogene in *Anaplasma phagocytophila* by bacterial RNA splicing: a novel mechanism for post-transcriptional regulation of a multigene family encoding immunodominant major outer membrane proteins. *Mol Microbiol* **46**, 135–145.

Zhi, N., Ohashi, N., Tajima, T., Mott, J., Stich, R. W., Grover, D., Telford, S. R., 3rd, Lin, Q. & Rikihisa, Y. (2002b). Transcript heterogeneity of the p44 multigene family in a human granulocytic ehrlichiosis agent transmitted by ticks. *Infect Immun* **70**, 1175–1184.

Interactions of *Yersinia pestis* with its flea vector that lead to the transmission of plague

B. Joseph Hinnebusch

Laboratory of Human Bacterial Pathogenesis, Rocky Mountain Laboratories, National Institute of Allergy and Infectious Diseases, National Institutes of Health, Hamilton, MT 59840, USA

YERSINIA PESTIS, A NEWLY EMERGED ARTHROPOD-BORNE AGENT

Bacterial pathogens transmitted by blood-feeding arthropods present serious public health problems worldwide and new ones continue to be recognized, for example the agents of Lyme disease, ehrlichiosis and cat scratch disease. Arthropod-borne transmission is relatively rare among the prokaryotes, but has evolved independently in a phylogenetically diverse group of eubacteria, including the rickettsiae, spirochaetes in the genus *Borrelia*, and the Gram-negative or proteobacteria *Yersinia pestis* and *Francisella tularensis*.

Y. *pestis*, the agent of plague, produces a cyclic infection of rodents and their fleas in many parts of the world. Maintenance of Y. *pestis* in nature depends on these rodent–flea–rodent transmission cycles, although it can also be transmitted in some cases by aerosol, direct contact or ingestion (Poland & Barnes, 1979). The disease produces periodic eruptive, rapidly spreading epizootics among certain highly susceptible rodents, followed by regression to focal areas (Barnes, 1982). Historic human plague pandemics exhibit this same pattern. Other species of wild rodents are more resistant to overt disease, and these species are thought to constitute ecologically important reservoir hosts (Kartman *et al.*, 1958). Human plague remains an international public health concern, and recent outbreaks in India and parts of Africa, where the disease had been dormant for decades, suggest a resurgence (Chanteau *et al.*, 1998).

SGM symposium 63: Microbe–vector interactions in vector-borne diseases.
Editors S. H. Gillespie, G. L. Smith & A. Osbourn. Cambridge University Press. ISBN 0 521 84312 X

More troubling is the isolation of multi-drug-resistant strains of Y. *pestis* during the current epidemic in Madagascar (Galimand *et al.*, 1997). These considerations, and the recognized potential of Y. *pestis* as a bioterrorism agent, underline the need for a detailed understanding of plague transmission and pathogenicity, which hopefully will bring about new effective vaccines and other control measures. No vaccine is currently available for plague.

Y. *pestis* is a newcomer to the list of arthropod-borne bacteria. Compared to the ancient relationship of rickettsiae and spirochaetes with their vectors, the vector relationship between Y. *pestis* and fleas is quite recent. Comparative genomics and population genetics analyses demonstrate that Y. *pestis* is a clonal variant of *Yersinia pseudotuberculosis* that diverged only within the last 1500 to 20 000 years (Achtman *et al.*, 1999; Hinchcliffe *et al.*, 2003). During this short evolutionary time period, two dramatic changes occurred. From the relatively benign Y. *pseudotuberculosis* recent ancestor, Y. *pestis* evolved increased invasiveness and virulence to become one of the most feared pathogens of humans, with an LD_{50} of less than 10 bacteria even from peripheral inoculation sites (Perry & Fetherston, 1997). Second, and no less remarkably, Y. *pestis* evolved an entirely new transmission mechanism using a blood-feeding insect. The extremely close genetic relationship between Y. *pseudotuberculosis* and Y. *pestis* implies that the genetic changes responsible for these radical changes are relatively few and discrete, thus providing an interesting and tractable model to study the evolution of arthropod-borne transmission. To recount this evolutionary story, it will be necessary to identify the specific genetic changes in Y. *pestis* that enabled it to adapt to the flea, and to characterize their mechanisms of action at a molecular level.

BIOLOGICAL TRANSMISSION OF *Y. PESTIS*: FOREGUT BLOCKAGE IN THE FLEA VECTOR

Four years after Alexandre Yersin and Shibasaburo Kitasato first isolated the plague bacillus from human patients during the last pandemic, Paul-Louis Simond (1898) demonstrated that it could be transmitted by fleas. Simond thought that contamination of the bite site with infected flea faeces was the likely mechanism, and it was not until 1914 that A. W. Bacot described the true biological mechanism of transmission (Bacot & Martin, 1914). Bacot observed that Y. *pestis* grew in the digestive tract of fleas in the form of large, dark-coloured gelatinous masses. In some fleas, these masses lodged in the proventriculus, the valve in the flea foregut that connects the oesophagus to the midgut. In fleas, the interior of the proventriculus is lined with spines covered with cuticle, the same material that makes up the insect exoskeleton. As the bacterial mass grew amongst the proventricular spines, it reduced the patency of the proventricular valve sufficiently to disrupt normal blood flow during feeding, and eventually completely blocked the passage of blood into the midgut. Such partially

Fig. 1. *Xenopsylla cheopis* flea, blocked with *Y. pestis*, examined immediately after attempting to feed. Fresh blood is present only in the oesophagus and not the midgut, which contains dark-coloured digestion products of previous blood meals and masses of bacteria.

or completely blocked fleas were unable to feed normally. In partially blocked fleas, the bacterial growth prevented the proventricular valve from closing fully, allowing blood mixed with *Y. pestis* from the midgut to leak out and flow back into the bite site. In completely blocked fleas, the oesophagus became distended with blood and was partially regurgitated into the bite site, carrying bacteria washed from the blocking mass along with it. Blocked fleas can be readily diagnosed by examining them microscopically after their attempts to feed. The midgut of uninfected fleas is filled with blood, whereas blocked fleas contain fresh blood only in the oesophagus (Fig. 1). The proventricular blockage-regurgitation model for flea-borne transmission has since been substantiated by several investigators, who went on to show that the development of proventricular infection and blockage is prerequisite for reliable transmission of disease to laboratory rodents (Burroughs, 1947; Pollitzer, 1954).

The retrograde route of delivery of *Y. pestis* by its flea vector is unusual. Most arthropod-borne agents, including borreliae, rickettsiae, plasmodia and arboviruses, use a different strategy. After entering their vector in a blood meal, they penetrate the gut epithelium, migrate through the haemocoel, invade the salivary glands, and are delivered in the saliva that is introduced into the skin when the vector feeds on a new host. These invasive processes do not harm the vector. Transmission via the saliva is efficient but obviously involves very complex vector–parasite interactions, indicative of long-standing coevolution. Compared to this sophisticated strategy, *Y. pestis* infection of the flea seems primitive and rather crude, perhaps a reflection of its very recent adaptation to the flea. Nevertheless, the basic transmission mechanism used by *Y. pestis* resembles that of at least one other arthropod-borne agent. Protozoan parasites in the genus *Leishmania* also infect a valve in the foregut of their sandfly vectors and disrupt the normal feeding mechanism, allowing parasites to exit through the same channel that they originally entered. This follows a complex development process in which the amastigotes ingested in the blood meal differentiate to motile promastigotes in the midgut, migrate to the foregut, multiply and further differentiate to haptomonad forms which attach to the stomodeal valve, and form a biological plug that is enclosed in a viscous-gel-like matrix. The valve itself is damaged during this process, allowing promastigotes behind the degenerating valve to swim forward into the bite site during blood feeding (Sacks & Kamhawi, 2001).

The *Y. pestis* plasminogen activator and the fibrin clot model of flea blockage

The first model to explain the mechanism of proventricular blockage proposed that the blocking mass was composed of bacteria embedded in a fibrin clot. Coagulation of the blood meal is observed in fleas, possibly due to trypsin-like endopeptidases secreted by the flea midgut epithelium (Cavanaugh, 1971). *Y. pestis* also had a coagulase activity, later found to be due to an outer-surface plasminogen activator (Pla) encoded on pPst, a 9·5 kb *Y. pestis* plasmid that is not present in *Y. pseudotuberculosis* (Sodeinde & Goguen, 1988). Coagulase activity of Pla is temperature-dependent. At temperatures below 28 °C the *Y. pestis* Pla produces a fibrin clot from rabbit plasma, but at 37 °C Pla is a potent plasminogen activator, with an opposite fibrinolytic activity (McDonough & Falkow, 1989). Based on these observations, Cavanaugh (1971) proposed the fibrin clot model, hypothesizing that at the low temperatures encountered in the flea gut, the coagulase activity of *Y. pestis* resulted in a fibrin matrix that embedded the bacteria in a blocking mass.

The fibrin clot model of flea blockage quickly became conventional wisdom and is often cited in infectious disease textbooks as the mechanism that enables flea-borne transmission. Several experimental results call the fibrin clot model into question, however. First, the *Y. pestis* coagulase activity is weak and is only observed with rabbit plasma, and not with plasma from mice, rats, guinea pigs, squirrels or humans, which, unlike rabbits, are normal hosts of *Y. pestis* (Beesley *et al.*, 1967; Jawetz & Meyer, 1944). More importantly, *Y. pestis* strains cured of the pPst plasmid, which contains the *pla* gene, show no defect in flea-blocking ability. All aspects of the flea infections caused by the two Pla⁻ strains examined, including infection rate, bacterial loads, formation of typical bacterial masses and proventricular blockage, were identical to those produced by the Pla⁺ parent strains (Hinnebusch *et al.*, 1998). In addition, the matrix of the bacterial masses that form in the flea gut is not degraded by proteases or by the fibrinolytic enzyme plasmin.

The true biologically relevant activity of Pla occurs after transmission to the mammal and is due to its potent plasminogen activator ability at 37 °C. Sodeinde *et al.* (1992) found that the *Y. pestis pla* gene is required for bacterial dissemination from a subcutaneous inoculation site in mice. Pla may also help *Y. pestis* to escape containment in the primary lymph nodes and spread systemically.

The biofilm model of flea blockage and the *Y. pestis hms* genes

An alternative model to account for the development of a transmissible infection arose from the discovery that the *Y. pestis hms* (haemin storage) genes are required for proventricular blockage (Hinnebusch *et al.*, 1996). The *hms* genes include the chromosomal operons *hmsHFRS* and the unlinked *hmsT* (Hare & McDonough, 1999; Jones *et al.*, 1999; Lillard *et al.*, 1997). All of them are required for the haemin storage or pigmentation phenotype based on the *in vitro* ability of *Y. pestis* colonies to bind haemin or the structurally analogous Congo red dye when grown at 28 °C or lower (Jackson & Burrows, 1956; Surgalla & Beesley, 1969). Because the low-temperature-dependent phenotype matched the flea better than the mammalian host temperature, the ability of *Y. pestis hms* mutants to infect and block fleas was examined. Hms⁻ mutants were able to infect and colonize the flea midgut at levels equivalent to the Hms⁺ isogenic parent strain. However, unlike the wild-type bacteria, the Hms⁻ mutants infected only the midgut – they were completely unable to colonize and block the proventriculus (Hinnebusch *et al.*, 1996). Subsequently, it was shown that the *hms* genes are not required for virulence in mice; they are specifically required to produce a transmissible infection in the vector (Hinnebusch *et al.*, 1996; Lillard *et al.*, 1999).

An important clue pertaining to the mechanism of the *hms* genes came from the recognition that *hmsR* and *hmsF* are similar to two genes in the *ica* (intercellular adhesion) operon of *Staphylococcus aureus* and *Staphylococcus epidermidis* that is required to synthesize the extracellular polysaccharide matrix of staphylococcal biofilms (Heilmann *et al.*, 1996; Lillard *et al.*, 1999). HmsR has 39 % identity and 58 % amino acid sequence similarity to IcaA, an N-acetylglucosamine transferase; and HmsF has 23 % identity and 42 % similarity to IcaB, an N-acetylglucosamine deacetylase. In addition, HmsT contains a GGDEF domain, which is present in several known bacterial proteins that regulate the biosynthesis of extracellular polysaccharide (Jones *et al.*, 1999). In *Acetobacter* and *Agrobacterium*, for example, GGDEF proteins

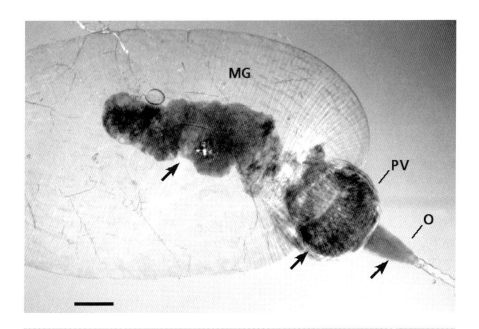

Fig. 2. Digestive tract dissected from a blocked *X. cheopis* flea. A large biofilm-like mass of *Y. pestis* embedded in a dark-coloured extracellular matrix (indicated by arrows) is lodged in the proventriculus (PV) and extends forward and backward into the oesophagus (O) and midgut (MG). Bar, 0·1 mm.

catalyse the formation of a cyclic diguanylate molecule that regulates extracellular cellulose production (Ausmees *et al.*, 2001; Ross *et al.*, 1987).

Many bacteria are able to adhere to a surface and grow in the form of a dense aggregate by forming a biofilm. While originally recognized in aquatic environments, it is now well established that many host-associated bacteria form biofilms *in vivo* which are important in establishing infections. Infection of oral microflora on dental surfaces, staphylococci on indwelling medical devices and heart valve tissue, and *Pseudomonas aeruginosa* in the lungs of cystic fibrosis patients all involve biofilm formation and are but a few of many examples of this phenomenon (Costerton *et al.*, 1999). A bacterial biofilm is operationally defined as a dense community of cells, usually adherent to a surface, which is enclosed within an extracellular matrix. The manner in which *Y. pestis* adheres to and blocks the flea proventriculus fits this definition (Fig. 2). The amino acid similarity of *Y. pestis hms* gene products to enzymes in other bacteria known to produce a biofilm-associated extracellular matrix suggests that they have a similar function in the flea. Further evidence for this hypothesis is the fact that the *hms* genes are required for the ability of *Y. pestis* to produce an adherent biofilm on the surface of glass flowcells, and for the production of amorphous extracellular material when growing on certain media (B. J. Hinnebusch, unpublished). Like pigmentation, the

Table 1. Temperature dependence of Hms phenotypes in *Y. pestis*

In vitro	21 °C	37 °C
Pigmentation on Congo red agar	+	−
Autoaggregation in liquid media	+	−
Biofilm produced on glass surface	+	−
Extracellular matrix produced	+	−
In the flea	**21 °C**	**30 °C**
Aggregation in the midgut	+	+
Extracellular matrix produced	+	−
Proventricular blockage	+	−

biofilm phenotype of *Y. pestis in vitro* is temperature-dependent, expressed only at room temperature and not at 37 °C (Table 1).

The foregoing observations suggest an alternative biofilm model to the fibrin clot model. According to this model, *Y. pestis* colonizes the hydrophobic, acellular surface of the spines in the flea proventriculus by forming an *hms*-dependent biofilm, and growth and expansion of the biofilm is responsible for blockage. Darby *et al.* (2002) have also shown that *Y. pseudotuberculosis* and *Y. pestis* growth on agar plates accumulates on the external mouthparts and interferes with the feeding of *Caenorhabditis elegans* nematodes placed on them, and proposed that this phenotype was due to the formation of an *hms*-dependent biofilm.

Y. PESTIS AT THE VECTOR–HOST INTERFACE

Plague is initiated during the brief connection between the triad of flea, *Y. pestis* and mammalian host when a blocked flea attempts to take a blood meal. Although the interaction at this interface has many important implications for pathogenicity, it has received little attention. The feeding mechanism of fleas determines the initial site of infection and dissemination route of *Y. pestis*. Two basic types of feeding mechanisms have been distinguished in blood-feeding arthropods (Lavoipierre, 1965). Vessel feeders, such as triatomine bugs, probe the dermis, cannulate a blood capillary or small venule with their mouthparts, and begin to feed. Pool feeders, such as ticks and tsetse flies, lacerate vessels with their mouthparts as they probe and feed from the resulting haemorrhage. Fleas and mosquitoes are considered to be largely vessel feeders, although careful observations suggest that a combination of pool and vessel feeding occurs (Deoras & Prasad, 1967; Lavoipierre, 1965). Fleas have a unique mechanism for perforating the skin which utilizes resilin, the same highly elastic protein used for the flea jumping mechanism (Wenk, 1980). The paired piercing stylets are alternately contracted, compressing a cushion of resilin. The stylet is forcibly thrust downward when the stored energy in the compressed resilin is suddenly released. Alternating, rapid movements of the left and right stylets pierce the skin, and mouthpart muscles can move the stylets laterally during probing, often causing haemorrhage. Saliva, which contains an apyrase anticoagulant activity, is introduced through a separate channel (Ribeiro *et al.*, 1990). The flea mouthparts are not long enough to penetrate through the dermis, so *Y. pestis* must be introduced intradermally, intravenously (directly into a venule), or into both compartments.

Inefficient transmission of *Y. pestis* by fleas

Surprisingly, the number of bacteria transmitted by a blocked flea has not been well established. The one reported study that attempted this quantification concluded that approximately 10^4 *Y. pestis* were transmitted (Burroughs, 1947). However, this estimate was based on the results from a single blocked flea and was calculated indirectly. We recently re-examined this issue by directly counting c.f.u. from skin biopsies of flea bite sites, and from the blood in an artificial feeding apparatus through which blocked fleas attempted to feed. Only half of the samples were positive, and the median number of c.f.u. transmitted was less than 100. Other data reinforce the conclusion that flea-borne transmission is inefficient. Infection and blockage of the flea is not very efficient in the first place. For example, the ID_{50} of *Y. pestis* for the rat flea *Xenopsylla cheopis*, usually cited as the plague vector par excellence, is about 10^4 (Hinnebusch *et al.*, 1996). Since the volume of a flea's blood meal is only $0 \cdot 1$–$0 \cdot 3 \, \mu$l, this means that high-density septicaemia [10^7–10^8 *Y. pestis* (ml peripheral blood)$^{-1}$] is required to result in a 50 % infection rate of the vector. Then, even if a chronic infection is successfully established in the flea gut, subsequent proventricular blockage develops only about half the time in *X. cheopis* (Burroughs, 1947). Furthermore, several previous investigations found that individual blocked fleas placed on a susceptible host transmit plague only about 50 % of the time (Burroughs, 1947; Pollitzer, 1954). The rather inefficient, erratic transmission of relatively low numbers of bacteria perhaps reflects the recent adaptation of *Y. pestis* to its vector. It also probably provided strong selective pressure for the emergence of more invasive and virulent strains, able to disseminate from a small infectious dose in the skin to eventually cause high-density septicaemia. Thus a mutually reinforcing linkage between the recent coevolution of both flea-borne transmission and increased virulence of *Y. pestis* can be suspected.

The transmission phenotype of *Y. pestis* and its implications for host interactions

In the flea foregut, *Y. pestis* exists as a biofilm enclosed within an extracellular matrix, and it is in this context that the bacteria are transmitted (Fig. 2). The infectious units that exit the flea and enter the mammal may consist not only of individual bacteria washed from the periphery of the blocking mass, but also bits of the biofilm itself, composed of clusters of *Y. pestis* within the extracellular matrix. The association of *Y. pestis* with an as yet undefined extracellular matrix in the flea has important implications for the initial encounter with the innate immune system of the host, because bacteria within a biofilm have been shown to be more resistant to uptake or killing by phagocytes (Donlan & Costerton, 2002). Known antiphagocytic factors such as the F1 protein capsule and the type III secretion system are not expressed at the low temperatures encountered in the flea (Straley & Perry, 1995). As it leaves the flea, therefore, *Y. pestis* is not protected by these virulence factors against uptake and killing by polymorphonuclear neutrophils. As has been found for other bacteria, the biofilm-like phenotype and the extracellular matrix associated with *Y. pestis* in the flea may provide initial protection from phagocytosis after transmission, until synthesis of the known antiphagocytic virulence factors is induced by the increased body temperature of the mammalian host. The biochemical composition of the extracellular matrix, its role if any in protecting the bacteria at the host–vector interface, and whether *Y. pestis* is transmitted in the form of small aggregates are important questions. The effect of flea saliva on the initial host response to transmission is also completely unknown.

The feeding mechanism of fleas suggests that some *Y. pestis* may be injected directly into a capillary or small venule. However, systemic infection does not usually develop until after local spread to the regional lymph nodes; and inflammation and infection at the flea bite site can occur, indicating that at least some bacteria are deposited in the extravascular dermis. The early events that lead to dissemination from the flea bite site have not been well characterized, but it is hypothesized that macrophages may transport the bacteria to the primary lymph node (Titball *et al.*, 2003).

The *Y. pestis* plasminogen activator has been shown to be required for the dissemination of *Y. pestis* from a subcutaneous injection site, and Pla has therefore been postulated to have the same function after flea bite transmission (Sodeinde *et al.*, 1992). The natural flea-borne route of transmission is different in several ways from needle and syringe injection of *in vitro*-grown bacteria, however. *Y. pestis* from laboratory cultures is not in the biofilm phenotype and is not surrounded by the flea-specific extracellular matrix. Fleas are unlikely to inject bacteria subcutaneously, and the effects of flea saliva are unknown. Thus the role of *pla* in dissemination from the flea bite site needs to be explicitly tested.

THE EVOLUTION OF ARTHROPOD-BORNE TRANSMISSION IN *Y. PESTIS*

Besides the *hms* operons and *pla*, the list of *Y. pestis* genes proven or likely to be important for flea-borne transmission (Table 2) includes the gene for *Yersinia* murine toxin (*ymt*). Ymt is an intracellular phospholipase D that is required for *Y. pestis* to

Table 2. *Y. pestis* genes important for flea-borne transmission

Gene	Location	Present in:		Function in *Y. pestis*
		Y. pestis	*Y. pstb*	
hmsHFSR,T	Chromosome	+	+	Extracellular matrix synthesis, biofilm formation in proventriculus
ymt	pFra	+	−	Phospholipase D, survival in flea midgut
pla	pPst	+	−	Plasminogen activator, dissemination from flea bite site

survive in the digestive milieu of the flea midgut (Hinnebusch *et al.*, 2002). Ymt is not required for blockage per se, but *Y. pestis* cells lacking this enzyme form spheroplasts and lyse within hours of being taken up in a blood meal. Therefore, Ymt⁻ *Y. pestis* strains are unable to establish an infection in the first place. The agent that causes rapid lysis of Ymt⁻ *Y. pestis* appears to be generated by digestion of a blood plasma component in the flea midgut, but the native substrate of the *Y. pestis* phospholipase D and its mechanism of protection remain to be discovered.

Inspection of the genetic factors listed in Table 2 suggests an evolutionary scenario that could account for the rapid evolutionary change to flea-borne transmission in this pathogen (Carniel, 2003). Because this transition occurred within the last 1500–20 000 years, the slow process of accumulation of random mutations was probably not the principal evolutionary mechanism. Two of the genes involved in the flea-borne route of transmission are located on plasmids, pPst and pFra, that are unique to *Y. pestis*. Thus acquisition of these two plasmids by horizontal exchange accounts for two important evolutionary steps. The third factor, the *hms* system, is also present in *Y. pseudotuberculosis* strains and thus pre-dates the divergence of *Y. pestis*. The Hms phenotype is different in *Y. pseudotuberculosis*, however. Most isolates are non-pigmented on Congo red agar (Brubaker, 1991; Carniel, 2003), but all *Y. pseudotuberculosis* strains tested, whether pigmented or not, were unable to cause proventricular blockage in *X. cheopis* fleas, even though they were able to establish a chronic midgut infection (Hinnebusch *et al.*, 2002; unpublished). This indicates that the Hms phenotype in *Y. pestis* was fine-tuned in some way to enable biofilm formation in the proventriculus, perhaps by a contributing change in some other outer-surface component.

CONCLUSIONS

As they cycle between vertebrate and arthropod hosts, vector-borne pathogens sense host-specific environmental cues and regulate gene expression appropriately. The biological function of many of these genes is specific to one or the other of the alternate hosts. For example, the *Y. pestis* transmission factors *hms* and *ymt* genes are required to produce a transmissible infection in the flea, but are not required for infection of the rodent; and the known *Y. pestis* virulence factors that have been tested are not required in the flea (Du *et al.*, 1995; Hinnebusch *et al.*, 1998; Lillard *et al.*, 1999). A shift in temperature from 37 °C to the lower, ambient temperature of the flea appears to be an important cue for *Y. pestis* to upregulate the flea-specific genetic programme. The expression of the *hms* and *ymt* gene products is upregulated at temperatures below 28 °C and is downregulated at 37 °C (Du *et al.*, 1995; Straley & Perry, 1995).

Y. pestis is unique among the enteric group of Gram-negative bacteria in having adopted an arthropod-borne route of transmission. *Y. pestis* populations worldwide constitute a highly uniform clone that diverged quite recently from *Y. pseudo-tuberculosis*, a typical enteric pathogen transmitted perorally through contaminated food and water (Achtman *et al.*, 1999). Comparative genomics analyses indicate that relatively few gene differences between *Y. pseudotuberculosis* and *Y. pestis* must account for the evolution of increased virulence and adaptation to the flea (Hinchcliffe *et al.*, 2003). The rapid evolutionary processes of horizontal gene transfer, and perhaps selective gene loss, explain some of the important changes that led to the newly acquired ability to produce a transmissible infection in the flea.

Given its recent common ancestry with an enteric pathogen, it is ironic that *Y. pestis* does not adhere to or invade the flea digestive tract epithelium. This puts the bacteria ingested in a blood meal at risk of being eliminated by the flea in the faeces. In fact, many fleas do clear themselves of infection in that manner even after feeding on highly septicaemic blood; infection of the flea by *Y. pestis* is not very efficient. The basic transmission strategy of *Y. pestis* is to form an adherent bacterial biofilm on the acellular, hydrophobic surface of the proventricular spines. Once this foothold is established, a chronic infection is ensured and a mechanism is instigated that enables *Y. pestis*, which is non-motile, to pass back out against the normal flow of blood when the flea feeds.

ACKNOWLEDGEMENTS

I thank Clayton Jarrett, Roberto Rebeil and Florent Sebbane for their participation in the work described in this review and for discussions about it. Research in my laboratory is supported in part by a New Scholars in Global Infectious Diseases award from the Ellison Medical Foundation.

REFERENCES

Achtman, M., Zurth, K., Morelli, G., Torrea, G., Guiyoule, A. & Carniel, E. (1999). *Yersinia pestis*, the cause of plague, is a recently emerged clone of *Yersinia pseudotuberculosis*. *Proc Natl Acad Sci U S A* **96**, 14043–14048.

Ausmees, N., Mayer, R., Weinhouse, H., Volman, G., Amikam, D., Benziman, M. & Lindberg, M. (2001). Genetic data indicate that proteins containing the GGDEF domain possess diguanylate cyclase activity. *FEMS Microbiol Lett* **204**, 163–167.

Bacot, A. W. & Martin, C. J. (1914). Observations on the mechanism of the transmission of plague by fleas. *J Hyg Plague Suppl 3* **13**, 423–439.

Barnes, A. M. (1982). Surveillance and control of bubonic plague in the United States. *Symp Zool Soc Lond* **50**, 237–270.

Beesley, E. D., Brubaker, R. R., Janssen, W. A. & Surgalla, M. J. (1967). Pesticins. III. Expression of coagulase and mechanism of fibrinolysis. *J Bacteriol* **94**, 19–26.

Brubaker, R. R. (1991). Factors promoting acute and chronic diseases caused by yersiniae. *Clin Microbiol Rev* **4**, 309–324.

Burroughs, A. L. (1947). Sylvatic plague studies. The vector efficiency of nine species of fleas compared with *Xenopsylla cheopis*. *J Hyg* **45**, 371–396.

Carniel, E. (2003). Evolution of pathogenic *Yersinia*, some lights in the dark. In *The Genus Yersinia. Entering the Functional Genomic Era*, pp. 3–11. Edited by M. Skurnik, J. A. Begoechea & K. Granfors. New York: Plenum.

Cavanaugh, D. C. (1971). Specific effect of temperature upon transmission of the plague bacillus by the oriental rat flea, *Xenopsylla cheopis*. *Am J Trop Med Hyg* **20**, 264–273.

Chanteau, S., Ratsifasoamanana, L., Rasoamanana, B., Rahalison, L., Randriambelosoa, J., Roux, J. & Rabeson, D. (1998). Plague, a reemerging disease in Madagascar. *Emerg Infect Dis* **4**, 101–104.

Costerton, J. W., Stewart, P. S. & Greenberg, E. P. (1999). Bacterial biofilms: a common cause of persistent infections. *Science* **284**, 1318–1322.

Darby, C., Hsu, J. W., Ghori, N. & Falkow, S. (2002). *Caenorhabditis elegans*: plague bacteria biofilm blocks food intake. *Nature* **417**, 243–244.

Deoras, P. J. & Prasad, R. S. (1967). Feeding mechanism of Indian fleas *X. cheopis* (Roths) and *X. astia* (Roths). *Indian J Med Res* **55**, 1041–1050.

Donlan, R. M. & Costerton, J. W. (2002). Biofilms: survival mechanisms of clinically relevant microorganisms. *Clin Microbiol Rev* **15**, 167–193.

Du, Y., Galyov, E. & Forsberg, A. (1995). Genetic analysis of virulence determinants unique to *Yersinia pestis*. *Contrib Microbiol Immunol* **13**, 321–324.

Galimand, M., Guiyoule, A., Gerbaud, G., Rasoamanana, B., Chanteau, S., Carniel, E. & Courvalin, P. (1997). Multidrug resistance in *Yersinia pestis* mediated by a transferable plasmid. *N Engl J Med* **337**, 677–680.

Hare, J. M. & McDonough, K. A. (1999). High-frequency RecA-dependent and -independent mechanisms of Congo red binding mutations in *Yersinia pestis*. *J Bacteriol* **181**, 4896–4904.

Heilmann, C., Schweitzer, O., Gerke, C., Vanittanakom, N., Mack, D. & Götz, F. (1996). Molecular basis of intercellular adhesion in the biofilm-forming *Staphylococcus epidermidis*. *Mol Microbiol* **20**, 1083–1091.

Hinchcliffe, S. J., Isherwood, K. E., Stabler, R. A. & 7 other authors (2003). Application of DNA microarrays to study the evolutionary genomics of *Yersinia pestis* and *Yersinia pseudotuberculosis*. *Genome Res* **13**, 2018–2029.

Hinnebusch, B. J., Perry, R. D. & Schwan, T. G. (1996). Role of the *Yersinia pestis* hemin storage (*hms*) locus in the transmission of plague by fleas. *Science* **273**, 367–370.

Hinnebusch, B. J., Fischer, E. R. & Schwan, T. G. (1998). Evaluation of the role of the *Yersinia pestis* plasminogen activator and other plasmid-encoded factors in temperature-dependent blockage of the flea. *J Infect Dis* **178**, 1406–1415.

Hinnebusch, B. J., Rudolph, A. E., Cherepanov, P., Dixon, J. E., Schwan, T. G. & Forsberg, Å. (2002). Role of Yersinia murine toxin in survival of *Yersinia pestis* in the midgut of the flea vector. *Science* **296**, 733–735.

Jackson, S. & Burrows, T. W. (1956). The pigmentation of *Pasteurella pestis* on a defined medium containing haemin. *Br J Exp Pathol* **37**, 570–576.

Jawetz, E. & Meyer, K. F. (1944). Studies on plague immunity in experimental animals. II. Some factors of the immunity mechanism in bubonic plague. *J Immunol* **49**, 15–29.

Jones, H. A., Lillard, J. W., Jr & Perry, R. D. (1999). HmsT, a protein essential for expression of the haemin storage (Hms$^+$) phenotype of *Yersinia pestis*. *Microbiology* **145**, 2117–2128.

Kartman, L., Prince, F. M., Quan, S. F. & Stark, H. E. (1958). New knowledge on the ecology of sylvatic plague. *Ann N Y Acad Sci* **70**, 668–711.

Lavoipierre, M. M. J. (1965). Feeding mechanism of blood-sucking arthropods. *Nature* **208**, 302–303.

Lillard, J. W., Jr, Fetherston, J. D., Pedersen, L., Pendrak, M. L. & Perry, R. D. (1997). Sequence and genetic analysis of the hemin storage (*hms*) system of *Yersinia pestis*. *Gene* **193**, 13–21.

Lillard, J. W., Jr, Bearden, S. W., Fetherston, J. D. & Perry, R. D. (1999). The haemin storage (Hms$^+$) phenotype of *Yersinia pestis* is not essential for the pathogenesis of bubonic plague in mammals. *Microbiology* **145**, 197–209.

McDonough, K. A. & Falkow, S. (1989). A *Yersinia pestis*-specific DNA fragment encodes temperature-dependent coagulase and fibrinolysin-associated phenotypes. *Mol Microbiol* **3**, 767–775.

Perry, R. D. & Fetherston, J. D. (1997). *Yersinia pestis* – etiologic agent of plague. *Clin Microbiol Rev* **10**, 35–66.

Poland, J. D. & Barnes, A. M. (1979). Plague. In *CRC Handbook Series in Zoonoses*, vol. 1, pp. 515–597. Edited by J. H. Steele. Boca Raton, FL: CRC Press.

Pollitzer, R. (1954). *Plague*. Geneva: World Health Organization.

Ribeiro, J. M. C., Vaughan, J. A. & Azad, A. F. (1990). Characterization of the salivary apyrase activity of three rodent flea species. *Comp Biochem Physiol* **95B**, 215–219.

Ross, P., Weinhouse, H., Aloni, Y. & 8 other authors (1987). Regulation of synthesis in *Acetobacter xylinum* by cyclic diguanylic acid. *Nature* **325**, 279–281.

Sacks, D. & Kamhawi, S. (2001). Molecular aspects of parasite-vector and vector–host interactions in leishmaniasis. *Annu Rev Microbiol* **55**, 453–483.

Simond, P.-L. (1898). La propagation de la peste. *Ann Inst Pasteur* **12**, 662–687.

Sodeinde, O. A. & Goguen, J. D. (1988). Genetic analysis of the 9·5-kilobase virulence plasmid of *Yersinia pestis*. *Infect Immun* **56**, 2743–2748.

Sodeinde, O. A., Subrahmanyam, Y. V., Stark, K., Quan, T., Bao, Y. & Goguen, J. D. (1992). A surface protease and the invasive character of plague. *Science* **258**, 1004–1007.

Straley, S. C. & Perry, R. D. (1995). Environmental modulation of gene expression and pathogenesis in *Yersinia*. *Trends Microbiol* **3**, 310–317.

Surgalla, M. J. & Beesley, E. D. (1969). Congo red agar plating medium for detecting pigmentation in *Pasteurella pestis*. *Appl Microbiol* **18**, 834–837.

Titball, R. W., Hill, J. & Brown, K. A. (2003). *Yersinia pestis* and plague. *Biochem Soc Trans* **31**, 104–107.

Wenk, P. (1980). How bloodsucking insects perforate the skin of their hosts. In *Fleas*, pp. 329–335. Edited by R. Traub & H. Starcke. Rotterdam: A. A. Balkema.

Transgenic malaria

Peter W. Atkinson[1] and David A. O'Brochta[2]

[1]Department of Entomology, University of California, Riverside, CA 92521, USA

[2]Center for Biosystems Research, University of Maryland Biotechnology Institute, College Park, MD 20742, USA

INTRODUCTION

The life cycle of the causative agent of human malaria, the protozoan *Plasmodium*, involves transmission through both humans and the mosquito vector. Malaria can therefore be tackled on three distinct biological fronts. The most explored is the development of vaccines or drugs that kill the parasite during its passage through the human host. Drugs such as chloroquine have been used with considerable success; however, the emergence of resistance to these drugs has compromised their effectiveness in many regions of the world. Considerable efforts have been directed towards the development of vaccines for malaria; however, the ability of the parasite to evade the human acquired immune response has so far frustrated these attempts. The publication of the complete genomic sequence of *Plasmodium falciparum* (Gardner *et al.*, 2002), the major malaria pathogen in sub-Saharan Africa, has rejuvenated efforts to develop drugs that will specifically target the parasite while minimizing side effects on the human host.

Development of malaria control or eradication strategies based on mosquito vector control has traditionally involved the use of chemical insecticides with or without elimination or reduction of excess standing water. These have been successful in terms of reducing or eliminating malaria from specific geographical regions. Regions of Europe, Asia, the Americas and Australia have been rendered malaria-free as a consequence of these area-wide strategies. The effectiveness of this approach has, however, diminished with the cessation of the application of DDT, the rapid emergence of resistance to insecticides amongst mosquito populations, and the overall reluctance

SGM symposium 63: Microbe–vector interactions in vector-borne diseases.
Editors S. H. Gillespie, G. L. Smith & A. Osbourn. Cambridge University Press. ISBN 0 521 84312 X ©SGM 2004

of communities to accept the spraying of any chemical insecticide. The short-term success of chemical insecticides combined with the management of excess standing water provided a graphic illustration of how successful vector control can be in reducing and even eliminating malaria from specific geographical regions. Modern genetic approaches to malaria control are being considered that use genetically modified mosquitoes rendered incapable of supporting *Plasmodium* development. These new approaches to combating malaria transmission by controlling the vector seek to expand on earlier successes using insecticides, but in a more sustainable and environmentally friendly manner. Here we review recent progress in the development and intended use of transgenic mosquitoes in fighting malaria. While there are no current plans for field-testing genetically modified mosquitoes, rapid developments in this field will likely lead to applications for these types of studies to be filed with regulatory agencies within the next few years.

TRANSGENIC MOSQUITOES ARE GENERATED IN THE LABORATORY USING TRANSPOSABLE ELEMENTS AS GENE VECTORS

Class II transposable elements are used to generate transgenic mosquitoes. Class II elements are transposable elements that transpose via a DNA intermediate. They are usually less than 5 kb in length and contain short inverted terminal repeats (ITRs) that are typically less than 30 bp in size. They encode a transposase that catalyses the transposition of the element from one genomic location to another. Five Class II transposable elements, four derived from insects and one from a bacterium, have been used to successfully transform mosquito species. These are the *Hermes* element from the housefly *Musca domestica* (Warren *et al.*, 1994), the *piggyBac* element from the cabbage looper, *Trichoplusia ni* (Cary *et al.*, 1989), the *Minos* element from *Drosophila hydei* (Franz *et al.*, 1994), the *Mos1 mariner* element from *Drosophila mauritiana* (Medhora *et al.*, 1988) and the Tn5 element from *Escherichia coli* (Rowan *et al.*, 2004). Physical details of each of these, together with the mosquito species that they have been used as gene transfer vectors in, are shown in Table 1.

The four elements from insects represent four different transposable element families but all appear to have similar mobility properties in mosquitoes with respect to both transformation frequencies and stability following transgenesis. The strategy for generating transgenic mosquitoes using Class II transposable elements is similar to that developed for *Drosophila melanogaster* (Rubin & Spradling, 1982). A mixture of two plasmids is co-injected into pre-blastoderm mosquito embryos such that approximately 1–10 million copies of each are introduced. One plasmid contains the transposable element vector into which a genetic marker, typically a fluorescent protein gene, is placed, thereby interrupting or replacing the open reading frame of the transposase

Table 1. Transposable elements used to genetically transform mosquito species

Transposable element	Species transformed	Effector gene used (if present)	Reference
Hermes	Aedes aegypti		Jasinskiene et al. (1998, 2000); Pinkerton et al. (2000); Moreira et al. (2000)
		Defensin A	Kokoza et al. (2000)
	Culex quinquefasciatus		Allen et al. (2001)
Mos1	Aedes aegypti		Coates et al. (1998, 1999, 2000); Jasinskiene et al. (2003); Wilson et al. (2003)
piggyBac	Aedes aegypti		Kokoza et al. (2001); Lobo et al. (2002)
	Anopheles gambiae		Grossman et al. (2001)
		Cecropin	Kim et al. (2004)
	Anopheles albimanus		Perera et al. (2002)
	Anopheles stephensi	Bee venom phospholipase	Moreira et al. (2002)
		SM1 peptide	Ito et al. (2002)
			Nolan et al. (2002)
Minos	Anopheles stephensi		Catteruccia et al. (2000)
Tn5	Aedes aegypti		Rowan et al. (2004)

gene. Active transposase must therefore be supplied *in trans*, and this is achieved by co-injecting the second 'helper' plasmid, which contains the appropriate transposase gene placed under the control of an inducible promoter. The injection site is the posterior pole of the embryo since it is here where presumptive germ cells are formed soon after fertilization. Following gene expression from the helper plasmid the transposase protein is synthesized and enters the pole cell nuclei where it mediates the transposition of the transposable element from the donor plasmid into the germ-line chromosomes. The immediate consequence of this recombination event is the creation of a chimeric embryo consisting of transgenic and non-transgenic cells. If some of the germ-line cells are transgenic they will produce transgenic sperm and ova. Typically, the chimeric insects developing from these injected embryos do not display any transgene-specific phenotype although this depends on the genes expressed on the gene vector and how they are regulated. Adults arising from injected embryos are backcrossed and their progeny examined for the expression of the vector-specific genetic marker. Transgenic lines are created and the molecular nature of the transposition event leading to transgenesis is characterized by standard molecular biological techniques.

The *Hermes* element was isolated from the housefly, *M. domestica*, and was found, using interplasmid transposition assays, to be mobile in somatic nuclei in a range of insect species, including the mosquitoes *Aedes aegypti*, *Anopheles gambiae* and *Culex quinquefasciatus* (Allen *et al.*, 2001; Sarkar *et al.*, 1997a, b; Zhao & Eggleston, 1998). Subsequently, it was used to transform *Ae. aegypti* and *Cx. quinquefasciatus* (Allen *et al.*, 2001; Jasinskiene *et al.*, 1998; Pinkerton *et al.*, 2000). Interestingly, *Hermes* was found to integrate differently into germ-line nuclei and somatic nuclei in *Ae. aegypti* (Jasinskiene *et al.*, 2000; O'Brochta *et al.*, 2003; Pinkerton *et al.*, 2000). In germ-line nuclei, the *Hermes* element and DNA from the flanking plasmid sequences were integrated into the chromosomes in a transposase-dependent reaction. This mode of integration into the germ line was also observed when an autonomous *Hermes* element was used to transform *Ae. aegypti* (O'Brochta *et al.*, 2003). In subsequent generations of this transgenic line, *Hermes* transpositions in the soma were by cut-and-paste transposition, with no flanking sequences involved. The molecular basis for this tissue-specific difference in the mode of transposition of the *Hermes* element in transgenic *Ae. aegypti* remains unknown. It has not been observed in transgenic lines of *D. melanogaster* containing the same autonomous *Hermes* element, in which the cut-and-paste mode of transposition occurred in both germ-line and somatic nuclei (Guimond *et al.*, 2003).

The *piggyBac* element was isolated from the cabbage looper, *T. ni*, and has proven to be a very successful gene vector of insects, and several mosquito species, including *Ae. aegypti* (Kokoza *et al.*, 2000), *An. gambiae* (Grossman *et al.*, 2001; Kim *et al.*, 2004) and *Anopheles albimanus* (Perera *et al.*, 2002). *piggyBac* integrates into both germ-line and somatic nuclei by cut-and-paste transposition. Z. N. Adelman & A. A. James (personal communication) have reported that flanking plasmid sequences transposed along with the *piggyBac* element in some transgenic lines of *Ae. aegypti*; however, it is not known how widespread this mode of integration is for *piggyBac* in mosquitoes. It is clear that cut-and-paste integrations of *piggyBac* in mosquitoes can be screened and selected for, which satisfies the demands of virtually all laboratory experiments in which this element is used to introduce transgenes into mosquitoes. The mobility of *piggyBac* in transgenic strains of *Ae. aegypti* was examined by O'Brochta *et al.* (2003), who found that this element was stable in transgenic mosquitoes in which a source of *piggyBac* transposase was absent. The ability of *piggyBac* to be re-mobilized by its transposase is currently being investigated.

The *Mos1 mariner* element was isolated from *D. mauritiana*. *Mariner* elements are abundant in their distribution throughout both genomes and species. However, only two active *mariner* elements have been described, *Mos1* and *Himar1*, a synthetic element constructed from *mariner* sequences present in the hornfly *Haemotoba irritans*. The *Mos1* element has been used to genetically transform *Ae. aegypti* (Coates *et al.*, 1998) and integrations were by cut-and-paste transposition. The mobility of *Mos1* in transgenic strains of *Ae. aegypti* was examined by Wilson *et al.* (2003) using the same strategy as outlined above for the *piggyBac* element. These transgenic *mariner* elements were found to be relatively immobile when present in heterozygotes that contained active *Mos1* transposase. These data were consistent with previous observations of *mariner* immobility in transgenic lines of *D. melanogaster* (Lohe *et al.*, 1995; Lozovsky *et al.*, 2002).

The *Minos* element was isolated from *D. hydei* and has been used to transform *D. melanogaster*, the Mediterranean fruit fly, *Ceratitis capitata*, and the mosquito *Anopheles stephensi* (Catteruccia *et al.*, 2000; Loukeris *et al.*, 1995a, b). *Minos* integrations into each of these species are largely by cut-and-paste transposition although integrations involving flanking plasmid sequences have been observed. Attempts to re-mobilize *Minos* in transgenic lines of *An. stephensi* have not been successful (O'Brochta *et al.*, 2003).

REQUIREMENTS FOR A TRANSGENIC-VECTOR-BASED TECHNOLOGY TO COMBAT THE TRANSMISSION OF MALARIA

There are three biological requirements for the development of a successful vector-based transgenic technology to combat malaria. One is the development of efficient gene vectors. As described above, vectors have been developed. However, there remains much to be done in both understanding how these transposable elements function in mosquitoes and in generating hyperactive forms of these elements that will have increased rates of transposition, and so will be better suited to driving genes through field populations of mosquitoes (see later). The second requirement is the development of effector genes whose expression will prevent transmission of the parasite. Important to the success of an effector gene is the confinement of its expression to those mosquito tissues that are directly involved in transmission. The third requirement is the development of a means of disseminating the effector gene through the field mosquito population. This may be directly dependent on the mobility properties of the transposable element used to introduce the gene into the mosquito since, at present, transposable elements remain the most likely genetic drive mechanisms to be developed in these pests.

Mobile gene vectors

The five transposable element vectors described above have performed to expectations in laboratory experiments. Even though transformation frequencies are, at best, modest, and, in the case of *Hermes*, integrations into the mosquito germ line occur by a mechanism other than cut-and-paste transposition, foreign genes or altered mosquito genes have been reintroduced into mosquitoes and the effects of these on the ability of the mosquito to transmit the malarial pathogen have been examined. Several of these studies have demonstrated a reduction in the number of parasites and are described in more detail later. As already noted, initial investigations into the mobility of the *Hermes*, *piggyBac* and *Mos1* elements post-integration show them to be quite immobile. While this is a desired characteristic for laboratory studies, it suggests that these elements may need to be modified if they are to be used to quickly spread beneficial genes through pest insect populations.

Tissue-specific promoters

During the course of infection of the adult female mosquito, the malaria parasite needs to transverse only a few tissues. These are the midgut and the salivary glands. It also encounters the haemocoel, which is produced by the fat body. In order for transmission to proceed the parasite must transverse each of these while evading the innate immune system of the mosquito. Current strategies aimed at preventing the transmission of malaria therefore seek to target expression of the transgene to these tissues only, and only at that time when the parasite is present. Because the parasite is injected with a blood meal, promoters that are activated either in response to the blood meal (such as the vitellogenin promoter), are active in the midgut (such as the carboxypeptidase promoter) or are active in salivary glands (such as the maltase and apyrase promoters) have been isolated and used to drive the expression of effector genes in transgenic mosquitoes. The localization of transgene expression to only a few specific tissues is important for two reasons. First it localizes the expression of the transgene to the sites of infection and so transgenic mosquitoes can be constructed in the knowledge that the desired effector protein, or RNA molecule if an RNAi-based strategy is used, will be present and active in those tissues of cells in which the parasite is also present, and at sufficiently high levels. Second, by localizing expression of the transgene to one or only a few tissues, the likelihood of any costs of fitness arising due to over- or inappropriate expression of the transgene throughout the mosquito is minimalized. To date only a handful of mosquito promoters have been identified and utilized to drive the expression of transgenes in mosquitoes. The completion of the *An. gambiae* genome project combined with microchip array analysis of the entire *Anopheles* transcriptosome will provide more promoters with perhaps even more specific tissue- and stage-specific regulation that can be tested for their ability to drive the targeted expression of transgenes in transgenic mosquitoes.

Effector genes in transgenic mosquitoes

A number of different types of effector mechanisms have been proposed or tested for their ability to prevent transmission of the malaria parasite. Nirmala & James (2003) have classified these into five groups: parasite ligands that target proteins of the parasite that are required for transmission through the mosquito; mosquito receptor proteins that the parasite utilizes to facilitate movement through the mosquito; biochemical targets within the malaria parasite itself; manipulation of the innate immune system of the mosquito; and the production of proteins by the mosquito that are toxic to *Plasmodium* (Nirmala & James, 2003). Of all these possibilities, only a handful have been directly tested in transgenic mosquitoes (Table 1). However, these tests have provided clear evidence that manipulation of the mosquito genotype can produce mosquito strains that reduce the transmission of malaria through the mosquito.

The first experiments involved placing the *Ae. aegypti* defensin gene into *Ae. aegypti* using the *Hermes* transposable element. These experiments sought to augment the innate immune response of mosquitoes to infection by *Plasmodium* and, in doing so, prevent transmission of the parasite through the female mosquito (Kokoza *et al.*, 2000). The defensin gene was placed under the control of the vitellogenin promoter. Transgenic lines were established and, following the blood meal, the defensin transcript was present in large amounts, as was defensin protein. Both the pattern and level of defensin expression were markedly different than in untransformed mosquitoes, in which low to moderate levels of defensin are produced only in response to bacterial or fungal infection. These transgenic mosquitoes were challenged with the Gram-negative bacterium *Enterobacter cloacae* and were found to be almost twice as resistant to infection as the non-transformed strain (Kokoza *et al.*, 2000). Defensins and other antimicrobial peptides of the insect immune system also possess anti-*Plasmodium* activity (Kim *et al.*, 2004; Shahabuddin *et al.*, 1998) and antimicrobial peptides have been demonstrated to be activated in *An. gambiae* upon infection by *P. falciparum* (Richman *et al.*, 1997). A prediction is, therefore, that increased expression of antimicrobial genes such as defensin and cecropin in response to the blood meal might confer resistance to *Plasmodium* infection and thus block transmission. This remains to be confirmed, however, since RNAi-mediated interference of defensin gene expression in *An. gambiae* does not affect *Plasmodium* development (Blandin *et al.*, 2002).

Another approach has been to prevent the transmission of the parasite through the female mosquito by targeting mosquito receptor molecules. A 12-amino-acid peptide, named SM1, was isolated from a phage display library based on its ability to bind to both salivary gland and midgut epithelial tissues, both of which are involved in *Plasmodium* transmission. Several repeated units of the SM1 coding sequence were then placed downstream of the carboxypeptidase promoter to target expression to the midgut. This chimeric gene was inserted into the *piggyBac* transposable element and transgenic lines of *An. stephensi* were generated (Ito *et al.*, 2002). Upon infection by *Plasmodium berghei*, the agent of rodent malaria, *Plasmodium* oocyst formation was found to be blocked by approximately 80 % relative to transgenic controls that did not express the SM1 peptide. As a consequence, sporozoite formation and vector competence were reduced in the transgenic lines expressing SM1 (Ito *et al.*, 2002).

Venom phospholipase A(2) also prevents *Plasmodium* oocyst formation within the blood meal (Zieler *et al.*, 2001). The precise mechanism by which this occurs is unknown but it has been proposed that this phospholipase modifies the midgut epithelial membrane properties, thereby preventing oocyst formation. The gene encoding honeybee venom phospholipase was placed under the control of the carboxypeptidase promoter and, once again, the *piggyBac* element was used as the gene vector to generate transgenic *An. stephensi*. Transgenic lines expressing honeybee phospholipase A(2) showed a reduction in *P. berghei* oocyst formation of approximately 87 % relative to transgenic controls (Moreira *et al.*, 2002). As a consequence, transmission of the parasite to naïve mice was greatly reduced, and, in many cases, no transmission occurred at all.

The feasibility of blocking the transmission of the parasite through the mosquito at the stage of salivary gland invasion has been shown using the Sindbis alpha virus expression system. A single-chain antibody fragment that was specifically targeted to the circumsporozoite protein of *Plasmodium* greatly reduced the number of *Plasmodium gallinaceum* sporozoites in salivary glands of *Ae. aegypti* expressing this virus-based transgene (de Lara Capurro *et al.*, 2000). Given that transposable-element-mediated transgenesis would be predicted to ensure that the expression of this transgene occurred in every cell, the efficacy of this approach would be expected to increase in transgenic mosquitoes expressing this gene.

Experimentally, therefore, the concept of genetically engineering mosquitoes to prevent the transmission of *Plasmodium* through the insect has been demonstrated for two of the animal models of human malaria. It has yet to be demonstrated for *P. falciparum*. However, based on the limited data reported, there is reason to believe that, at least in the laboratory, transmission of this parasite species can also be blocked. Two questions remain regarding whether these data will be able to be extended into the field. First, can 100 % refractoriness levels be obtained? If not, can refractoriness levels approaching this lead to a reduction in the incidence of malaria in human populations? Both outcomes are contingent on how rapidly and effectively these beneficial effector genes can be spread through a field mosquito population by genetic drive agents.

TRANSGENIC MOSQUITOES IN THE FIELD

The feasibility of extending transgenic-based strategies in mosquitoes from the laboratory to the field attracts much discussion. Several obvious problems need to be overcome such as the fitness of transgenic mosquitoes and discovery of a means by which these transgenes can be spread through mosquito populations. Questions of the feasibility of using the effector genes themselves have been recently discussed by Billingsley (2003) and the reader is referred to this review for a critique of these approaches beyond the laboratory.

The need for complete refractoriness?

The ability, in theory, for transgenic mosquitoes to spread beneficial effector genes linked to transposable elements through mosquito populations was examined recently (Boete & Koella, 2002). This model assumed that refractoriness was determined by a single gene and that the refractory, or effector, gene was linked to a transposable element. The model showed that if the efficacy of refractoriness was 100 % then the spread of malaria through human populations could be halted and in fact eradicated, provided that the effector gene was linked to a transposable element that, due to genetic drive, increased the probability that the transposable element effector gene would be passed on to following generations. The model assumed that this linkage was absolute (i.e. no recombination within the chimeric transposable element was permitted). The model also assumed that there was a cost in fitness of being transgenic and, in the case of infected females, a cost arising from harbouring the malaria parasite. Encouragingly, when absolutely linked to a transposable element, a refractory gene that was predicted to be 100 % effective in preventing transmission of the parasite through the mosquito was predicted to be capable of eliminating malaria in a human population. The question then becomes how feasible this is to attain in practice. While much progress has been made in demonstrating that some effector genes can dramatically reduce and even eliminate the transmission of malaria under laboratory conditions, expecting 100 % effectiveness in the field from a genotype as complex and potentially variable as the blocking of transmission of every individual parasite by all cells of the female mosquito in all transgenic mosquitoes is placing high expectations on the system. This can be overcome, in principle, by placing multiple, independent, refractory systems in the same transgenic mosquito so that more than one genetic system has the opportunity to prevent transmission of the parasite through the mosquito. The fitness costs of multiple systems simultaneously present in a single genetic strain may, however, be severe. In addition, ensuring that all of these transgenes remain genetically linked to each other will also be a challenge since it is likely that, given present technology, multiple transformations will need to be made in order to generate a single genetic strain with more than a single refractory system. The ability of transposable elements to spread through mosquito populations also remains unknown. Transposable elements have yet to be observed to spread through mosquito populations in either the laboratory or the field, and, within insects, the examples of *P* and *hobo* element spread through natural populations of *D. melanogaster* (Daniels *et al.*, 1990a, b) remain the paradigm upon which the assumption that transposable elements can provide the genetic drive to move effector genes through mosquito populations is based. At present the transformation frequencies of all five transposable elements used to transform mosquitoes are low and, as already discussed, their re-mobilization frequencies are lower than expected. Yet the abundance and distribution of transposable elements in insect genomes suggests that the elements not only invade but also propagate efficiently within genomes. The interactions between invading transposable elements, particularly those that contain transgenes, and the new host genome remain unknown. Potentially these could rapidly inactivate any new transposable element. Indeed the RNAi-based mechanism of single-strand RNA destruction has been proposed to be a natural immune system that is not just limited to invading viruses but that also applies to invading transposable elements (Vastenhouw *et al.*, 2003).

Fitness of transgenic mosquitoes

The mating competitiveness of strains of *An. stephensi* containing genetically modified *piggyBac* elements has been examined (Catteruccia *et al.*, 2003). Mating competitiveness was severely affected in all four transgenic lines investigated. None of these lines was designed for field release, and so extrapolations from the behaviour of these transgenic mosquitoes to transgenic mosquitoes specifically designed for release in genetic control strategies cannot be justified. A life table analysis on three transgenic strains of *Ae. aegypti* comparing fecundity, longevity and development times to the laboratory strain from which they each derived was more recently performed (Irvin *et al.*, 2004). All strains displayed severely reduced fitness relative to the laboratory strain. This strain was used to generate the transgenic lines and had been maintained in the laboratory for over 40 years. A transgenic strain containing autonomous *Hermes* elements was equally unfit as a transgenic strain containing non-autonomous *Hermes* elements. As for the *An. stephensi* studies of Catteruccia *et al.* (2003), none of these genetically engineered strains of *Ae. aegypti* were intended for field release. In both studies only the mating competitiveness or fitness of homozygous individuals was examined. Since in a field release of genetically modified mosquitoes it would be heterozygotes that spread the transgene through a natural population, it is the fitness of these individuals that is more relevant to the success of a genetic control programme. Nonetheless, these two preliminary studies show that the process of transgenesis can incur fitness costs. One of the attractions of transgenesis over traditional insect genetic modification strategies, such as the construction of sex-chromosome-linked translocations, was that the process of transposable element-mediated insertion of a transgene into the genome was precise and resulted in, at least at the genetic level, a small and defined change to the genome. In contrast, the generation of translocation strains resulted in large and undefined changes in the genome. An expectation was that genetically engineered strains might therefore be more fit than strains containing translocations. Based on these preliminary studies this may not be the case, and so more detailed research needs to be undertaken. For example, the genetic basis of the decline in fitness and/or mating competitiveness needs to be established. Green fluorescent protein was used as the genetic marker in both of the studies described above and it is possible that the expression of this leads to a decline in the fitness of mosquitoes that express it. Similarly the role of inbreeding depression in reducing fitness in transgenic mosquitoes needs to be examined and measured. All transgenic mosquitoes undergo an extreme population bottleneck in that each transgenic strain arises from a single injected embryo. The magnitude of inbreeding depression has only been examined in tree-hole breeding mosquitoes, where it was shown to be a significant factor in the viability of the population (Armbruster *et al.*, 2000). In the experiments conducted by Irvin *et al.* (2004), the laboratory line used had been maintained in the laboratory for over 40 years so the effects of inbreeding associated with these experiments were estimated to be minimal.

Spreading transgenes through mosquito populations using transposable elements

Transposable elements remain the only means by which foreign DNA can be introduced into mosquitoes and then passed onto successive generations. Endosymbionts such as *Wolbachia* have yet to be developed into gene vectors in mosquitoes, although the distribution of *Wolbachia* through populations of *Culex pipiens* indicates that, should it be successfully introduced into a naïve species such as *An. gambiae*, *Wolbachia* might also spread through this species. An attractive feature of transposable elements is their ability to move within genomes and to increase in copy number, which results in an enhanced vertical transmission potential relative to an immobile gene. This makes them, in principle, good candidates for spreading beneficial transgenes through mosquito populations.

Optimism for the feasibility of this strategy comes from the invasion of *P* elements, most probably from *Drosophila willistoni*, through wild populations of *D. melanogaster* over the past 70 years. Similarly, the distribution in insect genomes of other transposable elements such as *mariner* strongly suggests that several forms of this element have also invaded a large number of arthropod species. Using transposable elements to spread genes through mosquito populations will be contingent upon a number of factors. Fitness and mating competitiveness of the released transgenic mosquitoes are likely to be critical factors in determining whether the genotype remains and increases in frequency within a population. The structure of the recipient populations and the degree of gene flow among them will strongly influence the rate and degree to which transgenes can be spread. Characteristics of the transposable element will also play a role in determining the rates and efficiencies of spread. Transposition rate (jumps per genome per generation) will be an important determinant of an element's spreading potential. While transposition rate will be important, transmission efficiency (percentage of gametes containing the element) will ultimately determine whether the element will increase or decrease in frequency within a population. Transmission efficiency is a function of transposition rate but other aspects of the behaviour of an element are also critical. For example, the timing of transposition can strongly influence transmission efficiency. An element that transposes post-meiotically will have a lower transmission efficiency than an element that transposes pre-meiotically. Clearly pre-meiotic transposition can result in large clusters of gametes containing the transposition event. A transposition event occurring before blastoderm formation could lead to an entire germ line with the new transposition event (100 % efficient). So, while an element may have a transposition rate of one jump per genome per generation, its transmission efficiency will be largely determined by the timing of the event. The replicative potential of an element is also very important. Class II transposable elements move by a conservative cut-and-paste transposition mechanism. Increase in copy number can only occur if the element exploits replication functions of the host genome. This can be done by repairing gaps resulting from element excision by templated gap repair mechanisms. Alternatively, timing transposition to occur during S phase can result in a net copy number increase if the element transposes after it has been replicated and inserts into an unreplicated region of the genome. Elements that have little replicative potential will have low transmission efficiencies even if they have high rates of transposition. Target site preference can also influence transmission efficiency. An element that jumps to positions tightly linked to the original integration site will have low transmission efficiency relative to elements that prefer to integrate into unlinked regions of the genome. To date the properties of existing insect gene vectors have not been examined much beyond their rates of integration from a plasmid donor molecule to a germ-line chromosome. It is quite possible that none of the existing insect gene vectors will have transmission efficiencies high enough to make them adequate genetic drive systems.

As described above, the inability to efficiently re-mobilize *mariner* and *piggyBac* elements in transgenic lines of *Ae. aegypti* indicates that these elements may be of limited used as gene-spreading agents, at least in this species. The different behaviour of modified *Hermes* elements in the germ-line and somatic nuclei in transgenic lines of *Ae. aegypti* also raise questions about the utility of at least this form of *Hermes* in this species. Existing forms of these elements, while very successful as laboratory tools, may therefore be of limited use in spreading genes through mosquito populations.

Until a genetic drive system is developed, investigations into insect transposable elements and their development into gene vectors should continue. There are two approaches to these studies that can be envisaged. One is to survey natural populations of the target mosquito species for active transposable elements that might be harnessed to spread genes through these populations. The second is to modify the existing transposable elements used to transform mosquitoes with the aim of developing hyperactive elements that can spread quickly through mosquito populations.

The discovery of naturally occurring transposable elements from mosquito populations has not, until recently, produced any active elements. The completion of the *An. gambiae* genome project revealed the presence of members of many previously characterized families of transposable elements (Holt *et al.*, 2002; Sarkar *et al.*, 2003). We have discovered an active form of a member of the *hAT* family of transposable elements that includes the *Hermes* element described previously. This new element, called *Herves*, is present in three copies per genome in one of the strains examined and appears to have a discontinuous distribution in laboratory populations of *An. gambiae* derived from populations obtained from across Africa (P. Arensburger, D. O'Brochta & P. W. Atkinson, unpublished). If this distribution is found to be present in actual field populations *Herves* may be an example of a transposable element that is presently spreading through populations of *An. gambiae*. As such, *Herves* and related *hAT* elements, such as *Hermes* and *hobo*, might prove to be vehicles for achieving genetic drive in this species.

Experiments to generate hyperactive forms of *Hermes*, *mariner*, *Minos* and *piggyBac* with enhanced mobility properties in mosquitoes have yet to be conducted in earnest. An active transposable element would be expected to display suboptimal levels of mobility since high levels of activity would lead to high rates of mutations and genetic damage that would reduce the viability of the individual.

ISSUES ARISING FROM GENERATING AND RELEASING TRANSGENIC MOSQUITOES

The use of transgenic technology to benefit human well-being elicits a spectrum of emotions and arguments. Genetically engineered food is accepted by many as an improvement in agriculture since it results in plant varieties that are resistant to insect predation, or to herbicides used to kill insect pests. Opposition to transgenic crops has come from a number of directions. In some cases, health risks to consumers are cited as outweighing any economic gains that benefit mainly growers. Sometimes grower autonomy and increasing reliance on technologies developed in and imported from large industrialized countries are cited as reasons for opposition. Opposition can also take the form of ethical concerns and the desire to limit man's impact on nature and the environment. The ongoing discussions about the appropriateness of certain transgenic crop technologies for continental Africa are preambles to the discussions that will take place in response to proposals to release transgenic mosquitoes to combat malaria. Powerful arguments based on improving the health of a large portion of the population of the continent are likely to be met with equally powerful arguments about risks associated with sustainability, unintended effects on vectorial capacity of other parasites and pathogens and disruption of local ecologies.

Progress in developing useful transgenic insect technologies for field release is likely to be slow relative to transgenic plant technologies because there are essentially no data that address any of the major risk issues associated with this technology, such as transgene stability, horizontal gene transfer, pleiotropy associated with transgene integration and alterations in insect life history characteristics and biotic interactions. The absence of strong commercial economic incentives to develop and adopt these technologies means that development will depend on resources for 'public good'. Countries likely to benefit from transgenic insect technologies are likely to lack sufficient resources to subsidize the major efforts required to develop them. Consequently, development of this technology will proceed largely through government funding in the West and North. The challenges that lie ahead are tremendous but the possible benefits, in the case of malaria control, are enormous and the potential of this technology needs to be fully explored.

CONCLUSION

More than a decade lapsed between the development of *D. melanogaster* genetic transformation and its successful extension into mosquitoes. Both experimental strategies involve the use of Class II transposable elements as gene vectors and the techniques required for transformation are the same in both insect species. As for *Drosophila*, mosquito transformation has led to new approaches in genetics since modified genes can be quickly tested for their effects on phenotype. Unlike *Drosophila*, however, the technique of mosquito transformation has not been taken up by a large number of laboratories. This is because far fewer laboratories are engaged in mosquito research and because mosquito transformation is less robust than *Drosophila* transformation. Despite these limitations, engineered mosquitoes with inhibited ability to transmit pathogens have already been generated for model systems. Researchers can now test a range of effector genes and promoters, either in their own laboratories or via collaboration, for function in transgenic mosquitoes. Immense challenges remain. Mosquito transformation is not robust and, in the case of *An. gambiae*, has only been achieved in three laboratories over the course of 16 years (Grossman *et al.*, 2001; Kim *et al.*, 2004; Miller *et al.*, 1987). If multiple transgenic strategies are to be tested, improvements to genetic transformation of this species need to be made. As yet, no genetic drive mechanism for mosquitoes has been identified. If genes are to be efficiently spread through mosquito populations, genetic drive mechanisms need to be identified and developed. Clearly transposable elements are successful for introducing genes into mosquitoes in the laboratory. Of the five transposable elements so far used for this purpose, all appear to have limitations for use in the field. Re-mobilization frequencies are very poor, suggesting that these elements in their current form will not be efficient gene-spreading agents. Research into how each of these elements interacts with their new host genomes needs to be undertaken so that accurate predictions about their mobility post-transformation can be made. Understanding the molecular basis of transposable element movement in mosquitoes should enable hyperactive elements to be isolated since mutations in critical regions of the transposon can be screened for enhanced activity. In the same vein, estimates of transposition frequency, whether transposition is pre-meiotic or meiotic and whether the element moves to genetically unlinked sites also need to be acquired so that a time frame of population replacement by transposable element spread can be made. Finally, during the development of this technology (not at the end), surveys of the target populations of mosquitoes need to be made so that release strains are as compatible as possible with the strains they seek to replace. The goal of reducing or eliminating malaria using transgenic mosquitoes is an ambitious one. However, the pay off, reducing the disease burden on a large proportion of humanity, is immense and so justifies this long-term approach to mosquito-borne disease control.

ACKNOWLEDGEMENTS

The authors recognize the members, past and present, of both of their laboratories and funding from the National Institutes of Health grants GM48106, AI45741 and AI45743.

REFERENCES

Allen, M. L., O'Brochta, D. A., Atkinson, P. W. & LeVesque, C. S. (2001). Stable germ-line transformation of *Culex quinquefasciatus* (Diptera: Culicidae). *J Med Entomol* **38**, 701–710.

Armbruster, P., Hutchinson, R. A. & Linvell, T. (2000). Equivalent inbreeding depression under laboratory and field conditions in a tree-hole-breeding mosquito. *Proc R Soc Lond B Biol Sci* **267**, 1939–1945.

Billingsley, P. F. (2003). Environmental constraints on the physiology of transgenic mosquitoes. In *Ecological Aspects for Application of Genetically Modified Mosquitoes*, pp. 149–161. Edited by W. Takken & T. W. Scott. Dordrecht: Kluwer.

Blandin, S., Moita, L. F., Kocher, T., Wilm, M., Kafatos, F. C. & Levashina, E. A. (2002). Reverse genetics in the mosquito *Anopheles gambiae*: targeted disruption of the *Defensin* gene. *EMBO Rep* **3**, 852–856.

Boete, C. & Koella, B. (2002). A theoretical approach to predicting the success of genetic manipulation of malaria mosquitoes in malaria control. *Malaria J* **1**, 3–9.

Cary, L. C., Goebel, M., Corsaro, B. G., Wang, H. G., Rosen, E. & Fraser, M. J. (1989). Transposon mutagenesis of baculoviruses: analysis of *Trichoplusia ni* transposon IFP2 insertions within the FP-locus of nuclear polyhedrosis viruses. *Virology* **172**, 156–169.

Catteruccia, F., Nolan, T., Loukeris, T. G., Blass, C., Savakis, C., Kafatos, F. C. & Crisanti, A. (2000). Stable germline transformation of the malaria mosquito *Anopheles stephensi*. *Nature* **405**, 959–962.

Catteruccia, F., Godfray, H. C. & Crisanti, A. (2003). Impact of genetic manipulation on the fitness of *Anopheles stephensi* mosquitoes. *Science* **299**, 1225–1227.

Coates, C. J., Jasinskiene, N., Miyashiro, L. & James, A. A. (1998). *Mariner* transposition and transformation of the yellow fever mosquito, *Aedes aegypti*. *Proc Natl Acad Sci U S A* **95**, 3748–3751.

Coates, C. J., Jasinskiene, N., Pott, G. B. & James, A. A. (1999). Promoter-directed expression of recombinant fire-fly luciferase in the salivary glands of *Hermes*-transformed *Aedes aegypti*. *Gene* **21**, 317–325.

Coates, C. J., Jasinskiene, N., Morgan, D., Tosi, L. R. O., Beverley, S. M. & James, A. A. (2000). Purified *mariner* (*Mos1*) transposase catalyzes the integration of marked elements into the germ line of the yellow fever mosquito, *Aedes aegypti*. *Insect Biochem Mol Biol* **30**, 1003–1008.

Daniels, S. B., Peterson, K. R., Strausbaugh, L. D., Kidwell, M. G. & Chovnick, A. (1990a). Evidence for horizontal transmission of the *P* transposable element between *Drosophila* species. *Genetics* **124**, 339–355.

Daniels, S. B., Chovnick, A. & Boussy, I. A. (1990b). Distribution of the *hobo* transposable elements in the genus *Drosophila*. *Mol Biol Evol* **7**, 589–606.

de Lara Capurro, M., Coleman, J., Beerntsen, B. T., Myles, K. M., Olsen, K. E., Rocha, E., Krettli, A. U. & James, A. A. (2000). Virus expressed, recombinant single-chain antibody blocks sporozoite infection of salivary glands in *Plasmodium gallinaceum*-infected *Aedes aegypti*. *Am J Trop Med Hyg* **62**, 427–433.

Franz, G., Loukeris, T. G., Dialektaki, G., Thompson, C. R. & Savakis, C. (1994). Mobile *Minos* elements from *Drosophila hydei* encode a two-exon transposase with similarity to the paired DNA-binding domain. *Proc Natl Acad Sci U S A* **91**, 4746–4750.

Gardner, M. J., Hall, N., Fung E. & 42 other authors (2002). Genome sequence of the human malaria parasite *Plasmodium falciparum*. *Nature* **419**, 498–511.

Grossman, G. L., Rafferty, C. S., Clayton, J. R., Stevens, T. K., Mukabayire, O. & Bendedict, M. Q. (2001). Germline transformation of the malaria vector, *Anopheles gambiae*, with the *piggyBac* transposable element. *Insect Mol Biol* **10**, 597–604.

Guimond, N. D., Bideshi, D. K., Pinkerton, A. C., Atkinson, P. W. & O'Brochta, D. A. (2003). Patterns of *Hermes* element transposition in *Drosophila melanogaster*. *Mol Genet Genomics* **268**, 779–790.

Holt, R. A., Subramanian, G. M., Halpern, A. & 120 other authors (2002). The genome sequence of the malaria mosquito *Anopheles gambiae*. *Science* **298**, 129–149.

Irvin, N., Hoddle, M. S., O'Brochta, D. A., Carey, B. & Atkinson, P. W. (2004). Assessing fitness costs for transgenic *Aedes aegypti* expressing the green fluorescent protein marker and transposase genes. *Proc Natl Acad Sci U S A* **101**, 891–896.

Ito, J., Ghosh, A., Moreira, L. A., Wimmer, E. A. & Jacobs-Lorena, M. (2002). Transgenic anopheline mosquitoes impaired in transmission of a malaria parasite. *Nature* **417**, 387–388.

Jasinskiene, N., Coates, C. J., Benedict, M. Q., Cornel, A. J., Salazar-Rafferty, C., James, A. A. & Collins, F. H. (1998). Stable, transposon-mediated transformation of the yellow fever mosquito, *Aedes aegypti*, using the *Hermes* element from the house fly. *Proc Natl Acad Sci U S A* **95**, 3743–3747.

Jasinskiene, N., Coates, C. J. & James, A. A. (2000). Structure of *Hermes* integrations in the germline of the yellow fever mosquito, *Aedes aegypti*. *Insect Mol Biol* **9**, 11–18.

Jasinskiene, N., Coates, C. J., Ashikyan, A. & James, A. A. (2003). High efficiency, site-specific excision of a marker gene by the phage P1 cre-lox system in the yellow fever mosquito, *Aedes aegypti*. *Nucleic Acids Res* **31**, e147.

Kim, W., Koo, H., Richman, A. M., Seeley, D., Vizioli, J., Klocko, A. D. & O'Brochta, D. A. (2004). Ectopic expression of a cecropin transgene in the human malaria vector mosquito *Anopheles gambiae* (Diptera: Culicidae): effects on susceptibility to *Plasmodium*. *J Med Entomol* (in press).

Kokoza, V., Ahmed, A., Cho, W. L., Jasinskiene, N. A., James, A. A. & Raikhel, A. S. (2000). Engineering blood-meal activated systemic immunity in the yellow fever mosquito, *Aedes aegypti*. *Proc Natl Acad Sci U S A* **97**, 9144–9149.

Kokoza, V., Ahmed, A., Wimmer, E. A. & Raikhel, A. S. (2001). Efficient transformation of the yellow fever mosquito *Aedes aegypti* using the *piggyBac* transposable element vector pBac[3xP3-EGFPafm]. *Insect Biochem Mol Biol* **31**, 1137–1143.

Lobo, N. F., Hua-Van, A., Li, X., Nolen, B. M. & Fraser, M. J. (2002). Germ line transformation of the yellow fever mosquito, *Aedes aegypti*, mediated by trans-positional insertion of a *piggyBac* vector. *Insect Mol Biol* **11**, 133–139.

Lohe, A. R., Moriyama, E. N., Lidholm, D. A. & Hartl, D. L. (1995). Horizontal transmission, vertical inactivation, and stochastic loss of mariner-like transposable elements. *Mol Biol Evol* **12**, 62–72.

Loukeris, T. G., Arca, B., Livadaras, I., Dialektaki, G. & Savakis, C. (1995a). Introduction of the transposable element *Minos* into the germ line of *Drosophila melanogaster*. *Proc Natl Acad Sci U S A* **92**, 9485–9489.

Loukeris, T. G., Livadras, I., Arca, B., Zabalou, S. & Savakis, C. (1995b). Gene transfer into the Medfly, *Ceratitis capitata*, using a *Drosophila hydei* transposable element. *Science* **270**, 2002–2005.

Lozovsky, E. R., Nurminsky, D. I., Wimmer, E. A. & Hartl, D. L. (2002). Unexpected stability of *mariner* transgenes in *Drosophila*. *Genetics* **160**, 527–535.

Medhora, M. M., MacPeek, A. H. & Hartl, D. L. (1988). Excision of the *Drosophila* transposable element mariner: identification and characterization of the *Mos* factor. *EMBO J* **7**, 2185–2189.

Miller, L. H., Sakai, R. K., Romans, P., Gwadz, R. W., Kantoff, P. & Coon, H. G. (1987). Stable integration and expression of a bacterial gene in the mosquito, *Anopheles gambiae*. *Science* **237**, 779–781.

Moreira, L. A., Edwards, M., Adhami, F., Jasinskiene, N., James, A. A. & Jacobs-Lorena, M. (2000). Robust gut-specific expression in transgenic *Aedes aegypti* mosquitoes. *Proc Natl Acad Sci U S A* **97**, 10895–10898.

Moreira, L. A., Ito, J., Ghosh, A. & 7 other authors (2002). Bee venom phospholipase inhibits malaria parasite development in transgenic mosquitoes. *J Biol Chem* **277**, 40839–40843.

Nirmala, X. & James, A. A. (2003). Engineering *Plasmodium*-refractory phenotypes in mosquitoes. *Trends Parasitol* **19**, 384–387.

Nolan, T., Bower, T. M., Brown, A. E., Crisanti, A. & Catterucia, F. (2002). PiggyBac-mediated germline transformation of the malaria mosquito *Anopheles stephensi* using the red fluorescent protein dsRED as a selectable marker. *J Biol Chem* **277**, 8759–8762.

O'Brochta, D. A., Sethuramuran, N., Wilson, R. & 9 other authors (2003). Gene vector and transposable element behavior in mosquitoes. *J Exp Biol* **206**, 3823–3834.

Perera, O. P., Harrell, R. A. I. & Handler, A. M. (2002). Germ-line transformation of the South American malaria vector, *Anopheles albimanus*, with a piggyBac/EGFP transposon vector is routine and highly efficient. *Insect Mol Biol* **11**, 291–297.

Pinkerton, A. C., Michel, K., O'Brochta, D. A. & Atkinson, P. W. (2000). Green fluorescent protein as a genetic marker in transgenic *Aedes aegypti*. *Insect Mol Biol* **9**, 1–10.

Richman, A. M., Dimopoulos, G., Seeley, D. C. & Kafatos, F. C. (1997). *Plasmodium* activates the innate immune response of *Anopheles gambiae* mosquitoes. *EMBO J* **16**, 6114–6119.

Rowan, K. H., Orsetti, J., Atkinson, P. W. & O'Brochta, D. A. (2004). *Tn*5 as an insect gene vector. *Insect Biochem Mol Biol* (in press).

Rubin, G. M. & Spradling, A. C. (1982). Genetic transformation of *Drosophila* with transposable element vectors. *Science* **218**, 348–353.

Sarkar, A., Yardley, K., Atkinson, P. W., James, A. A. & O'Brochta, D. A. (1997a). Transposition of the *Hermes* element in embryos of the vector mosquito, *Aedes aegypti*. *Insect Biochem Mol Biol* **27**, 359–363.

Sarkar, A., Coates, C. J., Whyard, S., Willhoeft, U., Atkinson, P. W. & O'Brochta, D. A. (1997b). The *Hermes* element from *Musca domestica* can transpose in four families of cylorrhaphan flies. *Genetica* **99**, 15–29.

Sarkar, A., Sengupta, R., Krzywinski, J., Wang, X., Roth, C. & Collins, F. H. (2003). P elements are found in the genomes of nematoceran insects of the genus *Anopheles*. *Insect Biochem Mol Biol* **33**, 381–387.

Shahabuddin, M., Fields, I., Bulet, P., Hoffmann, J. A. & Miller, L. H. (1998). *Plasmodium gallinaceum*: differential killing of some mosquito stages of the parasite by insect defensin. *Exp Parasitol* **89**, 103–112.

Vastenhouw, N. L., Fischer, S. E. J., Robert, V. J. P., Thijssen, K. L., Fraser, A. G., Kamath, R. S., Ahringer, J. & Plasterk, R. H. (2003). A genome-wide screen identifies 27 genes involved in transposon silencing in *C. elegans*. *Curr Biol* **13**, 1311–1316.

Warren, W. D., Atkinson, P. W. & O'Brochta, D. A. (1994). The *Hermes* transposable element from the housefly, *Musca domestica*, is a short inverted repeat-type element of the *hobo*, *Ac*, and *Tam3* (*hAT*) element family. *Genet Res* **64**, 87–97.

Wilson, R., Orsetti, J., Klocko, A. K., Aluvihare, C., Peckham, E., Atkinson, P. W., Lehane, M. J. & O'Brochta, D. A. (2003). Post-integration behavior of a *Mos1 mariner* gene vector in *Aedes aegypti*. *Insect Biochem Mol Biol* **33**, 853–863.

Zhao, Y. & Eggleston, P. (1998). Stable transformation of an *Anopheles gambiae* cell line mediated by the *Hermes* mobile genetic element. *Insect Biochem Mol Biol* **28**, 213–219.

Zieler, H., Keister, D. B., Dvorak, J. A. & Ribeiro, J. M. (2001). A snake venom phospholipase A(2) blocks malaria parasite development in the mosquito midgut by inhibiting ookinete association with the midgut surface. *J Exp Biol* **204**, 4157–4167.

Vaccines targeting vectors

Geoffrey A. T. Targett

Department of Infectious and Tropical Diseases, London School of Hygiene & Tropical Medicine, London WC1B 3DP, UK

INTRODUCTION

It has been known for a long time that arthropods requiring blood meals induce pronounced immune responses in the host on which they feed (Trager, 1939). These, together with the wide range of haemostatic mechanisms that are employed by the host to avoid blood loss, present a formidable barrier to the arthropod's essential need for regular blood meals in order to moult or produce eggs.

The immune responses first recognized were in those vector–host interactions where the arthropod remains attached to its host for a long period – days or often weeks. This is true of many ticks. The much cited work of William Trager (1939) is recognized as one of the seminal studies on anti-tick resistance and, during the past 60 years, a great deal has been learned about the numerous underlying mechanisms.

It is not surprising that such a lengthy and intimate association leads to cellular and humoral responses to vector antigens. However, it is now equally clear that such responses can also develop to haematophagous arthropods that feed rapidly and then leave the host. Notable here are insects of medical importance such as mosquitoes, sandflies and black flies.

Extensive studies have shown that, in both the slow-feeding (tick) and rapid-feeding (insect) arthropods, there is an array of immunomodulatory factors to counter the induction of immune responses, to deflect the immunity away from mechanisms that are harmful to the arthropod, or to lessen the effect of these if they do occur.

SGM symposium 63: Microbe–vector interactions in vector-borne diseases.
Editors S. H. Gillespie, G. L. Smith & A. Osbourn. Cambridge University Press. ISBN 0 521 84312 X ©SGM 2004

Interest in the artificial induction by vaccination of immune responses that will interfere with blood feeding and/or be harmful to the arthropod has advanced most with the development of anti-tick responses, and there are effective commercial vaccines that have been in use for several years. By contrast, artificial active immunization to impair insect feeding has not progressed that far, though studies in recent years on the immunomodulatory role of saliva of the insects have given promise of immunological approaches to vector control.

Anti-arthropod vaccines are developed for two reasons. With ticks, an important objective of vaccination has been control of the ticks themselves; this then is not strictly development of an anti-vector vaccine but one directed against the tick as an ectoparasite. Ticks are nevertheless vectors of important pathogens such as *Borrelia*, *Theileria* spp. and arboviruses (Table 1), so the vaccination could in some cases serve a dual function. Vaccines directed against haematophagous insects are primarily intended to have an impact on the diseases of major importance that they transmit. Notable examples include the parasites causing malaria and dengue fever transmitted by mosquitoes, and leishmaniasis transmitted by sandflies. There is here an intricate relationship involving immune responses induced by the host to both the parasite and the vector, and their ability to modulate or evade its harmful effects.

Examples have been selected here to show the nature of the immune responses that need to be induced, what has been achieved, and the opportunities that exist for adding anti-vector vaccines to other control measures directed at the vectors or the disease-causing organisms they transmit.

HOST RESPONSES TO TICK FEEDING

As indicated, it is not surprising that there are immune responses induced by the feeding tick since larval, nymphal and adult stages feed slowly and the tick saliva contains an array of molecules that are antigenic. It should be borne in mind that most of what has been learned about these acquired immune responses has come from studies of host–vector relationships that are not natural. Nevertheless, it was important for economic reasons to try to establish how cattle might be protected immunologically against tick infestation, and the studies on tick feeding on experimental hosts such as mice and guinea pigs have allowed immune mechanisms to be dissected in detail and given clues to the immune effector mechanisms that need to be induced if vaccination is to be effective.

The consequence for the ticks of the development of resistance is reduced engorgement of blood and hence lower weight, longer periods of attachment and attempted feeding, reduced egg production, prevention of moulting, and, in some cases, death of the tick.

Table 1. Tick-borne diseases

Tick	Host (main)	Disease
Ixodes	Humans and wild and domestic animals	Lyme disease (*Borrelia burgdorferi*) Tick-borne encephalitis Babesiosis Erlichiosis Anaplasmosis
Boophilus	Cattle	Babesiosis Anaplasmosis
Hyalomma	Ruminants	Theileriosis (*Theileria annulata*)
Rhipicephalus	Cattle Wild ruminants	East Coast fever (*Theileria parva*)
Amblyomma	Cattle Wild ruminants	Heartwater (*Cowdria ruminantium*)

The immune responses that have been seen are diverse. Langerhans' cells carry tick immunogens from the skin to draining lymph nodes. Circulating antibodies to tick salivary gland molecules are induced but, perhaps more importantly, homocytotropic antibodies which will bind to cells infiltrating the feeding sites have been shown. Complement activation by both the classical and alternative pathways occurs.

There is a very pronounced and diverse cellular infiltration as a consequence of repeated feeding. Basophils and eosinophils are common and degranulation involving both tissue binding antibodies and complement will lead to release of molecules (such as histamine) which can inhibit tick feeding. CD4$^+$ T cells are the predominant infiltrating lymphocytes.

Cytokine responses are of particular interest, especially in relation to the relative importance of Type 1 (Th1) and Type 2 (Th2) responses. Mejri *et al.* (2001) demonstrated a predominantly Th2 response in BALB/c mice infested with *Ixodes ricinus* larvae. Higher levels of interleukin 4 (IL-4) mRNA than of interferon gamma (IFN-γ) mRNA were demonstrated in draining lymph nodes and this polarization was confirmed by *in vitro* re-stimulation of lymph nodes cells with tick saliva, salivary gland extract, and with more purified salivary extract fractions. Continued exposure, however, has shown a shift towards a Th1 response with greater production of IFN-γ and IL-2 (Wikel, 1999).

Ticks must be able to counter this array of immune responses in order to be able to feed effectively and to survive. Extensive studies have revealed a plethora of immuno-suppressive measures evolved by the tick to dampen down the acquired immune responses of the host. These include inhibition of different components of complement

(Valenzuela *et al.*, 2000), reduction of pro-inflammatory responses by macrophages, depressed Th1 cytokine responses (Ramachandra & Wikel, 1995) and T-cell proliferation (Schoeler & Wikel, 2001), lowered antibody responses and impaired NK cell function (Kopecky & Kuthejlova, 1998).

The extent of the immunosuppressive effects varies with the periods of infestation and it is important to remember that, as with most established host–parasite relationships, there is a balance in the protection afforded to both the host and its (ecto)parasite.

However, immunosuppression by the vector can have additional consequences in those situations where vector feeding also facilitates microbial transmission. We shall consider this in more detail in relation to other host–vector–parasite relationships but there is compelling evidence that the changes in host response induced by the tick salivary components, especially of cytokine production, can lead to greater susceptibility to various pathogens. Thus protection against *Borrelia burgdorferi* is mediated via tumour necrosis factor α (TNF-α), IFN-γ and IL-2, production of all of which is inhibited by tick salivary components (Zeidner *et al.*, 1996). This creates a site in which the spirochaetes can become established in the new host (Gillespie *et al.*, 2000). Salivary gland extracts of *Dermacentor reticulatus* ticks also serve to promote arbovirus transmission by suppressing production and action of interferon (Hajnicka *et al.*, 2000). As we have seen, ticks are vectors of a wide range of parasites so vaccine development can serve an important dual function.

VACCINATION AGAINST TICKS

While accepting that some of the mechanisms of acquired immunity described above come from unnatural host–parasite interactions, they do demonstrate a potential approach to vaccination by induction of the same responses artificially, and this is being investigated.

It can be argued that the balanced regulation that has most likely evolved in terms of the naturally occurring immune responses and the tick countermeasures may make these less suitable targets for a vaccination strategy. A better strategy would be immunization with tick antigens not naturally involved in immune reactions, so-called concealed antigens. This approach has also been investigated and, with ticks, has been more effective.

Jittapalapong *et al.* (2000) provide an example of both approaches, comparing vaccination with repeated infestation. They immunized dogs with salivary gland or midgut extracts of *Rhipicephalus sanguineus*. Salivary gland extracts induced host responses that reduced the numbers of ticks engorged, and led to an increase in the feeding period but with still a significantly lower engorgement weight. The midgut extract vaccine also had a significant impact on numbers feeding but its other effects that were harmful to the tick were in terms of its fecundity, with decreases in egg masses and increases in the pre-oviposition and oviposition periods.

Immunization of cattle with partially purified antigens derived from larval (Sharma *et al.*, 2001) and nymphal (Das *et al.*, 2000) antigens of *Hyalomma anatolicum* ticks led to rejection of larval nymph and adult stages, decrease in engorgement weight, and a decrease in the ability to moult.

The best characterized vaccines, notably Bm86, are based on midgut glycoproteins used to give protection of cattle from *Boophilus microplus*. A high level of expression was achieved in *Pichia pastoris* and forms the basis of valuable commercial vaccines (Gavac and Tick GARD), especially when appropriately combined with use of acaricides (Redondo *et al.*, 1999). Bm86 gives protection against tick species other than *B. microplus* (de Vos *et al.*, 2001; de la Fuente *et al.*, 1999) but sequence variations in Bm86 can affect protective efficacy (Garcia-Garcia *et al.*, 1999). Bm95, a homologue of Bm86, the gene for which is also expressed in yeast, was shown to give protection against strains of *B. microplus* that are resistant to Bm86 vaccine protection (Garcia-Garcia *et al.*, 2000). On vaccinated cattle, there are fewer engorged ticks and their weight and reproductive capacity are reduced (de la Fuente *et al.*, 1998). Synthetic peptides based on the Bm86 sequence have been produced and were highly effective at inducing resistance to tick challenge (Patarroyo *et al.*, 2002).

The Bm86 vaccine efficacy is associated with induction of antibody that is ingested with the blood meal and causes lysis of the tick gut cells (Kemp *et al.* 1989). Wang & Nuttall (1999) have shown, however, that ticks have another protective mechanism in the form of proteins that will bind immunoglobulins and then excrete them via the saliva. They suggest this will need to be blocked (possibly by immunological means) to allow ingested antibodies to be fully effective.

Vaccination with the concealed antigens, though advantageous in some respects, induces an immune response that can be boosted only by re-vaccination. In an attempt to retain the benefits of a vaccine-induced immunity to the tick midgut but also to have that immunity boosted when the ticks feed, Trimnell *et al.* (2002) selected a saliva antigen that cross-reacts with tick midgut antigens. Tick cement protein from *Rhipicephalus appendiculatus*, the vector of East Coast fever, anchors tick mouth parts to allow it to remain attached for several days. Guinea pigs were immunized with a recombinant form of the cement protein. Effects on ticks subsequently fed on the immunized animals included damage to the midgut and increased mortality.

The number of tick antigens that can be expressed in a (recombinant) form that makes them potential vaccine candidates is still small (Willadsen, 2001). Though commercial vaccines are in use, greater efficacy is likely to require combinations of antigens rather than the single tick molecules currently being used.

INSECTS OF MEDICAL IMPORTANCE

Limited attempts have been made to immunize against vectors of a range of microbial and parasitic organisms but the most revealing are the more sustained attempts to develop vaccines against sandflies and mosquitoes.

Sandflies and *Leishmania*

Phlebotomine sandflies occur extensively throughout the Old and New Worlds. Their medical importance derives from the microbial and parasitic infections they transmit, which include numerous species of *Leishmania* parasites. Female sandflies – predominantly *Phlebotomus* spp. in the Old World and *Lutzomyia* spp. in the New World – are the vectors responsible for the many different forms of leishmaniasis. Like many other insects they require blood meals every 4–5 days in order to be able to complete egg development. Remarkably this small insect has mouthparts that penetrate the skin through to the blood capillaries and lymphatics of the upper dermis. They are pool feeders, lacerating the capillaries to create a haemorrhagic pool which leads to formation of a haematoma.

Injected saliva facilitates the feeding process and it contains a range of pharmacologically active components that have anticoagulant, anti-platelet aggregation and vasodilatory properties. Kamhawi (2000) provides a summary of the molecules that have been identified; with a few exceptions these have been identified through the activities of the saliva. Most of the molecules are only partially characterized.

The molecule under most detailed investigation is a 63-amino-acid peptide named maxadilan. It has been cloned and shown to bind to cell surface receptors on both vascular and neural tissues, notably to the type 1 receptor for pituitary adenylate cyclase-activating neuropeptide (PACAP). It is found in *Lutzomyia* (but not *Phlebotomus*) sandflies and is a potent vasodilator (Lerner *et al.*, 1991). In passing, it has been shown that adenosine and 5′-AMP found in the saliva of *Phlebotomus* species have similar vasodilatory properties (Ribeiro *et al.*, 1999).

Salivary gland extracts from *Lutzomyia* females also have a significant immuno-modulatory effect when released as the sandfly feeds. They inhibit macrophage activation, impairing the production of TNF-α and both nitric oxide (NO) and oxygen-derived metabolites such as H_2O_2. The local immune responses assessed by cytokine production become more strongly Th2 rather than Th1. Importantly, the extracts also inhibit T-cell activation and proliferation and their effects have been estimated to persist for at least 4 days. Maxadilan contributes to these immunomodulatory effects; it has been shown to inhibit T-cell proliferation, prevent delayed-type hypersensitivity (DTH) reactions, inhibit TNF-α production by macrophages, and to lead to increased production of the Th2 cytokines IL-6 and IL-10 (Rogers & Titus, 2003).

These are all interesting observations but not in themselves a justification for trying to develop an anti-sandfly vaccine. However, when transmission of *Leishmania* parasites is factored into the feeding cycle, vaccine development becomes relevant and a potentially valuable addition to prevention and control measures for leishmaniasis.

Leishmaniasis is a spectrum of cutaneous, mucocutaneous and visceral forms of infection, the outcome being largely, though not entirely, dependent on the species of the infecting parasite. Most of the experimental studies on the role of salivary components on transmission of the parasite to mice have been with a cutaneous form of *Leishmania* (*Leishmania major*) which in humans is usually self-healing though producing disfiguring lesions. The protective immunity to this and to most *Leishmania* infections is not fully understood but is strongly associated with Th1 responses and with secretion of IFN-γ and IL-12. Th2 responses marked by the production of the cytokines IL-4, IL-5, IL-6 and IL-10 allow disease progression (Etges & Muller, 1998). This indicates how the immunomodulatory properties of saliva from the sandfly most likely permit and promote leishmanial infection.

The effects of salivary gland extracts or of maxadilan alone are dramatic. Increases in lesion size of 5–10 times are seen when compared with controls injected with the metacyclic (infective) stages of the parasite in the absence of salivary gland extracts. In its definitive host (e.g. human, rodent) *Leishmania* develops within phagocytic cells. When injected with salivary gland components (as would occur naturally) the metacyclic parasites give rise to lesions that contain up to 5000 times the number of intracellular parasites seen in the controls (Titus & Ribeiro, 1988; Gillespie *et al.*, 2000). Most of these experimental studies involve injection of large numbers of metacyclics, e.g. 10^5, which is far more than would be introduced when a sandfly feeds. It has been variously estimated that between 10^1 and 10^3 metacyclic forms will be introduced by the feeding sandfly, and the presence of the salivary gland components like maxadilan seems to be necessary with such numbers to ensure that the parasite becomes established. The parasites form a plug in the proventriculus (mouth part) of the sandfly which makes feeding more difficult. The fly probes more frequently, introducing more saliva as well as parasites.

Table 2. Vaccination with maxadilan and challenge with *Leishmania major* (adapted from Morris *et al.*, 2001)

Treatment	Challenge	*L. major*/footpad	P value
Maxadilan and adjuvant/saliva	*L. major*	9.9×10^5	–
Adjuvant/saliva alone	*L. major* and saliva	2.4×10^7	<0.01
Maxadilan and adjuvant/saliva	*L. major* and saliva	1.8×10^3	<0.001

It could be expected that immunity to sandfly saliva would confer protection from subsequent *Leishmania* infections since the immunomodulatory properties of the saliva would be neutralized (Belkaid *et al.*, 1998). Vaccination against maxadilan (or other components of saliva) should therefore impair sandfly feeding by inhibiting its vasodilatory properties, reducing the harmful immunomodulatory effects, and decreasing the infectivity of the *Leishmania* metacyclics.

Morris *et al.* (2001) vaccinated mice with 25 µg doses of maxadilan, given with first Freund's complete then Freund's incomplete adjuvant, followed by a boost of soluble maxadilan. The vaccine gave substantial protection when the challenge was parasites plus saliva (Table 2). Even when the challenge was parasites alone there was a significant effect on numbers, presumably as a result of loss of the immunomodulatory properties of maxadilan. Anti-maxadilan antibody was used as the primary readout of vaccinations but draining lymph node cells showed induced production of IFN-γ and NO. The responding cells were predominantly CD4$^+$.

Similar protection against *L. major* infection in mice was achieved by vaccination against a salivary protein of the Old World sandfly *Phlebotomus papatasi*. This is designated SP15. Vaccination with cDNA encoding SP15 had a marked protective effect on lesion size after challenge of the vaccinated mice with *Leishmania* parasites plus saliva. Though the vaccine induced both humoral and DTH responses, vaccination of B-cell-deficient mice showed that the protection was DTH rather than antibody-dependent (Valenzuela *et al.*, 2001).

An interesting recent study has shown that pre-exposure to the bites of uninfected sandflies gives protection in terms of lesion size and parasite numbers after subsequent exposure to infective bites (Kamhawi *et al.*, 2000). The pre-exposure induced up-regulation of IFN-γ and IL-12 locally, both of which would indicate activation of macrophages, presumably to a level where parasite control occurred. This observation, together with the identification of salivary components such as maxadilan and SP15, has encouraged the idea of sandfly saliva as the basis of a vaccine that indirectly would reduce the transmission of *Leishmania* (and other parasites transmitted by the sandflies) (Enserink, 2001).

Mosquitoes and malaria

The development of malaria parasites in the mosquito vector is initiated when male and female gametocytes are ingested from an infected patient when the female mosquito takes a blood meal. Very rapidly the male microgametocyte exflagellates to release usually eight microgametes, while the female macrogametocyte gives rise to a single macrogamete. The zygote formed by their fertilization transforms into an elongated, motile ookinete which moves from the midgut lumen, penetrates the peritrophic membrane that surrounds the blood meal, crosses the midgut wall, and settles as an oocyst on the outer side of the wall. It takes normally 10–14 days for several thousand sporozoites to form within each oocyst. Rupture of the oocyst releases the sporozoites into the haemolymph of the mosquito and they home to the salivary glands and become positioned to be injected when the mosquito next takes a blood meal.

The concept of a vaccine that targets the sexual and sporogonic cycle of the parasite in the mosquito is not only now well established but has reached the stage where phase I clinical trials are beginning. Such vaccines are intended to prevent the parasites from infecting the mosquito vectors. Consequently, as they reduce or interrupt transmission of malaria in human populations, they are called transmission-blocking vaccines (Carter, 2001).

The stages against which vaccines have been developed are the male and female gametes, zygotes and ookinetes. The end result is the same: the absence or marked reduction in numbers of oocysts and hence of infective sporozoites. Experimentally, a highly effective immunity has been shown. It is antibody-mediated and directed against surface antigens of either gametes (preventing fertilization) or the ookinetes. The antigens, designated Pfs48/45 and Pfs230, occur on the surface of both male and female gametes of *Plasmodium falciparum* and are being intensively studied as vaccine candidates. More progress has been achieved towards production of clinical grade formulations with antigens Ps25 and Ps28, which are expressed on the zygote and ookinete surfaces (i.e. after fertilization has occurred). It is recombinant forms of these that are now in clinical trials against both *P. falciparum* and *Plasmodium vivax* malaria (Carter, 2001; Carter *et al.*, 2000; Tsuboi *et al.*, 2003).

Transmission-blocking vaccines have considerable value in different ways. Malaria is a focal disease even in highly endemic regions of the world and it is in this context that measures to interrupt transmission should be considered. Mosquitoes from an active breeding site will interact with the human population within a relatively short range, and vaccination to stop transmission could be highly effective within such a population. Such defined areas of transmission may be more easily identified in areas of lower intensity transmission or in epidemic situations. Breaking the malaria cycle in this way is also recognized to have other benefits. The concept of multi-component vaccines is widely believed to be the only way to achieve high efficacy against malaria parasites (Carvalho *et al.*, 2002). Inclusion of transmission-blocking vaccine components would serve to preserve the usefulness of other vaccines (against blood or liver stages of the infection) by preventing or reducing the spread of parasites that have become vaccine resistant.

The factors that underlie the development of transmission-blocking vaccines directed against malaria parasites in the mosquito have been stressed here because they have relevance in terms of attempts to develop anti-mosquito vaccines. In fact, from a theoretical point of view, a good mosquitocidal vaccine could be better than an anti-parasite vaccine. Interestingly, it is in sub-Saharan Africa where this might work best since the principal vector, *Anopheles gambiae*, is highly anthropophilic, i.e. it feeds mainly or wholly on humans. The female mosquito needs blood meals every 3–4 days and survives 2–5 weeks (Service & Townson, 2002). Thus a single female mosquito will feed between 4 and maybe 12 times (Kileen *et al.*, 2000). Each mosquito is most unlikely to incubate more than one cycle of parasite development as it takes at least 10 days from ingestion of gametocytes before infective sporozoites appear in the salivary glands. Vaccination targeting the vector could thus give several 'hits' at the parasite life cycle. This approach also has the benefit that it would be effective against different malaria parasite species.

Feeding mosquitoes do induce antibody responses and these are being investigated particularly in relation to intensity of transmission (J. Mitchell, C. J. Drakeley & P. Billingsley; F. Simondon and others, personal communications). As shown with other arthropods, mosquito saliva also has immunomodulatory activity that affects T cells, cytokines, and other pro-inflammatory molecules (Gillespie *et al.*, 2000; Schoeler & Wikel, 2001). The active components have not yet been characterized and, while immunization against salivary components does have a blocking effect on transmission of malaria (Brennan *et al.*, 2000) and gives protective effects akin to those seen with maxadilan from sandflies, there is likely to be a better hope of success by modelling anti-mosquito vaccines on the anti-tick vaccines by use of so-called concealed antigens.

Immunization with haemolymph (Gulia *et al.*, 2002) and mosquito ovary (Gakhar *et al.*, 2001) reduces mosquito fecundity but, following extensive studies with a variety of tissues and organs, Almeida & Billingsley (1998) concluded that the midgut was the most obvious target organ for attempts to reduce mosquito survival and fecundity. In addition it is the organ to attack in order to disrupt malaria parasite development since the passage of ookinetes from the midgut lumen and their transition to become oocysts is a particularly vulnerable stage in the parasite life cycle. Polyclonal antibodies to these crude midgut antigens have been shown to reduce significantly the numbers of oocysts developing on the midgut when mosquitoes are fed on immunized and infected animals or through a membrane on gametocyte-containing blood suspended in sera from animals immunized with midgut homogenates (Lal *et al.*, 1994; Ramasamy & Ramasamy, 1990; Srikrishnaraj *et al.*1995; Suneja *et al.*, 2003).

Lal *et al.* (2001) made first attempts towards defining the midgut components that are particularly important as immunogens. They showed that monoclonal antibodies against midgut antigens of *Anopheles stephensi* would inhibit the development of oocysts of both *P. falciparum* and *P. vivax* in a variety of species of mosquito; in other words, it was not a mosquito-specific effect, which argues for them being developed against conserved molecules. There was also a demonstrable effect on mosquito survival and fecundity.

Almeida & Billingsley (2002) induced similar effects by immunizing with subcellular fractions of the midgut of *A. stephensi*. Interestingly, while antisera to a microvilli fraction reduced the numbers of *P. berghei* malaria oocysts, antiserum to the baso-lateral plasma membranes induced higher oocyst burdens in the feeding mosquitoes.

Foy *et al.* (2003) specified their target antigens more precisely, focusing on mosquito survival rather than effects of vaccination on parasites within the vector. Mice were immunized with a cDNA library constructed from mRNA derived from mosquito midguts, or with specific cDNAs encoding a peritrophic matrix protein and a mucin expressed on the luminal side of the midgut wall. They showed significantly increased mortality amongst mosquitoes fed on mice immunized with either the whole library or the mucin cDNA but not when mice were immunized with the peritrophic protein. The mosquitocidal effect was a pro-inflammatory Th1-type associated with raised TNF-α and IFN-γ, with negligible antibody. A booster dose of midgut protein given after the cDNA immunization shifted the immune responses to a Th2 type with raised IL-5 and IL-10 cytokines and much higher antibody titres. These altered immune responses then had no significantly harmful effect on the mosquitoes. It seems, therefore, that the harmful immune responses generated by this method of immunization rely on ingestion of immune effector cells and may be difficult to show other than in experimental situations like those employed.

FUTURE VACCINE STUDIES NEEDED

The purpose of an anti-vector vaccine is to break the life cycle of the parasite that the vector transmits. The anti-tick vaccines, though developed commercially for the purpose of reducing tick infestation, demonstrate an approach that could be applied for control of the infections of veterinary importance spread by these arthropods.

Developing vaccines that could conceivably be used in humans to combat major infectious diseases like malaria or the leishmaniases will inevitably be more difficult. However, the experimental results are again sufficiently encouraging to warrant designing experimental studies that more directly have vaccine testing in humans as their future objective. Taking malaria mosquitoes as the model, it has been shown that normally concealed antigens of the mosquito midgut will induce immune responses that significantly reduce mosquito survival and fecundity; these effects are not mosquito species specific. At the same time they reduce development of the malaria parasites within the vector. It is now vitally important to reproduce and enhance these effects with specific recombinant proteins or synthetic molecules since, even for experimental studies, the aim should be to produce and use candidate vaccines that would be suitable for later testing in humans.

Part of this process is to begin to apply some of the more recent approaches to vaccine strategy that have not so far been done. For example, a prime–boost strategy is being widely investigated for vaccination against a range of infectious disease agents. This often requires that an antigen used for priming is presented in a form that is different from that used for boosting. Thus future vaccine studies should plan to produce any candidate antigen in different forms, for example, as a DNA vaccine, expressed in viral vectors, as soluble protein, or as a synthetic molecule, so that combinations of these may be tested for priming and boosting.

This may all seem very ambitious but, if anti-vector vaccines are to be taken seriously and not just viewed as being of experimental interest, the approach to their development, whether for veterinary or medical use, must reflect this.

REFERENCES

Almeida, A. P. & Billingsley, P. F. (1998). Induced immunity against the mosquito *Anopheles stephensi* Liston (Diptera: Culicidae): effects on mosquito survival and fecundity. *Int J Parasitol* **28**, 1721–1731.

Almeida, A. P. & Billingsley, P. F. (2002). Induced immunity against the mosquito *Anopheles stephensi* (Diptera: Culicidae): effects of cell fraction antigens on survival, fecundity, and *Plasmodium berghei* (Eucoccidiida: Plasmodiidae) transmission. *J Med Entomol* **39**, 207–214.

Belkaid, Y., Kamhawi, S., Modi, G., Valenzuela, J., Noben-Trauth, N., Rowton, E., Ribeiro, J. M. & Sacks, D. L. (1998). Development of a natural model of cutaneous leishmaniasis: powerful effects of vector saliva and saliva preexposure on the long-term outcome of *Leishmania major* infection in the mouse ear dermis. *J Exp Med* **188**, 1941–1953.

Brennan, J. D. G., Kent, M., Dhar, R., Fujioko, H. & Kumar, N. (2000). *Anopheles gambiae* salivary gland proteins as putative targets for blocking transmission of malaria parasites. *Proc Natl Acad Sci U S A* **97**, 13859–13864.

Carter, R. (2001). Transmission blocking malaria vaccines. *Vaccine* **19**, 2309–2314.

Carter, R., Mendis, K. N., Miller, L. H., Molyneaux, L. & Saul, A. (2000). Malaria transmission-blocking vaccines – how can their development be supported? *Nat Med* **6**, 241–244.

Carvalho, L. J. M., Daniel-Ribeiro, C. T. & Goto, H. (2002). Malaria vaccine: candidate antigens, mechanisms, constraints and prospects. *Scand J Immunol* **56**, 327–343.

Das, G., Ghosh, S., Khan, M. H. & Sharma, J. K. (2000). Immunization of cross-bred cattle against *Hyalomma anotolicum anatolicum* by purified antigens. *Exp Appl Acarol* **24**, 645–659.

de la Fuente, J., Rodriguez, M., Redondo, M. & 16 other authors (1998). Field studies and cost-effectiveness analysis of vaccination with Gavac^TM against the cattle tick *Boophilus microplus*. *Vaccine* **16**, 366–373.

de la Fuente, J., Rodriguez, M., Montero, C. & 11 other authors (1999). Vaccination against ticks (*Boophilus* spp.): the experience with the Bm86-based vaccine Gavac^TM. *Genet Anal* **15**, 143–148.

de Vos, S., Zeinstra, L., Taoufik, O., Willadsen, P. & Jongejan, F. (2001). Evidence for the utility of the Bm86 antigen from *Boophilus microplus* in vaccination against other tick species. *Exp Appl Acarol* **25**, 245–261.

Enserink, M. (2001). Sandfly saliva may be key to new vaccine. *Science* **293**, 1028.

Etges, R. & Muller, I. (1998). Progressive disease or protective immunity to *Leishmania major* infection: the result of a network of stimulatory and inhibitory interactions. *J Mol Med* **76**, 372–390.

Foy, B. D., Magalhaes, T., Injera, W. E., Sutherland, I., Devenport, M., Thanawastien, A., Ripley, D., Carnenas-Freytag, L. & Beier, J. C. (2003). Induction of mosquito-cidal activity in mice immunized with *Anopheles gambiae* midgut cDNA. *Infect Immun* **71**, 2032–2040.

Gakhar, S. K., Jhamb, A., Gulia, M. & Dixit, R. (2001). Anti-mosquito ovary antibodies reduce the fecundity of *Anopheles stephensi* (Diptera: Insecta). *Jpn J Infect Dis* **54**, 181–183.

Garcia-Garcia, J. C., Gonzalez, I. L., Gonzalez, D. M. & 9 other authors (1999). Sequence variations in the *Boophilus microplus* Bm86 locus and implications for immunoprotection in cattle vaccinated with this antigen. *Exp Appl Acarol* **23**, 883–895.

Garcia-Garcia, J. C., Montero, C., Redondo, M. & 10 other authors (2000). Control of ticks resistant to immunization with Bm86 in cattle vaccinated with the recombinant antigen Bm95 isolated from the cattle tick, *Boophilus microplus*. *Vaccine* **18**, 2275–2287.

Gillespie, R. D., Mbow, M. L. & Titus, R. G. (2000). The immunomodulatory factors of blood feeding arthropod saliva. *Parasite Immunol* **22**, 319–331.

Gulia, M., Suneja, A. & Gakhar, S. K. (2002). Effects of anti-mosquito hemolymph antibodies on fecundity and on the infectivity of malarial parasite *Plasmodium vivax* to *Anopheles stephensi* (Diptera: Insecta). *Jpn J Infect Dis* **55**, 78–82.

Hajnicka, V., Kocakova, P., Slovak, M., Labuda, M., Fuchsberger, N. & Nuttall, P. A. (2000). Inhibition of the antiviral action of interferon by tick salivary gland extract. *Parasite Immunol* **22**, 201–206.

Jittapalapong, S., Stich, R. W., Gordon, J. C., Wittum, T. E. & Barriga, O. O. (2000). Performance of female *Rhipicephalus sanguineus* (Acari: Ixodidae) fed on dogs exposed to multiple infestations or immunization with tick salivary gland or midgut tissues. *J Med Entomol* **37**, 601–611.

Kamhawi, S. (2000). The biological and immunomodulatory properties of sand fly saliva and its role in the establishment of *Leishmania* infections. *Microbes Infect* **2**, 1765–1773.

Kamhawi, S., Belkaid, Y., Modi, G., Rowton, E. & Sacks, D. (2000). Protection against cutaneous leishmaniasis resulting from bites of uninfected sand flies. *Science* **290**, 1351–1354.

Kemp, D. H., Pearson, R. D., Gough, J. M. & Willadsen, P. (1989). Vaccination against *Boophilus microplus*: localization of antigens on tick cells and their interaction with the host immune system. *Exp Appl Acarol* **7**, 43–58.

Kileen, G. F., McKenzie, F. E., Foy, B. D., Schieffelin, C., Billingsley, P. F. & Beier, J. C. (2000). A simplified model for predicting malaria entomologic inoculation rates based on entomologic and parasitologic parameters relevant to control. *Am J Trop Med Hyg* **62**, 535–544.

Kopecky, J. & Kuthejlova, M. (1998). Suppressive effect of *Ixodes ricinus* salivary gland extract on mechanisms of natural immunity. *Parasite Immunol* **20**, 169–174.

Lal, A. A., Schriefer, M. E., Sacci, J. B., Goldman, I. F., Louis-Wileman, V., Collins, W. E. & Azad, A. F. (1994). Inhibition of malaria parasite development in mosquitoes by anti-mosquito-midgut antibodies. *Infect Immun* **62**, 316–318.

Lal, A. A., Patterson, P. S., Sacci, J. B., Vaughan, J. A., Paul, C., Collins, W. E., Wirtz, R. A. & Azad, A. F. (2001). Anti-mosquito midgut antibodies block development of *Plasmodium falciparum* and *Plasmodium vivax* in multiple species of *Anopheles* mosquitoes and reduce vector fecundity and survivorship. *Proc Natl Acad Sci U S A* **98**, 5228–5233.

Lerner, E. A., Ribeiro, J. M. C., Nelson, R. J. & Lerner, M. R. (1991). Isolation of Maxadilan, a potent vasodilatory peptide from the salivary glands of the sand fly *Lutzomyia longipalpis*. *J Biol Chem* **17**, 11234–11236.

Mejri, N., Franscini, N., Rutti, B. & Brossard, M. (2001). Th2 polarization of the immune response of BALB/C mice to *Ixodes ricinus* instars, importance of several antigens in activation of specific Th2 subpopulations. *Parasite Immunol* **23**, 61–69.

Morris, R. V., Shoemaker, C. B., David, J. R., Lanzaro, G. C. & Titus, R. G. (2001). Sand fly Maxadilan exacerbates infection with *Leishmania major* and vaccinating against it protects against *L. major* infection. *J Immunol* **167**, 5226–5230.

Patarroyo, J. H., Portela, R. W., De Castro, R. O., Couto Pimentel, J., Guzman, F., Patarroyo, M. E., Vargas, M. I., Prates, A. A. & Dias Mendes, M. A. (2002). Immunization of cattle with synthetic peptides derived from the *Boophilus microplus* gut protein (Bm86). *Vet Immunol Immunopathol* **88**, 163–172.

Ramachandra, R. N. & Wikel, S. K. (1995). Effects of *Dermacentor andersoni* (Acari: Ixodidae) salivary gland extracts on *Bos indicus* and *Bos taurus* lymphocytes and macrophages: *in vitro* cytokine elaboration and lymphocyte blastogenesis. *J Med Entomol* **32**, 338–345.

Ramasamy, M. S. & Ramasamy, R. (1990). Effect of anti-mosquito antibodies on the infectivity of the rodent malaria parasite *Plasmodium berghei* to *Anopheles farauti*. *Med Vet Entomol* **4**, 161–166.

Redondo, M., Fragoso, H., Ortiz, M. & 9 other authors (1999). Integrated control of acaricide-resistant *Boophilus microplus* populations on grazing cattle in Mexico using vaccination with Gavac^{TM} and amidine treatments. *Exp Appl Acarol* **23**, 841–849.

Ribeiro, J. M., Katz, O., Pannell, L. K., Waitumbi, J. & Warburg, A. (1999). Salivary glands of the sand fly *Phlebotomus papatasi* contain pharmacologically active amounts of adenosine and 5' AMP. *J Exp Biol* **11**, 1551–1559.

Rogers, K. A. & Titus, R. G. (2003). Immunomodulatory effects of Maxadilan and *Phlebotomus papatasi* sand fly salivary gland lysates on human primary *in vitro* immune responses. *Parasite Immunol* **25**, 127–134.

Schoeler, G. B. & Wikel, S. K. (2001). Modulation of host immunity by haematophagous arthropods. *Ann Trop Med Parasitol* **95**, 755–771.

Service, M. W. & Townson, H. (2002). The *Anopheles* vector. In *Essential Malariology*, 4th edn, pp. 59–84. Edited by D. A. Warrell & H. M. Gilles. London: Arnold.

Sharma, J. K., Ghosh, S., Khan, M. H. & Das, G. (2001). Immunoprotective efficacy of a purified 39 kDa nymphal antigen of *Hyalomma anatolicum anatolicum*. *Trop Anim Health Prod* **33**, 103–116.

Srikrishnaraj, K. A., Ramasamy, R. & Ramasamy, M. S. (1995). Antibodies to *Anopheles* midgut reduce vector competence for *Plasmodium vivax* malaria. *Med Vet Entomol* **9**, 353–357.

Suneja, A., Gulia, M. & Gakha, S. K. (2003). Blocking of malaria parasite development in mosquito and fecundity reduction by midgut antibodies in *Anopheles stephensi* (Diptera: Culicidae). *Arch Insect Biochem Physiol* **52**, 63–70.

Titus, R. G. & Ribeiro, J. M. C. (1988). Salivary gland lysates from the sand fly *Lutzomyia longipalpis* enhance *Leishmania* infectivity. *Science* **239**, 1306–1308.

Trager, W. (1939). Acquired immunity to ticks. *J Parasitol* **25**, 57–81.

Trimnell, A. R., Hails, R. S. & Nuttall, P. A. (2002). Dual action ectoparasite vaccine targeting 'exposed' and 'concealed' antigens. *Vaccine* **20**, 3560–3568.

Tsuboi, T., Tachibana, M., Kaneko, O. & Torii, M. (2003). Transmission-blocking vaccine of vivax malaria. *Parasitol Int* **52**, 1–11.

Valenzuela, J. G., Charlab, R., Mather, T. N. & Ribeiro, J. M. C. (2000). Purification, cloning and expression of a novel salivary anticomplement protein from the tick, *Ixodes scapularis*. *J Biol Chem* **275**, 18717–18723.

Valenzuela, J. G., Belkaid, Y., Garfield, M. K., Mendez, S., Kamhawi, S., Rowton, E. D., Sacks, D. L. & Ribeiro, J. M. C. (2001). Towards a defined anti-*Leishmania* vaccine targeting vector antigens: characterization of a protective salivary protein. *J Exp Med* **194**, 331–342.

Wang, H. & Nuttall, P. A. (1999). Immunoglobulin-binding proteins in ticks: new target for vaccine development against a blood-feeding parasite. *Cell Mol Life Sci* **56**, 286–295.

Wikel, S. K. (1999). Tick modulation of host immunity: an important factor in pathogen transmission. *Int J Parasitol* **29**, 851–859.

Willadsen, P. (2001). The molecular revolution in the development of vaccines against ectoparasites. *Vet Parasitol* **101**, 353–368.

Zeidner, N., Dreitz, M., Belasco, D. & Fish, D. (1996). Suppression of acute *Ixodes scapularis*-induced *Borrelia burgdorferi* infection using tumor necrosis factor α, interleukin-2 and interferon-γ. *J Infect Dis* **173**, 187–195.

INDEX

References to tables/figures are shown in italics

Aedes aegypti
  defensins 202, 211, *214*, 216, 351
  dengue viruses (DENV)
    infection 150, 164–165
    vaccine 122–123
  genetics 149–151
  geographic distribution 165
  midgut 109, *143*
  peritrophic membrane 205
  Sindbis virus (SINV) *107*, *143*, 158
  transgenic lines 347–349, *347*, 351, 354–355,
    356
  trypsin genes 109
  vector competence 148, 150
  Venezuelan equine encephalitis virus
    (VEEV) 157
  viral gene effects 121–122
  yellow fever virus (YFV)
    17D vaccine 2–3, 122, 123, 124–125
    infection 2–3, 104, 150
    neurotropic vaccine 2
Aedes albopictus
  antiviral proteins 115
  Bunyamwera virus 95, 96
  climate effects 183–184
  geographic distribution 183–184
  minireplicon system 96–98, *96*, *97*
  virus receptors 111
  Wolbachia superinfection 290
Aedes triseriatus, see Ochlerotatus triseriatus
African horse sickness virus (AHSV)
  climate effects 184–187
  Culicoides spp. 184–187
  El Niño/southern oscillation (ENSO) 188
  novel vectors 192
  overwintering mechanism *186*
African trypanosomiasis (sleeping sickness)
    10–11
Anaplasma phagocytophilum
  human granulocytic ehrlichiosis (HGE) 7–8,
    301–302
  morphology 302
  neutrophils
    apoptosis inhibition 316–318
    colonization 303, 307–312
    cytokines 318–319
    NADPH oxidase inhibition 312–316, *313*
    P-selectin glycoprotein ligand-1 (PSGL-1)
      307–309

  p44 gene family 304–306, *304*
  survival strategies (in neutrophils) 309–316
  ultrastructure *302*
Anopheles gambiae
  antimicrobial compounds (AMPs) 211–213,
    214
  infection susceptibility 151
  killing factors 238–239
  prophenoloxidase (PpO) 228–229
  RNA silencing 69
  transposable elements *357*
  trypsin genes 109
  vaccine 372
Anopheles stephensi 169, 209, 218–219, 351, 354,
    373
antimicrobial peptides (AMPs) 211–213,
    214–215, 216–218, 237
aphids 48–49
arboviruses, *see also* individual viruses
  circulative–propagative transmission 46–47
  environmental effects *166*
  evolution 123–124
  host virulence 166–167
  mosquitoes *140*
  mutations 157
  reassortment 158
  stages of infection 105–107
  temperature effects 184
  transmission 140–144
  transovarial transmission *143*
  vector virulence 167–169
  virus families 103, *104*
arthropods, *see also* individual species
  diseases *200–201*
  feeding behaviour 153–156
  host preferences 153–156
  host-seeking behaviour 153
  innate immune defences *204*
  intracellular signalling pathways *233*
  oviposition 155–156
  vaccines 364

Bluetongue virus (BTV)
  climate effects 184–187
  Culicoides spp. 126, 159, 184–187
  epidemiology 189
  European outbreak 1998–2003 189–193, *190*
  genome segment reassortment 159
  novel vectors 192

Old World vectors 189
vector competence 159, 192
Bm86 vaccine 367
Boophilus microplus, *see* ticks
Borrelia spp.
    Borrelia burgdorferi
        disease-causing species 76
        diversity 31–33
        evolution 31–35
        geographic distribution 33–35, *34*
        Lyme borreliosis 9–10, 31, 76, 77–78, 81, 82, 85
        relapsing fever *76*, 77–78, *80*, 81, 82, 84, 85
        ticks 9–10, 32, *34*, 78–81, 79, *80*, 82, 82–84
        transovarial transmission *80*, 81–82
        vertebrate host range 77
Bunyamwera virus
    Aedes albopictus 95, 96
    host protein synthesis 98–99, 99
    minireplicon system 96–98, 96, 97
bunyaviruses
    characteristics 91–92
    genome segment reassortment 94–95, *94*
    minireplicon system 96–98, 96, 97
    orthobunyaviruses 92–93, *92*
    reassortant 93
    replication effects on host 95–96, 95
    vector competence 93

*C*aulimoviruses 51–53, *52*
cecropins 11, 202, 211–213, *214–215*, 216, 217, *347*
circulative virus transmission 45–50
    circulative–non-propagative 47–50
    circulative–propagative 46–47
Class II transposable elements
    Hermes 347–348, *347*, 351, 352, 357
    Himar1 347, 348–349
    Minos 347, 349
    Mos1 mariner 347, 348–349
    piggyBac 347, 348, 351, 354
climate effects
    Aedes albopictus 183–184
    African horse sickness virus (AHSV) 184–187
    bluetongue virus (BTV) 184–187
    Culicoides spp. 184–187
    insect vectors 182, 183–184
    tick-borne flaviviruses 27–31, *29*, *30*
    Western tick-borne encephalitis virus (WTBEV) 24–25
cucumber mosaic virus (CMV) 51, *52*, 66, 67
Culex pipiens 106, 114, 117, 120–121, 163, 164

Culicoides spp.
    African horse sickness virus (AHSV) 184–187
    bluetongue virus (BTV) 126, 159, 184–187
    climate effects 184–187
    geographic distribution 189–192, *190*
    vector competence 192–193

*D*efensins 202, 211, *214–215*, 216, 351
dengue fever 3–4
dengue virus (DENV) 3–4, 119, 122–123, 150, 164–165

*E*hrlichiosis 7–8
El Niño/southern oscillation (ENSO) 188
evolution
    arboviruses 123–124
    Borrelia burgdorferi sensu lato 31–35
    tick-borne flaviviruses 21–31
    virus adaptation 157–159
    Yersinia pestis 339–340

*F*leas, *see Yersinia pestis*
flock house virus (FHV)
    B2 protein 67–68
    genome organization 66
    RNA silencing 65–68

*G*eographic distribution
    Aedes aegypti 165
    Aedes albopictus 183–184
    Borrelia burgdorferi sensu lato 33–35, *34*
    Culicoides spp. 189–192, *190*
    tick-borne encephalitis viruses 26–27
    vector competence 161–162, 165
Glossina spp., *see* tsetse flies
grapevine fanleaf virus (GFLV) 269, 275, 279, *279*

*H*uman granulocytic ehrlichiosis (HGE) 7–8, 301–302

*I*xodes spp., *see* ticks

*L*a Crosse virus (LACV) 109, 117, 120–121, 150, 158, *158*
Leishmania 160, *208*, 334, 368–370, *370*
louping ill virus (LIV) 25–26, 28–29, *30*
Lumpy skin disease virus (LSDV) 54
luteoviruses 48–49
Lutzomyia spp., *see* sandflies
Lyme borreliosis 9–10, 31, 76, 77–78, 81, 82, 85
lymphatic filariasis 13–14

Mahalanobis distance 28, *29*
malaria
    mosquito vaccine 371–373
    overview 12–13
    transgenic vectors 345–346, 353, 357
    transmission determinants 159–160
    vector competence modelling 146
maxadilan 368–370, *370*
mosquitoes, *see also* individual species *and*
       transgenic mosquitoes
    anatomy *107*, *141*
    antimicrobial compounds (AMPs) 211–213,
       214–215
    arbovirus transmission *140*
    genetics 124–126
    gut infection *142*
    immune response 211–216, *212*
    infection
        arboviruses 105–107
        deleterious effects 120–121
        flaviviruses 122–123
        orthobunyaviruses 92–93, *92*
        parasite-associated factors *208*
        virus genetics and 121–122
    infection sites *107*
        digestive tract 108–111, *168*
        fat body 116
        haemocoel 115
        Malpighian tubules 118
        muscle 116
        nervous tissues 116
        reproductive tissues 116–118
        salivary glands 118–120
    midgut escape barrier (MEB) 114–115, 122
    midgut infection barrier (MIB) 46, 110, 115
    minireplicon system 96–98, *96*, *97*
    nitric oxide 218–219
    oviposition 156
    peritrophic matrix 108
    RNA silencing 69
    saliva 119–120
    superinfection 111
    transmission-blocking vaccines 371–372
    transovarial transmission 116–118, 121
    vaccines 371–373
    vector competence 106, 124–126
    virus dissemination 113–115
    virus replication 112–113

Nanoviruses 47–50
nematodes
    feeding behaviour 276–277
    isolation 272
    ultrastructure *264*

    virus retention sites 272–276, *273*
    Wolbachia infection 291–293, *292*
nematode-transmitted plant viruses
    disease characteristics 267–268, *270*
    disease control 268–271
        avoidance 268
        chemical methods 268–269, *270*
        cultural methods 269
        plant resistance 271
    nepoviruses
        genome organization 279
        longidorid vectors 265–266, *264*, *266*
        transmission determinants 279
    retention sites 272–276, *273*
    specificity 278–279
    tobraviruses
        2b protein 281
        genome organization *280*
        transmission determinants 279–282
        trichodorid vectors 264–265, *264*, *265*,
        274
nepoviruses
    genome organization 279
    longidorid vectors 265–266, *264*, *266*
    transmission determinants 279
New World trypanosomiasis (Chagas' disease)
    11–12
non-circulative virus transmission
    beetle transmission 53
    capsid strategy 51, *52*
    helper strategy 51–53, *52*
    mechanical transmission 53–54

*O*chlerotatus triseriatus 121, 150, 158, *158*
orthobunyaviruses 92–93, *92*

*P*aratrichodorus spp., *see* nematodes
pea early-browning virus (PEBV) 274–275
Phlebotomus spp., *see* sandflies
phleboviruses 91, 93
plague 8–9, 331–332
Plasmodium spp.
    antimicrobial peptides (AMPs) 211–216, 237
    malaria 12–13
    nitric oxide killing 218–219
    NOS 238–239
    ookinetes 210, 224, 371
    transgenic mosquitoes 351–352
potyviruses 51–53, *52*

*Q*uantitative trait loci (QTL) 149, 150, 224
quasispecies 55–56

Rabbit hemorrhagic disease virus (RHDV) 53–54

relapsing fever 76, 77–78, *80*, 81, 82, 85

Rhodnius spp. 219, 223

Rickettsia spp. 6–7, 161

Rift valley fever virus (RVFV) 106, 114, 117, 120

RNAi (RNA interference), *see* RNA silencing

RNA silencing 63–70, *65, 67*

    antiviral role 64, 68, 69–70, 115

    cucumber mosaic virus (CMV) 67

    Drosophila cells 68

    flock house virus (FHV) 65–68

    mosquito cells 69

    suppression 66–68

    virus infection and 64

RNA-silencing-based antiviral response (RSAR) 66, 68, 69–70

Rocky Mountain spotted fever 6–7

*S*andflies *208*, 334, 368–370, *370*

Sindbis virus (SINV) 106, *107*, *143*, 158

snails *208*, 227, 239–240

spotted fever 6–7

stable fly (*Stomoxys*) 54, 217–218

superinfection

    arboviruses 111

    Wolbachia 290

symbionin 49

*T*ick-borne flaviviruses

    climate effects 27–31, *29, 30*

    evolution 21–31

    louping ill virus (LIV) 25–26, 28–29, *30*

    phylogeny 21–23, *22*

    tick-borne encephalitis virus (TBEV) 5–6, 26–27

    Western tick-borne encephalitis virus (WTBEV) 23–25, 29

ticks

    Bm86 vaccine 367

    Borrelia spp. 9–10, 32, *34*, 78–81, *79, 80, 82*, 82–84

    cytokine responses 365

    diseases 365

    HGE 301

    host immune responses 364–366

    immunosuppression by 365–366

    Lyme borreliosis 9–10, 31, 76, 77–78, 81, 82, 85

    tick-borne encephalitis virus (TBEV) 5–6, 26–27

    vaccination 366–368

    Western tick-borne encephalitis virus (WTBEV) 23–25

Tobacco mosaic virus (TMV) 53

tobacco rattle virus (TRV) 265, *270*, 271, 274, 280, *280*, 281–282

tobraviruses

    2b protein 281

    genome organization *280*

    transmission determinants 279–282

    trichodorid vectors 264–265, *264, 265*, 274

transgenic mosquitoes, *see also* mosquitoes *and* individual species

    Class II transposable elements

        Hermes 347–348, *347*, 351, 352, 357

        Himar1 *347*, 348–349

        Minos *347*, 349

        Mos1 mariner *347*, 348–349

        piggyBac *347*, 348, 351, 354

    defensins 351

    effector genes 350–352

    field studies 352–357

    fitness 35

    generation 346–349

    Herves 357

    mobile gene vectors 349–350

    Plasmodium spp. 351–352

    risks 358–359

    SM1 peptide 351

    tissue-specific promoters 350

    transgene spread 355–356

    venom phospholipase A2 352

transovarial transmission

    arboviruses 143

    Borrelia spp. *80*, 81–82

    insects 46–47

    La Crosse virus (LACV) 117

    mosquitoes 116–118

    Rift valley fever virus (RVFV) 117

    virus persistence 151

Trichodorus spp., *see* nematodes

trypanosomes 10–11, *208*, 209, 217, 219, 223

trypanosomiasis

    African (sleeping sickness) 10–11

    New World (Chagas' disease) 11–12

tsetse flies 10, 108, 155, 206, *208*, 216–217, 237, 239

typhus 6

*V*accines

    anti-arthropod 364

    Bm86 367

    flavivirus 122–123

    malaria 371–373

    maxadilan 368–370, *370*

    sandflies 368–370, *370*

    ticks 366–368

transmission-blocking 371–372
vector immunity 202
yellow fever virus
 17D 2–3, 122, 123, 124–125
 neurotropic 2
vector competence
 Aedes aegypti 148, 150
 bluetongue virus (BTV) 159, 192
 bunyaviruses 93
 Culicoides spp. 192–193
 disease prevalence 148
 flaviviruses 150
 genetics 149–152
 geographic effects 161–162, 165
 La Crosse virus (LACV) 158
 mathematical models 146–148, *147*
 mosquitoes 106, 124–126
 Plasmodium 151–152
 rainfall/humidity 164–165
 temperature effects 161–164, *166*
vector invasion
 arthropod immune response *204*
 gut barrier 205–220, *207*
  bacterial symbionts 210–211
  digestive enzymes 206–208
  glycoprotein receptors 210
  immune gene upregulation 211–218, *212*
  lectins 209–210
  nitric oxide killing 218–219
  peptides 209
  prophenoloxidase (PpO) 220, 223–227, *225*
 haemolymph
  cellular reactions 221–227
  encapsulation 223–227
  humoral factors 227–236
   lectins 234–235
   pattern recognition receptors (PRRs) 230–236, *233*
   prophenoloxidase (PpO) 228–230
  killing factors
   antimicrobial peptides (AMPs) 211–213, *214–215*, 216–218, 236–238
   reactive nitrogen intermediates 238–240
   reactive oxygen species 238–240
  phagocytosis 222–223
 peritrophic membrane 205, *207*

physicochemical barriers 203–205
salivary glands 240–241
viral dissemination index 114

Western equine encephalomyelitis virus (WEEV) 110–111, 120, 163
Western tick-borne encephalitis virus (WTBEV) 23–25, 29
West Nile virus (WNV) 4–5
Wolbachia
 Aedes albopictus superinfection 290
 antibiotic effects 292–293
 arthropod infection
  cytoplasmic incompatibility 288–289
  feminization 289
  male killing 289–290
  parthenogenesis 289
  pathogenic effects 290
  superinfection 290
 disease control strategies 293–294
 filariasis 14
 nematode infection 291–293, *292*
 phylogeny *288*
 transgene spread (mosquitoes) 355
 ultrastructure *292*
 vector control strategies 294

Xenopsylla cheopsis (fleas), *see Yersinia pestis*
Xiphenema spp., *see* nematodes

Yellow fever virus (YFV)
 17D vaccine 2–3, 122, 123, 124–125
 Aedes aegypti infection 2–3, 104, 150
 neurotropic vaccine 2
Yersinia pestis
 fleas
  feeding mechanism 337–338
  foregut blockage 332–337, *333*, *336*
   biofilm model 335–337
   fibrin clot model 334–335
  transmission efficiency 338
  transmission phenotype 338–339
 hms genes 335, *337*
 plague 8–9
 transmission 332–340
  evolution 339–340
  genetics 339–340, *340*
Yersinia pseudotuberculosis 332, 341